NATIVE PLANTS
of the
SYDNEY REGION

ALAN FAIRLEY & PHILIP MOORE

NATIVE PLANTS *of the* SYDNEY REGION

From Newcastle to Nowra and west to the Dividing Range

ALLEN&UNWIN

Photograph names:
p. ii: *Dillwynia rudis*
p. v: *Persoonia oblongata*
p. viii: *Eucalyptus paniculata*
p. x: *Hibbertia monogyna*

First published in 1989 by Kangaroo Press Pty Ltd.
Second edition published in 2002 by Kangaroo Press, an imprint of Simon & Schuster.
This revised third edition published in 2010 by Allen & Unwin.

Copyright © Alan Fairley and Philip Moore 2010

All rights reserved. No part of this book may be reproduced or transmitted in
any form or by any means, electronic or mechanical, including photocopying,
recording or by any information storage and retrieval system, without prior
permission in writing from the publisher. The Australian *Copyright Act 1968*
(the Act) allows a maximum of one chapter or 10 per cent of this book, whichever
is the greater, to be photocopied by any educational institution for its educational
purposes provided that the educational institution (or body that administers it) has
given a remuneration notice to Copyright Agency Limited (CAL) under the Act.

Jacana Books,
an imprint of Allen & Unwin
83 Alexander Street
Crows Nest NSW 2065
Australia
Phone: (61 2) 8425 0100
Fax: (61 2) 9906 2218
Email: info@allenandunwin.com
Web: www.allenandunwin.com

Cataloguing-in-Publication details are available
from the National Library of Australia
www.librariesaustralia.nla.gov.au

ISBN 978 1 74175 571 8

MIX
Paper from responsible sources
FSC® C124385
www.fsc.org

Internal design by Avril Makula
Maps by Mapgraphics
Illustrations by Edwina Riddell
Printed in China at Everbest Printing Co

10 9 8

Contents

ABBREVIATIONS	ix
INTRODUCTION	1
THE SYDNEY ENVIRONMENT	3
MAJOR VEGETATION TYPES	11

FORK FERNS — 16
Psilotaceae Fork Fern — 16

CLUB MOSSES — 18
Lycopodiaceae Club Moss — 18
Selaginellaceae Club Moss — 19

FERNS — 20
Osmundaceae King Fern — 20
Schizaeaceae Comb Fern — 22
Gleicheniaceae Coral, Fan Fern — 23
Hymenophyllaceae Filmy Fern — 25
Cyatheaceae Tree Fern — 25
Dicksoniaceae Tree Fern — 26
Dennstaedtiaceae Ground Fern, Bracken — 26
Lindsaeaceae Screw Fern — 28
Pteridaceae Brake — 28
Adiantaceae Maidenhair Fern — 29
Davalliaceae Hare's Foot Fern — 31
Thelypteridaceae — 32
Aspleniaceae Spleenwort — 33
Athyriaceae Lady Fern — 34
Dryopteridaceae Shield Fern — 34
Polypodiaceae Elkhorn, Kangaroo Fern — 36
Grammitidaceae Finger Fern — 36
Blechnaceae Fishbone, Water, Rasp Fern — 38
Marsileaceae Nardoo — 41

CYCADS — 42
Zamiaceae Cycad — 42

PINES (CONIFERS) — 44
Cupressaceae Cypress Pine — 44
Podocarpaceae Podocarpus — 45
Araucariaceae Dwarf Pine, Wollemi Pine — 47

FLOWERING PLANTS — 48
DICOTYLEDONS — 49
Winteraceae Pepperbush — 49
Eupomatiaceae — 49
Monimiaceae Sassafras — 50
Lauraceae Laurel — 52
Cassythaceae Dodder — 54
Piperaceae Pepper Vine — 55
Peperomiaceae — 55
Ranunculaceae Buttercup — 55
Menispermaceae — 57
Ulmaceae Native Peach — 58
Moraceae Fig — 59
Urticaceae Nettle — 61
Casuarinaceae She-oak — 62
Aizoaceae Pigface — 65
Portulacaceae Purslane — 66
Caryophyllaceae Starwort — 67
Amaranthaceae — 68
Chenopodiaceae Saltbush — 68
Polygonaceae Knotweed — 70
Dilleniaceae Guinea Flower — 72
Clusiaceae St John's Wort — 80
Elaeocarpaceae Oliveberry — 80
Sterculiaceae Rusty-petals, Flame Tree — 82
Malvaceae Hibiscus — 87
Flacourtiaceae — 88
Violaceae Violet — 89
Passifloraceae Passionfruit — 92
Ericaceae Epacris, Styphelia — 93
Ebenaceae — 113
Sapotaceae — 113
Myrsinaceae River Mangrove — 114
Primulaceae Brookweed — 115
Euphorbiaceae Euphorbia — 115
Thymelaeaceae Rice-flower — 124
Droseraceae Sundew — 126
Cunoniaceae Christmas Bush, Coachwood — 127
Eucryphiaceae Leatherwood — 129
Baueraceae Dog Rose — 130
Grossulariaceae — 131
Pittosporaceae Pittosporum — 132
Crassulaceae Stonecrop — 135

Rosaceae Brambles, Blackberry	136
Mimosaceae Wattle	138
Caesalpiniaceae Cassia, Senna	164
Fabaceae Pea Flower	165
Proteaceae Banksia, Grevillea, Hakea	204
Myrtaceae Eucalyptus, Tea-tree, Bottlebrush	235
Onagraceae Willow Herb	308
Haloragaceae Raspwort	309
Sapindaceae Hop Bush	311
Meliaceae Cedar	316
Rutaceae Boronia, Philotheca, Zieria	317
Geraniaceae Geranium	341
Polygalaceae Matchhead	344
Tremandraceae Black-eyed Susan	345
Araliaceae Star-hair	349
Apiaceae Flannel Flower	352
Olacaceae	358
Santalaceae Native Cherry, Native Currant	359
Loranthaceae Mistletoe	362
Viscaceae Mistletoe	366
Icacinaceae	367
Celastraceae	368
Stackhousiaceae	369
Rhamnaceae Pomaderris	370
Vitaceae Native Grape	379
Oleaceae Olive	380
Caprifoliaceae Elderberry	380
Loganiaceae	382
Asclepiadaceae Milk Vine	383
Apocynaceae Silkpod	384
Menyanthaceae Marshwort	385
Rubiaceae Woodruff	386
Convolvulaceae Bindweed	390
Boraginaceae	393
Solanaceae Kangaroo Apple, Nightshade	394
Scrophulariaceae Speedwell	398
Bignoniaceae Wonga Vine	402
Gesneriaceae Fieldia	403
Lentibulariaceae Bladderwort	405
Myoporaceae Boobialla	406
Acanthaceae	408
Plantaginaceae Plantain	409
Verbenaceae	410
Avicenniaceae Grey Mangrove	411
Chloanthaceae	411
Lamiaceae Mintbush, Westringia	412
Campanulaceae Bluebell	426
Lobeliaceae Lobelia	428
Stylidiaceae Trigger Plant	432
Goodeniaceae Goodenia, Dampiera, Fan flower	433
Brunoniaceae Blue Pincushion	442
Asteraceae Daisies	443
MONOCOTYLEDONS	466
Alismataceae	466
Hydrocharitaceae Eel-weed	466
Juncaginaceae Water Ribbons	468
Potamogetonaceae Pond Weed	468
Arecaceae Palm	470
Araceae Settlers Flax, Spoon Lily	471
Commelinaceae	472
Xyridaceae Yellow-eye	474
Eriocaulaceae	477
Flagellariaceae Whip Vine	477
Restionaceae Rush	479
Centrolepidaceae	485
Juncaceae Rush	486
Cyperaceae Sedge, Rush	489
Poaceae Grass	516
Sparganiaceae	526
Typhaceae Bulrush	526
Zingiberaceae Native Ginger	527
Iridaceae Native Iris	527
Liliaceae Christmas Bell, Fringe Lily	530
Xanthorrhoeaceae Grass Tree, Lomandra	544
Smilacaceae Sarsaparilla	554
Dioscoreaceae Native Yam	556
Agavaceae Gymea Lily	556
Philydraceae Frogmouth	557
Amaryllidaceae Swamp Lily	558
Haemodoraceae Bloodroot	558
Hypoxidaceae	558
Orchidaceae Orchid	561
GLOSSARY	592
LEAF AND FLOWER SHAPES	596
COMMON NAME INDEX	597
SCIENTIFIC NAME INDEX	604
FURTHER READING	614

Abbreviations

Geographical features

bch	= beach
Blue Mtns	= Blue Mountains
ck	= creek
e.	= east
Gt Div. Ra.	= Great Dividing Range
hd	= head
h'lands	= highlands
l'out	= lookout
mtns	= mountains
n., nth	= north
NP	= National Park
NR	= Nature Reserve
pantrop.	= pantropical
Pk	= Park
Pl.	= Plain
Plat.	= Plateau
R.	= River
SCA	= State Conservation Area
se.	= south-east
SF	= State Forest
s., sth	= south
Sthn H'lands	= Southern Highlands
Sthn T'lands	= Southern Tablelands
t'land	= tableland
trop.	= tropical
V	= Valley
w.	= west

States

ACT	= Australian Capital Territory
N Terr	= Northern Territory
NSW	= New South Wales
Qld	= Queensland
SA	= South Australia
Tas	= Tasmania
Vic	= Victoria
WA	= Western Australia

NSW regions

CC	= Central Coast
CT	= Central Tablelands
CWS	= Central Western Slopes
FWP	= Far Western Plains
NC	= North Coast
NT	= Northern Tablelands
NWP	= North Western Plains
NWS	= North Western Slopes
SC	= South Coast
ST	= Southern Tablelands
SWP	= South Western Plains
SWS	= South Western Slopes
WP	= Western Plains
WS	= Western Slopes

Other regions

Afr.	= Africa
Amer.	= America
Eur.	= Europe
LHI	= Lord Howe Island
N Cal.	= New Caledonia
Norfolk Is.	= Norfolk Island
NZ	= New Zealand
Pac. Iss.	= Pacific Islands
PNG	= Papua New Guinea

Others

±	= more or less
<	= less than
>	= greater than
aff.	= affinis, affinities with
cm	= centimetre
diam.	= diameter
Dist.	= Distribution
e.g.	= for example
esp.	= especially
Fl.	= Flowering
m	= metre
mm	= millimetre
subsp.	= subspecies
syn.	= synonym
var.	= variety

Introduction

The Sydney district is one of the great wildflower regions of Australia, containing over 2000 native plant species. It compares favourably with other recognised wildflower areas, such as the sand plains of south-west Western Australia, the Grampians in Victoria, the Wallum heathlands in Queensland and the arid Centre after rains. The amazing variety shown in the following pages is proof that the Sydney district is home to a significant and enduring part of the nation's floral wealth.

The impact of the plant life on early Europeans is evident in the naming of Botany Bay by James Cook in April 1770: botanists Joseph Banks and Daniel Solander, who accompanied Cook, collected enough plants in their short stay to require 2400 sheets of drying paper, and Banks' collection aroused such interest in Europe that Linnaeus suggested that the newly discovered southern land be called Banksia. Throughout the nineteenth century, Australia was a new world open to botanical enquiry, attracting many scientists and enthusiasts who worked with boundless energy exploring, collecting, drawing and classifying.

Sydney's flora still evokes a special fascination. Increasing numbers of people visit national parks and other natural areas to experience our native plants. There is a trend in both public and private landscaping towards the growing of native plants. Public involvement in bush regeneration and environmental study courses is increasing. While many of us enjoy the special feeling of 'being in the bush' and are aware of its nuances and beauty without needing to know the names of particular plants, for others a more detailed understanding of the plants provides an added dimension. There is much pleasure to be had by the discovery of a 'new' plant or the rediscovery of an old favourite. Often without noticing, we develop an insight into plant variety and relationships and begin to see order in the apparent confusion.

Classification and arrangement

The naming of families and the segregation of genera into families are based on the classification systems of Takhtajan and of Cronquist and consequently differ in some respects from other publications which use alternative classifications.

We have used common terms to describe the six major categories of plants recognised by modern taxonomy and present in the Sydney region. These are all of equal taxonomic rank. Those exhibiting features (character states) considered by botanists to be primitive are presented ahead of those considered to be advanced. The first three groups (with others) are popularly called ferns and fern allies. Cycads and pines (with others) are sometimes called gymnosperms and the flowering plants may be called angiosperms. Although not part of current formal naming, these terms are useful and convenient in referring to the major groups.

Geographical coverage of this book

This book covers an area of some 23,000 square kilometres stretching from Newcastle to the Shoalhaven River and from the coast to the watershed of the Great Dividing Range. It includes most of the geological unit known as the Sydney Basin and the catchment area of the Nepean–Hawkesbury rivers (except the upper Wollondilly River). It also represents the botanical subdivisions of Central Coast (CC) and the Central Tablelands (CT) east of the Great Dividing Range. In the individual plant descriptions, these are referred to as 'coast' and 'tablelands'; 'central' is assumed.

Boundaries in the north and west are determined by dividing ranges. The Watagan Mountains and the complex of sandstone ridges and plateaus running westwards through Bucketty along the

Hunter Range north of Howes Valley to Mount Coricudgy (1254m) separate the north-flowing streams of the Hunter system from the south-flowing streams of the Hawkesbury system. This natural boundary has been extended in the north-east to include the southern portion of Newcastle and those small creeks and slow-moving rivulets which flow into Lake Macquarie. From Mount Coricudgy to west of Wombeyan Caves, the Great Dividing Range forms a natural boundary. The area bordered by Marulan, Goulburn, Crookwell and Taralga in the upper reaches of the Wollondilly River is in the Southern Tablelands (ST) botanical subdivision and is excluded. On the coast, the Shoalhaven River forms a convenient southern boundary.

The coastal plain is surrounded by a semi-circular belt of rugged sandstone plateaus, which are eroded into a complex of walled gorges, steep slopes and infertile ridges. This rough country effectively hemmed in the early colony and, because it was unsuitable for agriculture or grazing, was mostly left uncleared. Today's road and rail links with the rest of the State have been built along interconnecting ridges or by massive construction feats cutting deep into sandstone ridges or infilling gullies. National parks and water catchment areas have been established over much of the sandstone belt, preserving most of the plant species discovered by the first European explorers and botanists.

MAP 1: Botanical divisions and subdivisions of New South Wales

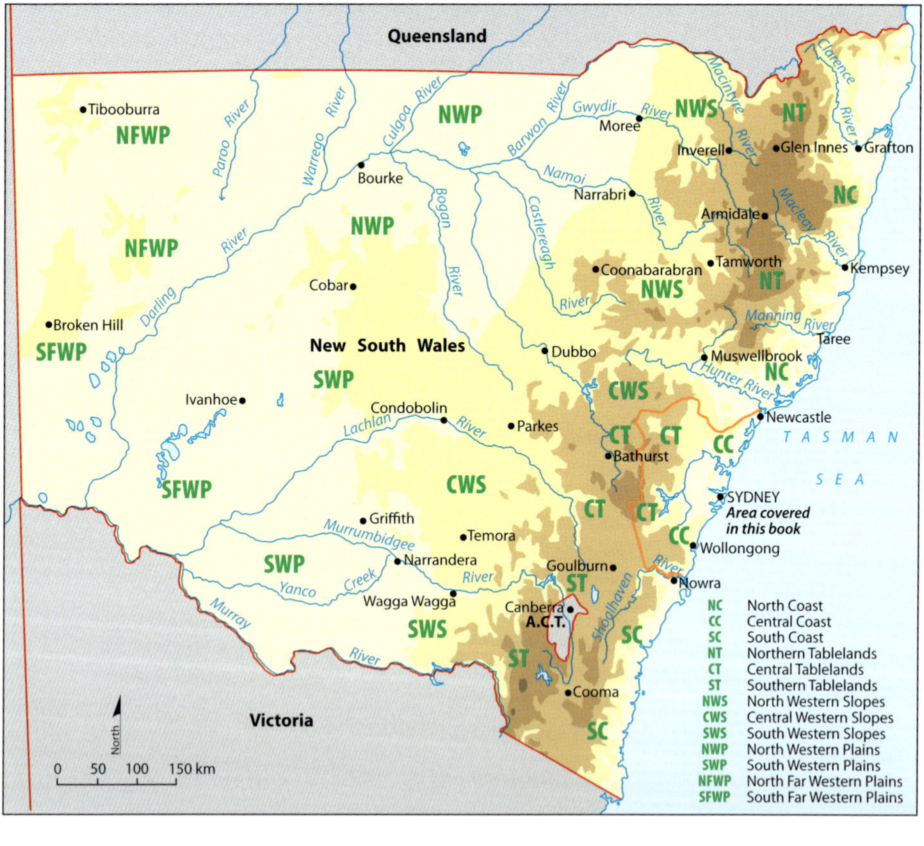

The Sydney environment

The Sydney district can be divided into a number of zones based on rock types and topography. Because the distribution and variety of plants are linked to the nature of the soils, topography and climate, these zones are also convenient botanical subdivisions.

Cumberland Plain

The Cumberland Plain is a gently undulating basin dipping westwards from Bankstown to the Nepean River and from Wilberforce southwards to Picton and Menangle. It consists of a deep, almost unbroken layer of Wianamatta shale at altitudes ranging from 20 to 100m above sea level. Poorly consolidated Tertiary alluvial deposits, mainly clays and silts, often laterised, occupy the area from Penrith to Windsor and Riverstone. Recent sand and clay alluvium occurs along the banks of the Nepean–Hawkesbury River, along South and Eastern creeks and around Liverpool.

The heavy clay soils of the Plain are often poorly drained and poorly aerated, hence swampy depressions are common. Summer temperatures are usually higher than nearer the coast and winter frosts can be severe. Rainfall is considerably less than along the coastal margin: in some parts it is about half that for Sydney, where the average is 1209mm per year. It is not surprising then that many plants that are adapted to the conditions of this zone are absent from the surrounding sandstone areas.

On the hills around Cecil Park and The Oaks, there are stands of Spotted Gum (*Corymbia maculata*) and Grey Box (*Eucalyptus moluccana*) with undershrubs such as *Bursaria spinosa* and *Dillwynia sieberi*. Cabbage Gum (*E. amplifolia*) is common on alluvial flats of tributary creeks, and the presence of Swamp Oak (*Casuarina glauca*) indicates a saline influence in the ground water. Forest Red Gum (*E. tereticornis*) occurs on better-drained slopes.

The clay flood plains of Eastern Creek near Doonside and South Creek near Shanes Park support a number of uncommon herbs and shrubs, including *Murdannia graminea*, *Zornia dyctiocarpa* and *Eremophila debilis*. On the western margin, on the river bank and flood plain of the Nepean River at Bents Basin, there is a stand of *Eucalyptus benthamii*, a tree once much more widespread but now rare because of clearing and the flooding of part of its range by the Warragamba Dam.

Near Agnes Banks is an unusual area of white sand of Pliocene or Pleistocene age. It originally covered an area of about 600ha, but sand extraction has reduced it to less than half that size. The sand forms a low stable set of east-west dunes with an average height of 3–4m. Swampy depressions lie between the main dunes. This area is botanically interesting because of its floristic and structural affinities with the sand dune vegetation of the coast. Here, for example, is *Banksia aemula*, a coastal species not found south of Botany Bay. Dominant trees of the sand deposit are *Eucalyptus sclerophylla* and *E. parramattensis*. On the clays and silts surrounding the sand, the larger trees are *E. fibrosa* and *Melaleuca decora*. A small portion of this sand deposit is protected by a nature reserve.

The Castlereagh Nature Reserve is also of interest since it contains the best remaining open forest community of *Eucalyptus fibrosa* and the most extensive area of undisturbed clay soil vegetation on the Cumberland Plain. Here are found the rare *Allocasuarina glareicola*, the only stand of *Dodonaea falcata* in the Sydney district, and such clay-loving species as *Chorizema parviflorum*, *Grevillea juniperina* and *Macrozamia spiralis*.

Further south at Razorback Mountain, between Camden and Picton, a protected west-facing slope of Wianamatta shale supports a dry rainforest containing a number of unusual species. There is a disjunct occurrence of the woody climber *Deeringia amaranthoides*, the yellow-flowered *Abutilon oxycarpum*, the rare climber *Cissus opaca* and the leafless shrub *Spartothamnella juncea*.

4 NATIVE PLANTS OF THE SYDNEY REGION

MAP 2: Major river systems and geographical zones of the Sydney region

THE SYDNEY ENVIRONMENT 5

MAP 3: Major locations

Hornsby Plateau

The area north of Sydney, known as the Hornsby Plateau, rises gradually from Port Jackson like the rim of a giant saucer to an elevation of 640m at Mount Warrawolong, the highest point in the Watagan Mountains on the Hunter–Hawkesbury divide. The general height of the plateau is 200–300m and the major outcropping rock is Triassic Hawkesbury sandstone. This has been eroded by numerous streams to form deep V-shaped valleys, a maze of rugged, narrow ridges and tree-clad slopes. The Hawkesbury River carves deep valleys through the plateau before emptying into Broken Bay. Feeding the Hawkesbury River are streams such as the Macdonald River, Mangrove Creek and Mooney Mooney Creek which have cut through the Hawkesbury sandstone to expose the underlying strata of the Narrabeen group of rocks, also of Triassic age. More recent deposits of silt and sand have been built up along the major watercourses. In the lower reaches of the Hawkesbury, the valleys have been flooded by rising sea-levels, forming 'drowned valleys' such as Cowan Waters, Berowra Creek and Mooney Mooney Creek, distinctive features of Broken Bay.

Pockets of heath and low open woodland are commonly found on rocky platforms and ridge tops. Taller open forest occurs on deeper plateau soils and on slopes. Soil removed from the slopes accumulates in the valleys where it may be relatively deep and contain a high percentage of humus. Where the Narrabeen group has been exposed in the deeper valleys along the western shores of Pittwater and along the eastern rim of the plateau, the eroded shales form deep clay-rich soils with a resultant change in the vegetation. In deeper cool valleys, there is a lush growth of tall trees, rainforest shrubs, ferns and herbs. Here emergent trees may be up to 30m tall and there is often a dense canopy cover. Trees such as *Eucalyptus deanei*, *Syncarpia glomulifera* and *Acacia elata* are mixed with rainforest elements.

Volcanic soils are uncommon in the area, although there are a few small intrusions of Tertiary basalt such as at Mount Wareng (Howes Valley), Basalt Hill (Kulnura) and Dillons Crater (Brisbane Water National Park) which have weathered to dark clay soils rich in iron and magnesium. A feature of the tidal inlets of Broken Bay is the presence of mangrove communities backed by reedlands and salt-marsh flats subject to frequent tidal inundation. On higher ground of the alluvial flats, the main trees are Swamp Oak (*Casuarina glauca*) and Prickly-leaved Paperbark (*Melaleuca styphelioides*). Further north, the Watagan State Forests give a degree of protection to valuable natural areas. Here are plants not found elsewhere in the Sydney district, often species that are common further north.

Northern Coast

In the north-eastern corner, the Hornsby Plateau is eroded to expose the more fertile Narrabeen shales and sandstones. These extend from Brisbane Water northwards and for the most part have been cleared of native vegetation. The southern section along the coastline of Bouddi National Park has a rugged cliff-line of Narrabeen sandstone and extensive eroded rock platforms. Higher ridges are capped by Hawkesbury sandstone. Further north the landscape is dominated by a series of shallow lakes—Tuggerah Lake, Munmorah Lake and Lake Macquarie—around which are extensive sand and alluvial deposits. Sands transported by the currents along the shore have been deposited as beaches and long strips of sands which trap the waters of the lakes. In some areas these sands support unique examples of native vegetation, such as the *Angophora costata* forest at The Entrance North, magnificent coastal heaths and swampy gullies dominated by Cabbage Tree Palms and sedges. Slow-moving streams such as Ourimbah Creek, Wyong Creek and Dora Creek are fringed in parts by remnant rainforest vegetation.

A feature of the Gosford–Ourimbah area is the presence of warm temperate rainforests, sometimes just narrow strips along creeks, or larger areas in Strickland, Ourimbah and Wyong State Forests which containing significant stands with a diversity of species. They are part of a large biogeographical complex extending from Broken Bay northwards and are important natural museums contributing to the preservation of rainforest species in the Sydney district. No significant coastal rainforest stands occur between Broken Bay and the Narrabeen outcroppings of the upper Hacking River (Royal National Park), so it is unsurprising that a number of rainforest species do not extend south beyond Broken Bay.

Colo Plateau–Wollemi

Colo Plateau is a high area of Hawkesbury sandstone eroded in the east by the Macdonald River and in the west by the Colo River and Wollemi Creek. Each river has a dense network of tributary creeks eating into the sandstone and producing a maze of gorges and ridges. The main north-south ridge, known as Mellong Range, carries the Putty Road, and branching from it like stubby fingers are Howes Range, Womerah Range and Culoul Range. The general height of the plateau is 200–300m on the eastern edge, rising to 600m on the western edge.

Remnants of Wianamatta shale occur on the higher ridge-tops, especially along the Putty Road north of Colo Heights and on Culoul Range where richer soils on protected southern slopes support a tall open forest of Mountain Mahogany (*Eucalyptus notabilis*), Grey Ironbark (*E. paniculata*) and Grey Gum (*E. punctata*). A deposit of alluvial sands on the Mellong Range has a clayey subzone which retains water near the surface. These conditions produce a distinctive plant community dominated by Narrow-leaved Apple (*Angophora bakeri*) and Drooping Red Gum (*E. parramattensis*).

West of the Colo Plateau, Narrabeen sandstones are up to 800m thick in parts and form the spectacular cliff-lines on the western rim of Wollemi National Park where the Wolgan and Capertee rivers cut deep gorges on their way eastwards from Newnes and Glen Davis.

Northern Blue Mountains

This region extends from Lithgow and Mount Wilson northwards across the Newnes Plateau to the old shale-mining settlements of Newnes and Glen Davis. The north-western corner is bounded by the Great Dividing Range sweeping in a broad arc to encircle the upper reaches of the Capertee River. It includes some of the highest parts of the Blue Mountains, rising to over 1000m on the Newnes Plateau (e.g. Birds Rock, 1181m). The dissected plateau above Glen Davis is 750–850m high, and temperatures are often very low in winter. To the east, a complex of streams cuts into the Narrabeen Series forming long spurs, deep gullies, cliff-lined canyons and isolated rocky outcrops. It is within this maze of gorges that the unique Wollemi Pine (*Wollemia nobilis*) was discovered in 1994. Eroded sandstone formations known as 'pagodas' standing conspicuously above the surrounding bushland are a feature of the landscape. Another feature of the plateau is the presence of headwater swamps, formed in broad shallow valleys on a clay base which impedes drainage.

In the north-west, erosion and undercutting have produced many prominent topographical features such as Baal Bone Point, Genowlar Mountain and isolated flat-topped mountains of Narrabeen sandstone, such as Mount Airly (1034m) and Pantoney's Crown (1000m). Beneath the cliff-lines, the talus slopes consisting of sandstone blocks, Permian soils and scree material from the cliffs above constitute a specialised habitat for a number of interesting plants including the rare *Prostanthera cryptandroides*.

The cold higher plateau has woodland dominated by *Eucalyptus dives* and *E. radiata*, with Snow Gum (*E. pauciflora*) common in some areas. Mallee eucalypts, *E. stricta* and *E. gregsoniana*, are common in heath on exposed shallow sandy soils. The high swamps contain disjunct populations of Snow Daisy (*Celmisia longifolia*) and the endemic *Boronia deanei*. *Banksia penicillata* is scattered in open forest on the plateau. Because of the rainshadow effect of the surrounding plateau and ranges, the valleys of the Capertee and Newnes receive less than 900mm of rain a year. A combination of low rainfall and infertile, skeletal soil produces woodland communities of trees such as *Eucalyptus albens*, *E. melliodora* and *Brachychiton populneus*, which are more common further west. A number of acacias are typically western slopes species. With such a variety of soil types, drainage and rainfall patterns, altitude, exposure and topography, it is not surprising that the complex of plant communities within the zone contains a number of rare or endemic species.

Blue Mountains

A steep monocline marks the eastern edge of the Blue Mountains, which rise gradually from 200m above sea level at Glenbrook to 1094m at Mount Piddington. The western rim is a series of cliff-lines above the Hartley, Megalong and Kanimbla valleys.

The lower Blue Mountains from Lapstone to Woodford consists of Hawkesbury sandstone. Further west this is replaced by rocks of the Narrabeen group which form the major part of the central and upper Blue Mountains plateau and the cliffs at Wentworth Falls, Katoomba and Blackheath. This sandstone is highly eroded and in places giant pillar-like fragments, such as the Three Sisters and Orphan Rock at Katoomba, are left standing separated from the main plateau surface. The cliff-fringed remnant of Mount Solitary and the long fingers of Narrow Neck and Kings Tableland are further evidence of the erosion processes affecting the Narrabeen group.

Residual Wianamatta shale caps occur along the Bells Line of Road from Kurrajong to Mount Tomah, but much of their native vegetation has been cleared for the growing of apples. Where the native vegetation is preserved, the well-drained clay soils support an open forest of Blue Mountain Mahogany (*Eucalyptus notabilis*), White Stringybark (*E. globoidea*) and Turpentine (*Syncarpia glomulifera*), with occasional examples of Mountain Blue Gum (*E. deanei*).

The Blue Mountains landscape is dotted with prominent peaks capped by basalt remnants. Mounts Wilson, Tomah, Irvine, Caley, Hay and Banks all have a thin basalt topping. This is 80m thick at Mount Tomah and covers an area of 240ha. Due to a high rainfall, rainforest and fern communities are present on the sheltered hillsides and gullies of Mount Wilson and Mount Tomah. Unusual species, such as Plume Bush (*Calomeria amaranthoides*) and *Prostanthera caerulea*, occur on these richer soils.

The Blue Mountains are a refuge for many rare or threatened plants. *Acrophyllum australe* is known chiefly from cliff-lines in the Woodford–Linden area. *Isopogon fletcheri* grows only on the cliffs at Govetts Leap near Blackheath. *Eucalyptus burgessiana* is restricted to ridges near Linden and Mount Tomah. *Darwinia fascicularis* subsp. *oligantha*, *Almaleea incurvata* and *Acacia ptychoclada* are all endemic in the Blue Mountains. Some species have localised populations in the mountains separated from other occurrences. *Darwinia peduncularis* occurs on Kings Tableland and in the Hornsby–Hawkesbury area. *Eucalyptus baeuerlenii* grows on south-facing cliff-lines at Wentworth Falls and near Braidwood. *Eucalyptus pulverulenta* is found along the Coxs River and at Bredbo in the Southern Tablelands.

Southern Blue Mountains–Nattai

The Southern Blue Mountains is the least visited part of the Sydney district. Rugged terrain and restricted access to water catchment areas require visitors to make long journeys to reach scenic spots such as Kanangra Walls, Yerranderie, Wombeyan Caves and the Nattai Plateau. The Boyd Plateau, a dominant feature of the area, reaches elevations of over 1300m in its highest parts and covers more than 13,000ha. Although its forests have been extensively logged, *Eucalyptus fastigata*, *E. radiata*, *E. dalrympleana* and *E. dives* are still common. Other cold-country eucalypts such as *E. pauciflora*, *E. ovata* and *E. stellulata* occur frequently. Alpine swamps such as those along Luthers Creek provide habitats for alpine sedges and herbs. Other species found here and sometimes in sub-alpine regions further south include *Oxylobium ellipticum*, *Grevillea imberbis* and *Hakea dohertyi*. A number of rare or uncommon plants occur in the Yerranderie–Kowmung area, including *Acacia clunies-rossiae*, *A. flocktoniae*, *Grevillea kedumbensis* and *Seringia arborescens*, as well as an unnamed species of *Kunzea*.

The geology south of the Kowmung River is complex, with outcroppings of Ordovician and Silurian rocks overlain by Permian sandstones near Yerranderie and small intrusions of granite. Due to a rainfall of only about 800mm a year, the hillsides, especially on the exposed northern and eastern slopes, are dry and open, supporting a woodland of Grey Box (*Eucalyptus moluccana*) and Yellow Box (*E. melliodora*), with a low prickly understorey. Caves occur in limestone outcrops at Jenolan, Colong, Billys Creek and Wombeyan. *Acacia chalkeri* from Wombeyan is restricted to the limestone habitat.

East of the Wollondilly River the high plateau country forming the catchment of the Nattai, Wingecarribee and Little Rivers consists predominantly of Hawkesbury sandstone. Several peaks, such as Mount Jellore, Mount Wanganderry and Mount Flora, are of Tertiary volcanic origin. Tall River Oaks (*Casuarina cunninghamiana*) line the banks of the Wingecarribee, Nattai and Wollondilly rivers.

Woronora Plateau–Illawarra

The Woronora Plateau forms the southern part of the sandstone basin surrounding Sydney. It ascends gradually from Botany Bay to link up with the richer shale and basalt of the 700m high Robertson Plateau in the Southern Highlands. Along the coast, the narrow Illawarra Plain widens to about 25km near Shellharbour. The Illawarra Escarpment, which forms the western margin of the plain, is a dramatic cliff-line of weathered Hawkesbury sandstone above exposed softer rocks of the talus slopes, benches and foothills. West of the escarpment, the sandstone dips gently towards the Wianamatta shales between Bargo and Campbelltown. Streams such as the Woronora, Cataract, Cordeaux and Avon rivers have their headwaters near the escarpment but flow westwards into the Nepean and Georges river systems.

A mixture of rainforest types occurs in the Illawarra. Warm-temperate rainforest is the most widespread, occurring in gullies, along creeks on soils derived from the Illawarra coal measures and on talus slopes above the benches of the escarpment. These rainforests, at altitudes of 150–450m, receive about 1600mm of rain a year. The dominant trees are Coachwood (*Ceratopetalum apetalum*), Sassafras (*Doryphora sassafras*), Brown Beech (*Cryptocarya glaucescens*) and Lillypilly (*Acmena smithii*). Subtropical rainforest elements occur in gully bottoms at Minnamurra Falls and near Jamberoo where the soils are enriched by latite from the Gerringong volcanics.

Other features of the region are:
- The chocolate-brown soils of the coast south of Royal National Park which support a remnant of littoral rainforest and a number of plants more at home in the rainforests north of Sydney.
- Sedgeland where drainage of shallow upland valleys is poor (such as at Maddens Plains). These areas are dominated by sedges and rushes, and contain a number of rare plants such as *Euphrasia collina*.
- Bass Point with its littoral rainforest on Gerringong volcanics and rainforest patches on Tertiary basalt cappings of Saddleback Mountain.
- Cappings of Wianamatta shale between Menai and Liverpool where a high number of uncommon plant species occur including *Grevillea diffusa*, *Melaleuca deanei*, *Rulingia dasyphylla* and an unnamed *Hibbertia*.
- The stabilised sand spit at the mouth of Minnamurra Sands which supports depauperate rainforest containing *Celtis paniculata* at its southern limit, *Crinum pedunculatum*, *Myoporum boninense*, the rare sedge *Cyperus enervis* and large trees of *Endiandra sieberi*.
- The Shoalhaven delta consisting of a beach sand-ridge system, backed by alluvial floodplains and terraces containing swamps and low-lying marshy ground.

Southern Highlands

The Southern Highlands is a fertile plateau of more than 2600 square kilometres, with peaks rising to between 600–870m above sea level. It covers the area between Mittagong, Robertson and Bundanoon, with fingers extending to Wingello and Barren Grounds. Its fertile nature is mainly a result of an extensive cover of Wianamatta shale which has weathered into a dark grey soil enriched from basalt hills scattered throughout the area. Although many of the richer parts have been cleared, an outcrop of volcanic trachyte rock at Mount Gibraltar (863m) between Mittagong and Bowral retains much of its original vegetation, and the Robertson Nature Reserve (45ha) preserves a mixed warm-temperate/cool-temperate rainforest.

The plateau is besieged on all sides by streams which are eating away at the fertile capping. In the west beyond Berrima, the Wingecarribee River cuts deep gorges into the rolling landscape. In the north, tributaries of the Nepean have exposed the Hawkesbury sandstone and in the south-east the Kangaroo River has eroded its way down to the Shoalhaven group below. However, it is in the south that the most spectacular changes occur, for here the plateau ends abruptly in a series of cliffs and deep V-shaped gorges. Streams such as Bundanoon Creek, Yarrunga Creek and Barrengarry Creek start in small swamps on the plateau, then plunge in spectacular waterfalls over the cliff-line before making their way in a series of cascades down narrow valleys to the Shoalhaven River. These gorges and falls such as Fitzroy, Belmore and Carrington are major scenic features of the area.

Extensive swamps occur on the plateau, especially Wingecarribee Swamp, and in Barren Grounds Nature Reserve and Budderoo National Park. Heaths also occur on poorly drained soils derived from Hawkesbury sandstone. Rare plants in these areas include *Grevillea rivularis*, *Viola silicestris*, *Oxylobium arborescens* and *Almaleea paludosa*.

The area between Penrose and Tallong has a drier climate and the plant communities show relationships with those of the Southern Tablelands. A number of species from this area are absent from, or uncommon in, the rest of the Sydney district. These include *Grevillea molyneuxii*, *Hibbertia praemorsa*, *Rulingia prostrata*, *Acacia jonesii*, *Actinotus gibbonsii* and *Asterolasia buckinghamii*.

Major vegetation types

Heath

Heaths are among the richest plant communities in the world, and over 750 species occur in the coastal heaths around Sydney. If defined by structure, a heath community is one dominated by shrubs less than 2m tall having a foliage cover of 30–70% (open heath) or 70–100% (closed heath).

Heaths are characterised by their variety of hard-leaved plants whose size and toughness are a response to shallowness of soil and low nutrient levels. They are found on coastal sands exposed to onshore wind and sea spray and on other exposed sites with shallow sandy soils. Closed heaths are common in the Blue Mountains in shallow headwater valleys where poorly drained, sandy, acid peat soils have developed.

There are fine examples at West Head, Kurnell, Jibbon, Barren Grounds, Narrow Neck (Katoomba) and Clarence–Newnes.

Scrub

Scrub communities are associated with sandy, poorly-drained soils on a wide range of topography but most often on ridge tops and slopes and along drainage lines bordering swampy sedgelands. They consist of shrubs 2–8m tall without tree cover, often with single-species thickets and dense growth of prickly-leaved plants.

Scrub-lined swamps are a feature of Blue Mountains plateaus where shrub height and density are influenced by drainage, soil type and fire frequency. *Leptospermum* species are common in such areas. Although large trees are absent, small trees from neighbouring woodland may invade the more open, better-drained areas in scrubland.

Woodland

Woodlands are open areas dominated by trees (usually eucalypts in Australia), whose canopy covers 10–30% of the ground area. Because the scattered nature of woodland trees is often a result of fairly dry conditions, woodlands are common in the western valleys of the ranges, on plateau tops and slopes and in low-rainfall parts of the Cumberland Plain.

On the coast, low open woodlands occupy less exposed sites than heath and grade into low open forest. *Angophora costata*, *Eucalyptus haemastoma*, *Corymbia gummifera* and *Banksia serrata* are common trees in coastal woodlands.

The understorey is rich in species. Woodlands of *Eucalyptus moluccana* and *E. tereticornis* were once common on the clay soils of the Cumberland Plain, but clearing has greatly reduced this community. Ground cover on these clays is generally low and open, with scattered shrubs and native grasses.

Both the structure and floristic composition of woodland varies considerably. The woodland of *Eucalyptus longifolia* and *Angophora floribunda* found on the Illawarra Plain, for example, bears little resemblance to the *E. rossii* woodland on the dry steep slopes of Wolgan Valley. A specialised woodland with dominant tall trees of *Casuarina cunninghamiana* and *Angophora floribunda* occurs on the banks and flats of some streams and rivers.

Open forest

Open forest is the predominant vegetation type over much of the Sydney district. It consists of trees to 30m tall, with a canopy cover of 30–70% of the ground area, and a dry understorey of smaller flowering plants.

In the coastal zone, open forest occurs on deeper sands, in gullies and on sheltered hillsides, especially with southern and eastern aspects. Here the main trees are *Eucalyptus piperita*, *Angophora costata*, *Corymbia gummifera* and *Banksia serrata*. *Eucalyptus sieberi* may be locally common on coastal plateaus. *Eucalyptus fibrosa* forests occur on alluvial soils in the western part of the Cumberland Plain, often with other ironbarks. *Eucalyptus sieberi/piperita* forests dominate the sandstone plateaus and ridges of the Blue Mountains between Wentworth Falls and Bell; stringybarks may also be present.

Many of the characteristic and most beautiful of Sydney's wildflowers are found in the understorey of these forests. Waratahs, Gymea Lilies or grass trees can be locally common, and wattles, boronias, grevilleas, peas, native irises, lilies and ground orchids all add their colour and magic to the understorey.

MAJOR VEGETATION TYPES   13

Tall open forest

This community occurs on the richer fertile soils of basalt caps and eroded necks, on enriched soils of sheltered slopes and in deep sheltered gullies. It consists of trees 30m or more tall, usually with straight trunks, with a canopy cover of 30–70%. Rainforest elements are often present in the understorey.

The slopes of basaltic peaks like Mount Wilson support tall trees of *Eucalyptus viminalis*, *E. cypellocarpa* and *E. fastigata*. *Eucalyptus oreades* is common on more rocky slopes on the edge of the basalt. In the eroded dykes of the lower Blue Mountains there are beautiful trees of *E. deanei* and tall *Syncarpia glomulifera*. At Blue Gum Forest in the Grose Valley, *E. deanei* is associated with *Backhousia myrtifolia* and an open grassy understorey. In sheltered gullies of the North Shore, *E. saligna* may dominate while on the upper slopes of the Illawarra Escarpment, on benches between rainforest or on foothill spurs, the main trees are *E. smithii*, *E. quadrangulata*, *E. muelleriana* and *Allocasuarina torulosa*. The understorey of these taller, moist eucalypt forests consists of climbers and scramblers, tall broad-leaved shrubs and a ground cover of ferns and herbs.

Rainforest

Rainforest is defined as a tree-dominated community with a closed canopy (70–100% foliage cover) where the upper leaves are in close and intermingled contact. In such an environment, trees tend to be straight-stemmed, flowers and fruit are borne high in the canopy, leaves are dark and broad and there are plentiful vines and epiphytes. Since little light reaches the ground, only scattered shrubs and ferns are present in the understorey. Suitable weather, soil and shelter occur in a number of places in the Sydney district.

The basaltic soils of Mount Wilson, Mount Tomah and the Southern Highlands support warm temperate rainforest, as do the soils of the Gerringong volcanics and coal measures along the Illawarra Escarpment. These are often dominated by *Ceratopetalum apetalum*, *Acmena smithii* and *Doryphora sassafras*. There are significant warm temperate rainforests in the Gosford–Wyong district, Palm Jungle–Hacking River area and in valleys of the Blue Mountains. Remnant stands of subtropical rainforest occur in the Illawarra and Kiama districts. There are significant differences in species composition between the littoral rainforest of Palm Jungle, Bass Point and Minnamurra Spit, the dry rainforest of Razorback Mountain, the mixed warm temperate and subtropical rainforest of Minnamurra Falls and the temperate rainforest on the basalt of Mount Wilson.

Sedgeland swamp

Swamplands dominated by members of Cyperaceae and Restionaceae occur commonly in plateau areas, especially in headwater valleys where impervious sandstone or clay layers create waterlogged conditions and a build-up of organic matter in the soil. As places of permanent moisture, they support specialised plants such as insectivorous sundews and bladderworts, yellow-flowered *Xyris*, *Leptospermum juniperinum*, *Epacris obtusifolia*, and *Sprengelia incarnata*. *Gahnia* sedge-thickets are common along drainage lines. The presence of alpine genera such as *Celmisia*, *Ranunculus* and *Sphagnum* in many higher mountain swamps indicates a former continuity between these areas and the alpine bogs of the Snowy Mountains.

Fresh water

There are few fresh water lakes, lagoons or river overflows in the Sydney district. Thirlmere Lakes, Mountain Lagoon, Marley Lagoon, Kings Waterhole (Putty), some swamps on the Cumberland Plain, scattered farm dams and river backwaters provide a few suitable sites for fresh water plants but pollution and infilling have destroyed much of this habitat.

Some fresh water plants such as *Ottelia ovalifolia*, *Triglochin procerum* and *Nymphoides geminata* are rooted in mud and have leaves which float on the water surface. A few, like *Vallisneria gigantea*, are completely submerged. Lagoons and swamps are often bordered by rushes (*Baumea*, *Eleocharis* spp.) and such wetland species as *Persicaria decipiens*, *Typha orientalis*, *Philydrum lanuginosum*, *Villarsia exaltata* and *Damasonium minus*.

Mangrove and salt marsh

Estuarine vegetation occurs in protected coastal areas along the tidal zone or on alluvial flats subject to regular tidal inundation but not exposed to wave action. This complex consists of a sequence of zones determined by salinity levels and the frequency and duration of flooding. Two species of mangrove occur on the seaward edge of the mudflats. The River Mangrove is usually on the landward side of the Grey Mangrove and often extends further up estuaries where the water is less saline. There is often a zone of salt marsh dominated by low succulent-stemmed Samphire (*Sarcocornia quinqueflora*) and Seablite (*Suaeda australis*) behind the mangroves. These salt marsh areas range from small (e.g. Patonga Creek, Salt Pan Creek) to extensive (e.g. Towra Peninsula). Behind the salt marsh, in areas infrequently covered by salt water, there is often a zone of rushes or reeds. The rush *Juncus kraussii* dominates boggy or shallow saline areas. *Phragmites australis* is found in less saline habitats. Here too may be found the small herbs *Wilsonia backhousii*, *Tetragonia tetragonoides* and *Samolus repens*. Swamp Oak (*Casuarina glauca*) and Swamp Mahogany (*Eucalyptus robusta*) occupy slightly drier ground on the landward side.

Coastal dunes

The area between beach and hind-dune woodland is an inhospitable place for plants. Frontal dunes are usually unstable and have poor water-holding capacity. On the seaward slope, the first plants to stabilise the sand are the grasses *Spinifex sericeus* and *Austrofestuca littoralis*. *Lomandra longifolia* and *Acacia sophorae* occur, often with the exotic *Hydrocotyle bonariensis*, on more stable sand. On the inner dune slope away from salt winds *Leptospermum laevigatum*, clumps of *Hibbertia scandens* and the climber *Stephania japonica* may occur. Shallow depressions which retain moisture often form between the dunes and some areas develop small lagoons immediately behind the dune area.

Fork ferns

ONLY TWO SMALL genera, *Psilotum* and *Tmesipteris*, survive as descendants of a group of land plants which was abundant during the Devonian and Silurian periods, 360–440 million years ago, possibly the earliest and most primitive of the vascular plants. They lack true roots and leaves, the functions of these being performed by the stems which provide photosynthesis in the upper part and anchor the plant and absorb water in the lower. Reproduction is by spores borne in sporangia, which are fused into structures called synangia.

PSILOTACEAE

Psilotum
Plants with much-branched stems. Synangia capsule-like and 3-lobed. One species in Sydney district.

Psilotum nudum

Psilotum nudum
Skeleton Fork Fern
Small plant with erect to trailing 3-ribbed yellowish stems to 60cm long, forked repeatedly near tips. Synangia 2–3mm across, yellow, sessile, on upper stems. **HABITAT:** Crevices of sandstone rocks in open forest. **DIST:** Mainly coast; also Colo, Bundanoon. NC, NT, WS, Qld, Vic, WA, NTerr, NZ, pantrop.

Tmesipteris
Plants with unbranched stems. Synangia elongated and 2-lobed. Four species in Sydney district, all uncommon, 1 less so.

Tmesipteris truncata
Small plant with stems to 25cm long. Lower stems ridged. Leaves about 20 x 2–4mm, apex truncate; midvein ending in fine point. Synangia pointed at both ends. **HABITAT:** Rock crevices, humus, trees, tree ferns and *Todea barbara* (King Fern) in rainforest and moist gullies. **DIST:** Widespread. Coast and ranges. Also NC, SC, ST, Qld, LHI.

FORK FERNS

Tmesipteris truncata

Club mosses

THE CLUB MOSSES are a much larger group than the fork ferns. Their ancestors appeared somewhat later, in the Devonian period about 350 million years ago. Although still showing many primitive characters, they are more highly developed than the fork ferns. They have small leaves with a midvein. The sporangia (kidney-shaped spore-producing structures) are at the ends of branches or scattered along them and are subtended by tiny leaves called sporophylls, the whole forming a cone-like structure known as a strobilus.

LYCOPODIACEAE

Lycopodiella
Plants with erect unbranched stems, rarely with 1–2 branches. If more branched then strobili pendent.

Lycopodiella cernua
Scrambling Clubmoss
Main stems creeping or pendulous, 1–3m long; aerial stems erect, to 1m tall, much branched in upper part. Leaves crowded, curved, 5mm long, acuminate. Strobili terminating on short side-branches, about 5mm long, pale, nodding or pendulous. **HABITAT:** Shaded moist forested areas, rock cuttings, swamp margins, cliff faces. **DIST:** Coast. Also NC, NT, Qld, WA, NTerr, pantrop.

Lycopodiella lateralis
Slender Clubmoss
Main stems underground; aerial stems erect, 10–40cm tall, branched once or twice. Leaves 4–7mm long, narrow, finely pointed. Strobili lateral at intervals along stems or terminal on short branches. Sporophylls broader than foliage leaves. **HABITAT:** Bogs, swamps, wet heaths. **DIST:** Coast and Blue Mtns. Also NC, SC, NT, Qld, Vic, Tas, SA, NZ, NCal.

Lycopodiella cernua

CLUB MOSSES

Lycopodiella lateralis

Lycopodium deuterodensum

Lycopodium
Plants with erect stems branching more than twice; strobili erect at ends of branches.

Lycopodium deuterodensum
Bushy Clubmoss
Aerial stems, 20–30cm tall (to 1m in sheltered sites), erect, branched. Leaves of 2 types: scattered near base of stem; dense, small (2mm long) and crowded near top. Strobili terminal, 1–2cm long. **HABITAT:** Sandy moist places, heath, woodland. **DIST:** Coast and ranges. Also NC, SC, NT, ST, Qld, Vic, Tas, SA, Norfolk Is, NCal.

SELAGINELLACEAE

Selaginella
Leaves in 4 rows, with tiny raised gland called a ligule on upper surface near base. Male and female spores produced in different sporangia.

Selaginella uliginosa

Selaginella uliginosa
Swamp Selaginella
Aerial stems to 30cm tall from branched underground rhizome, simple or branched. Leaves in 4 rows, to 3mm long, ovate, spreading, pointed, keeled, green turning bronze in sunny locations. Sporophylls similar to leaves but smaller. **HABITAT:** Damp soil in heath or open forest. **DIST:** Coast and ranges. Also NC, SC, NT, all States (except SA), LHI.

Ferns

THE FERNS ARE a group of vascular plants whose ancestry stretches back nearly 350 million years. They produce spores from sporangia occurring on the underside of ordinary fronds or on specialised fertile fronds. The sporangia usually occur in small clusters called sori which often have a characteristic size, colour, shape and arrangement. In some cases, the sorus is protected by an outgrowth of tissue known as an indusium. There are possibly 10,000 species of fern, with 400 in Australia and about 120 in the Sydney district, some rare, localised and seldom seen.

OSMUNDACEAE

Leptopteris
Terrestrial ferns with membranous mature fronds. Sporangia clustered in patches on pinnae near the base, less abundant than those of *Todea*.

Leptopteris fraseri
Crepe Fern
Terrestrial fern with erect rhizome forming short trunk on older plants. Fronds mostly bipinnate-pinnatifid, to 1m long, arching, delicate; stipes with soft woolly brown hairs; sporangia in light brown patches. **HABITAT:** Wet, cool shady places, rainforest caves, near waterfalls. **DIST:** Chiefly Blue Mtns; also Minnamurra Falls to Belmore Falls. NC, NT, ST, Qld.

Todea
Terrestrial ferns with leathery mature fronds. Sporangia along minor veinlets, covering most of the lower surface of pinnules. A genus of only 2 species (1 native), considered to be among the most primitive of the modern ferns.

Leptopteris fraseri

Todea barbara
King Fern
Terrestrial fern with short dark trunk and 2 or more crowns of bipinnate fronds to 2m long; pinnules narrow with serrated margins. Sporangia initially on veins of pinnules, later on most of underside of pinnules, velvety brown. **HABITAT:** Shady creek banks, sheltered moist crevices. **DIST:** Common. Coast to WS. Also Qld, Vic, Tas, SA, NZ, S. Afr.

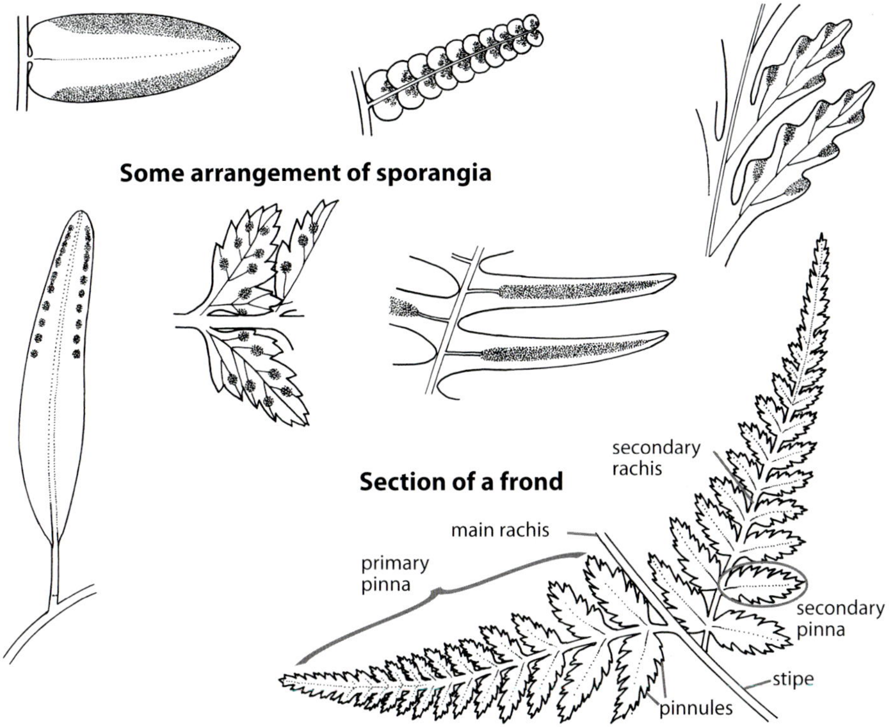

FERNS

Some arrangement of sporangia

Section of a frond

- primary pinna
- main rachis
- secondary rachis
- secondary pinna
- stipe
- pinnules

Todea barbara

Schizaea dichotoma

Schizaea rupestris

Schizaea bifida

SCHIZAEACEAE

Schizaea
Small terrestrial ferns with erect fronds. Fertile segments pinnate at apex, resembling a cock's comb.

Schizaea bifida
Forked Comb Fern
Fronds narrow, wiry, 10–25cm tall. Fertile fronds divided once or rarely twice, pinnae comb-like, 1–2cm long. **HABITAT:** Open forest, scrub, heath, on sandy soils. **DIST:** Coast and ranges. Also NC, SC, NT, Qld, Vic, Tas, SA, NZ, NCal.

Schizaea dichotoma
Branched Comb Fern
Fronds erect to 30cm tall. Fertile fronds divided 3–6 times, scabrous, with comb-like segments. Sterile fronds smaller, with 20 or more branches. **HABITAT:** Open forest, scrub, heath, on sandy soils. **DIST:** Coast nth from Royal NP. Also NC, Qld, WA, NTerr, tropics.

Schizaea rupestris
Tiny mat-forming fern with unbranched flat sterile fronds to 12cm x >1mm. Fertile fronds few, to 20cm x <1mm. **HABITAT:** Damp mossy rocks near waterfalls, wet sandstone rocks. **DIST:** Coast and ranges (e.g. Royal NP, Lawson, Robertson). Also NC, SC, ST.

GLEICHENIACEAE

Gleichenia
Terrestrial ferns with slender creeping scaly rhizome, often forming large tangled masses. Fronds branched several times with dormant apical bud in each fork. Pinnules at right angles to rachises. Sori without indusia, consisting of 2–5 sporangia on ultimate segments.

Gleichenia dicarpa
Pouched Coral Fern
Fronds to 2m long; pinnules narrow-linear, 3–5cm long; ultimate segments rounded, dull green, recurved, forming a pouch on lower surface holding 2 sporangia. **HABITAT:** Damp sunny places around creeks, swamps and cliff bases. **DIST:** Common. Coast and ranges. Also NC, SC, NT, ST, Qld, Vic, Tas, NZ, Pac. Iss.

Gleichenia microphylla
Scrambling Coral Fern
Ultimate segments about 2mm long, obliquely saw-toothed, dark green, paler beneath, flat or slightly recurved; sori with 3–4 sporangia. **HABITAT:** Damp sunny places around swamps and at cliff bases. **DIST:** Coast and ranges. Also NC, SC, all other States, NZ, SE Asia.

Gleichenia rupestris
Fern with glossy brown glabrous stipes and rachises, forking 1–3 times. Pinnules about 4cm long with flat rounded-triangular ultimate segments 2mm long, green above, glaucous beneath. Sporangia 3–4 near upper margin. **HABITAT:** Wet sandstone crevices. **DIST:** Coast and ranges. Also NC, SC, ST, Qld.

Gleichenia dicarpa

Gleichenia microphylla

Gleichenia rupestris

Sticherus flabellatus

Sticherus lobatus

Sticherus
Erect terrestrial ferns with long-creeping scaly rhizome, often forming large masses. Fronds forking 2–4 times with a dormant apical bud in each fork. Ultimate segments flat, 5–45 x 1.5–3.5mm. Sori numerous, in single row on each side of midvein.

Sticherus flabellatus
Umbrella Fern
Fronds to 1m tall; stipe green or brown. Pinnules absent between lowermost fork and first fork of lateral branches. Ultimate pinnules attached to rachises at about 45°, 25–35mm long, with finely serrated margins, shiny green above, paler and glabrous/glaucous beneath. **HABITAT:** Moist gullies, creek banks. **DIST:** Common. Coast and ranges. Also NC, SC, ST, Qld, Vic, NZ, NCal.

Sticherus lobatus
Spreading Fan Fern
Fronds to 2m high; stipe brown. Pinnules lobed or unlobed between lowermost fork and first fork of branches. Ultimate pinnules attached to rachises at 80–90°, 20–45mm long, with entire margins, dull light green above, paler and glabrous beneath. **HABITAT:** Moist slopes of open forest, rainforest margins. **DIST:** Chiefly Blue Mtns, Bundanoon; rarer on coast. Also NC, SC, ST, NT, Qld, Vic.

Sticherus urceolatus*
Similar to *S. lobatus* except lateral axes have fringed scales and pinnules are at 50–65°, smaller (15–25 x 2–3mm) and have hairs and scales on the lower surface. **HABITAT:** Wet forest, creek banks, seepage areas. **DIST:** Chiefly Blue Mtns. Also NC, NT, Vic, Tas.

  * Formerly included in *S. tener* but pinnules broader and at smaller angle to axis.

Sticherus urceolatus

Hymenophyllum cupressiforme

Cyathea australis

HYMENOPHYLLACEAE

Hymenophyllum
Delicate small epiphytic or terrestrial fern with creeping rhizome forming mats on damp rocks, decaying logs or sheltered tree trunks, including tree ferns. Fronds with a membranous lamina 1 cell thick. Sporangia enclosed in a marginal 2-lipped indusium. Seven species in the Sydney area.

Hymenophyllum cupressiforme
Common Filmy Fern
Small creeping fern; rhizome wiry, dark, with red-brown hairs when young. Fronds to 9cm long, lamina pinnate-pinnatifid; glabrous, with toothed margins. **HABITAT:** Rainforest, moist forest, on rocks and tree trunks. **DIST:** Coast and ranges. Also NC, SC, NT, ST, CWS, Qld, Vic, Tas.

CYATHEACEAE

Cyathea leichhardtiana

Cyathea
Tree ferns reaching several metres in height. Base of fronds covered with scales. A genus of over 600 species, 11 in Australia, 2 native to Sydney district.

Cyathea australis
Rough Tree Fern
Common tree fern with erect trunk 2–6m tall and 40cm diam. Upper trunk covered by persistent bases of old fronds. Base of frond stalks (stipes) covered with flat, shiny, red-brown scales. Fronds to 4m long. Sori round, in single row on each side of midvein of pinnules. **HABITAT:** Forested gullies, moist hillsides. **DIST:** Coast and ranges. Also NC, SC, NT, ST, WS, Qld, Vic, Tas.

Cyathea leichhardtiana
Prickly Tree Fern
Tree fern with erect slender trunk to 7m tall (usually much less) and 15cm diam. Upper trunk covered by prickly persistent bases of old fronds. Base of frond stalks (stipes) covered with whitish scales. Fronds to 3m long. Sori round, in single row on each side of midvein of pinnules. **HABITAT:** Rainforest gullies, creeks, sheltered slopes. **DIST:** Coast, Blue Mtns. Also NC, SC, NT, Qld, Vic.

Cyathea cooperi is now regarded as naturalised in Sydney area. It resembles *C. australis*, but its trunk is more slender and patterned with large oval scars left by detached frond bases.

DICKSONIACEAE

Calochlaena
Terrestrial fern, sometimes placed in Culcitaceae. Two species in Australia, 1 in Sydney district. The local species is common, often forming large banks in protected open areas where its soft light green fronds contrast with the shiny dark green of bracken (*Pteridium esculentum*).

Calochlaena dubia
Common Ground Fern
Softly hairy terrestrial fern with stout creeping rhizome, brown at base of stipe, otherwise green. Fronds to 1.5m long with broadly triangular tripinnate-pinnatifid lamina 30–80cm long. Sori small, in 2 rows along margins. **HABITAT:** Slopes of open forest, sheltered gullies. **DIST:** Common. Coast and ranges. Also NC, SC, NT, ST, CWS, Qld, Vic, Tas.

Dicksonia
Arborescent ferns, sori numerous, marginal on the ends of veinlets. Three Australian species, with 1 common in eastern Australia in cool moist mountain areas.

Dicksonia antarctica
Soft Tree Fern
Tree fern usually 1–6m tall. Trunk stout with persistent stipe bases on upper part, the bases and upper trunk densely covered with terete red-brown hairs. Mature fronds to 4m long, dark green above, paler beneath. Sori 1mm diam. in cup-shaped or 2-valved indusium; indusia on pinnule lobes. **HABITAT:** Rainforest creeks, moist gullies. **DIST:** Illawarra, Blue Mtns. Also NC, SC, NT, ST, CWS, Vic, Tas.

DENNSTAEDTIACEAE

Dennstaedtia
Terrestrial ferns similar to common bracken (*Pteridium*), but with circular sori and cup-like indusium. One species in Australia.

Dennstaedtia davallioides
Lacy Ground Fern
Fern with large delicate finely divided fronds about 1m tall. Stipes shiny, reddish brown, rising at intervals along creeping rhizome; stipes and rachises grooved and hairy on upper surface. Lamina 3–4-pinnate, broadly triangular, dark green, soft, lacy. Sori mostly in forks between lobes. **HABITAT:** Rainforest margins, creek banks in sheltered forest. **DIST:** Illawarra (e.g. Hacking R.), upper Blue Mtns (e.g. Mt Tomah, Mt Wilson). Also NC, Qld, Vic, Norfolk Is.

Histiopteris
Terrestrial ferns with rhizome covered with scales and/or hairs. Sori linear, marginal. One species in Sydney district.

Histiopteris incisa
Bat's-wing Fern
Fern with fronds to 1.5m tall. Stipes smooth, reddish brown at base with a few hairs, shiny green and glabrous higher up. Lamina pale green, glabrous, often glaucous, soft. Pinnae with pair of stipule-like leaflets at base. Sori

Calochlaena dubia

Dicksonia antarctica

Dennstaedtia davallioides

Histiopteris incisa

Hypolepis muelleri

along margins of ultimate segments.
HABITAT: Forms thickets in wet gullies, rainforest margins. **DIST:** Coast and ranges. Also NC, SC, NT, ST, all States (except WA), pantrop.

Hypolepis
Terrestrial ferns with circular sori; indusium mostly absent. Three species in Sydney district.

Hypolepis muelleri
Harsh Ground Fern
Fern with fronds to 80cm tall, not sticky-hairy. Stipes nearly glabrous, reddish brown at base, brown to yellow higher up. Secondary rachises with stiff non-glandular hairs which curve upwards. **HABITAT:** Along creeks or rainforest margins. **DIST:** Chiefly coast; rarely t'lands (e.g. Bundanoon). Also NC, SC, ST, CWS, Qld, Vic, Tas.

Pteridium esculentum

Lindsaea linearis

Pteridium
Common terrestrial ferns. Rhizome covered with hairs. Sori linear, marginal. Three species in Australia, 1 very common in all States. Poisonous to horses and cattle when eaten in large quantities.

Pteridium esculentum
Common Bracken
Fronds leathery, glossy, dark green above, whitish with fine appressed hairs beneath. Lamina tripinnate to bipinnate, usually with a small lobe where each branch (pinna) joins the main rachis. Sori inconspicuous, continuous around the slightly recurved margins. **HABITAT:** Open forest, grassy clearings, roadsides, pastures. **DIST:** Coast and ranges. Most of NSW, all States, NZ, Pac. Iss.

LINDSAEACEAE

Lindsaea
Small terrestrial ferns with slender stipes and wedge-shaped pinnules; sori marginal.

Lindsaea linearis
Screw Fern
Small fern with 1-pinnate fronds to 20cm x 9–12mm, often forming small colonies but not tufted. Barren fronds shorter and broader than fertile fronds. Stipe dark, shiny; pinnae triangular, longer than broad; outer margin crenate. Sori continuous along outer margin. **HABITAT:** Forest, heath, on rocks, near swamps. **DIST:** Coast and ranges. Also NC, SC, NT, ST, all States, NZ, Pac. Iss.

Lindsaea microphylla

Lindsaea microphylla
Lacy Wedge Fern
Tufted fern to 40cm tall with short-creeping brown, scaly rhizome. Stipe slender, yellowish brown, shorter than lamina. Lamina 2–3-pinnate; delicate, pale green; ultimate lobes cuneate. Sori on outer margin of lobes. **HABITAT:** Open forest, in sheltered areas. **DIST:** Coast and ranges. Also NC, SC, NT, Qld, Vic.

PTERIDACEAE

Pteris
Terrestrial ferns with linear marginal sori and indusium formed by the recurved margin of the lamina.

Pteris tremula
Tender Brake
Tufted fern to 1.5m tall. Stipe brown, shiny, deeply grooved; rachis brown; lamina usually

FERNS

Pteris tremula

Pteris umbrosa

Adiantum aethiopicum

ADIANTACEAE

Adiantum
Terrestrial ferns with lacy fronds on thin shiny stipes. The popular name Maidenhair comes from the specific name of the European species *A. capillus-veneris*.

Adiantum aethiopicum
Common Maidenhair Fern
Fern forming extensive patches from creeping rhizomes with yellow, transparent scales. Fronds 10–30cm tall. Stipes thin, reddish brown, smooth, shiny. Ultimate segments with irregularly rounded upper margin and symmetrical base. Sori reniform, along upper margin. **HABITAT:** Moist situations with open tree cover. **DIST:** Common. Coast and ranges. Also NC, SC, NT, ST, NWS, all States, cosmop.

Adiantum atroviride is similar to *A. aethiopicum* but has darker stipe and lamina, and brown opaque rhizome scales with fringed margins.

3-pinnate, glabrous, light green; ultimate segments narrow-oblong, to 20mm long, finely toothed. Sori linear, on narrow fertile segments. **HABITAT:** Moist rocky gullies, creek banks. **DIST:** Blue Mtns and Cambewarra. Also NC, SC, NT, WS, SWP, all States (except WA), NZ, Pac. Iss.

Pteris umbrosa
Jungle Brake
Fern with erect fronds 1m or more tall. Stipes grooved, reddish or yellowish brown. Lamina dark green, deeply divided to partially 2-pinnate. Sori conspicuous along margins. **HABITAT:** Rainforest, usually on rocky slopes, esp. basalt. **DIST:** Illawarra, Blue Mtns, Robertson, Bundanoon. Also NC, SC, NT, Qld, Vic.

Adiantum formosum

Adiantum formosum
Giant Maidenhair Fern
Fern to 80cm tall, sometimes covering large areas. Stipes black, shiny, slightly rough. Lamina 3–4 pinnate; ultimate segments lobed and irregularly toothed on upper margin, the base asymmetric. Sori in shallow depressions on upper margin. **HABITAT:** Rainforest, covering extensive areas in mid-shade. **DIST:** Coast and ranges. Also NC, SC, NT, CWS, Qld, Vic, NZ.

Adiantum hispidulum
Rough Maidenhair Fern
Fern to 50cm tall. Rhizome short-creeping with dark brown scales. Stipes stout, black, slightly rough. Stipes and rachises with a few fine white hairs. Ultimate segments with soft white hairs, often pink when young, oblique, broader than long, attached to short stalk at a lower corner. Sori along upper and outer margins. **HABITAT:** Rainforest, tall forests, often in rock crevices. **DIST:** Coast and ranges. Also NC, SC, NT, NWS, Qld, Vic, NZ, Afr., SE Asia, Pac. Iss.

Cheilanthes
Terrestrial ferns with lamina 2–3 pinnate. Sori marginal, circular. Often known as Rock Ferns but they frequently occur away from rock outcrops.

Cheilanthes distans
Bristly Cloak Fern
Fronds erect, hairy, to 30 x 3cm; stipe shiny, with brown scales; lamina 2-pinnate, with scales on rachises and lower surface of pinnae, densely hairy on upper surface. **HABITAT:** Open forest, esp. on shale and basalt. **DIST:** Coast, Jenolan Caves. Most of NSW; also Qld, Vic, SA, WA, NZ, Pac. Iss.

Cheilanthes sieberi
Poison Rock Fern
Fronds erect to 30 x 3cm; stipe dark brown, shiny, glabrous; lamina 2–3 pinnate, glabrous. **HABITAT:** Rocky hillsides in open forest and grassland, esp. on shale and basalt. **DIST:** Coast and ranges. Most of NSW. All States (except Tas), NZ, Pac. Iss.

Cheilanthes austrotenuifolia has fronds to 55 x 20cm with a glabrous triangular lamina which is sometimes sparsely hairy and scaly below. It occurs between Mittagong and Goodmans Ford in the Sydney district and in all southern states.

Pellaea
Terrestrial ferns with fronds 1-pinnate. Sori linear, continuous as band along margin of pinna.

Pellaea falcata
Sickle Fern
Hardy fern resembling the common fishbone fern (*Nephrolepis*) but with smooth, bright green fronds to 60cm tall. Stipe shiny dark brown clothed with long narrow scales. Pinnae ovate, 2–5cm long, on stalks less than 1mm long. **HABITAT:** Moist shady open forest and rainforest. **DIST:** Coast and ranges. Also NC, SC, NT, ST, WS, Qld, Vic, Tas, Asia, Pac. Iss.

Pellaea nana. Similar to *P. falcata* but smaller and with pinnae 6–20mm long. A rainforest species.

Pellaea paradoxa. Pinnae longer and wider (to 9 x 4cm) than those of *P. falcata*. Chiefly Wyong–Ourimbah rainforests.

Adiantum hispidulum

FERNS

Cheilanthes distans

Davallia solida var. *pyxidata*

Cheilanthes sieberi

Pellaea falcata

Arthropteris tenella

DAVALLIACEAE

Arthropteris
Small rainforest epiphytes. Rhizomes creeping, fronds 1-pinnate.

Arthropteris tenella
Jointed Fern
Rhizome wiry, clothed with brown scales. Fronds about 30cm long, with a slender rachis and 6–20 well-spaced alternate stalked pinnae. Pinnae 3–10cm long, shiny, green. Sori spaced around margin at end of forked veinlet.
HABITAT: On trees and rocks in rainforest. **DIST:** Mainly coast; also Robertson, Blackheath, NC, SC, NT, NWS, Qld, NZ, Pac. Iss.

Arthropteris beckleri. Small epiphyte from darker areas of rainforest. Pinnae to 15mm long, hairy.

Davallia
Epiphytes with conspicuous creeping rhizome covered with reddish brown scales. Indusium cup-like, facing outwards.

Davallia solida var. *pyxidata*
Hare's Foot Fern
Rhizome to 12mm diam., robust, densely scaly (furry). Stipes in 2 rows along rhizome, straw coloured. Fronds to 50cm long. Lamina glossy green, 2- or 3-pinnate; ultimate segments deeply lobed. **HABITAT:** Rock crevices in eucalypt forests or rainforest on trees. **DIST:** Coast. Also NC, SC, NT, WS, Qld, Vic, Tas.

Christella dentata

THELYPTERIDACEAE

Christella
Terrestrial ferns with 1-pinnate fronds. Lower surface of lamina with fine white hairs along veins.

Christella dentata
Tussock-forming fern. Rhizomes creeping, densely covered with brown scales. Fronds to 80cm tall, dark green; basal pinnae reduced in size, 3–5cm long. Sori circular, indusium hairy. **HABITAT:** Along streams in shady forest or rainforest. **DIST:** Coast, lower Blue Mtns. Also NC, Qld, SA, WA, pantrop.

Cyclosorus
Terrestrial ferns; fronds 1–2 pinnate. Lower pinnae not reduced in size.

Cyclosorus interruptus
Rhizome long-creeping. Fronds erect to 1m tall. Stipe to 40cm with at least 10 pairs of opposite pinnae. Pinnae lobes ovate-triangular in shape. Sori in crowded rows near lobe margins. **HABITAT:** Fresh water swamps, often forming large colonies. **DIST:** Coast, but rare in Sydney district (e.g. Kurnell, Scarborough Pk). Also NC, Qld, WA, NTerr, NZ, trop.

Cyclosorus interruptus

ASPLENIACEAE

Asplenium
Terrestrial and epiphytic ferns with over 650 species worldwide. Fronds simple, lobed or pinnate. Sori linear. The common name Spleenwort derives from the medieval belief that some species could cure illness of the spleen.

Asplenium australasicum
Bird's Nest Fern
Large nest-like epiphytic. Fronds spreading from thick rhizome, to 2m x 20cm, simple, entire. Sori numerous on undersurface, forming oblique lines parallel to veins. **HABITAT:** On trees and rocks in rainforest. **DIST:** Coast and ranges. Also NC, SC, NT, CWS, Qld, Vic, Pac. Iss.

Asplenium australasicum

Asplenium bulbiferum
Mother Spleenwort
Fronds to 1m tall from a tufted rhizome, divided 2–3 times (usually twice), soft, occasionally bearing tiny plantlets on pinnae. Sori elliptic, 2–4 on secondary pinna. **HABITAT:** On rocks and trees in rainforest creeks and gullies. **DIST:** Upper Blue Mtns (e.g. Mt Tomah, Mt Wilson). Also NC, SC, ST, Qld, Vic, Tas, SA, NZ.

Asplenium difforme. Fern with thick fronds occurring in rock crevices along coast from Barrenjoey to La Perouse.

Asplenium flabellifolium

30–100cm long; pinnae 4–6cm long, toothed, broad at base, tapering gradually to a drawn-out apex. Sori numerous, linear, parallel with veinlets. **HABITAT:** Rainforest, on trees, logs, rocks. **DIST:** Coast, Blue Mtns, rare (e.g. Hacking R., Minnamurra Falls, Mt Wilson). Also NC, SC, NT, ST, Qld, NZ, SE Asia, Pac. Iss.

Asplenium flabellifolium
Necklace Fern
Small trailing fern forming colonies. Fronds 1-pinnate, decumbent, to 20cm long, rooting from the tips of the fronds. Pinnae up to 20 pairs, fan-shaped, pale green. **HABITAT:** Open forest, rainforest, in soil, rock crevices or on tree trunks. **DIST:** Coast and ranges. Most of NSW, Qld, Vic, Tas, NZ.

Asplenium polyodon
Willow Spleenwort
Epiphyte with weeping fronds to 80cm long. Stipe dark and shiny. Lamina pinnate,

Asplenium bulbiferum

Asplenium polyodon

ATHYRIACEAE

Diplazium
Terrestrial ferns; stipes clustered; rachises grooved above. Sori elongated.

Diplazium australe
Austral Lady Fern
Uncommon fern, tufted or with small trunk. Fronds to 1.5m tall, bright green, fine, tender, bipinnate with pinnules divided almost to midvein. Sori in 2 rows on veinlets on each side of midvein. **HABITAT:** Rainforest, esp. damp basalt soils and creek banks. **DIST:** Illawarra, upper Blue Mtns. Also NC, ST, CWS, Qld, Vic, Tas.

Diplazium australe

DRYOPTERIDACEAE

Lastreopsis
Terrestrial ferns, rhizomes usually creeping. Fronds 2- to 4-pinnate. Main rachis with 2 ridges; hairy in channel of rachis.

Lastreopsis acuminata
Shiny Shield Fern
Tufted fern with short rhizome. Fronds 40–60cm long, 2- or 3-pinnate. Axes covered with soft spreading hairs. Lamina pale green, shiny, upper pinnae not fused at base to rachis. **HABITAT:** Rainforest, tall damp forest, often near creeks. **DIST:** Coast, upper Blue Mtns. Also NC, SC, NT, WS, Qld, Vic, Tas, SA.

Lastreopsis decomposita
Trim Shield Fern
Fern, often covering large areas. Rhizome thick, short-creeping; stipes crowded. Fronds to 50–80cm long, 3- or 4-pinnate. Lamina dull green, hairy, upper pinnae finely pointed at apex; basal lobes fused to rachis. **HABITAT:** Rainforest, tall damp forest. **DIST:** Coast, Blue Mtns. Also NC, SC, NT, CWS, Qld, Vic.

Lastreopsis hispida. Rachises with dark brown bristle-like scales. Rare in Blue Mtns rainforest.

Lastreopsis decomposita

Lastreopsis acuminata

Lastreopsis microsora

Polystichum australiense

Lastreopsis microsora
Creeping Shield Fern
Rhizome long-creeping, with stipes spaced along it about 3cm apart. Fronds to 30–80cm long, 3- or 4-pinnate. Rachises shiny above, pubescent below. Lamina soft, pale green. Upper pinnae broadly rounded at apex, finely toothed, with basal lobes fused to rachis.
HABITAT: Rainforest, tall damp forest. **DIST:** Coast and ranges. Also NC, SC, NT, Qld, Vic, NZ.

Polystichum
Terrestrial ferns. Rhizome erect or short-creeping, with broad scales. Fronds clustered, often bearing new growth (proliferous buds) at tip.

Polystichum australiense
Rhizome and base of stipe with narrow dull brown scales. Lamina 40–80cm long, narrow, dark green, often with pale brown proliferous bud at apex. Ultimate segments 10–15mm long, distinctly aristate. **HABITAT:** Open forest, rainforest margins. **DIST:** Coast and ranges. Also NC, SC, CWS.

Polystichum proliferum
Mother Shield Fern
Rhizome erect, thick, with glossy pale-edged scales. Main rachis densely scaly. Fronds 2- or 3-pinnate, to 80cm long, often with proliferous buds. Ultimate segments toothed but not aristate. **HABITAT:** Open forest, rainforest margins, mainly on basalt soils. **DIST:** Upper Blue Mtns, Minnamurra Falls. Also NC, SC, NT, ST, NWS, Vic, Tas.

Polystichum proliferum

POLYPODIACEAE

Microsorum
Epiphytes or lithophytes. Fronds variable; veins reticulate; sori large, in 2 rows.

Microsorum pustulatum
Kangaroo Fern
Rhizomes creeping, to 6mm wide, with appressed deciduous shiny brown scales. Fronds to 50cm long, slightly leathery and unscented. Lamina entire or divided with <9 pairs of well-spaced lobes. **HABITAT:** Rainforest, on rocks, tree trunks. **DIST:** Upper Blue Mtns, Sthn H'lands. Also NC, SC, NT, ST, Qld, Vic, Tas, NZ, Norfolk Is.

Microsorum scandens
Fragrant Fern
Rhizomes creeping, to 4mm wide, with spreading, persistent, shiny brown scales. Fronds to 50cm long, thin textured, with elusive musky smell. Lamina entire or divided with up to 20 lobe pairs. **HABITAT:** Rainforest, on rocks, tree trunks. **DIST:** Coast and ranges. Also NC, SC, NT, ST, CWS, Qld, Vic, NZ, LHI.

Platycerium
Large bracket epiphytes or lithophytes, with 2 types of fronds: broad sterile fronds protecting the rhizome and anchoring the plant, and long fertile fronds.

Platycerium bifurcatum
Elkhorn
Fertile fronds to 90cm long, erect or pendulous, divided 2–5 times like deer antlers. Sporangia in brown velvety masses on lower parts of segments. **HABITAT:** Rainforest, occasionally swamp and moist tall forest, on trees and rocks. **DIST:** Coast (e.g. Gosford, Illawarra). Also NC, SC, Qld, PNG, LHI.

Pyrrosia
Creeping epiphytes or lithophytes. Lamina simple; surface covered with stellate hairs.

Pyrrosia rupestris
Rock Felt Fern
Fern with creeping rhizome and 2 types of fronds. Sterile fronds circular to 3cm across. Fertile fronds narrow-lanceolate, to 10cm long.

Microsorum scandens

Sori in several rows on either side of midrib covering most of undersurface. **HABITAT:** Rainforest, moist forest, common on rocks and tree trunks. **DIST:** Coast and ranges. Also NC, SC, NT, ST, WS, Qld, PNG.

Pyrrosia confluens. A rare epiphyte, differing from *P. rupestris* in having sori only at the tip of fertile fronds. Occurs Gosford–Watagan Mtns.

GRAMMITIDACEAE

Grammitis
Creeping epiphytes or lithophytes. Fronds mostly simple. Sori circular.

Grammitis billardierei
Finger Fern
Rhizome short-creeping, with brown scales. Stipes densely hairy. Fronds simple, 5–15 x 1cm, with finely wavy margins. Sori in rows on either side of midrib. **HABITAT:** Rainforest, on rocks, tree trunks, near waterfalls. **DIST:** Blue Mtns, rarely coast. Also NC, SC, ST, NT, ST, Qld, Vic, Tas, NZ.

Grammitis stenophylla. Stipe glabrous; fronds <5cm long.

Microsorum pustulatum

Platycerium bifurcatum

Pyrrosia rupestris

Grammitis billardierei

BLECHNACEAE

Blechnum
Terrestrial ferns with fertile fronds usually much narrower than sterile fronds. Sori elongated, on both sides of midvein. Eleven species in Sydney district.

Blechnum ambiguum
Rhizome creeping, with reddish brown scales. Fronds pinnate, 30–50cm long, thin, dull, mid-green. Pinnae of sterile fronds 6–10cm x 15–25mm, shortly stalked. Pinnae of fertile fronds 4–9cm x 3–10mm. **HABITAT:** Open forest, on wet cliff faces and rock ledges. **DIST:** Coast and ranges. Also SC, Qld.

Blechnum camfieldii. Lower sterile pinnae reduced in size, ovate. Chiefly along margins of saline rivers and fresh water swamps.

Blechnum cartilagineum
Gristle Fern
Rhizome short-creeping. Fronds erect, 50–100cm long, usually pink when young, but leathery when aged. Fertile and sterile fronds similar. Pinnae with fine serrated margins and attached to rachis by broad bases. Lower pinnae longest, to 150 x 15mm. **HABITAT:** Open forest, rainforest, on sheltered slopes and creek banks. **DIST:** Common. Coast and ranges. Also NC, SC, ST, CWS, Qld, Vic, Tas, Malesia.

Blechnum indicum
Swamp Water Fern
Clumped fern. Fronds erect, 30–70cm tall, shiny green, margins of pinnae finely toothed for entire length. Sterile and fertile fronds similar. Pinnae distinctly stalked. **HABITAT:** Brackish swamps. **DIST:** Coast nth of Jervis Bay (e.g. Deep Ck, Kurnell, Berry). Also NC, Qld, NTerr, Malesia, Pac. Iss, Cent. Amer.

Blechnum minus. Lower sterile pinnae reduced in size, circular. Chiefly upper Blue Mtns in gorges near waterfalls.

Blechnum nudum
Fishbone Water Fern
Rhizome erect, tufted, sometimes developing into short trunk. Fronds erect to 80cm tall. Stipe smooth, shiny, black. Lamina pinnately divided; pinnae attached to rachis by broad bases. Fertile fronds shorter, with very narrow pinnae. **HABITAT:** Rocky creek beds and banks in rainforest and cool eucalypt forest. **DIST:** Coast and ranges. Common, esp. Blue Mtns. Also NC, SC, NT, ST, Qld, Vic, Tas, SA.

Blechnum patersonii subsp. *patersonii*
Strap Fern
Rhizome erect. Fronds semi-erect or pendent, 20–40cm long, strap-shaped or lobed, often on same plant. New fronds pinkish, older ones dark green, glossy. Sterile fronds 10–25mm wide; fertile fronds 2–4mm wide. **HABITAT:** Rainforest, slopes near streams, in deep shade. **DIST:** Coast and ranges. Also NC, SC, ST, CWS, Qld, Vic, Tas, NZ, PNG.

Blechnum ambiguum

FERNS 39

Blechnum cartilagineum

Blechnum indicum

Blechnum nudum

Blechnum patersonii subsp. *patersonii*

Blechnum wattsii

Blechnum wattsii
Hard Water Fern
Rhizome creeping; scales dark brown to black with finely toothed margins. Fronds pinnate, 30–70cm long; new growth pinkish, older fronds leathery. Pinnae of sterile fronds shortly stalked, 10–20mm wide. Pinnae of fertile fronds linear, 2–3mm wide. **HABITAT:** Creek banks in rainforest. **DIST:** Upper Blue Mtns, Barren Gardens. Also NC, SC, NT, ST, Qld, Vic, Tas, SA.

Doodia
Terrestrial ferns reproducing by horizontal runners, rooting at nodes. Sori in two rows on either side of midvein.

Doodia aspera
Rasp Fern
Fronds 20–40cm long, fishbone-shaped. Pinnae attached to rachis by broad base, stiff, often pink when young; margins prickly.
HABITAT: Rainforest, moist open forest, among

Doodia aspera

rocks or sheltered creek banks; sometimes in large colonies. **DIST:** Coast and ranges. Also NC, SC, NT, ST, WS, Qld, Vic.

Doodia caudata
Small Rasp Fern
Fronds erect to decumbent, to 30cm long, with 2 forms on the one plant. Pinnae opposite, broad or narrow, shortly stalked, toothed on

Marsilea mutica

Doodia caudata

margins, lobed at base, with rounded apex.
HABITAT: Rainforest, moist open forest, among rocks or on sheltered creek banks. **DIST:** Coast, upper Blue Mtns. Also NC, SC, NT, ST, all States (except WA), NZ, Pac. Iss.

Doodia linearis (formerly *D. caudata* var. *laminosa*). Frond undivided but with a few pairs of pinnae at base. Occurs mainly nth of Sydney.

Doodia australis (formerly *D. media* subsp. *australis*). Similar to *D. caudata*, but with fronds all alike. Pinnae sessile near middle of lamina; lower 1–3 pairs of pinnae stalked. Widespread.

MARSILEACEAE

Marsilea
Semi-aquatic ferns. Sterile fronds with long stipe and 2 pairs of leaflets (pinnae) resembling a 4-leaf clover.

Marsilea mutica
Nardoo
Rhizome long-creeping, branched, rooted in mud. Rhizome and fronds glabrous. Sporocarp globose, not toothed. Leaflets 10–50 x 20–40mm, rounded at apex, usually with brownish band separating 2 different shades of green.
HABITAT: Permanent fresh water swamps, wet places in floodplain. **DIST:** Chiefly Cumberland Pl. (e.g. Pitt Town, Mt Annan). Also NC, SC, ST, CT, NWP, all other States, NCal.

Marsilea hirsuta. Rhizomes and leaflets hairy. Leaflets 5–20 x 5–12mm. Sporocarp with 2 teeth, hairy. Uncommon, on Cumberland Pl. (e.g. Doonside, Yarramundi).

Cycads

THE CYCADS ARE an ancient group of palm-like or fern-like plants which flourished throughout much of the Mesozoic Era before they were largely displaced by the conifers and flowering plants about 80 million years ago. They range from small plants with a few short fronds to massively trunked trees with a dense crown of pinnate leaves or fronds several metres long. Plants are wind-pollinated and unisexual with either male (pollen-producing) or female (egg-producing) sexual organs borne on modified leaves called sporophylls which are formed into large ovoid or cylindrical cones or are clustered among the leaf bases at the top of the trunk. Although formerly included in the gymnosperms with conifers and several other groups, cycads have developed separately from conifers for nearly 300 million years and are not closely related.

ZAMIACEAE

Macrozamia
Palm-like plants with stem forming a trunk or mainly subterranean. Male and female cones on different plants. Male cones cylindrical, female cones ovoid. Seed with a bright-coloured outer layer. Three species occur in the Sydney district with another 3 on the northern (*M. pauli-guilielmi*) and north-western (*M. secunda, M. reducta*) outskirts.

Macrozamia communis
Burrawang
Trunk buried or shortly erect, bearing 20–100 fronds to 1.5m long, each with many rigid, sharp-pointed pinnae 15–35cm long, shortening towards base. Cones 20–50cm long, growing from centre, especially after fire; male cones reddish, cylindrical, 6–12cm diam.; female cones orange-red, 10–20cm diam. **HABITAT:** Woodland, forest, on sandy soils. **DIST:** Coast (e.g. Kurnell, Woy Woy, Colo Hts). Also NC, SC.

Macrozamia elegans has pungent pinnae 9–14mm wide with pink bases. It is restricted to Wheeny Ck–Mountain Lagoon area.

Macrozamia spiralis
Small plant with buried trunk and 2–12 fronds <1m long. Rachis spirally twisted about half a turn. Pinnae 10–20cm x 5–9mm, non-pungent, not shortening towards base of frond. Male plants with 1–4 cylindrical cones to 20 x 6cm; females usually with a solitary ovoid cone to 20 x 9cm. **HABITAT:** Dry open forest, woodland, on clay-shale soils. **DIST:** Western Sydney (Wallacia to Putty); Lucas Hts–Bankstown to Goulburn R. Valley.

Macrozamia communis

Macrozamia spiralis

Pines (Conifers)

PINES, LIKE CYCADS, differ from the ferns and fern allies in producing seeds (as distinct from spores) and from the flowering plants in bearing their seeds on the surface of cone scales on sporophylls (as distinct from bearing seeds in the ovary of a flower). In pines, pollen is carried by wind from male sporophylls to naked ovules on the same or a different plant and fertilization and reproductive processes are more streamlined than in cycads. As the seeds develop, the female cones enlarge and become woody in most species. Seeds are often winged since dispersal relies largely on wind.

CUPRESSACEAE

Callitris
Conifers with scale-like leaves in whorls of 3, decurrent on stems. Female cones globular or conical, with 6 woody scales opening from the top to release seeds and reveal the columella, a distinctively shaped 3-lobed projection of the axis.

Callitris endlicheri
Black Cypress Pine
Tree to 10m tall with green or glaucous spreading branches. Leaves 2–4mm long. Cones ovoid to globular, 15–20mm diam.; scales with small conical point near apex; alternate scales much reduced in size. Columella short, 3-lobed or with 3–4 parts. **HABITAT:** Chiefly on rocky hillsides. **DIST:** Blue Mtns, Sthn H'lands (e.g. Mt Victoria, Berrima). Also inland NSW, Qld, Vic.

Callitris endlicheri

PINES

Callitris muelleri

Callitris rhomboidea

Microstrobos fitzgeraldii

Callitris muelleri
Mueller's Cypress
Tree or shrub to 6m tall with erect branches. Leaves 4–9mm long including decurrent part. Cones globular, 20–30mm diam.; scales with small conical point near apex; alternate scales smaller, pointed. Columella 1–4mm long, 3-lobed. **HABITAT:** Chiefly on rocky hillsides. **DIST:** Coast and ranges (e.g. Mt Victoria, Bundanoon, Gosford). Also SC.

Callitris rhomboidea
Port Jackson Pine
Tree to 15m tall, but usually a narrow shrub to 7m with erect branches drooping at ends. Leaves 2–3mm long. Cones globular, 12–20mm diam.; scales with short wide protuberance near middle; broadening above, abruptly angled into obtuse apex; alternate scales reduced in size. Columella short with 3 parts. **HABITAT:** Open forest, woodland. On sandy or rocky slopes and ridges. **DIST:** Coast and ranges. Also NC, SC, NT, Qld, Vic, Tas, SA.

PODOCARPACEAE

Microstrobos
Shrubs with leaves <4mm long, closely overlapping. Male and female cones terminal. A genus of 2 species, one (*M. niphophilus*) restricted to alpine areas of Tasmania.

Microstrobos fitzgeraldii
Dwarf Mountain Pine
Shrub to 1m tall with weak drooping or erect branches. Leaves crowded, spreading, pointed, 2–4 x 1mm. Male and female cones solitary on separate plants; males purplish brown, 6mm long; females with a few scales, 3mm long. **HABITAT:** Wet rocks and ledges beneath south-facing waterfalls. **DIST:** Endemic. Restricted to 7 locations between Wentworth Falls and Katoomba.

Podocarpus elatus

Podocarpus

Trees and shrubby plants. Male and female cones on separate plants; males at ends of branches, pollen-bearing, short, cylindrical; mature female cones with expanded fleshy peduncle.

Podocarpus elatus
Plum Pine
Tree seldom more than 5m tall. Leaves linear-oblong, to 14cm x 16mm, dark green; new growth light yellow-green. Female cones solitary on stalks in leaf axils, the mature receptacle about 20mm diam., blue-black, glaucous. **HABITAT:** Rainforest. **DIST:** Uncommon. Coast (e.g. Whispering Gully, Minnamurra Spit, Gosford). North from Jervis Bay to Qld.

Podocarpus spinulosus
Spreading shrub with straggling branches to 1m tall. Leaves 3–6cm long, rigid, narrow-linear, dark green above, paler below; apex acute. Female cones solitary on stalks; mature receptacle blue-black, 10mm diam. **HABITAT:** Open forest, in sheltered areas on sandy soils. **DIST:** Coast, from Woy Woy to Ulladulla. Also Glen Davis, NC, CT, Qld.

Podocarpus spinulosus

ARAUCARIACEAE

Wollemia
The discovery of the Wollemi Pine in September 1994 in a remote canyon in Wollemi NP attracted immense public and scientific interest. Although the location of these trees is a closely held secret, specimens are now being sold in nurseries.

Wollemia nobilis
Wollemi Pine
Tree to 40m tall with dark brown nodular bark. Leaves crowded, of 3 types: narrow-oblong to 4cm long on adult horizontal branches, linear to 8cm long on juvenile shoots, 1cm long, pungent on vertical shoots. Male cones cylindrical; female cones spherical, about 8cm diam. **HABITAT:** Deep sheltered gorges, with Coachwood and Sassafras. **DIST:** Endemic. Forty adult trees known from 2 sites in Wollemi NP.

Wollemia nobilis

Flowering Plants (Angiosperms)

THE FLOWERING PLANTS are the most recently evolved and the largest major group (both in number of species and number of individuals) and constitute the dominant vegetation on Earth. They appear to have developed from a group of Mesozoic pteridosperms (seed ferns) about 130 million years ago and diversified rapidly during the Cretaceous and early Tertiary periods. Their most obvious attribute is the possession of flowers.

A flower consists of a number of whorls. The outer whorl (calyx) protects the developing flower. The second whorl (corolla) consists of petals to attract pollinating insects and animals. The third whorl is of stamens which produce pollen. The inner whorl consists of one or more carpels, each consisting of an ovary topped by a column (style) and a pollen receptor (stigma). In the various families and genera, these four whorls are modified in different ways and frequently some are reduced in size or absent entirely.

The possession of flowers provides a more efficient and varied means of pollination than wind alone, which is relied upon by less advanced groups. In many species the seeds are dispersed by mammals and birds which feed on the fruit formed from the ovary or ovaries or accidentally carry the fruit on their bodies.

The flowering plants are divided into 2 classes, the monocotyledons and the dicotyledons. Monocot leaves usually have parallel venation, leaf bases which sheath the stem and leaves that are not clearly differentiated into petiole (stalk) and lamina (blade). Dicots have leaves with reticulate venation, are rarely sheathed at the base and usually have a petiole. Floral parts of monocots are typically in sets of 3; in dicots they are usually in sets of 4 or 5. As indicated by their names, the two groups differ in the number of cotyledons (the first leaves of the embryo or germinated seed).

Dicotyledons

WINTERACEAE

Tasmannia
PEPPERBUSH
Shrubs. Leaves with oil glands and peppery taste. Male and female flowers usually on separate plants. Male flowers with many stamens. Inflorescence a terminal umbel.

Tasmannia insipida
Brush Pepperbush
Shrub to 3m tall. Stems often reddish. Leaves oblanceolate, blunt at base, 8–16 x 1.5–3cm, green, glossy. Flowers with 2 yellow-white petals. Fruit a purplish berry mottled with red. **HABITAT:** Shaded gullies, rainforest. **DIST:** Coast and ranges. Also NC, SC, NT, ST, Qld. **FL:** Sept–Dec.

Tasmannia insipida

Tasmannia lanceolata

Tasmannia lanceolata
Mountain Pepperbush
Shrub to 3m tall. Similar to *T. insipida*, but leaves tapering to short stalk, lanceolate, 5–8 x 1–2cm. **HABITAT:** Sheltered gullies. **DIST:** Upper Blue Mtns (e.g. Blackheath, Kanangra). Also ST, Vic, Tas. **FL:** Sept–Dec.

EUPOMATIACEAE

Eupomatia
Glabrous shrubs. Flowers creamy, waxy, not differentiated into sepals and petals. Perianth fused into cap which falls off to expose rows of stamens and staminodes. Fruit a berry with numerous seeds. Flowers pollinated by beetles which feed on staminodes.

Eupomatia laurina
Bolwarra
Large shrub with weak straggling branches. Leaves alternate, 2-ranked, oblong-elliptic, mostly 7–10 x 2–5cm, shiny. Flowers to 25mm across, solitary. Berry with flat top and raised rim. **HABITAT:** In or near rainforests, sheltered gullies. **DIST:** Coast and ranges. Also NC, SC, NT, ST, CWS, Qld, Vic, PNG. **FL:** Dec–Jan.

Eupomatia laurina

Atherosperma moschatum

MONIMIACEAE

Atherosperma
Monotypic genus. Flowers solitary, enclosed by 2 bracts in bud; male flowers above single female flower on each branch. Fruiting receptacle cup-shaped, enclosing hairy seeds.

Atherosperma moschatum
Southern Sassafras
Tree. Leaves opposite, 3–10cm long, entire, greyish underneath, fragrant when crushed. Flowers solitary in leaf axils, cup-shaped, creamy, turning pink with age. **HABITAT:** Deep gullies, along streams. **DIST:** Upper Blue Mtns (e.g. Gordon Falls, Katoomba Falls). Also NT, ST, Vic, Tas. **FL:** Aug–Oct.

Doryphora
Trees with aromatic bark and leaves. Inflorescences axillary, 3-flowered. Flowers bisexual; buds enclosed by 2 bracts; stamens 6; staminodes 6–12. Fruiting receptacle ribbed, enclosing fruit; seeds hairy.

Doryphora sassafras
Sassafras
Medium to tall tree. Leaves opposite, dark green, glossy, 6–10cm long, serrated. Flowers

Doryphora sassafras

white. **HABITAT:** In or near rainforests, sheltered gullies. **DIST:** Coast and ranges. Also NC, SC, NT, ST, CWS, Qld. **FL:** July–Sept.

FLOWERING PLANTS | 51

Hedycarya angustifolia

Palmeria scandens

Hedycarya
Dioecious shrubs or trees. Fruit compound, resembling a mulberry.

Hedycarya angustifolia
Native Mulberry
Large shrub. Leaves ovate, 5–10cm long, toothed, strongly veined, paler beneath. Male flowers with numerous stamens on a flat disc. Fruit yellow. **HABITAT:** Rainforests, sheltered gullies. **DIST:** Coast and ranges (e.g. Royal NP, Kiama, Bilpin, Robertson). Also NC, SC, NT, ST, Qld, Vic, Tas. **FL:** Aug–Sept.

Palmeria
Woody climbers. Flowers in axillary racemes, dioecious. Male flowers with 5 lobes; female flowers smaller and globular.

Palmeria scandens
Anchor Vine
Climber, becoming woody with age. Leaves opposite, broadly elliptic, 5–13cm long, dark green above, rough. Flowers yellowish. Fruit of 3–7 red drupes in fleshy receptacle about 20mm diam. **HABITAT:** In or near rainforests. **DIST:** Coast (e.g. Minnamurra Falls, Bulli Pass, Gosford). Also Robertson, NC, SC, NT, Qld. **FL:** June–Aug.

Wilkiea huegeliana

Wilkiea
Dioecious shrubs or trees. Leaves toothed, with small oil dots. Fruit an ovoid drupe.

Wilkiea huegeliana
Veined Wilkiea
Large shrub. Leaves dark green, leathery, to 15 x 6cm, serrated, strongly veined. Flowers insignificant. Fruits clustered, 12mm long, black. **HABITAT:** Rainforests, sheltered gullies, esp. on volcanic soils. **DIST:** Coast (e.g. Bass Pt, Mt Keira, Palm Jungle). From Jervis Bay to Qld. **FL:** Dec–Feb.

LAURACEAE

Cinnamomum
Trees and shrubs. Leaves usually alternate (rarely opposite), entire, dotted with oil glands. Perianth segments in 2–3 whorls; stamens often 9. Fruit a drupe or berry.

Cinnamomum oliveri
Oliver's Sassafras
Medium to large tree. Leaves ± opposite, lanceolate, 8–15cm long, glossy green above, paler below. Flowers to 6mm long, hairy, in panicles. Leaves aromatic when crushed. Fruit ovoid, 12mm long, black. **HABITAT:** Rainforest gullies. **DIST:** Coast (e.g. Saddleback Mt, Ourimbah Ck, Wambina NR). Also NC, Qld. **FL:** Oct–Nov, irregular.

Cryptocarya
Trees or shrubs. Leaves mostly alternate. Flowers small, with 6 perianth segments and 9 perfect stamens in 2 rows, the inner row alternating with staminodes. Fruit a drupe surrounded by succulent receptacle and persistent floral parts.

Cryptocarya glaucescens
Jackwood, Brown Beech
Tree to 25m tall. Leaves elliptic, 6–12cm long; undersurface midvein yellow. Flowers in terminal clusters. Fruit hanging in bunches, depressed-globose, black. **HABITAT:** Moist forests, rainforests. **DIST:** Coast (e.g. Budderoo NP, Royal NP, Wyong). Also Robertson, NC, SC, NT, Qld. **FL:** Oct–Nov.

Cryptocarya glaucescens

Cryptocarya microneura
Murrogun, Brown Jack
Tree similar to *C. glaucescens*. Leaves green on both surfaces, paler beneath; midvein whitish. Fruit ± globose, pointed, black. **HABITAT:** Rainforests. **DIST:** Coast (e.g. Minnamurra Falls, Bass Point, Wambina NR). Also SC, NC, Qld. **FL:** Sept–Dec.

Cryptocarya obovata (Ourimbah) has leaves broadest above the middle, lower surface grey-hairy and new growth velvety brown.

Cryptocarya rigida (Bouddi–Wyong) has ovate leaves 8–12cm long and new growth with pale brown hairs.

Endiandra
Trees and shrubs, usually with alternate leaves. Flowers in short axillary panicles; perianth segments 6–8; stamens 3; staminodes 3. Fruit a large drupe.

Endiandra sieberi
Corkwood
Medium tree with distinctive corky bark. Leaves narrow-elliptic, 5–8cm long; midrib yellow beneath. Flowers 2mm long, finely hairy. Fruit a shiny black ovoid drupe to 25mm long. **HABITAT:** Littoral rainforest. **DIST:** Coast (e.g. Stanwell Park, Royal NP, Wyong). Also NC, SC, Qld. **FL:** June–Oct.

Endiandra discolor is a North Coast species extending to Wyong and Bouddi NP. It has large domatia on the leaf undersurface.

Cinnamomum oliveri

FLOWERING PLANTS 53

Cryptocarya microneura

Endiandra sieberi

Litsea
Dioecious trees with distinctly veined alternate leaves. Flowers with 6 perianth segments and 9 stamens in 2 rows. Fruit a drupe on a cup-like receptacle.

Litsea reticulata
Bolly Gum
Tree to 25m tall, with grey mottled bark marked by shallow oval indentations. Leaves leathery, 4–8cm long, with prominent net veins. Flowers insignificant. Fruit a purple-black drupe about 12mm long. **HABITAT:** Rainforests. **DIST:** Coast (e.g. Whispering Gully, Mt Keira, Ourimbah). Also NC, SC, Qld. **FL:** May–June.

Litsea reticulata

Neolitsea
Dioecious trees with alternate leaves, 3-veined from base. Flowers in axillary clusters, small, with 4 perianth segments and 6 stamens in 2 rows. Fruit a drupe.

Neolitsea dealbata
White Bolly Gum
Shrub or tree to 10m tall. Leaves elliptic-obovate, 10–20 x 2–8cm; undersurface whitish to blue-grey. Branchlets, buds and underleaf veins rusty-hairy. **HABITAT:** Rainforests. **DIST:** Illawarra (e.g. Mt Keira, Wyong–Ourimbah). Also NC, Qld. **FL:** Mar–June.

Neolitsea dealbata

Cassytha glabella

Cassytha pubescens

CASSYTHACEAE

Cassytha
DEVIL'S TWINE
Leafless twining parasites attaching to host plants by stem suckers. Stems greenish yellow. Sometimes included in Lauraceae because of the structure of the tiny white flowers and berry-like fruit. Flowers and fruit similar to members of Lauraceae.

Cassytha glabella
Twiner with glabrous stems to 0.5mm diam. Flowers and fruits sessile in crowded heads. Fruit ovoid, about 3mm diam. **HABITAT:** Open forest, heath. **DIST:** Coast, lower Blue Mtns. Also NC, SC, Qld, Vic, Tas, SA, WA. **FL:** Sept–Mar.

Cassytha pubescens (incl. *C. paniculata*)
Twiner with softly hairy stems, 0.5–1.5mm thick. Flowers in spikes, racemes or heads. Fruit globular, hairy, 8–10mm long, with 6 raised ridges. **HABITAT:** Open forest, scrub, heath. **DIST:** Coast and ranges. Most of NSW. Also Qld, Vic, SA, NZ. **FL:** Sept–Dec.

Piper novae-hollandiae

Peperomia blanda var. *floribunda*

PIPERACEAE

Piper
Climbers with small flowers in terminal spike and fruit embedded in the fleshy axis. Includes Pepper (*P. nigrum*) and Kava (*P. methysticum*). About 2000 species worldwide.

Piper novae-hollandiae
Pepper Vine
Large climber with stems to 40m long. Leaves ovate, to 12cm long, with 5–7 conspicuous veins from the base. Male flowers in axillary spikes as long as leaves. **HABITAT:** On trees in rainforest. **DIST:** Coast (e.g. Mt Keira, Palm Jungle). Also NC, SC, NT, Qld. **FL:** June–Sept.

PEPEROMIACEAE

Peperomia
Succulent herbs with small flowers in cylindrical spikes and fruit embedded in the fleshy axis.

Peperomia blanda var. *floribunda*
Herb to 30cm tall. Leaves opposite, ovate, 1–3cm long, softly hairy. Flower spikes 4–10cm long. **HABITAT:** On rocks in damp gullies and rainforest. **DIST:** Coast (e.g. Razorback Mtn, Gosford, Woy Woy). Also NC, Qld, Asia, Pac. Iss. **FL:** Nov–Dec.

Peperomia tetraphylla (coast and Jenolan Caves) has glabrous leaves in whorls of 4.

Clematis aristata

RANUNCULACEAE

Clematis
OLD MAN'S BEARD
Climbers with opposite compound leaves having 3 leaflets. Flowers often unisexual with female flowers developing into clusters of fruit with distinctive long silky appendages. Flowers with 4 petal-like sepals; petals absent or minute.

Clematis aristata
Traveller's Joy
Robust climber. Leaflets firm, irregularly toothed. Flowers abundant, white to 4cm across. Stamens with appendage projecting above anthers; appendage as long as anthers. **HABITAT:** Forested gullies, shaded slopes. **DIST:** Coast and ranges. Also NC, SC, NT, ST, SWS, Qld, Vic, Tas, WA. **FL:** Oct–Nov.

Clematis glycinoides
Headache Vine
Robust climber. Differs from *C. aristata* in having leaflets which are thin and mostly entire, smaller flowers (to 3cm) and stamens lacking an appendage (or <1mm long). **HABITAT:** Open forests, shaded slopes. **DIST:** Coast and ranges. Also NC, SC, NT, ST, WS, Qld, Vic. **FL:** Aug–Oct.

Ranunculus
BUTTERCUP
Herbs with leaves mostly lobed or divided. Flowers with 5 sepals, 5 petals and numerous stamens. Fruit an achene. Eight native species, some rare and localised, and 5 naturalised species in Sydney region.

Ranunculus inundatus
River Buttercup
Glabrous herb, often forming large patches. Leaves with long slender stalks; lamina to 35mm diam., divided into several narrowly lobed segments. Flowers to 15mm across. **HABITAT:** Ponds, swamps, in wet mud. **DIST:** Coast and ranges. Also NC, SC, NT, ST, SWS, SWP, Vic, SA. **FL:** Nov–Jan.

Ranunculus lappaceus
Common Buttercup
Softly hairy plant with erect branched flower stems to 50cm tall. Leaves divided into 3

Clematis glycinoides

Ranunculus inundatus

Ranunculus lappaceus

FLOWERING PLANTS | 57

Ranunculus plebeius

Legnephora moorei

wedge-shaped lobes; each lobe toothed or divided at apex. Flowers 2–3cm across. Achene smooth, topped by a slender coiled beak. **HABITAT:** Damp sites in grassland and open forest, on shale or clay soils. **DIST:** Coast and ranges. Most of NSW. Also Qld, Vic, Tas, SA. **FL:** Aug–Dec.

Ranunculus plebeius
Hairy Buttercup
Herb to 30cm tall. Leaves hairy; lamina to 7cm long, divided into 3 lobes, each with 3 broad teeth. Flowers 10–15mm diam.; petals golden yellow; sepals reflexed against stalk. Achene smooth, with short strongly recurved beak. **HABITAT:** Damp woodland, swamps, creek banks. **DIST:** Coast and ranges. Also NC, SC, ST, Qld, Vic. **FL:** Chiefly Nov–Feb.

MENISPERMACEAE

Legnephora
Dioecious climbers with rounded leaves. Male flowers with 6 petals and 6 stamens; female flowers without petals; staminodes 6; carpels 3. Fruit a drupe.

Legnephora moorei
Round-leaf Vine
Tall woody climber. Leaves circular, to 20cm diam., green and glabrous above, ash-grey beneath. Drupe ± globose, 6–8mm long, red. **HABITAT:** Upper canopy of rainforest. **DIST:** Coast. Also NC, SC, CWS, Qld. **FL:** Dec–Jan.

Sarcopetalum
Climber with ovate leaves. Flowers unisexual, in racemes on main stems; petals 2–6; males with staminal filaments united into column; females with 3–6 carpels. Fruit a flattened drupe.

Sarcopetalum harveyanum
Pearl Vine
Woody climber. Leaves ovate, to 12 x 9cm on stalk to 9cm long; lamina leathery, glabrous, 7-veined from base; base of lamina deeply cordate. Flowers pink-yellow in racemes to 7cm long. Drupes 5–7mm diam., red. **HABITAT:** Rainforest, sheltered gullies. **DIST:** Coast and ranges. Also NC, SC, CWS, Qld, Vic, PNG. **FL:** Oct–Dec.

Sarcopetalum harveyanum

Stephania japonica var. *discolor*

Celtis paniculata

Stephania
Climbers with peltate leaves. Flowers in umbels; petals 3–5; males with staminal filaments united into a column; females with 1 carpel. Fruit a compressed drupe.

Stephania japonica var. *discolor*
Snake Vine
Slender climber. Leaves ovate to circular, to 12 x 8cm, glabrous and green above, pale greyish green below. Drupes obovoid, 2–5mm long, green turning orange then red. **HABITAT:** In or near rainforest, coastal dunes. **DIST:** Coast. Also Kowmung, NC, SC, NT, ST, WS, Qld, Asia. **FL:** Dec–Feb.

ULMACEAE

Trema tomentosa var. *viridis*

Celtis
Shrubs or trees. Leaves 3-veined from base. Inflorescence axillary. Flowers unisexual or bisexual, axillary. Fruit a drupe. One species in Sydney district, at the southern limit of its range.

Celtis paniculata
Native Celtis
Tree to 8m tall. Leaves elliptic, 5–8 x 2–4cm, glabrous, dull green. Perianth <2mm long; male flowers in panicle, female flowers solitary or 2–3 together. Fruit a purplish black drupe 8–10mm long. **HABITAT:** Littoral rainforest on sand. **DIST:** Coast (e.g. Bass Point, Grays Point). Also NC, Qld, Pac. Iss, Malesia. **FL:** Jan–Feb.

Trema
Trees or shrubs. Leaves finely toothed. Flowers unisexual and bisexual mixed in axillary cymes. Fruit a drupe.

Trema tomentosa var. *viridis* (formerly *T. aspera*)
Native Peach
Shrub or small tree with pubescent young branches. Leaves lanceolate, sandpapery, to 8 x 3cm, toothed. Perianth 4–5 lobed. Drupe globose, black, 4mm diam. **HABITAT:** Moist forests, rainforest margins. **DIST:** Coast and ranges. Most of NSW. Also Vic, Qld, NTerr. **FL:** Nov–Jan.

FLOWERING PLANTS | 59

Ficus obliqua

Ficus coronata

MORACEAE

Ficus
FIG
Trees with minute male and female flowers borne on inner wall of the structure called a 'fig'. Pollinated by wasps which lay eggs inside.

Ficus coronata
Sandpaper Fig
Shrub or tree to 10m tall. Leaves elliptic with asymmetric base, 5–14 x 2–5cm, upper surface sandpapery. Fig ovoid, 15mm long, densely covered with rough hairs. **HABITAT:** Creek banks, rainforest, sheltered rocky areas. **DIST:** Coast and ranges. All along NSW coast, t'lands, WS. Also Qld, Vic, NTerr. **FRUIT:** Ripe Dec–June.

Ficus macrophylla
Moreton Bay Fig
Large spreading 'strangler' tree to 40m tall, often with root buttressing. Leaves elliptic, to 25cm long, glabrous, usually red-brown on undersurface; stalks to 10cm long. Fig solitary, 25mm diam., orange turning purple with white dots; stalks to 15mm long. **HABITAT:** Rainforest. **DIST:** Illawarra to Shoalhaven R. Widely planted in Sydney parks. Also NC, Qld. **FRUIT:** Ripe Feb–May.

Ficus obliqua
Small-leaved Fig
'Strangler' tree to 40m tall, with spreading buttresses and aerial roots. Leaves elliptic, 4–7cm long, smooth, with stalks 6–15mm. Figs paired, 6–8mm diam., yellow turning orange; stalks to 4mm long. **HABITAT:** Rainforest. **DIST:** Coast (e.g. Jamberoo, Bola Ck, Ourimbah). Also NC, SC, CWS, Qld, Malesia. **FRUIT:** Ripe Apr–July.

Ficus macrophylla

Ficus rubiginosa
Port Jackson Fig, Rusty Fig
Shrub or spreading tree to 40m tall, also small scrambling shrub on rock faces. Young stems rusty. Leaves 6–10 x 5–6cm, rusty below; stalks to 3cm long. Figs paired, 12–20mm diam., yellow turning red, marked with warts; stalks to 5mm long. **HABITAT:** Open forest and dry rainforest gullies, rocky slopes. **DIST:** Coast and ranges. Also NC, SC, NT, WS, NWP, Qld. **FRUIT:** Ripe Feb–July.

Ficus superba var. *henneana*
Deciduous Fig
'Strangler' tree to 40m tall with buttressed trunk. Leaves ovate, 8–12cm long, green and smooth on both surfaces, deciduous in late winter; stalks 2–5cm long. Figs solitary, globular, 25mm diam., yellow turning red-purple, spotted; stalks 5–10mm long. **HABITAT:** Riverine and littoral rainforest, often on rocky slopes. **DIST:** Coast (e.g. Jamberoo, Burning Palms). Also NC, Qld, NTerr. **FRUIT:** Ripe Dec–June.

Maclura
Dioecious woody climber or straggly shrub with flowers in axillary globose heads. Compound fruit globose.

Maclura cochinchinensis
Cockspur Thorn
Untidy plant, forming thickets. Stems with stout thorns. Leaves alternate, elliptic, 3–8cm long. Fruit orange, 12–16mm diam., more noticeable than smaller flower heads. **HABITAT:** In or near rainforest. **DIST:** Coast, esp. Illawarra. Also Cobbitty, Calga, SC, NC, Qld, Malesia. **FL:** Nov–Jan.

Ficus rubiginosa

Ficus superba var. *henneana*

FLOWERING PLANTS

Maclura cochinchinensis

Streblus brunonianus

Streblus
Dioecious trees with milky latex. Fruit a drupe topped by a bifid style.

Streblus brunonianus
Whalebone Tree
Tall shrub or tree. Leaves alternate, lanceolate, to 7 x 4cm, toothed, often scabrous. Male flowers in dense cylindrical spikes to 3cm long; female flowers in shorter spikes. Fruit yellow, ovoid, 6mm long. **HABITAT:** Rainforest, on dry rocky slopes. **DIST:** Coast, coastal ranges. Also NC, SC, CT, Qld. **FL:** Oct–Feb.

URTICACEAE

Dendrocnide
STINGING TREE
Trees and shrubs with stinging hairs on young branches, leaf stalks and underside of leaves.

Dendrocnide excelsa
Giant Stinging Tree
Tree to 30m tall. Leaves heart-shaped, to 25 x 20cm, pale green; stalk to 15cm long. Flowers yellowish in panicles to 12cm long. Pedicel of fruit swollen, fleshy, becoming pink with age. **HABITAT:** Rainforest, on fertile soils. **DIST:** Coast (e.g. Royal NP, Jamberoo, Calga). Also Kanangra Walls, SC, NC, NT, WS, Qld. **FL:** Dec–Apr.

Dendrocnide excelsa

Elatostema reticulatum

Elatostema
Dioecious herbs with alternate 2-ranked leaves. Flowers in dense heads.

Elatostema reticulatum
Spreading plant to 70cm tall with fleshy stems. Leaves lanceolate, to 15 x 6cm, toothed, with asymmetric base. Male flowers in dense heads to 20mm diam. on peduncles 7cm long; female heads smaller, sessile. **HABITAT:** Wet rocks and near streams in rainforest. **DIST:** Illawarra, Sthn H'lands. Also SC, NC, Qld. **FL:** Nov–Jan.

Urtica
STINGING NETTLE
Herbs with rigid stinging hairs. Leaves opposite, serrated. Inflorescence axillary. Flowers with 4 perianth lobes. Fruit a nut enclosed by perianth.

Urtica incisa
Scrub Nettle
Monoecious herb to 1m tall. Leaves opposite, lanceolate, to 10cm long, irregularly toothed; stalk to 6cm long. Flowers in slender panicles or racemes longer than leaf stalks. **HABITAT:** Rainforest and damp open forest. **DIST:** Coast and ranges. Most of NSW. Also Qld, Vic, Tas, SA, Pac. Iss. **FL:** Aug–Dec.

Urtica incisa

The introduced species, *Urtica urens*, occurs as an annual weed in disturbed areas. Its flowers are clustered on the short branches of a panicle.

CASUARINACEAE

Allocasuarina
SHE-OAK
Wind-pollinated trees or shrubs. Male flowers in spikes at ends of branchlets; female flowers clustered on lower branches (monoecious) or on separate plants (dioecious). Branchlets cylindrical with regular nodes, each carrying a whorl of reduced leaves forming a ring of triangular teeth. 'Seeds' (samaras) red-brown to black, shiny, embedded in cones.

Allocasuarina diminuta subsp. mimica
Dioecious shrub to 2m tall. Branchlets erect, with waxy bloom. Leaf teeth 6–10. Male spikes to 5cm long. Cones truncate, 20 x 12mm. **HABITAT:** Woodland, heath. **DIST:** Endemic. Scattered but uncommon (e.g. Kingsford, Lucas Hts, Leura). **FL:** Mar–Aug.

Subsp. *diminuta* has 6–7 teeth. It occurs Putty–Capertee, Berrima, WS.

Allocasuarina diminuta subsp. *mimica*

Allocasuarina distyla

Allocasuarina distyla
Scrub She-oak
Spreading dioecious shrub to 3m tall. Leaf teeth 6–8. Male spikes to 7cm long, abundant, giving whole plant a tan appearance. Cones cylindrical with pointed ends, 20–30mm long. **HABITAT:** Woodland, heath, scrub. **DIST:** Coast and ranges (e.g. Royal NP, Woodford, Bundanoon). Also NC, SC, ST, Vic. **FL:** Apr–Oct.

Allocasuarina littoralis
Black She-oak
Dioecious tree to 10m tall (a stunted shrub on stony heath). Leaf teeth 6–8. Male spikes to 20mm long. Cones cylindrical, truncate, to 30mm long. **HABITAT:** Open forest, woodland, scrubby heath. **DIST:** Coast and ranges. Also NC, SC, NT, ST, WS, NWP, Qld, Vic, Tas. **FL:** Apr–Oct.

Allocasuarina nana
Dioecious shrub 1–2m tall. Leaf teeth 5–6. Male spikes deep tan, 5–15mm long, erect on short branchlets. Cones oblong, to 20mm long. **HABITAT:** Heath, exposed ridges. **DIST:** Coast sth from Cowan; common in upper Blue Mtns. Also SC, ST, Vic. **FL:** June–Oct.

Allocasuarina littoralis

Allocasuarina nana

Allocasuarina paludosa
Spreading monoecious shrub to 2m tall. Branchlets softly hairy in furrows. Leaf teeth 6–8. Male spikes to 18mm long. Cones oblong, to 20mm long. **HABITAT:** Wet heath, woodland. **DIST:** Coast, mainly sth of Sydney, Sthn H'lands, Bilgola, Narrabeen Head. Also SC, ST, Vic, Tas, SA. **FL:** Nov–Feb.

Allocasuarina torulosa
Forest Oak
Graceful dioecious tree to 20m tall, with furrowed bark. Leaf teeth 4–5. Male spikes 2–4cm long. Cones ± globular, to 25mm long. **HABITAT:** Forest, on moderately fertile soils. **DIST:** Coast and ranges. Also NC, SC, NT, WS, Qld. **FL:** Mar–Sept.

Allocasuarina verticillata
Drooping She-oak
Spreading dioecious tree with drooping branches. Leaf teeth 9–13. Male spikes 5–10cm long. Cones 15–35mm long, cylindrical, tapering to base. **HABITAT:** Coast on clay-shale; inland on rocky outcrops. **DIST:** Coast (e.g. Garie, Bungan Head); upper Blue Mtns (e.g. Nullo Mtn). Also SC, ST, WS, WP, Vic, Tas, SA. **FL:** May–Aug.

Allocasuarina glareicola (Castlereagh SF). Rare endemic. Shrub with 5–7 teeth; cones 13 x 8mm.

Allocasuarina portuensis (Nielsen Park). Rare endemic. Shrub with 7–8 teeth; cones 15 x 10mm.

Casuarina cunninghamiana

Casuarina
SHE-OAK
Wind-pollinated trees similar to *Allocasuarina*, but differing in having pale dull seeds (samaras).

Casuarina cunninghamiana
River Oak
Stately dioecious tree to 25m tall. Leaf teeth 8–10. Male spikes to 15mm long. Cones ± globular to ± cylindrical, 10 x 8mm. **HABITAT:** Banks of rivers and streams. **DIST:** Coast and ranges (e.g. Macquarie R., Nepean R., Coxs R.). Most of NSW. Also Qld. **FL:** Flowers all year.

Allocasuarina paludosa

Allocasuarina torulosa

Allocasuarina verticillata

FLOWERING PLANTS | 65

Casuarina glauca

Carpobrotus glaucescens

Casuarina glauca
Swamp Oak
Dioecious tree to 20m tall (a stunted shrub on exposed headlands). Leaf teeth 12–16. Male spikes to 40mm long. Cones cylindrical, 9–18 x 7–9mm. **HABITAT:** Estuarine flats, brackish streams. **DIST:** Coast, Cumberland Pl. Also NC, SC, CWS, Qld. **FL:** June–Sept.

AIZOACEAE

Carpobrotus
Prostrate perennials with trailing stems and fleshy leaves. Fruit succulent with numerous seeds embedded in flesh.

Carpobrotus glaucescens
Pigface
Leaves glaucous, 3-angled, to 10 x 1.5cm. Flowers to 6cm across; staminodes numerous, petal-like, purple with white base; stamens 300–400, yellow. Fruit purple-red. **HABITAT:** Coastal dunes, sea-cliffs. **DIST:** Coast, from Vic to Qld. **FL:** June–Jan.
 Fruit and leaves roasted and eaten by Aborigines, perhaps adding salt to their diet.

Tetragonia
Sprawling herbs or small shrubs. Leaves alternate, entire. Flowers in axillary clusters, green or yellow, sessile or pedunculate.

Tetragonia tetragonioides
New Zealand Spinach
Scrambling herb with thick succulent stems and leaves bearing small glistening blisters. Leaves rhombic, to 8 x 5cm. Flowers yellow;

Tetragonia tetragonioides

perianth lobes 4–5, 2mm long; stamens 4–many. **HABITAT:** Salt marsh edges, coastal rock crevices. **DIST:** Coast. Also WS, WP, all States, Pac. Iss. **FL:** Aug–Mar.
 Leaves of this species were cooked as a spinach-like vegetable by early European explorers in Australia.

PORTULACACEAE

Calandrinia

Succulent herbs. Leaves mainly basal, smaller on stems. Inflorescences few to many-flowered cymes. Flowers with 4–10 petals, free, mostly on erect pedicles. Fruit a capsule.

Calandrinia calyptrata
Pink Purslane

Stems to 20cm long. Leaves cylindrical, 1–5cm long, often pink-red. Flowers pink, on stalks in upper leaf axils; stamens 5–7. Seeds lenticular, smooth or nearly so, brown. **HABITAT:** Dry rocky sites. **DIST:** Cumberland Pl., Sthn H'lands. Also WS, SWP, Vic, Tas, SA, WA. **FL:** Sept–Oct.

Calandrinia calyptrata

Calandrinia eremaea
Pink Purslane

Succulent herb with stems to 20cm long. Leaves narrow-elliptic, sessile, to 5cm long. Inflorescence straight (not twining). Flowers deep pink; petals 3–7mm long; stamens 7–20. Seeds reniform, 3–5mm long, pimply with coppery lustre. **HABITAT:** Swampy grassland, woodland. **DIST:** Uncommon. Upper Blue Mtns. Also most of NSW, other States (except Tas). **FL:** Aug–Nov.

Calandrinia pickeringii is similar to *C. calyptrata* but has rough, pimply dark-brown to black seeds.

Calandrinia eremaea

Stellaria flaccida

CARYOPHYLLACEAE

Stellaria
Annual and perennial herbs with opposite or basal leaves and star-like white flowers (*stella*, 'star').

Stellaria flaccida
Forest Starwort
Stems trailing or scrambling over other plants. Leaves ovate, soft, 7–16mm long; stalk 2–6mm long, fringed. Petals 5, longer than sepals, deeply divided giving the appearance of 10. Stamens 10. **HABITAT:** Rainforest edges, sheltered gullies. **DIST:** Coast and ranges. Also NC, SC, NT, ST, Qld, Vic, Tas, SA. **FL:** Oct–Feb.

Stellaria pungens
Prickly Starwort
Stems weak, angular, forming mats or spreading over other plants and rocks. Leaves narrow-lanceolate, pungent, 6–8mm long. Petals 5, as long as sepals, deeply divided and appearing as 10. Stamens 10. **HABITAT:** Damp tall forests. **DIST:** Upper Blue Mtns. Also SC, NT, ST, WS, Vic, Tas, SA. **FL:** Oct–Dec.

Stellaria pungens

Alternanthera denticulata

Nyssanthes erecta

AMARANTHACEAE

Alternanthera
Weed-like herbs with opposite leaves and inflorescence axillary or terminal; flowers in spikes or sometimes clustered. The genus includes the pest species Alligator Weed (*A. philoxeroides*) and Khaki Weed (*A. pungens*).

Alternanthera denticulata
Lesser Joyweed
Glabrous annual herb to 30cm tall. Leaves variable in shape, finely toothed, to 40 x 6mm. Flowers with papery bracts, appearing as sessile chaffy clusters 4–8mm diam. in leaf axils. **HABITAT:** Grassy woodlands, disturbed clay areas, wetland edges. **DIST:** Cumberland Pl. (e.g. Yallah, Mt Annan). Most of NSW, all States. **FL:** Oct–Feb.

Nyssanthes
Herbs with opposite leaves. Bracts spreading spinescent. Flowers with 4 rigid segments, the outermost spinescent. Fruit enclosed in perianth.

Nyssanthes erecta
Erect herb to 50cm tall with spreading upper branches. Leaves elliptic, 2–7 x 1–2cm. Flowers with 4 stamens in sessile clusters. Spines to 10mm long. **HABITAT:** Beside creeks, disturbed areas. **DIST:** Cumberland Pl., Razorback Mtn, Kiama. Also NC, SC, CWS, Qld. **FL:** Jan–Mar.

CHENOPODIACEAE

Einadia
Perennial shrubs and herbs. One of a group commonly known as saltbushes. Leaves broad hastate. Inflorescences spike-like.

Einadia hastata
Saloop
Spreading perennial to 50cm tall. Leaves stalked, lamina 10–25mm long. Flowers in short clusters. Fruit red, succulent, to 3mm across. **HABITAT:** Woodland, grassland, on clay-shale and volcanics. **DIST:** Widespread. Most of NSW, Qld, Vic. **FL:** Oct–Feb.

Einadia nutans. Twining plant with leaves linear or sagittate.

Einadia polygonoides. Prostrate or climbing herb. Fruit dry. On clay soils of Cumberland Pl.

Einadia trigonos. Similar to *E. hastata*, but with dry fruit. Scattered in clay-shale and basalt soils.

Einadia hastata

FLOWERING PLANTS | 69

Enchylaena tomentosa

Suaeda australis

Enchylaena
Shrubs. Leaves alternate, entire. Flowers solitary, axillary. Fruit succulent.

Enchylaena tomentosa
Ruby Saltbush
Weak perennial to 1m tall with soft grey hairs on stems and leaves. Leaves terete, 6–16mm long. Fruit flattened globular, 5–8mm across, yellow to red, finally black. **HABITAT:** Coastal sands. **DIST:** Illawarra (e.g. Minnamurra Spit), Terrigal. Also NC, SC, drier inland areas, all mainland States. **FL:** Sept–Dec.

Rhagodia
Dioecious shrubs. Leaves simple, usually with bladder-like hairs. Fruit a fleshy berry.

Rhagodia candolleana
Coastal Saltbush
Weak shrub, about 1m tall. Leaves elliptic, to 25mm long, covered with grey sheen below. Flowers clustered in terminal panicles to 15cm long, mealy white. Fruit red, globular, 2–3mm across. **HABITAT:** Coastal dunes. **DIST:** Coast (e.g. Werrong, Minnamurra). Also NC, SC, Vic, Tas, SA, WA. **FL:** Most of year.

Sarcocornia
Perennials spreading horizontally in mud with erect, succulent, jointed branches. Leaves reduced to scales at nodes. Inflorescence terminal, spike-like. Fruiting perianth dry. Often covering large areas of salt marsh.

Rhagodia candolleana

Sarcocornia quinqueflora

Sarcocornia quinqueflora
Samphire
Branches to 30cm tall, green, often with reddish tinge. **HABITAT:** Coast, estuaries, salt marsh. **DIST:** Coast. Also NC, SC, Qld, Vic, Tas, SA, WA, LHI. **FL:** Nov–Feb.

Suaeda
Herbs or shrubs from saline habitats. Leaves alternate. Flowers bisexual, in axillary clusters or spikes near ends of branches.

Suaeda australis
Seablite
Bushy glabrous herb to 80cm tall. Leaves soft, succulent, semi-terete, to 3cm long. Flowers in sessile axillary clusters of 1–3 along branches, tiny, 5-lobed, succulent. Seeds glossy, red-brown, 1mm diam. **HABITAT:** Salt marsh, landward side of mangroves. **DIST:** Coast, NSW, all States (except NTerr). **FL:** Sept–Dec.

POLYGONACEAE

Persicaria
Herbs of wet places. Stems with sheath (ocrea) encircling node at base of each leaf stalk. Flowers in terminal or axillary spike-like inflorescences. Perianth segments 4–5, white, pink or green. Fruit a nut.

Persicaria decipiens
Slender Knotweed
Decumbent or ascending glabrous plant to 30cm tall. Ocrea with long fine bristles. Leaves lanceolate, to 12cm x 12mm, often with purple blotch. Flowers pink, in slender spikes to 6cm long. **HABITAT:** Shallow water, wet stream banks. **DIST:** Coast, Sthn H'lands. Also NC, SC, NT, ST, NWS, all States, Europe, Asia, Pac. Iss. **FL:** Jan–May.

Persicaria elatior. Rare plant to 90cm tall with glandular hairs on ocrea. Recorded from Thirlmere Lakes.

Persicaria decipiens

Persicaria hydropiper
Water Pepper
Weak glabrous plant to 80cm tall. Ocrea with long fine bristles. Leaves narrow-ovate, to 12cm long, with a peppery taste. Flowers greenish white, on slender drooping spikes to 9cm long; perianth dotted with glandular hairs. **HABITAT:** Shallow water, wet stream banks. **DIST:** Coast to slopes and plains. Also Qld, Vic, WA. **FL:** Jan–May.

Persicaria lapathifolia

Persicaria lapathifolia
Pale Knotweed
Erect glabrous plant to 1.5m tall. Ocrea without bristles. Leaves narrow-elliptic, 20cm long, gland dotted underneath. Flowers on dense drooping spikes to 6cm long, pink. Perianth dotted with glandular hairs. **HABITAT:** Shallow water, wet stream banks. **DIST:** Widespread (e.g. Doonside, Albion Park, Coxs R.). Also most of NSW, other States (except WA), Asia, Afr. **FL:** Dec–Mar.

Persicaria hydropiper

FLOWERING PLANTS

Persicaria orientalis
Princes Feathers
Erect densely hairy plant to 3m tall. Ocrea hairy with winged collar. Leaves ovate to 20 x 12cm. Flowers deep pink in dense spikes to 8cm long. **HABITAT:** Swampy margins. **DIST:** Cumberland Pl., Illawarra. Also NC, SC, WS, WP, Qld, NTerr, Asia. **FL:** Jan–May.

Persicaria strigosa
Spotted Knotweed
Decumbent plant with stiff hairs on stems. Ocrea brown with short bristles. Leaves hastate, to 7cm long. Inflorescences 2–4-branched, with small terminal flower clusters 5–10mm long. Flowers white. **HABITAT:** Shallow water, wet stream banks. **DIST:** Coast to ranges. Also NC, Qld, Malesia. **FL:** Dec–Jan.

Persicaria subsessilis
Hairy Knotweed
Decumbent hairy shrub. Ocrea membranous, pale brown, 10–15mm long with fine bristles. Leaves lanceolate, to 15cm long, hairy. Flowers white in dense spikes to 5cm long. **HABITAT:** Shallow water, wet stream banks. **DIST:** Rare. Cumberland Pl. (e.g. Shanes Park, Narellan). Also NC, SC, Qld, Vic, Tas, WA, PNG, NCal. **FL:** Dec–Jan.

Persicaria praetermissa. Creeping plant with hairy stems, hastate leaves and pink flowers in spikes. Widespread.

Persicaria prostrata. Small plant with hairy stems, toothed ocrea, and cream flowers in small dense spikes. Widespread.

Persicaria orientalis

Persicaria subsessilis

Persicaria strigosa

DILLENIACEAE

Hibbertia

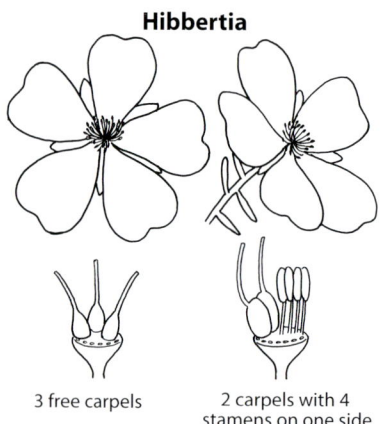

3 free carpels | 2 carpels with 4 stamens on one side

Hibbertia
GUINEA FLOWER
Shrubs or climbers. Flowers with 5 yellow spreading petals. Stamens few to numerous; carpels more or less free, 2–5. Fruit a follicle.

Hibbertia acicularis
Prickly Guinea Flower
Glabrous shrub to 1m tall. Leaves linear with recurved margins, to 15mm long, pungent. Flowers about 20mm across on slender stalks to 20mm long; stamens 4–8 on one side of 2 hairy carpels. **HABITAT:** Woodland, rocky areas. **DIST:** Coast and ranges. Also NC, SC, NT, ST, WS, Qld, Vic, Tas, SA. **FL:** Sept–Dec.

Hibbertia acicularis

Hibbertia aspera subsp. *aspera*

Hibbertia aspera subsp. *aspera*
Rough Guinea Flower
Hairy shrub to 60cm tall. Leaves obovate with recurved margins, to 18mm long, dark green and rough above. Flowers 10–15mm across on slender stalks 12–15mm long. Stamens 6 on one side of 2 hairy carpels. **HABITAT:** Open forest, woodland, heath. **DIST:** Coast and ranges. Also NC, SC, NT, Qld, Vic. **FL:** Aug–Dec.

Hibbertia bracteata
Erect shrub to 1m tall with glabrous branches. Leaves oblong, to 25mm long, shiny green above, margins flat or recurved, with a minute point. Flowers terminal, sessile, surrounded by bracts; sepals silky. Stamens 16 on one side of 2 hairy carpels. **HABITAT:** Moist open forests, sheltered slopes. **DIST:** Endemic. Coast and ranges (e.g. Mt Colah, Maroota, Lawson). **FL:** July–Dec.

Hibbertia bracteata

Hibbertia cistiflora
Shrub to 80cm tall. Leaves linear, to 10 x 1mm, margins revolute, with tiny yellowish point and warty surface. Flowers sessile, about 16mm across; sepals glabrous; stamens 6 on one side of 2 glabrous carpels. **HABITAT:** Woodland, heath, rocky areas. **DIST:** Coast and ranges (e.g. Warrah, Maroota, Mt Banks). Also ST, Vic. **FL:** July–Sept.

Hibbertia dentata
Twining Guinea Flower
Twiner with wiry stems. Leaves ovate, to 7 x 3cm, toothed. Flowers about 3cm across; stamens >30 surrounding 3 glabrous carpels. Fruit splitting to reveal orange-red seeds. **HABITAT:** Sheltered tall forest, coastal scrub. **DIST:** Coast and ranges. Also NC, SC, NT, Qld, Vic. **FL:** Aug–Nov.

Hibbertia cistiflora

Hibbertia dentata

Hibbertia diffusa
Low to prostrate shrub, often mat-forming. Leaves variable, usually obovate with a notched apex, to 25mm long, glabrous. Flowers sessile, about 20mm across; stamens 20–25 surrounding 3 glabrous carpels. **HABITAT:** Woodland, heath, esp. on clay-shales and alluvium. **DIST:** Widespread (e.g. Bouddi, Agnes Banks, Bargo). Also NC, SC, ST, NT, Qld, Vic. **FL:** Aug–Feb.

Hibbertia empetrifolia
Trailing Guinea Flower
Decumbent shrub to 40cm tall. Leaves oblong, 5–7mm long, margins recurved, upper surface minutely warty or rough. Flowers about 10mm across on slender stalks to 10mm long. Stamens 6–12 on one side of 2 hairy carpels. **HABITAT:** Open forest, heath. **DIST:** Coast and ranges. Also NC, SC, WS, Qld, Vic, Tas. **FL:** Chiefly Sept–Mar.

Hibbertia fasciculata
Bundled Guinea Flower
Diffuse shrub to 40cm tall. Leaves clustered, linear, 5–7mm long. Flowers axillary, sessile, 16–20mm across; stamens 8–12 surrounding 3 glabrous carpels. **HABITAT:** Open forest, scrub, heath. **DIST:** Coast and ranges. Also NC, SC, NT, ST, Qld. **FL:** July–Dec.

Hibbertia linearis
Diffuse to erect shrub to 80cm tall with glabrous stems. Leaves variable, linear to obovate, 8–30 x 1–4mm, pointed or rounded at tip. Flowers about 20mm across; stamens 15–25 surrounding 3 glabrous carpels. **HABITAT:** Woodland, scrub, heath. **DIST:** Chiefly coast. Also upper Blue Mtns, NC, SC, NT, CWS, Qld. **FL:** May–Nov.

Hibbertia fasciculata

Hibbertia monogyna
Erect glabrous shrub to 50cm tall. Leaves wedge-shaped with broad notched apex, to 12 x 5mm, folded along midvein. Flowers sessile; stamens 10–15 surrounding 1 glabrous carpel. **HABITAT:** Open forest, woodland. **DIST:** Coast and ranges (e.g. Newnes SF, Bundanoon). Also SC, ST, WS. **FL:** Aug–Dec.

Hibbertia nitida
Shining Guinea Flower
Erect glabrous shrub to 1m tall; branchlets often reddish. Leaves oblanceolate with acute or mucronate tip, to 25 x 8mm, glossy. Flowers sessile, about 20mm across; sepals glabrous; stamens 11 on one side of 2 hairy carpels. **HABITAT:** Sheltered slopes, often near salt water inlets. **DIST:** Endemic. Coast (e.g. Roseville, Audley, O'Hares Ck). **FL:** Sep–Nov.

Hibbertia obtusifolia
Erect or diffuse shrub to 60cm tall; stems and leaves with soft grey hairs. Leaves oblanceolate to spathulate with notched or round tips, 10–30 x 3–10mm. Flowers sessile, 20–30mm across; stamens 30–40 surrounding 3 glabrous carpels. **HABITAT:** Open forest, woodland. **DIST:** Coast and ranges. Most of NSW; also Qld, Vic. **FL:** Nov–May.

A prostrate form occurs along the Colo–Howes Swamp Road.

Hibbertia diffusa

Hibbertia empetrifolia

FLOWERING PLANTS

Hibbertia linearis

Hibbertia nitida

Hibbertia monogyna

Hibbertia obtusifolia

Hibbertia riparia

FLOWERING PLANTS

Hibbertia praemorsa

Hibbertia pedunculata
Stalked Guinea Flower
Prostrate or rarely erect diffuse shrub. Leaves linear, 4–6mm long, hairy; margins recurved. Flowers on slender stalks twice as long as leaves, about 12mm across; stamens 15–20 surrounding 2 hairy carpels. **HABITAT:** Open forest, on alluvial clay soils. **DIST:** Cumberland Pl. (e.g. Rookwood, Berkshire Park, Casula). Also NC, NT, ST, WS, Qld, Vic. **FL:** Nov–Mar.

Hibbertia praemorsa
Silky Guinea Flower
Spreading shrub to 1.3m tall with villous branchlets. Leaves oblong, 5–10 x 3–4mm, apex obtuse, margins recurved; both surfaces densely covered with stellate and long soft hairs. Flowers single, sessile, 20mm across; stamens 7–9 on one side of 2 hairy carpels. **HABITAT:** Open forest, in sandy soil or near rocks. **DIST:** Endemic. Sthn H'lands (e.g. Bundanoon, Wingello SF). **FL:** Oct–Feb.

Hibbertia riparia
Erect Guinea Flower
Variable shrub to 60cm tall. Leaves linear with revolute margins, 8–15 x 1–2mm, scabrous.

Hibbertia pedunculata

Flowers about 22mm across; sepals hairy; stamens 6–16 on one side of 2 hairy carpels. **HABITAT:** Woodland, heath. **DIST:** Chiefly coastal. Most of NSW, all other States (except NTerr). **FL:** May–Dec.

It is expected that some varieties will be raised to new species status.

Hibbertia rufa

Hibbertia rufa
Spreading shrub to 30cm tall; older stems reddish. Leaves linear with revolute margins, 5–8mm long, acute. Flowers 14mm across, on slender stalks to 12mm long. Stamens 4 on one side of 2 glabrous carpels. **HABITAT:** Wet heath, damp sedgy sites. **DIST:** Scattered; uncommon (e.g. Barren Grounds, Leura, Penrose). Also NC, NT, ST, Qld, Vic. **FL:** Aug–Nov.

Hibbertia saligna
Shrub to 2m tall; new growth softly hairy. Leaves sessile, narrow-elliptic, to 10cm x 15mm, glabrous above, pubescent beneath. Flowers on short branches, to 40mm across; stamens 20–35 surrounding 3 glabrous carpels. **HABITAT:** Sheltered forests, gullies. **DIST:** Blue Mtns (e.g. Kurrajong Heights, Lawson, Erskine Ck). Also SC, ST, CWS. **FL:** Aug–Nov; Apr–May.

Hibbertia scandens
Climbing Guinea Flower
Climber with long scrambling stems. Leaves stem-clasping, obovate, to 120 x 40mm. Flowers 5–6cm across; stamens >30 surrounding 3–7 glabrous carpels. Fruit splitting to reveal red seeds. **HABITAT:** Coastal scrub, dunes, rainforest edges. **DIST:** Coast. Also NC, SC, NT, CT, Qld. **FL:** Mar–Nov.

Hibbertia serpyllifolia
Decumbent or prostrate shrub. Leaves linear, 3–8mm long, upper surface scabrous, margins revolute. Flowers terminal, sessile, about 12mm across; stamens 15–20 surrounding 3 hairy carpels. **HABITAT:** Heath, scrubland, on sandy or rocky sites. **DIST:** Chiefly Blue Mtns; rare on coast. Also NC, SC, NT, ST, NWS, Vic, Tas. **FL:** Oct–Feb.

FLOWERING PLANTS

Hibbertia saligna

Hibbertia scandens

Hibbertia vestita

Hibbertia serpyllifolia

Hibbertia vestita
Hairy Guinea Flower
Diffuse shrub with spreading hairy branches, often prostrate. Leaves oblong, 4–7mm long, hairy; margins recurved. Flowers terminal, sessile, to 20mm across. Stamens about 30 surrounding 3 hairy carpels.
HABITAT: Coastal heath. **DIST:** Coast north of Wyong (e.g. Munmorah SCA, Lake Macquarie). Also NC, NT, NWS, Qld. **FL:** Sept–Nov.

The genus *Hibbertia* is currently being revised. It is likely that a number of forms and newly discovered populations will be described as new species. Some of those in the Sydney district are:
- *Hibbertia* (Menai): erect hairy shrub with 6–8 stamens on one side of 2 hairy carpels.
- *Hibbertia* (Maddens Plains): decumbent glabrous shrub with 5–7 stamens on one side of 2 hairy carpels.
- *Hibbertia* (East Heathcote): spreading shrub with 16–22 stamens surrounding 3 hairy carpels.
- *Hibbertia* (Minnehaha Falls): sprawling shrub with 8 stamens surrounding 3 hairy carpels.
- *Hibbertia* (Megalong Valley): sprawling shrub with 6 stamens on one side of 2 hairy carpels.
- *Hibbertia* (Howes Swamp): sprawling shrub with 8 stamens on one side of 2 slightly hairy carpels.

CLUSIACEAE

Hypericum
Perennials, with yellow 5-petalled flowers and numerous stamens. Two native species.

Hypericum gramineum
Small St John's Wort
Erect perennial to 30cm tall. Leaves stem-clasping, opposite, ovate, 5–25mm long, revolute. Flowers usually in terminal cymes; each flower 5–12mm diam., with 20–50 stamens. **HABITAT:** Open forest, grassland. **DIST:** Coast and ranges. Most of NSW, all other States. **FL:** Sept–Feb.

Hypericum japonicum is a prostrate or decumbent herb to 15cm tall with clusters of sessile flat leaves <8mm long.

ELAEOCARPACEAE

Elaeocarpus
Trees or shrubs. Leaves alternate, simple or 1-foliolate. Flowers in axillary racemes; petals often with linear lobes and appearing fringed. Fruit a blue or black drupe.

Elaeocarpus holopetalus
Black Oliveberry
Small to tall tree. Leaves lanceolate to obovate, 3–7 x 1–3cm, finely and sharply toothed, strongly veined, with dense pale hairs on undersurface. Petals white, not fringed, 6mm long. Fruit ovoid, 9 x 6mm, black. **HABITAT:** Cool mountain gullies. **DIST:** Upper Blue Mtns (e.g. Gordon Falls); near Robertson. Also SC, NT, ST, Vic. **FL:** Oct–Nov.

Elaeocarpus kirtonii
Pigeonberry Ash
Tree to 15m tall. Leaves narrow-elliptic with long tapering acute apex, to 20 x 4cm, finely toothed, with prominent small veins; stalk 15–50mm long. Flowers bell-like, pendulous, with fringed white petals 10mm long. Fruit ovoid, 10–13mm long, pale blue. **HABITAT:** Rainforests. **DIST:** Illawarra (e.g. Mt Keira), Gosford area. Also NC, SC, NT, CT, Qld. **FL:** Jan–Mar.

Hypericum gramineum

Elaeocarpus holopetalus

Elaeocarpus reticulatus

Elaeocarpus kirtonii

Elaeocarpus obovatus

Elaeocarpus reticulatus (flower)

Elaeocarpus obovatus
Hard Quandong
Small to tall tree. Leaves obovate, 5–9cm long, with shallow teeth in upper part and 1–5 domatia in angles of lateral veins in lower part. Flowers white, fringed, bell-like, in narrow racemes as long as leaves. Fruit 6–12mm long, blue. **HABITAT:** Rainforest margins, sheltered sites. **DIST:** Coast nth from Gosford (e.g. Bateau Bay, Nth Entrance). Also NC, CWS, Qld. **FL:** Sept–Nov.

Elaeocarpus reticulatus
Blueberry Ash
Small tree. Leaves elliptic with pointed apex, to 12 x 3cm, finely toothed, discolorous; reticulate veins prominent; leaf stalk 5–20mm long. Flowers white (occasionally pink), fringed, bell-like, pendulous. Fruit ± globose, 9–11mm long, blue. **HABITAT:** Sheltered forests. **DIST:** Widespread. Also NC, SC, ST, CWS, Qld, Vic, Tas. **FL:** Oct–Dec.

Sloanea

Trees. Leaves alternate, 1-foliolate, petiole often with swelling at junction with lamina of leaflet. Flowers solitary or in racemes or panicles. Fruit a capsule with bristles.

Sloanea australis
Maiden's Blush
Large spreading tree, often leaning and with buttressed trunk. Leaves to 30 x 10cm, toothed; new leaves reddish pink. Flowers to 2cm across, white. Fruit orange, turning light brown. **HABITAT:** Rainforest along creeks. **DIST:** Coast: Illawarra, Gosford (e.g. Hacking R., Ourimbah). Also NC, NT, Qld. **FL:** Oct–Nov.

STERCULIACEAE

Brachychiton
Trees, often with swollen trunks, rarely shrubs. Leaves simple, often palmately lobed and veined. Flower consisting of a tubular calyx with 5 lobes. Fruit a follicle resembling a pod.

Brachychiton acerifolius
Illawarra Flame Tree
Deciduous tree to 30m tall. Leaves rhombic to ovate, 10–25cm long, entire or 3-lobed, on petiole 10–20cm long. Flowers massed, red, bell-shaped, appearing when tree loses all or some of its leaves. 'Pod' boat-shaped, glabrous, black, about 15cm long. **HABITAT:** Rainforests. **DIST:** Illawarra, Gosford–Wyong. Also NC, CWS, Qld. **FL:** Dec–Jan.

Sloanea australis

Brachychiton acerifolius

Commersonia fraseri

Commersonia
Shrubs or small trees. Flowers bisexual with 5-lobed calyx and 5 petals as long as calyx; stamens 5, united at base, alternating with 5 staminodes. Fruit a capsule covered with bristles.

Commersonia fraseri
Brush Kurrajong
Bushy shrub to 3m tall. Leaves ovate, 5–17cm long, toothed, lower surface grey-hairy. Flowers white, massed in loose panicles. **HABITAT:** Rainforest margins, sheltered gullies, disturbed roadsides. **DIST:** Illawarra, Ourimbah, lower Blue Mtns (Erskine Ck). Also NC, SC, ST, Qld, Vic. **FL:** Sept–Feb.

Keraudrenia
Shrubs with rusty-hairy branches. Leaves simple, entire or toothed. Flowers in cymes, either terminal or opposite leaves. Calyx 5-lobed enlarging and colouring after flowering; petals absent. Fruit a capsule with long hairs.

Keraudrenia corollata var. *denticulata*

Keraudrenia corollata var. *denticulata*
Velvet-flower
Compact shrub to 2m tall. Leaves ovate, 6–12cm long, toothed, green and wrinkled above, paler and hairy below. Flowers white-pink in short cymes of about 5 flowers. **HABITAT:** Open forest, edge of floodplains. **DIST:** Localised, rare (e.g. Colo R. along Putty Rd). Also NC, Qld. **FL:** Oct–Dec.

Lasiopetalum ferrugineum var. *cordatum*

Lasiopetalum joyceae

Lasiopetalum ferrugineum var. *ferrugineum*

Lasiopetalum
Shrubs ± covered with rusty hairs. Flowers bisexual, mostly with 5 petal-like calyx lobes and 5 stamens. Petals tiny. Fruit a capsule without bristles.

Lasiopetalum ferrugineum var. *cordatum*
Rusty-petals
Shrub to 1m tall. Leaves to 12cm long, 1–4cm wide, lobed at base, lower surface rusty-hairy. Flowers in dense many-flowered cymes; calyx creamy and hairy inside, rusty-hairy outside. **HABITAT:** Open forest. **DIST:** Chiefly Blue Mtns. Also Pennant Hills, Thirlmere, Wollemi, NC, ST, NWS. **FL:** Aug–Nov.

Lasiopetalum ferrugineum var. *ferrugineum*
Rusty-petals
Similar to var. *cordatum*, but with narrow-elliptic leaves <10mm wide. **HABITAT:** Open forest, heath. **DIST:** Widespread, esp. on coast. Also NC, SC, Vic. **FL:** Aug–Nov.

The 2 varieties of *L. ferrugineum* intergrade (possibly hybrids).

Lasiopetalum joyceae
Joyce's Rusty-petals
Slender shrub to 1m tall. Leaves linear, to 80 x 5mm, green above, white to rusty-hairy below. Calyx lobes pink, 8–12mm long, densely hairy. **HABITAT:** Open forest, heath. **DIST:** Endemic. Hornsby Plat. (e.g. Mt Colah, near Terrey Hills). **FL:** Aug–Oct.

Lasiopetalum macrophyllum
Shrubby Rusty-petals
Erect shrub to 2m tall. Flowers in short dense cymes. Similar to *L. ferrugineum* var. *cordatum* but inside of calyx glabrous; outside with rusty-white hairs; bracteoles lanceolate. **HABITAT:** Open forest, coastal headlands. **DIST:** Coast and ranges. Also NC, SC, ST, CWS, Vic, Tas. **FL:** Sept–Oct.

Lasiopetalum parvifolium
Small Rusty-petals
Shrub to 80cm tall. Leaves narrow-oblong, to 60 x 6mm, green above, sparsely pubescent becoming glabrous, grey-rusty-hairy below. Flowers in short dense cymes. Bracteoles narrow, brown. Calyx lobes white, hairy outside but glabrous inside, about 3mm long. **HABITAT:** Open forest, on clay-shale soils. **DIST:** Coast, Sthn H'lands. Also NC, SC. **FL:** Aug–Nov.

Lasiopetalum rufum
Red Rusty-petals
Erect shrub to 80cm tall. Leaves linear, to 6cm x 2–4mm, green above, grey-hairy below. Flowers in open drooping cymes. Calyx lobes 4–5mm long, hairy, reddish inside. **HABITAT:** Woodland, heath. **DIST:** Endemic. Coast, between Woy Woy, Putty, Waterfall and Cataract Dam. **FL:** Aug–Nov.

Lasiopetalum macrophyllum

Lasiopetalum parvifolium

Lasiopetalum rufum

Rulingia dasyphylla

Rulingia

Shrubs. Flowers with 5 petals and 5-lobed calyx. Stamens 5, opposite petals, alternating with 5 linear-lanceolate staminodes. Fruit a capsule with bristles.

Rulingia dasyphylla
Kerrawang
Erect shrub 1–2m tall, with softly hairy stems. Leaves ovate to lanceolate, 3–10cm long; margins toothed, often scabrous and wrinkled above; hairy beneath. Flowers clustered in loose cymes, white, 5–7mm across. **HABITAT:** Open forest, esp. on sheltered slopes. **DIST:** Uncommon. Coast and ranges. Also NC, SC, NT, WS, Qld, Vic. **FL:** Aug–Oct.

Rulingia hermanniifolia
Sprawling or trailing shrub <50cm tall. Leaves ± lanceolate, 5–20mm long on stalk <3mm long, dark green above, tomentose below, very wrinkled; margins irregularly toothed. Flowers in short dense cymes, often abundant. **HABITAT:** Coastal heath, cliff tops. **DIST:** Coded 'rare species' but locally common after fire. Coast (e.g. Kurnell, Royal NP, Bouddi NP). Also SC. **FL:** Aug–Sept.

Rulingia prostrata is a mat-forming shrub from swamp margins at Thirlmere Lakes and Penrose SF.

Seringia

Shrub. Flowers with 5 lobes, with matted short hairs. Petals absent but calyx petal-like. Stamens 5 alternating with 5 staminodes. Fruit a hairy capsule.

Seringia arborescens
Shrub 3–5m tall; new growth rusty-hairy. Leaves ovate, toothed, 5–15cm long, with rusty hairs underneath. Flowers in short leaf-opposed cymes, greenish white. **HABITAT:** Open forests, gullies. **DIST:** Scattered. Uncommon (e.g. Nortons Basin, Yerranderie). Also NC, SC, Qld, PNG. **FL:** Oct–Nov.

Rulingia hermanniifolia

FLOWERING PLANTS

Seringia arborescens

MALVACEAE

Abutilon
Shrubs or herbs. Leaves alternate, mostly broad, cordate, lobed. Flowers solitary on axillary peduncles. Calyx 5-lobed, petals usually yellow. Ovary and fruit with 5 cells or more.

Abutilon oxycarpum

Abutilon oxycarpum
Lantern-bush
Shrub to 2m tall with velvety stems and leaves. Leaves cordate, 2–8cm long. Flowers on peduncles 2cm long. Capsule about 8mm across, flat on top with a star pattern, mericarps 8–12, awned, exceeding calyx. **HABITAT:** Dry rainforest margins, on clay-shales. **DIST:** Illawarra, sw. Sydney (e.g. Razorback Mtn, Albion Park). Also NC, SC, NT, WS, WP, Qld, SA, WA, NTerr. **FL:** Jan–Apr.

Hibiscus
Trees, shrubs and herbs with stellate hairs. Leaves often lobed or toothed. Flowers solitary, axillary, with epicalyx of 5–12 bracts; petals 5; staminal column with many filaments; style divided into 5 branches. Fruit a 5-valved capsule.

Hibiscus heterophyllus
Native Rosella
Tall shrub or tree to 5m; stems prickly. Leaves ovate or 3–5-lobed, to 15cm long; finely toothed. Epicalyx with 10–12 linear segments; petals white or pale pink, 6–7mm long. Capsule densely hairy, 2cm long. **HABITAT:** Sheltered forest, rainforest margins. **DIST:** Coast, esp. Illawarra. Also Colo R., NC, NT, Qld. **FL:** Nov–Jan.

Hibiscus diversifolius (coast). Tall shrub. Flowers yellow with a red spot. Rare.

Hibiscus splendens (Bulli area). Tall shrub. Flowers pink. Rare.

Hibiscus sturtii var. *sturtii* (Wollondilly Valley). Small shrub usually from WS and WP. Flowers pink.

Hibiscus heterophyllus

Howittia

Monotypic genus. Shrub. Flowers solitary, axillary. Epicalyx absent. Staminal column short; stamens numerous. Ovary 3-celled.

Howittia trilocularis
Shrub to 2m tall. Stems, leaf stalks and buds with yellow stellate hairs. Leaves ovate, 2–10 x 1–5cm, entire or toothed, lower surface with white or yellow hairs. Flowers purple with yellow stamens; petals 10–12mm long. Capsule shorter than calyx, 8mm across. **HABITAT:** Sheltered forest, rainforest margins. **DIST:** Coast and ranges. Also NC, SC, NT, CWS, Vic, SA. **FL:** Aug–Nov.

FLACOURTIACEAE

Scolopia
Trees or shrubs. Flowers in racemes, bisexual; sepals and petals similar, usually 4–6; stamens numerous in several rows. Fruit a berry with withered flower at base.

Scolopia braunii
Flintwood
Glabrous shrubby tree with tan new growth. Leaves alternate, 4–9 x 1.5–3.5cm, glossy above; young leaves ± rhombic-ovate, angular; older

Howittia trilocularis

leaves lanceolate; stalk 4–10mm long. Racemes 2–4cm long; petals greenish white, 1–1.5mm long. Berry ovoid, 12 x 10mm, red turning black. **HABITAT:** In and near rainforests. **DIST:** Coast (e.g. Palm Jungle, Patonga, Olney SF). Also NC, SC, Qld. **FL:** Oct–Nov.

Scolopia braunii

VIOLACEAE

Hybanthus
Slender herbs or shrubs. Differs from *Viola* in having lower petal much larger than other 4 and leaves ± sessile. Fruit a capsule.

Hybanthus monopetalus
Slender Violet
Stems weak to 40cm tall. Leaves linear, 5–90mm long, upper ones opposite; margins recurved. Flowers in racemes longer than leaves; lower petal blue-mauve, 7–20mm long. **HABITAT:** Woodland, heath, on drier sites. **DIST:** Coast and ranges. Most of NSW. Also Qld, Vic, SA. **FL:** Sept–Dec.

Hybanthus monopetalus

Hybanthus vernonii subsp. *vernonii*
Glabrous plant to 80cm tall. Similar to *H. monopetalus* but with solitary flowers on short stalks in axils and lower petal 9–13mm long. **HABITAT:** Woodland, heath. **DIST:** Coast, lower Blue Mtns. Also NC, SC, ST, Vic. **FL:** Aug–Jan.

Hymenanthera
Shrubs with alternate or clustered leaves. Flowers axillary, solitary or clustered, often on older stems; petals 5, ± equal; stamens with appendage. Fruit a berry.

Hymenanthera dentata
Tree Violet
Rigid glabrous shrub to 4m tall, often much smaller; side-branches short, spinescent. Leaves ± sessile, narrow-obovate, entire or bluntly toothed, to 5cm long. Flowers pale yellow, bell-like; petals 3–5mm long, recurved near apex. Berry 4–5mm diam., purple-black. **HABITAT:** Near rainforests, riverbanks. **DIST:** Coast and ranges. Also NC, SC, NT, ST, WS, Vic, Tas, SA. **FL:** Sept–Nov.

Hybanthus vernonii subsp. *vernonii*

Hymenanthera dentata

Viola banksii

Viola

Herbs, often spreading by stolons. Leaves basal, on long stalks; lamina usually oblate to reniform. Flowers solitary, on long scapes; lowest of 5 petals spurred or pouched at base. Fruit a capsule.

Viola banksii
Banks' Violet

Leaves broad-reniform, 18–25 x 30–45, toothed, with deep V at base. Flowers on scapes to 15cm long, strikingly bicoloured, violet and white. Anterior petal ovate, widest in middle. Seeds glossy, purple-black. **HABITAT:** Swamps, rainforest edges, sheltered forest. **DIST:** Coast (e.g. Garie Bch, Kurnell). Ulladulla to NC and se. Qld. **FL:** Sept–Mar.

Similar to *V. hederacea* which is less robust and has pale flowers with the anterior petal widest near the apex.

Viola betonicifolia
Showy Violet

Stems short, erect. Leaves with triangular lamina 2–6cm long on stalk 2–8cm long. Flowers on scapes to 20cm long, violet, lateral petals bearded inside. **HABITAT:** Woodland, esp. shaded areas. **DIST:** Coast and ranges. Also NC, SC, ST, WS, Qld, Vic, Tas, SA. **FL:** Sept–Dec.

Viola betonicifolia

Viola hederacea

Viola silicestris

Viola sieberiana

Viola sieberiana
Leaves tufted at nodes of stolons; lamina ovate-rhombic, about 12 x 8mm, toothed, on stalk 1–3cm long. Flowers on scapes exceeding leaves, pale mauve; lateral petals 6–9mm long, glabrous. **HABITAT:** Open damp sites. **DIST:** Coast and ranges (e.g. Royal NP, Fitzroy Falls, Mt Tomah). Also SC. **FL:** Sept–Dec.

Viola silicestris
Sandstone Violet
Herb with rosettes of leaves at nodes of stolons; some stems scrambling to 40cm long. Leaves broad-reniform, 6–10 x 10–25mm, with small even teeth. Flowers uniformly pale mauve-blue. Seeds glossy, purple-black. **HABITAT:** Woodland, scrub, on sandy soils. **DIST:** Jervis Bay–Bundanoon (e.g. Carrington Falls), Blue Mtns (e.g. Bilpin, Newnes Plat.). Also NC, Qld (Lamington NP). **FL:** Oct–Feb.

Viola caleyana. Herb with lax stems, leaves with heart-shaped laminae, white flowers. Occurs mainly in upland swamps.

Viola hederacea
Ivy-leaved Violet
Leaves tufted; lamina ± oblate to reniform, variously toothed or lobed, dark green above, paler beneath; stalks 2–6cm long. Flowers on scapes exceeding leaves, pale mauve and white. Anterior petal widest towards apex. Seeds dull brown. **HABITAT:** Moist forests. **DIST:** Coast and ranges. Also SC, ST, CWS, Vic, Tas, SA. **FL:** Sept–Mar.

This 'species' is a widespread, variable complex of forms.

Passiflora cinnabarina

Passiflora herbertiana

PASSIFLORACEAE

Passiflora

Climbers with characteristic passion flowers and fruit. Leaves with axillary tendrils, alternate, 1–5-lobed; petiole at least half as long as lamina. Two native and 7 weed species in the Sydney region.

Passiflora cinnabarina
Red Passionflower

Glabrous climber. Leaves 3-lobed with ± entire margins; lamina 6–10cm long; lobes broadly elliptic with pointed apex. Flowers red, 4–7cm diam. Fruit ovoid, greenish grey, 2–3cm long.
HABITAT: Sheltered forest, esp. on volcanic soils. **DIST:** Rare. Illawarra, upper Blue Mtns. Also SC, NT, ST, WS, Vic. **FL:** Aug–Nov.

Passiflora herbertiana
Native Passionfruit

Stems and leaf undersurface finely hairy. Leaves usually 3-lobed, (rarely 1- or 5-lobed); lobes shallow; lamina 6–12cm long; petiole with 2 dark glands near lamina. Flowers white to yellow-orange with green centre, about 6cm diam. Fruit subglobose, green, about 5 x 4cm.
HABITAT: Sheltered forest. **DIST:** Coast, lower Blue Mtns. Also NC, SC, NT, Qld. **FL:** Aug–Nov.

ERICACEAE

Acrotriche
Rigid shrubs with flowers in short spikes or crowded on older stems. Flowers greenish, with hairs at tips of spreading lobes. (The name means 'tip-hair'.) Fruit a drupe.

Acrotriche divaricata
Ground-berry
Spreading dense shrub 1–2m tall. Leaves at about 90° to stems, lanceolate, rigid, 6–15 x 2–4mm, pungent. Flowers in clusters of 3–5 in short axillary spikes. Drupe 3mm diam., red. **HABITAT:** Sheltered forest, rainforest margins. **DIST:** Coast and ranges. Also NC, SC, ST, Vic. **FL:** July–Sept.
  Acrotriche aggregata (Blue Mtns) has larger leaves and flowers in clusters of 5–10.

Acrotriche divaricata

Astroloma
Flowers tubular, with hair tufts inside below middle; lobes 5, hairy inside; stamens inserted at throat but not projecting beyond tube. Fruit a drupe.

Astroloma humifusum (photograph overleaf)
Cranberry Heath
Diffuse mat-forming shrub to 50cm tall. Leaves narrow-lanceolate, 5–12mm long, sharply pointed, minutely fringed. Flowers red, 7–12mm long, hidden in foliage. Drupe 5–6mm long. **HABITAT:** Open forest, woodland, on shale-clay soils. **DIST:** Coast and ranges. Also NC, SC, ST, WS, SWP, Vic, Tas, SA. **FL:** Feb–June.

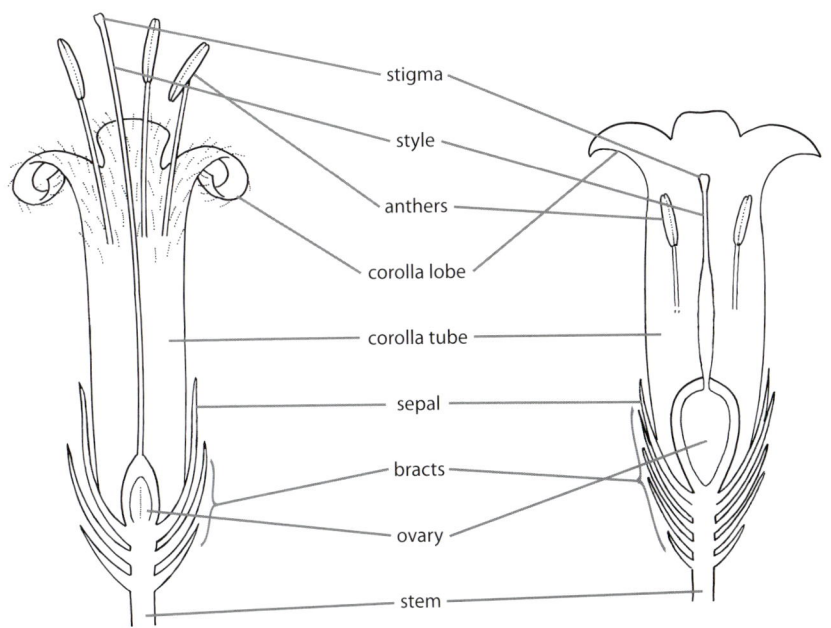

Styphelia flower (section)   **Epacris flower (section)**

Astroloma humifusum

Brachyloma daphnoides

Brachyloma
Shrubs with flower tube glabrous, the throat closed by hairs which descend between the anthers. Fruit a drupe.

Brachyloma daphnoides
Daphne Heath
Erect green-grey shrub to 1m tall. Leaves ovate-elliptic, 4–10 x 2–4mm. Flowers axillary, often grouped, white, 6mm long, sweet-scented, broadest near base. **HABITAT:** Open forest, woodland, heath. **DIST:** Coast and ranges. Most of NSW. Also Qld, Vic, Tas, SA. **FL:** June–Nov.

Astroloma pinifolium
Pine Heath
Shrub to 60cm tall. Leaves narrow-linear, 10–20mm long, crowded. Flowers red with yellow-green tip, 11–20mm long. Drupe 10–15mm long. **HABITAT:** Woodland, heath. **DIST:** Chiefly coast. Also NC, SC, ST, WS, SWP, Tas, SA, WA. **FL:** May–Sept.

Dracophyllum
Stems with annual leaf scars and long sheathing leaves. Flowers with cylindrical tube and small spreading lobes. Fruit a capsule.

Dracophyllum secundum
Erect to pendulous shrub, mostly 30–60cm tall. Leaves 5–14cm x 5–12mm, long-pointed. Buds and flowers with long brown bracts. Flowers in 1-sided panicles, tubular, 5–8mm long, pink or white (sometimes all red). **HABITAT:** Wet cliffs, damp rock ledges. **DIST:** Coast and ranges. Also NC, SC, ST. **FL:** July–Sept.

Astroloma pinifolium

Dracophyllum secundum

Epacris calvertiana var. *versicolor*

Epacris
Shrubs with glabrous tubular or bell-shaped flowers. Stamens with filaments attached to wall of tube, free part very short; anthers in throat of tube. Style in pit of ovary summit. Fruit a capsule.

Epacris calvertiana var. calvertiana
Shrub to 1m tall with woolly branchlets. Leaves elliptic, subulate and inrolled at tip, 1–2cm x 2–3mm; margins finely fringed. Flowers white or yellowish, 10–15mm x 3–5mm, with swollen tube and recurved lobes. **HABITAT:** Forest, on sheltered rocky sites. **DIST:** Sthn H'lands, Blue Mtns. Also NC, SC, ST. **FL:** Aug–Nov.

Epacris calvertiana var. *calvertiana*

Epacris calvertiana var. versicolor
Shrub to 1m tall, similar to var. *calvertiana* but with young leaves having ciliate margins and flowers pink-red with white lobes*.
HABITAT: Sheltered creek gullies. **DIST:** Restricted. Sthn H'lands (e.g. Budderoo NP, Barren Grounds). **FL:** Aug–Sept.

  * Unlike *E. longiflora*, the sepals are about half the length of floral tube.

Epacris coriacea

Epacris coriacea
Bushy shrub to 2m tall. Leaves broad-elliptic-obovate, flat, thick, 4–10 x 3–7mm; stalk 1–2.5mm long. Flowers with 5 spreading lobes ± as long as floral tube. **HABITAT:** Sandstone cliffs. **DIST:** Endemic. Illawarra (e.g. O'Hares Ck, Bulli L'out). Also CT (near Rylstone). **FL:** Aug–Nov.

Epacris crassifolia

Epacris crassifolia
Variable shrub with trailing stems; some forms upright to 20cm tall. Small leaf/flower form with leaves obovate, hairy, 5 x 2mm, and flowers 5–8 x 2mm. Large leaf/flower form with glabrous obovate leaves, 13 x 5mm, and flowers 15–22 x 7mm. **HABITAT:** Damp sandstone ledges and cliffs. **DIST:** Endemic. Coast and ranges (e.g. Blackheath, Fitzroy Falls, Brisbane Water NP). **FL:** Nov–Jan.

Epacris longiflora
Fuchsia Heath
Shrub to 1.5m tall. Leaves crowded, 5–17 x 3–6mm, apex acuminate, base obtuse or cordate. Flowers pendulous, tubular, to 2cm long, red with white spreading lobes (rarely completely white). **HABITAT:** Forest, woodland, heath. **DIST:** Coast and adjacent plateaus. Also NC, NT, Qld. **FL:** Mar–Nov.

Epacris longiflora

Epacris microphylla var. *microphylla*
Coral Heath
Shrub 30–60cm tall. Leaves ovate, 3–4mm long, concave, with short point. Buds and flowers numerous along upper stems, white, often tinged pink; floral tube broad, 2mm long. **HABITAT:** Woodland, heath. **DIST:** Coast and ranges. Also NC, SC, NT, ST, SWS, Qld, Vic. **FL:** June–Nov.

Epacris muelleri
Mueller's Heath
Weak decumbent shrub to 30cm tall, with glabrous branchlets. Leaves elliptic, 2–4mm long, thick, obtuse. Flowers axillary along upper stems, white or cream, 5–6mm diam.; stalks 2–4mm long. **HABITAT:** Damp rock faces and crevices. **DIST:** Endemic. Upper Blue Mtns (e.g. Woodford, Newnes SF, Mt Hay); also Bonnum Pic. **FL:** Oct–Nov.
   Similar to *E. rigida* but that sp. has softly hairy branchlets and leaves with a thick lower midrib.

Epacris obtusifolia
Blunt-leaf Heath
Erect shrub to 80cm tall with pubescent branches. Leaves erect, elliptic, 7–11mm long, blunt-tipped. Flowers in long leafy inflorescences on upper stems, white or cream, narrow bell-shaped, 4–8mm long. **HABITAT:** Damp heath, sedgeland. **DIST:** Coast and ranges. Also NC, SC, NT, ST, Qld, Vic, Tas. **FL:** June–Jan.

Epacris microphylla var. *microphylla*

Epacris muelleri

Epacris obtusifolia

NATIVE PLANTS OF THE SYDNEY REGION

Epacris paludosa

Epacris paludosa
Swamp Heath
Bushy shrub to 1m tall. Leaves ascending, narrow-lanceolate, acuminate, 7–10mm long. Flowers crowded at ends of branches, white, with a narrow floral tube and obtuse spreading lobes. **HABITAT:** Wet heath, swampy sites. **DIST:** Woronora Plat., upper Blue Mtns. Also SC, ST, Vic, Tas. **FL:** Aug–Dec.

Epacris pulchella
Slender shrub to 1m tall with woolly branchlets. Leaves spreading, ovate, sessile, 4–6mm long, concave, acuminate. Flowers white, often pinkish; floral tube to 6mm long; sepals and bracts finely pointed. **HABITAT:** Woodland, heath. **DIST:** Coast and ranges. Also NC, SC, NT, ST, Qld. **FL:** Jan–June (mainly autumn).

Epacris purpurascens var. *purpurascens*
Erect shrub to 1m tall with hairy branchlets. Leaves ovate, 8–12mm long, with aristate tip. Flowers white or pink-mauve; floral tube to 6mm long; sepals as long as floral tube. **HABITAT:**

Epacris purpurascens var. *purpurascens*

Scrub, in poorly drained clay soils. **DIST:** Endemic. Rare. Coast (e.g. Rookwood, Dural, Wilton). **FL:** July–Sept.

Var. *onosmiflora* (upper Blue Mtns) has white flowers and floral tube (to 7mm) longer than sepals.

FLOWERING PLANTS

Epacris pulchella

Epacris reclinata

Epacris reclinata
Shrub with weak decumbent branches to 50cm tall. Leaves heart-shaped, 4–6mm long. Flowers tubular, 8–12mm long, uniformly red or pink. **HABITAT:** Sheltered sandstone cliff faces. **DIST:** Upper Blue Mtns. Also Budderoo NP, CWS. **FL:** Aug–Nov.

Epacris rigida
Erect shrub to 80cm tall, with softly hairy branchlets. Leaves erect, crowded, almost sessile, elliptic, obtuse, thick, 2–4mm long, with raised lower midrib. Flowers in upper axils; white or cream; tube 2mm long; lobes 2–3mm long; stalk 2mm long. **HABITAT:** Open rocky heath. **DIST:** Endemic. Central Blue Mtns (e.g. Kings T'land, Mt Hay). **FL:** Sept–Oct.

Epacris rigida

Epacris sparsa
Shrub 30–80cm tall. Branches finely pubescent, with leaf scars. Leaves erect, elliptic to ovate, flat, glabrous, 11–16 x 3–4mm, apex acute. Flowers in upper axils, creamy white; tube to 18mm long; sepals 3mm long. **HABITAT:** Near creeks, often among rocks. **DIST:** Endemic. Lower Grose R. and feeder creeks. **FL:** May–June.

Epacris sparsa

Leucopogon amplexicaulis

Leucopogon appressus

Leucopogon
WHITE-BEARD, BEARD-HEATH

Flowers few to many in solitary or axillary spikes, usually white (except for *L. fraseri* in Sydney district), with hairs on lobes and inside of floral tube. Stamens inserted in throat with anthers partly enclosed by tube. Fruit a drupe.

Leucopogon amplexicaulis
Scrambling undershrub with hairy stems to 1m long. Leaves stem-clasping, broadly heart-shaped, 10–30 x 4–20mm, fringed. Spikes to 35mm long with up to 12 flowers. Corolla lobes longer than tube. **HABITAT:** Sheltered woodland, shaded rock faces. **DIST:** Coast (e.g. Muogamarra, Oatley Park, Audley). Also NC, SC. **FL:** Aug–Oct.

Leucopogon appressus
Wiry shrub to 60cm tall. Leaves erect, pressed against stem, narrow-ovate, 4–10 x 1–2.5mm, concave, tapering to a fine point. Flowers 1–2 in axils at ends of branchlets, inconspicuous, white; corolla tube and lobes each 1.5mm long. **HABITAT:** Open forest, heath. **DIST:** Endemic. Coast and ranges (e.g. Bouddi NP, Newnes SF). **FL:** Nov–Feb.

Leucopogon ericoides
Slender shrub, 60–80cm tall. Leaves spreading, oblong, to 8mm long, recurved, pungent. Flowers erect, in short spikes of 1–5 or more from leaf axils; corolla tube white to pinkish, to 4mm long, longer than lobes. **HABITAT:** Open forest, heath, old dunes. **DIST:** Coast and ranges. Also NC, SC, ST, CWS, Vic, Tas. **FL:** July–Oct.

Leucopogon esquamatus
Erect shrub to 80cm tall. Leaves ascending, elliptic, 15 x 3.5mm, flat with slight twist, finely pointed. Flowers 1–2 in upper axils on peduncle 1mm long; tube 2mm long, shorter than lobes. **HABITAT:** Woodland, damp heath. **DIST:** Widespread, esp. coast. Also NC, SC, ST, Vic, Tas. **FL:** Aug–Sept.

Leucopogon exolasius
Shrub to 1m tall with pubescent branchlets. Leaves oblong-linear, to 15 x 2mm, recurved, convex above, lined below, tapering to pungent point. Flowers 1–3 in upper axils, erect to pendulous on stalks 2–4mm long; tube 4mm long, shorter than lobes, villous outside, softly hairy inside. **HABITAT:** Rocky river banks. **DIST:** Endemic. Upper Georges R., Woronora R. **FL:** Aug–Sept.

Leucopogon fletcheri subsp. fletcheri
Shrub to 2m tall with scabrous branchlets. Leaves linear, to 8 x 3mm, convex above, lined below, tip pungent. Flowers pendent and solitary in upper axils on peduncles 1mm long; tube to 5mm long, longer than lobes. **HABITAT:** Lateritic soils in woodland. **DIST:** Western Sydney, lower Blue Mtns (e.g. Annangrove, Springwood). **FL:** Aug–Sept.

Leucopogon fraseri
Prostrate or spreading shrub to 40cm tall. Leaves elliptic-obovate, 6–10 1.5–2.5mm, ± flat, with 3 longitudinal veins beneath, pungent-pointed. Flowers 1–2 on peduncle 1mm long; tube to 5mm long, longer than lobes, pink-red, with woolly white lobes. **HABITAT:** Woodland, heath. **DIST:** Sthn H'lands (e.g. Wingello SF), upper Blue Mtns. Also NC, SC, NT, ST, Vic, Tas, NZ. **FL:** July–Oct.

Leucopogon juniperinus
Prickly Beard-heath
Dense shrub to 1m tall with pubescent branchlets. Leaves ± oblong, 5–10 x 1–2mm, convex above, lined below, pungent-pointed. Flowers mostly solitary on peduncles 1mm long; tube narrow, to 7mm long; lobes 2mm long, sparsely hairy inside. **HABITAT:** Open forest, esp. on clay soils. **DIST:** Coast. Also NC, SC, NT, CT, Qld, Vic. **FL:** June–Oct.

Leucopogon ericoides

Leucopogon esquamatus

Leucopogon exolasius

Leucopogon fletcheri subsp. *fletcheri*

Leucopogon juniperinus

Leucopogon fraseri

Leucopogon lanceolatus var. *lanceolatus*
Lance-leaf Beard-heath
Bushy glabrous shrub to 2m tall. Leaves narrow-lanceolate, to 40 x 10mm, with about 5 pale longitudinal veins. Flower spikes clustered in upper axils, longer than leaves, with 6–20 or more flowers. **HABITAT:** Sheltered forest, rainforest margins. **DIST:** Coast and ranges. Also NC, SC, NT, ST, NWS, Qld, Vic, Tas, SA. **FL:** Sept–Dec.

Leucopogon lanceolatus var. *lanceolatus*

Leucopogon microphyllus var. *microphyllus*

Leucopogon microphyllus var. *microphyllus*
Small-leaved Beard-heath
Wiry shrub to 80cm tall with fine hairs on branchlets. Leaves erect, ovate, 2–4 x 1–2mm, concolorous, margins often hairy, convex above, veined below. Flowers in short compact spikes clustered at branch ends; corolla tube 1–2mm long, mostly obscured by brown sepals 2.5mm long. **HABITAT:** Open forest, heath. **DIST:** Common. Coast and ranges. Also NC, SC, NT, ST, CWS, Qld. **FL:** May–Nov.

Leucopogon muticus
Blunt Beard-heath
Shrub to 1.5m tall, often with reddish stems. Leaves erect, obovate, to 20 x 4mm, dull green, discolorous, with 3–5 longitudinal veins. Flowers in short spikes of 4–10; corolla tube 2.5mm long, longer than lobes, densely hairy inside. **HABITAT:** Open forest, heath. **DIST:** Blue Mtns, Hornsby Plat., Burragorang V. Also NC, SC, NT, ST, WS, Qld. **FL:** Aug–Oct.

Leucopogon parviflorus
Coast Beard-heath
Erect shrub, usually 1–2.5m tall. Leaves sessile, elliptic, 10–25 x 2–6mm, acute but not pungent, convex above, paler and lined below. Flowers in dense spikes to 30mm long; corolla tube <2mm long, shorter than lobes. **HABITAT:** Sand dunes, heath, woodland near sea. **DIST:** Coast (e.g. Jibbon Bch). Also NC, SC, all States, LHI. **FL:** Aug–Sept.

Leucopogon setiger
Spreading shrub to 1.5m tall. Leaves sessile, linear-lanceolate, 12 x 3mm, flat or convex above, finely lined below, tapering to pungent point. Flowers in short spikes of 1–2, pendulous along stems on stalks 2–8mm long, bell-shaped; tube 2–3mm long, shorter than lobes; lobes conspicuously hairy. **HABITAT:** Open forest. **DIST:** Coast and ranges. Also NC, SC, CWS. **FL:** Aug–Oct.

Leucopogon virgatus
Slender erect shrub to 50cm tall with pubescent branchlets. Leaves erect, ± appressed to stems, ovate, to 15 x 1–3mm, concave above, finely pointed. Flowers in short spikes of 4–7 in upper axils; corolla tube 2mm long, slightly shorter than lobes. **HABITAT:** Open forest, woodland, old dunes. **DIST:** Coast, Sthn H'lands (e.g. Long Bay, Agnes Banks NR, Bouddi NP). Also NC, SC, NT, ST, WS, Qld, Vic, SA. **FL:** Sept–Oct.

FLOWERING PLANTS

Leucopogon parviflorus

Leucopogon muticus

Leucopogon setiger

Leucopogon virgatus

Lissanthe

Flowers in short terminal and axillary inflorescences, stalked, white; floral tube cylindrical to urceolate, hairy inside above middle; lobes erect or spreading, triangular, glabrous or hairy. Anthers at throat, visible between lobes. Fruit a drupe.

Lissanthe sapida
Native Cranberry
Spreading shrub to 1m tall. Leaves stalked, lanceolate, 15–25 x 2–4mm, ribbed beneath, pointed. Flowers pendent on slender peduncles and pedicels in clusters of 1–4; corolla tube 6mm long; lobes much shorter, reflexed. Drupe red, 5–6mm long, with persistent style. **HABITAT:** Open forest, on cliff edges and along creeks. **DIST:** Endemic. Lower Blue Mtns (e.g. Glenbrook, Mountain Lagoon). **FL:** May–Sept.

Lissanthe strigosa subsp. subulata
Peach Heath
Erect shrub to 60cm tall. Leaves linear with finely pointed apex, 6–16 x 1–2mm, strongly 1–3-ribbed on undersurface. Buds often with pink tinge. Flowers white to pink; tube 3–5mm long; lobes much shorter, erect. **HABITAT:** Woodland, on sandy and clayey soils. **DIST:** Coast and ranges (e.g. Loftus, Lawson, Hill Top). Also NT, ST, WS, Qld, Vic, Tas, SA. **FL:** July–Sept.

Subsp. *strigosa* (Cumberland Pl.) has convex leaves <6mm long with only the midrib visible on the undersurface.

Melichrus

Shrubs with leaves crowded at ends of branches. Flowers axillary, solitary, sessile on lower part of stems; corolla tube short and broad, with 5 glandular scales alternating with stamens. Fruit a drupe.

Melichrus procumbens
Jam Tarts
Shrub to 20cm tall or forming mats. Leaves upwards-pointing, 12–25 x 1–4mm long, hairy, finely pointed but not pungent, with fringed margins. Flowers creamy, hidden in foliage and facing the ground, usually with nectar; lobes spreading, 10–12mm across. Drupe red, 2–4mm long. **HABITAT:** Open forest, heath. **DIST:** Coast nth of East Hills, Hornsby Plat. Also NC, NT, Qld. **FL:** Aug–Sept.

Melichrus urceolatus
Urn Heath
Stiff erect shrub to 1m tall. Leaves spreading to reflexed, narrow-lanceolate, 10–25 x 2–6mm glabrous or scabrous, pungent-pointed. Flowers urceolate, creamy green; tube 4–5mm long, nearly as long as lobes; lobes ± erect. Drupe greenish, 4mm long. **HABITAT:** Woodland, scrub. **DIST:** Upper Blue Mtns, Wedderburn–Appin. Most of NSW. Also Qld, Vic. **FL:** Aug–Nov.

Monotoca
BROOM-HEATH
Flowers in short axillary spikes, rarely in terminal racemes, bell-shaped, glabrous. Ovary 1-celled (hence *mono-*, 'single'; *tokos*, 'birth'). Fruit a drupe.

Monotoca elliptica
Tree Broom-heath
Shrub or small tree to 5m tall. Leaves oblong-elliptic, to 12–25 x 3–6mm; undersurface pale, often glaucous, finely lined. Racemes usually crowded at ends of branches, longer than leaves, with 5–15 flowers; tube 3–5mm long. Drupe 3–4mm long, orange to red. **HABITAT:** Coastal dunes, open forest near salt water. **DIST:** Common. Coast and

Lissanthe sapida

Lissanthe strigosa subsp. *subulata*

FLOWERING PLANTS | 107

Melichrus urceolatus

Melichrus procumbens

Monotoca elliptica

estuaries. Also Barren Grounds, Leura, Newnes, NC, SC, ST, Qld, Vic, Tas. **FL:** July–Sept.

Monotoca ledifolia, a rare small shrub with erect oblong leaves, occurs sth of Sydney on the edge of rocky sites.

Monotoca scoparia
Prickly Broom-heath
Rigid shrub to 1m tall. Leaves narrow-oblong, 8–10mm long, abruptly pointed, glabrous, convex above, paler and lined below. Spikes very short, with 2–5 flowers; corolla tube bell-shaped, white, 1–2.5mm long, longer than lobes. Drupe 3mm long, yellow to orange. **HABITAT:** Open forest, heath. **DIST:** Coast and ranges. Also NC, SC, NT, ST, WS, Qld, Vic, Tas, SA. **FL:** Apr–Sept.

Monotoca scoparia

Rupicola

Flowers solitary in axils; corolla tube cylindrical, shorter than sepals, glabrous; lobes spreading, glabrous, longer than tube; stamens attached at base of tube; anthers exserted from tube, close together around long style but not fused. Fruit a capsule.

Rupicola apiculata
Shrub to 50cm tall with hairy branchlets. Leaves crowded, spreading, ovate with long callus tip, 4–8 x 2–6mm, concave above. Flowers white; corolla tube 1.5–3.5mm long with spreading lobes 2.5–5mm long; anthers orange. **HABITAT:** Damp rock ledges. **DIST:** Endemic. Blue Mtns. **FL:** Oct–Dec.

Rupicola ciliata
Decumbent shrub with hairy branchlets. Leaves crowded, spreading, elliptic-ovate, 8–14 x 3–4mm, flat, hairy, with minutely fringed margins. Flowers creamy white, 7–8mm across with spreading lobes 5–6mm long. **HABITAT:** Rock faces, crevices and ledges. **DIST:** Lower Blue Mtns (Kurrajong–Bilpin). **FL:** Oct–Dec.

Rupicola sprengelioides
Shrub to 2m tall with hairy branchlets. Leaves crowded, ± erect, narrow-elliptic, 10–25 x 2–3mm, rigid, concave above, scabrous, acute. Flowers creamy white; tube 1.5–2mm long with spreading lobes 4–6mm long; anthers orange. **HABITAT:** Cliff faces. **DIST:** Endemic. Restricted to McMahons L'out (Kings T'land) and Burragorang L'out. **FL:** Sept–Dec.

Rupicola apiculata

Rupicola ciliata

Rupicola sprengelioides

Sprengelia monticola

Sprengelia
Slender glabrous shrubs with leaf sheaths overlapping on stems. Flowers with short tube and long spreading lobes or ± free petals; anthers cohering around style. Fruit a capsule.

Sprengelia incarnata
Pink Swamp Heath
Shrub to 1.5m tall. Leaves ovate, 10–20 x 3–6mm, reflexed above sheath, tapering to a long point. Flowers in clusters in upper axils, pink, sometimes white; tube much shorter than lobes; lobes, spreading, 3–5mm long, acute, glabrous. **HABITAT:** Swampy areas, esp. heaths. **DIST:** Coast and ranges. Also NC, SC, NT, ST, Vic, Tas, SA. **FL:** July–Oct.

Sprengelia incarnata

Sprengelia monticola
Rock Sprengelia
Diffuse shrub to 50cm tall. Leaves ovate, stem-sheathing, 3–6mm long. Flowers solitary, white, star-like; tube ± absent; anthers orange-brown. **HABITAT:** Sheltered damp rocks and cliffs. **DIST:** Endemic. Blue Mtns (e.g. Katoomba, Blackheath). **FL:** Aug–Dec.

Sprengelia sprengelioides

Sprengelia sprengelioides
Erect shrub usually 30–50cm tall. Leaves ovate, crowded, stem-sheathing, 4–10 x 2–3mm, pointed. Flowers solitary, white, star-like; tube shorter than lobes; sepals broad, green, leaf-like. **HABITAT:** Swampy areas, esp. sedgy heaths. **DIST:** Ku-ring-gai Chase NP (e.g. Salvation Ck). Also NC, Qld. **FL:** June–Oct.

NATIVE PLANTS OF THE SYDNEY REGION

Styphelia angustifolia

Styphelia longifolia

Styphelia
FIVE-CORNERS
Shrubs with rigid, pungent leaves. Flowers with cylindrical floral tube with hairy throat and lobes and 5 stamens projecting beyond the tube. Fruit a drupe.

Styphelia angustifolia
Shrub to 1.5m tall. Leaves on stalk to 1mm long, narrow-lanceolate, to 30 x 5mm, acute, with tiny serrations along margins. Flowers pendent on stalk 1–2mm long; tube pale green (pink-cream at Audley), 15–20mm long. **HABITAT:** Open forest on rocky outcrops. **DIST:** Rare. Audley, lower Blue Mtns, Hill Top. Also NC, ST, NWS. **FL:** Dec–Feb.

Styphelia laeta
Robust shrub 1–2m tall with silky branchlets. Leaves erect, elliptic-ovate, 15–35 x 6–15mm, pungent-pointed, finely toothed or fringed. Flowers not pendent; tube yellow-green or pinkish; 15–26mm long. **HABITAT:** Open forest, heath. **DIST:** Widespread. Coast and ranges. Also NC, NWS. **FL:** Mar–July.

Styphelia longifolia
Long-leaf Five-corners
Erect shrub to 2m tall; branchlets with silky hairs. Leaves lanceolate, 25–45 x 3–5mm, concave, tapering to a fine point. Flowers pale green. **HABITAT:** Open forest. **DIST:** Rare. Coast (Waterfall to Woy Woy). Also NC. **FL:** May–June.

Styphelia triflora
Bushy shrub to 2m tall with glabrous branchlets. Leaves elliptic, to 30 x 7mm, acute, flat to concave above, margins entire; stalk 2–4mm long. Flowers with pink tube and yellow lobes; calyx light green. **HABITAT:** Open forest, woodland. **DIST:** Coast, upper Blue Mtns (e.g. Royal NP, Nullo Mtn). Also NC, SC, NT, ST, WS, NWP, Qld. **FL:** July–Dec.

Styphelia laeta

FLOWERING PLANTS | 111

Styphelia triflora

Styphelia viridis subsp. *viridis*

Styphelia tubiflora

Styphelia tubiflora
Red Five-corners
Much-branched undershrub to 80cm tall with pubescent branchlets. Leaves spreading, oblong-lanceolate, to 20 x 3mm, acute. Flowers spreading or drooping; tube 15–25mm long, red; calyx greenish yellow. **HABITAT:** Woodland, heath. **DIST:** Coast and ranges. Also NC, SC, ST. **FL:** Apr–Aug.

Styphelia viridis subsp. *viridis*
Green Five-corners
Shrub to 2m tall with finely pubescent branchlets. Leaves crowded, oblong-obovate, 10–25 x 4–7mm, obtuse, mucronate; stalk 1–2mm long. Flowers green with brown anthers. **HABITAT:** Coastal heath, old dunes near sea. **DIST:** Coast. Royal NP to Port Jackson (e.g. Bundeena, La Perouse). **FL:** Apr–Aug.

Trochocarpa
Shrubs or trees with flowers in spikes. Corolla lobes overlapping in bud. Fruit drupe-like with stone (endocarp) splitting into 8–10 single-seeded nutlets (pyrenes).

Trochocarpa laurina
Tree Heath
Shrub or tree to 10m tall (usually 2–5m) with pinkish or tan new leaves. Leaves alternate or whorled, elliptic, to 7 x 3cm, with 5–7 longitudinal veins, glossy and dark green above, paler beneath. Flowers in terminal spikes 2–4cm long, white; tube 2–3mm long. Fruit blue-black, 6–8mm diam. **HABITAT:** Gullies, rainforest margins. **DIST:** Coast. Also NC, SC, CT, Qld. **FL:** Jan–Apr.

Trochocarpa laurina

FLOWERING PLANTS

Woollsia pungens

Woollsia
Monotypic genus. Flowers solitary and sessile in axils; bracts overlapping and passing into sepals; corolla tube exceeding sepals; lobes overlapping and twisted in bud. Fruit a capsule.

Woollsia pungens
Shrub to 2m tall. Leaves crowded, ovate, to 12 x 6mm, tapering to pungent point. Flowers white, often pink-tinged (a deep red-flowered form occurs at Jervis Bay), tube narrow, 6–12mm long; lobes spreading, wrinkled, 5mm long. **HABITAT:** Hind-dunes, open forest, heath. **DIST:** Coast and ranges. Also NC, SC, ST, Qld. **FL:** May–Sept.

EBENACEAE

Diospyros
Dioecious trees or shrubs. Branchlets often zig-zagged. Leaves alternate, 2-ranked. Fruit a berry with persistent enlarged calyx.

Diospyros australis
Black Plum
Small tree. Leaves oblong-elliptic to 8 x 4cm, dark green and glossy above, yellowish green beneath; apex obtusely rounded. Flowers creamy green, 5mm long, with 4 petals. Fruit a black shiny berry 2cm diam. with a single seed surrounded by purple flesh. **HABITAT:** Rainforest. **DIST:** Coast (e.g. Calga, Bola Ck, Kangaroo R.). Also NC, SC, NT, Qld. **FL:** Oct–Dec.

Diospyros pentamera occurs rarely in rainforest gullies. It has lanceolate pointed leaves, zig-zag branchlets and 5 petals.

SAPOTACEAE

Planchonella
Trees and shrubs. Flowers bisexual, axillary; sepals 5; petals 5, fused to halfway. Fruit a berry containing 1–6 seeds. This genus is sometimes known as *Pouteria*.

Planchonella australis
Black Apple
Tree with milky sap. Leaves ovate, 8–12 x 2–4cm, thick, leathery; apex obtuse. Flowers in small axillary clusters, white, bell-shaped, 5mm long. Fruit black, plum-like, to 5cm long; seeds 3–5, 2cm long, shiny brown. **HABITAT:** Rainforest. **DIST:** Coast (e.g. Ourimbah, Royal NP, Jamberoo). Also NC, Qld. **FL:** Nov–Jan.

Planchonella australis

Diospyros australis

MYRSINACEAE

Aegiceras
Trees or shrubs from saline habitats. Flowers in umbels or racemes, with 5-lobed calyx, 5 petals and 5 stamens. Fruit a cylindrical curved capsule germinating on the tree.

Aegiceras corniculatum
River Mangrove
Glabrous shrub to 4m tall. Leaves alternate, obovate, 4–8 x 2–4cm, leathery, often with salt crystals on upper surface. Flowers in umbels of 10–30, white. Fruit 3–4cm long, pointed. **HABITAT:** Mud flats of salt water estuaries, tidal rivers. **DIST:** Coast nth of Merimbula to Qld. Also WA, NTerr, Asia, Pac. Iss. **FL:** June–Nov.

Aegiceras is the smaller of the 2 species of mangrove in the Sydney district. It occurs on the upstream or landward side of the taller Grey Mangrove (*Avicennia marina*).

Myrsine howittiana

Myrsine variabilis

Myrsine
Shrubs or small trees with flowers in clusters on old wood. Fruit a small globular drupe. Formerly known as *Rapanea*.

Myrsine howittiana
Brush Muttonwood
Tall shrub with rusty-hairy new growth. Leaves obovate to 12 x 4cm, shiny green above, paler and dull beneath, undulate; stalk 1–2cm long. Flowers tiny, 5-lobed; corolla divided nearly to base. **HABITAT:** Rainforest margins. **DIST:** Coast and ranges. Also NC, SC, ST, CWS, Vic. **FL:** Aug–Nov.

Myrsine variabilis
Muttonwood
Tall shrub with pale brown hairs on new growth. Leaves ± obovate, to 10 x 3cm, entire or toothed, shiny above, dull beneath, lateral and marginal veins distinct; stalk 3–7mm long. Flowers tiny, usually 4-lobed; corolla divided <halfway to base. **HABITAT:** Open forest, littoral rainforest. **DIST:** Coast, lower Blue Mtns. Also NC, SC, NT, CWS, Qld. **FL:** June–Oct.

Aegiceras corniculatum

FLOWERING PLANTS

Samolus repens

Alchornea ilicifolia

PRIMULACEAE

Samolus
Glabrous herbs. Flowers with 5-lobed calyx and corolla and 5 stamens alternating with 5 staminodes. Fruit a capsule with 5 valves at apex.
  This genus is sometimes placed in families Theophrastaceae or Samolaceae.

Samolus repens
Creeping Brookweed
Erect to creeping herb with warty stems. Leaves basal and alternate on stems, variable, <8mm wide. Flowers with floral bract >5mm long at base of pedicel; petals white, 6mm long. **HABITAT:** Salt marsh, mangroves, rock ledges near sea. **DIST:** Coast. Also NC, SC, other States (except NTerr), Pac. Iss, S. Amer. **FL:** Oct–Feb.
  Samolus valerandi has smooth stems, leaves 10–20mm wide, and occurs on shaded creek banks.

EUPHORBIACEAE

Alchornea
Dioecious shrubs or trees. Male flowers with 4 perianth segments and 8 stamens; female flowers with 4 perianth segments and 3 stigmas.

Alchornea ilicifolia
Native Holly
Shrub to 2m tall. Leaves alternate, 4–8 x 3–5cm, with spine-tipped lobes on margins and spiny apex, glabrous, stiff, prominently veined. Flowers in short axillary spikes, tiny, greenish. Fruit a 3-lobed capsule about 8mm across. **HABITAT:** Dry rainforest, on clay and volcanic soils. **DIST:** Rare. Illawarra coast, Razorback Mtn. Also NC, CWS. **FL:** Nov.

Amperea
Perennial shrubs with alternate reduced leaves, often appearing leafless. Flowers unisexual, clustered at nodes and surrounded by brown bracts, tiny, with 4 or 5 perianth segments.

Amperea xiphoclada var. xiphoclada
Broom Spurge
Monoecious plant. Stems erect, stiff, angular, to 40cm tall. Leaves oblong, 5–10mm long, often with 1 or 2 teeth, becoming smaller and scale-like on upper stems. Flowers 2–3mm long. Fruit an ovoid capsule 3–4mm long. **HABITAT:** Open forest, woodland, old dunes. **DIST:** Coast and ranges. Also NC, SC, NT, ST, CWS, Qld, Vic, SA, Tas. **FL:** Sept–Jan.

Amperea xiphoclada var. *xiphoclada*

Baloghia inophylla

Baloghia
Shrubs or small trees. Leaves alternate or opposite, often with 2 marginal glands near leaf base. Flowers in terminal racemes, unisexual.

Baloghia inophylla
Brush Bloodwood
Shrub or tree. Trunk with thin brownish bark exuding pale sap quickly turning deep red when cut. Leaves opposite, oblong-elliptic, 7–13 x 2–5cm, thick, strongly veined, dark green above, paler beneath; stalk 6–10mm long. Petals 5, white or cream to pale pink, 10–12mm long. Capsule 12–18mm diam., slightly 3-lobed, with 1 large seed in each compartment. **HABITAT:** Rainforest. **DIST:** Illawarra. Rare near Wyong. Also NC, SC, NT, CWS, Qld, Pac. Iss. **FL:** Sept–Nov.

Bertya
Monoecious shrubs. Leaves well-developed. Flowers with 5 perianth segments; males with fused column of stamens; females with 3-celled ovary topped by branched styles. Fruit a small oval capsule.

Bertya pomaderroides
Shrub to 1m tall with soft white to rusty hairs. Leaves ovate to oblong, to 3 x 1cm, obtuse, dark green and glabrous above, whitish tomentose beneath. Capsule acute, glabrous, 10mm long; stalk 15mm long. **HABITAT:** Gullies, creek banks. **DIST:** Uncommon. Woronora Plat. (e.g. Woronora R., Douglas Park). Also NC, SC, ST, CWS. **FL:** June–Nov.

Beyeria
Shrubs or small trees. Leaves alternate, entire, often sticky. Flowers axillary; sepals 5, petals 0 or 5. Male flowers in small clusters, with numerous stamens; female flowers solitary, with 3-celled ovary topped by sessile stigma. Fruit a capsule.

Beyeria viscosa
Pinkwood
Shrub to 3m tall. Leaves ± elliptic, to 50 x 10mm, glabrous, sticky with resin. Flowers yellowish, sepals 3–4mm long, petals absent. Capsule globular, 8mm diam. **HABITAT:** Open forest, shallow soils. **DIST:** Scattered. Uncommon (e.g. upper Georges R., Nepean R., Kowmung R.). Also SC, NT, ST, WS, WP, Qld, WA, NTerr. **FL:** Sept–Dec.

Bertya pomaderroides

Beyeria viscosa

Breynia
Monoecious shrubs. Leaves alternate, entire. Flowers with 6 perianth segments; males with 3 stamens fused into a column; females with a 3-locular ovary and 3 styles. Fruit a berry.

Breynia oblongifolia
Coffee Bush
Spreading shrub to 2m. Leaves in 2 rows, alternate, oblong-elliptic, 20–30 x 10–15mm, glabrous, green above, paler beneath; stalk 1–3mm long. Flowers small, hanging on slender stalks to 10mm long. Berry globose, 6mm diam., orange turning black. **HABITAT:** Near rainforest, damp forest, dunes. **DIST:** Coast and ranges. Also NC, SC, WS, Qld, PNG. **FL:** Oct–Dec.

Breynia oblongifolia

Claoxylon
Tall shrub or small tree. Flowers with 3 or 4 perianth segments; males clustered in terminal racemes; females solitary.

Claoxylon australe
Brittlebush
Small tree with pale brittle branches. Leaves alternate, elliptic, 8–16 x 2–6cm, toothed, soft, thin, ± glabrous, veins prominent; stalk 2–4cm long with 2–4 raised glands at junction with blade. Flowers greenish. Capsule purple-black, 6mm diam. **HABITAT:** Rainforest. **DIST:** Coast. Also NC, SC, CT, ST, CWS, Qld. **FL:** Nov–Dec.

Claoxylon australe

Croton

Shrubs or small trees with alternate entire or toothed leaves bearing 2 raised glands at junction of stalk and blade. Flowers unisexual with males and females in same raceme; male flowers with 4 or 5 sepals and petals and numerous stamens; females with smaller petals or petals absent.

Croton verreauxii
Native Cascarilla
Tree to 6m tall with purplish branchlets. Leaves elliptic, to 12 x 3cm, toothed; glabrous, green, often turning orange before falling; glands stalked. Racemes 3–5cm long. Flowers yellow-green. Capsule ± globular, 6mm diam., orange-brown. **HABITAT:** In and near rainforest. **DIST:** Coast. Illawarra, rare near Gosford. Also SWC, NC, NTerr. **FL:** Nov–Dec.

Glochidion

Monoecious shrubs or trees. Leaves alternate, simple, entire, with stipules. Flowers in clusters. Fruit a capsule with numerous lobes, each with 2 seeds.

Glochidion ferdinandi var. ferdinandi
Cheese Tree
Shrub or tree to 15m tall. Leaves glabrous, elliptic, 5–10 x 2–4cm, dark green and glossy above, paler and yellowish beneath. Male flowers usually in clusters of 3 in axils; females solitary in axils. Fruit glabrous, to 20mm across. **HABITAT:** Rainforest margins, along tidal inlets. **DIST:** Coast. Also NC, SC, NWS, Qld, WA, NTerr. **FL:** Aug–Dec.

Var. *pubens* has hairy leaves and fruit. It is less common.

Micrantheum

Monoecious shrubs with leaves in groups of 3 alternate on stems. Flowers solitary or a few in axils, small; perianth segments 4–6. Fruit a 3-lobed capsule with 2 seeds in each lobe.

Micrantheum ericoides
Inconspicuous shrub to 70cm tall. Leaves linear, 8 x 1mm. Flowers solitary, 2mm across; stamens mostly 3. Capsule 6mm long. **HABITAT:** Open forest, heath. **DIST:** Coast and ranges. Also NC, SC, Qld. **FL:** Aug–Nov.

Micrantheum hexandrum
Erect shrub to 2m tall. Leaves narrow-oblanceolate, to 15 x 3mm. Flowers 1–3 in axils, white, 5mm across; stamens usually 6. Capsule 3-lobed, 7mm long. **HABITAT:** River banks, among rocks. **DIST:** Upper Georges R., Sthn H'lands, Colo R., Lett R. Also NT, ST, Qld, Vic, Tas. **FL:** July–Nov.

Croton verreauxii

Glochidion ferdinandi var. *ferdinandi*

Micrantheum hexandrum

Micrantheum ericoides

Monotaxis

Monoecious shrubs. Leaves alternate or opposite, entire, with minute stipules. Flowers in dense clusters of several males and 1 female; perianth segments 4 or 5; male flowers with 8–10 stamens. Fruit a 3-angled capsule.

Monotaxis linifolia

Glabrous subshrub with many slender stems to 30cm tall. Leaves opposite, distant, entire or with 1 or 2 teeth, 10–20 x 1–4mm, acutely pointed. Flowers in terminal cymose clusters of 10–15 males surrounding 1 female flower, white. **HABITAT:** Wet sandy heath. **DIST:** Coast (e.g. La Perouse, Royal NP). Also Robertson, SC, ST, Qld. **FL:** Oct–Jan.

Omalanthus (syn. Homalanthus)

Glabrous monoecious shrubs or small trees with milky sap. Leaves alternate, triangular to ovate, often turning red with age; stipules prominent. Inflorescence a dense, terminal raceme with numerous male flowers and a few stalked female flowers at the base. Fruit a capsule.

Monotaxis linifolia

Omalanthus populifolius
Bleeding Heart

Shrub, usually 3–5m tall. Leaf blade broadly ovate, 8–13 x 3–10cm; stalk 3–10cm long. Capsule compressed-globose, 8–10mm wide, glaucous. **HABITAT:** Rainforest margins. **DIST:** Coast and ranges. Also NC, SC, NT, NWS, Qld, LHI. **FL:** Sept–Dec.

Omalanthus stillingifolius is a smaller shrub with leaf blades 2–5cm long and non-glaucous capsules 5mm wide. It is rare in rocky gullies (e.g. Wheeny Ck, Whispering Gully).

Phyllanthus

Shrubs or herbs, usually monoecious. Leaves alternate, often distichous, entire. Flowers tiny, with glands at base of the 6 perianth segments; males in axillary clusters; females solitary. Fruit a capsule.

Phyllanthus gasstroemii
Blunt Spurge

Glabrous shrub to 1.5m tall with spreading branches. Leaves elliptic-obovate, to 20 x 10mm, obtuse. Male flowers in clusters of 3–4; stamens 3, filaments united. Female flowers on stalks lengthening to 6mm in fruit; styles 3. Capsule ± globular, glabrous, 4mm diam. **HABITAT:** Sheltered forest, creek banks. **DIST:** Coast and ranges. Also NC, SC, NT, ST, WS, Qld. **FL:** Aug–Nov.

Phyllanthus hirtellus
Thyme Spurge

Shrub to 40cm tall with soft hairs. Leaves 2–6 x 1–2.5mm, with broad obtuse or truncate or notched or mucronate apex. Male flowers in clusters of 2–3; stamens 3, filaments free. Capsule hairy, 4mm diam. **HABITAT:** Open forest, heath. **DIST:** Coast and ranges. Also NC, SC, NT, ST, WS, SWP, Qld, Vic. **FL:** July–Nov.

Omalanthus populifolius

FLOWERING PLANTS | 121

Phyllanthus gasstroemii

Phyllanthus hirtellus

Poranthera corymbosa

Poranthera
Monoecious herbs or small shrubs with flowers in short dense racemes or in terminal corymbs. Flowers small, white, with 5 sepals and 5 petals; males with 5 stamens; females with 3 bifid styles. Fruit a capsule.

Poranthera corymbosa
Glabrous shrub to 60cm tall. Leaves crowded, linear, 3–5cm x 1–4mm; margins strongly recurved. Inflorescence a large open corymb 10–15cm across. **HABITAT:** Woodland, open forest. **DIST:** Coast and ranges. Also NC, SC, NT, WS. **FL:** Sept–Nov.

Poranthera ericifolia

Poranthera microphylla

Poranthera ericifolia
Heath-leaved Poranthera
Shrub to 25cm tall with a few branches on upper stems, often minutely hairy. Leaves crowded, sessile, linear, 10–12 x 1–2mm, revolute. Corymbs 3–8cm across. **HABITAT:** Woodland, heath. **DIST:** Coast and ranges. Also NC, SC, ST. **FL:** Sept–Nov.

Poranthera microphylla
Small Poranthera
Diffuse annual herb to 10cm tall with soft glabrous branches. Leaves obovate, 2–10 x 1–2mm. Corymbs <1cm across. **HABITAT:** Woodland, open forest, in shady areas. **DIST:** Coast and ranges. Also NC, SC, NT, ST, WS, Qld, Vic, Tas. **FL:** Sept–Dec.

Pseudanthus
Monoecious shrubs. Flowers in upper axils; males clustered on short peduncles; females solitary; perianth segments 6.

Pseudanthus pimeleoides
Much-branched glabrous shrub to 1m tall. Leaves crowded, linear to narrow-ovate, 8–12mm long, acutely pointed. Flowers clustered at ends of branches, white, segments 2–12mm long; female flowers inconspicuous. Capsule 5mm long. **HABITAT:** Creek beds, rocky hillsides. **DIST:** Coast to ranges (e.g. Heathcote Ck, Patonga, Mt Hay Rd). Also NC, CWS. **FL:** July–Oct.

Ricinocarpos
Monoecious shrubs, usually with alternate entire leaves. Flowers solitary or in racemes; calyx 4–6-lobed; petals 4–6, free; male flowers with fused stamens; females with 3-celled ovary and 3 bifid styles. Fruit a capsule.

Ricinocarpos pinifolius
Wedding Bush
Glabrous shrub about 2m tall. Leaves linear, 20–40 x 2–3mm, revolute. Inflorescences terminal, with 3–6 male flowers and 1 female flower. Petals 10–15mm long, white; stamens yellow. Fruit globular, 12mm diam., densely spiny. **HABITAT:** Open forest, heath. **DIST:** Coast, lower Blue Mtns. Also NC, SC, NT, CT, Qld, Vic, Tas, NTerr. **FL:** Aug–Nov.

Pseudanthus pimeleoides

Ricinocarpos pinifolius

THYMELAEACEAE

Pimelea
Mostly shrubs with tough fibrous bark. Leaves usually opposite or nearly so, simple, entire. Inflorescence usually a condensed head subtended by leaf-like bracts, rarely spike-like. Floral tube with 4 sepal lobes; petals absent; stamens 2, exserted from tube.

Pimelea curviflora
Curved Rice-flower
Hairy shrub 20–150 cm tall. Variable species with 4 varieties recognised in Sydney area. Leaves to 20 x 8mm. Heads with 6–20 flowers. Floral tube red (var. *curviflora*) or green-yellow. **HABITAT:** Woodland, often on clay-shale soils. **DIST:** Coast and ranges. Also SC, NC, NT, ST, WS, Qld, Vic, Tas, SA. **FL:** Oct–Apr.

Pimelea glauca
Smooth Rice-flower
Glabrous shrub to 1m tall. Leaves narrow-elliptic, to 20 x 5mm, ± glaucous. Heads with 10–35 creamy white flowers; bracts 4, the inner pair fringed with hairs. **HABITAT:** Grassland, on clay-shale soils. **DIST:** Scattered, rare (e.g. Camden, Wallerawang, Bowral). Most of NSW, Qld, Vic, Tas. **FL:** Aug–Jan.

Pimelea latifolia subsp. *hirsuta*
Hairy spreading shrub to 30cm tall. Leaves elliptic, 2–10 x 3–6mm, with tangled spreading hairs on both surfaces. Heads with 1–7 greenish yellow flowers. **HABITAT:** Open forest, hillsides. **DIST:** Hornsby Plat., Blue Mtns (e.g. West Pittwater, Kurrajong Heights). Also NC, SC, ST. **FL:** Sept–Mar.

Pimelea curviflora

Pimelea ligustrina subsp. *ligustrina*
Tall Rice-flower
Shrub to 2m tall. Leaves ovate-lanceolate, 2–6 x 1–2cm. Heads about 4cm across, with 15 to many white flowers 10–15mm long; peduncles glabrous. Heads usually elongating and bracts becoming reflexed in fruit. **HABITAT:** Moist forests, rainforest margins. **DIST:** Chiefly upper Blue Mtns. Also Albion Park, NC, SC, NT, ST, Qld, Vic, Tas, SA. **FL:** Sept–Nov.

Subsp. *hypericina* is similar but has hairy peduncles. It occurs in Royal NP and Illawarra.

Pimelea linifolia subsp. *linifolia*
Slender Rice-flower
Glabrous scrambling to erect slender shrub to 1m tall. Leaves narrow-elliptic, to 30 x 4mm. Heads with about 10–40 white flowers 10–20mm long; peduncles glabrous; bracts 4, ovate. **HABITAT:** Coastal dunes to forests. **DIST:** Common. Coast and ranges. Most of NSW. Also Qld, Vic, Tas, SA. **FL:** Mainly July–Oct.

A pink-flowered form occurs at Bundanoon.

Pimelea glauca

Pimelea latifolia subsp. *hirsuta*

Pimelea ligustrina subsp. *ligustrina*

Pimelea linifolia subsp. *linifolia*

Pimelea linifolia subsp. *linoides*
Leafy shrub to 2m tall. Leaves oblong-elliptic, to 35 x 6mm. Flowers white, 30–60 per head; bracts often reddish. **HABITAT:** Cliffs, scrubby slopes. **DIST:** Upper Blue Mtns (e.g. Wentworth Falls, Echo Point). Also NC, CWS, Vic. **FL:** Aug–Nov.

Subsp. *collina* has leaves and bracts with prominent intermarginal veins. It occurs in upper Blue Mtns and at Bundanoon.

Pimelea linifolia subsp. *linoides*

Pimelea spicata
Slender shrub to 40cm tall. Leaves narrow-elliptic, 5–20 x 2–8mm. Flowers in racemes, white, often pinkish, lacking bracts. **HABITAT:** Grassy areas on shale soils. **DIST:** Endemic. Cumberland Pl. (e.g. Lansdowne, Mt Annan, Narellan), Illawarra (e.g. Bass Point). **FL:** Dec–Apr.

Pimelea spicata

DROSERACEAE

Drosera
Small carnivorous herbs growing in nitrogen-deficient waterlogged soils and possessing special adaptations for trapping and digesting insects. Leaves with fine hairs producing a sticky fluid.

Drosera binata
Forked Sundew
Erect plant with short basal leaves and erect leaves to 30cm tall, divided into narrow segments with golden and reddish glandular hairs. Flowering stems 1 or 2, exceeding leaves, with several white flowers 25mm across. **HABITAT:** Swamps, wet cliffs. **DIST:** Coast and ranges. Also NC, SC, NT, ST, Qld, Vic, SA, Asia. **FL:** Nov–Feb.

Drosera peltata
Pale Sundew
Plant to 30cm tall. Leaves in basal rosette (which may die away) and on stems; stem leaves alternate, fan-shaped, with sticky reddish hairs. Flowers in 1-sided terminal inflorescence, usually white, terminal; sepals hairy. **HABITAT:** Damp sites in grassland forest and heath. **DIST:** Coast and ranges. Most of NSW. All other States (except NTerr), Asia. **FL:** Aug–Dec.

Drosera auriculata. Similar to *D. peltata* but sepals glabrous and flowers white or pink.

Drosera pygmaea
Pigmy Sundew
Plant with rosette of glandular leaves 12mm diam., silvery hairy in centre. Lamina circular, 2mm diam. on slender stalks. Flowers solitary, white or pink, on stem to 5cm tall. **HABITAT:** Moist peaty soils, in full sun. **DIST:** Coast, Cumberland Pl., Kanangra Walls. Also NC, SC, all other States (except NTerr), NZ. **FL:** Oct–Apr.

Drosera peltata

Drosera binata

Drosera pygmaea

FLOWERING PLANTS

Drosera spatulata

Acrophyllum australe

Drosera spatulata
Common Sundew
Plant with rosette of reddish spathulate glandular leaves 8–20mm long. Inflorescences 1-sided with up to 15 pink or white flowers 6–10mm across; styles 3, divided to the base. **HABITAT:** Wet heath, scrub, in full sun. **DIST:** Coast and ranges. Also NC, SC, Qld, Vic, Tas, NZ, Asia. **FL:** Oct–Feb.

Drosera burmannii. Similar to *D. spathulata* but flowers with 5 styles. Uncommon in Cumberland Pl; common in Megalong V.

Drosera glanduligera. Rosette of leaves with distinct stalks and circular leaf blade. Flowers many, red or orange. Uncommon. Castlereagh SF; more frequent WS.

CUNONIACEAE

Acrophyllum
Rare shrub with simple leaves in whorls. Flowers with 5 sepals, 5 petals, 10–15 stamens and 2-locular superior ovary with 2 styles. Fruit a capsule topped by persistent styles. Monotypic genus.

Acrophyllum australe
Spreading glabrous shrub. Leaves ± sessile in whorls of 3 (rarely 4), ovate, 4–10 x 1–4cm, rigid, glabrous above, glaucous beneath; margins serrated. Flowers in several dense cymes in axils of reduced leaves at end of branches; petals white, 3–4mm long. **HABITAT:** Drip ledges of sandstone cliffs. **DIST:** Endemic. Mid Blue Mtns (e.g. Linden). **FL:** Oct–Nov.

Aphanopetalum
Straggling climbers with opposite simple leaves. Flowers axillary; calyx lobes 4; petals 4 or absent. Fruit a nut surrounded by enlarged calyx.

Aphanopetalum resinosum
Gum Vine
Glabrous climber; stems with raised lenticels. Leaves ovate-elliptic, 4–10 x 2–3cm, shiny, toothed; stalk 2–5mm long. Flowers few in short cymes; petals 3mm long, greenish cream. Nut 2–3mm diam., exceeded by sepals. **HABITAT:** Rainforests, wet gullies. **DIST:** Scattered (e.g. Royal NP, Razorback Mtn, Katoomba, Robertson). Also NC, SC, NT, ST, CWS, Qld. **FL:** Aug–Nov.

Aphanopetalum resinosum

Callicoma serratifolia

Callicoma
Trees or shrubs with hairy branchlets. Flowers in dense globular heads. Fruit a capsule splitting into 2 or 3 carpels.

Callicoma serratifolia
Black Wattle
Leaves opposite, elliptic, 5–10 x 3–5cm, acutely pointed, serrated, dark green and glabrous above, white- or rusty-hairy beneath. Heads axillary, 1–2cm diam., cream-yellow, on peduncles 1–3cm long. Capsules clustered. **HABITAT:** Creek banks, sheltered gullies. **DIST:** Coast and ranges. Also NC, SC, Qld. **FL:** Oct–Dec.

Stems used to construct wattle-and-daub buildings in early Sydney.

Ceratopetalum
Shrubs and trees with opposite compound leaves. Petals small or absent; sepals enlarging and colouring as they mature; stamens 10. Fruit dry, surrounded by persistent sepals.

Ceratopetalum apetalum
Coachwood
Small to tall tree, with smooth straight trunk blotched whitish grey. Leaves 1-foliolate; leaflet elliptic, 8–16 x 3–5cm, finely toothed, pale green on undersurface; stalk 1–2cm long. Flowers in loose cymes; sepals 5, cream, turning pink. **HABITAT:** Sheltered gullies, watercourses. **DIST:** Common. Coast and ranges. Also NC, SC, Qld. **FL:** Nov–Jan.

Ceratopetalum apetalum

Ceratopetalum gummiferum

Ceratopetalum gummiferum
Christmas Bush
Shrub or small tree. Leaves 3-foliolate; leaflets lanceolate, 4–10 x 1–2.5cm, finely toothed, glabrous, pale green beneath. Flowers in loose cymes; sepals 5, white turning pink or red. **HABITAT:** Slopes and gullies of open forest. **DIST:** Coast and ranges. Also NC, SC. **FL:** Oct–Nov.

Schizomeria
Tree with opposite simple toothed leaves. Flowers in terminal or axillary cymes; calyx lobes 4–6, not enlarging in fruit; petals small; stamens 10. Fruit a drupe.

Schizomeria ovata
Crabapple
Tree to 30m tall with deeply furrowed bark. Leaves ovate-elliptic, 8–18 x 2–5cm, leathery, finely toothed to nearly entire, paler beneath; stalk 10–25mm long. Calyx lobes white, 2mm long. Drupe creamy white, 12mm diam. **HABITAT:** Rainforest gullies. **DIST:** Coast (e.g. Gosford, Royal NP, Mt Keira). Also NC, SC, NT, Qld. **FL:** Sep–Nov.

EUCRYPHIACEAE

Eucryphia
Shrubs or trees. Flowers solitary or few in leaf axils; sepals 4, petals 4; stamens numerous. Fruit a capsule with winged seeds.

Eucryphia moorei
Plumwood, Pinkwood
Leaves pinnate with 3–13 leaflets, the terminal leaflet largest; leaflets oblong, 20–60 x 5–15mm, leathery, dark green above, white-tomentose beneath. Flowers white, fragrant, 2–3cm diam., petals overlapping. Capsule ovoid, 10–15mm long. **HABITAT:** Sheltered rocky sites, plateaus on volcanic soils. **DIST:** Illawarra, Sthn H'lands (e.g. Bundanoon, Knapsack Hill). Also SC, ST, Vic. **FL:** Feb–Mar.

Schizomeria ovata

Eucryphia moorei

BAUERACEAE

Bauera
Small to medium shrubs. Leaves sessile, opposite, 3-foliolate, often appearing as whorls of 6. Flowers with 4–10 sepals and petals, and many stamens. Fruit a capsule.

Bauera microphylla
Weak trailing plant to 30cm tall. Leaflets 5–9mm long. Flowers usually white, sometimes pinkish; sepals and flower stalks ± glabrous. **HABITAT:** Sandy wet heath. **DIST:** Coast, from Wollongong to NC. **FL:** Aug–May.

Bauera rubioides
River Rose, Dog Rose
Tangled scrambling shrub to 2m tall, often sprawling 3–4m. Leaflets 9–15mm long. Flowers pink, with 8–10 petals and numerous stamens; sepals and flower stalks hairy. **HABITAT:** Wet situations, esp. creek banks. **DIST:** Coast, upper Blue Mtns. Also Bargo, ST, Qld, Vic, Tas, SA. **FL:** July–Dec.

Bauera microphylla

Bauera rubioides

GROSSULARIACEAE

Abrophyllum
Shrubs or small trees. Leaves alternate, simple, toothed. Inflorescences paniculate. Calyx lobes, petals and stamens 5; ovary 5-locular; stigma 5-lobed. Fruit a berry with many seeds.
   This genus is also placed in Escalloniaceae and at least 3 other families.

Abrophyllum ornans
Native Hydrangea
Shrub to 8m tall with hairy new growth and inflorescences. Leaves elliptic, tapering to both ends, to 20 x 8cm, marginal teeth often with callus point, veins distinct on pale undersurface. Panicles shorter than leaves; petals 4–5mm long, yellowish. Berry, about 10 x 6mm, purplish black. **HABITAT:** Cool moist sites, rainforests, gullies. **DIST:** Uncommon. Coast and ranges (e.g. Bola Ck, Erskine Ck, Colo R.). Also NC, SC, Qld. **FL:** Nov–Jan.

Abrophyllum ornans

Polyosma
Trees or shrubs. Leaves ± opposite, rarely whorled, simple, entire or toothed. Inflorescence a terminal raceme. Flowers with 4 petals, joined into a tube; stamens 4. Fruit a 1-seeded berry.

Polyosma cunninghamii
Featherwood
Tree to 15m tall with hairy new growth. Leaves elliptic, tapering to both ends, to 9 x 3cm, soft, glossy, distantly toothed. Flowers greenish cream, strongly scented. Berry ovoid, ribbed, black. **HABITAT:** Rainforest. **DIST:** Coast (e.g. Strickland SF, Mt Keira). Also Robertson, NC, SC, CT, Qld. **FL:** Apr–Oct.

Polyosma cunninghamii

Quintinia
Trees and shrubs. Leaves alternate, simple, usually entire. Flowers with 5-toothed calyx, 5 petals, 5 stamens and 3–5-lobed stigma. Fruit a capsule with many seeds.

Quintinia sieberi
Possumwood
Tree to 25m tall. Branchlets often with red-brown glands; terminal shoots and leaf stalks pinkish. Leaves elliptic, 5–10 x 2–5cm. Inflorescence with many small white flowers in terminal panicles to 12cm long. **HABITAT:** Rainforest gullies, sheltered forests. **DIST:** Chiefly Blue Mtns, Sthn H'lands. Rarer on coast. Also NC, SC, NT, ST, Qld. **FL:** Oct–Nov.

Quintinia sieberi

PITTOSPORACEAE

Billardiera
Twining plants or subshrubs with bell-shaped flowers and succulent cylindrical berries.

Billardiera mutabilis
Climbing Apple Berry
Slender climber, with silky-hairy new shoots and leaves. Leaves elliptic, 4–6cm long. Flowers, pendulous on stalks to 40mm long, green-yellow. Fruit, cylindrical, to 2cm long, glabrous; seeds in flesh. **HABITAT:** Moist sites in forests. **DIST:** Coast sth from Royal NP; uncommon in Blue Mtns. Also SC, ST, Vic, Tas. **FL:** Sept–Dec.

Billardiera scandens

Billardiera scandens
Hairy Apple Berry
Scrambling shrub with some branches climbing; new shoots silky-hairy. Leaves narrow-ovate, 20–50 x 5–15mm. Flowers 1–3 in axils, pendulous on stalks 10mm long; tube 12–25mm long, greenish yellow to cream, hairy. Fruit cylindrical, to 3cm long, greenish; seeds in separate dry cells. **HABITAT:** Open forest, woodland. **DIST:** Coast and ranges. Also NC, SC, NT, ST, WS, Qld, Vic. **FL:** Sept–Dec.

Bursaria
Rigid much-branched shrubs, often with spiny branches. Leaves alternate or clustered. Flowers small, solitary or clustered in racemes or panicles. Fruit a thin-walled capsule.

Bursaria spinosa subsp. *spinosa*
Native Blackthorn
Shrub to 5m tall. Leaves in clusters, linear-obovate, 10–35 x 3–10mm, ± glabrous. Flowers mostly in terminal panicles 10–25cm long, a few in axillary clusters, white, 5-petalled, fragrant. **HABITAT:** Woodland, on clay-shale soils. **DIST:** Coast and ranges. Also NC, SC, ST, WS, Qld, Vic, Tas, SA. **FL:** Jan–Aug.

Bursaria longisepala (upper Blue Mtns). Similar to *B. spinosa* but with sepals to 5mm long and flowers along branches among the spines.

Billardiera mutabilis

Bursaria spinosa subsp. *spinosa*

FLOWERING PLANTS

Citriobatus pauciflorus

Hymenosporum flavum

Citriobatus
Shrubs with rigid thorny branches. Leaves alternate, entire or lobed, glabrous. Flowers solitary in axils, ± sessile; sepals free; petals fused in lower half. Fruit a globose berry.

Citriobatus pauciflorus
Orange Thorn
Intricate wiry shrub about 1m tall with fine thorns along branches. Leaves broad-obovate to circular, to 12 x 8mm, usually finely toothed towards apex. Flowers few, petals 4mm long, white. Berry to 10mm diam., orange. **HABITAT:** Rainforest, on clay-shales and basalts. **DIST:** Coast, Robertson. Also NC, SC, NT, CWS, SWP, Qld. **FL:** Sept–Nov.

Hymenosporum
Shrubs or trees. Leaves alternate or clustered, stalked, soft-textured, glabrous, discolorous. Flowers conspicuous in terminal panicles; petals fused in lower half, forming 5 spreading lobes in upper half. Fruit a capsule with numerous winged seeds. Monotypic genus.

Hymenosporum flavum
Native Frangipani
Leaves oblanceolate with pointed tip, to 16 x 5cm. Flowers cream to yellow, 5cm diam., scented, silky-hairy inside. **HABITAT:** Rainforest margins. **DIST:** Uncommon. Lower Blue Mtns (e.g. Wheeny Ck, Grose Vale). Also NC, NT, CT, CWS, Qld, PNG. **FL:** Oct–Nov.

Pittosporum revolutum

Pittosporum undulatum

Pittosporum undulatum (fruit)

Pittosporum

Shrubs or trees with alternate or clustered petiolate leaves. Flowers bell-shaped, with 5 free sepals, 5 partly fused petals spreading or recurved in upper part, and 5 stamens. Fruit a capsule; seeds in sticky pulp.

Pittosporum revolutum
Yellow Pittosporum
Shrub to 3m tall. Leaves obovate, to 15 x 5cm, rusty-hairy beneath when young. Flowers in terminal clusters, fragrant; petals yellow, 9–12mm long. Fruit about 25mm diam., splitting into 3 to reveal bright red seeds. **HABITAT:** Sheltered forest, rainforest margins. **DIST:** Coast and ranges. Also NC, SC, ST, CWS, Qld, Vic. **FL:** Sept–Oct.

Pittosporum undulatum
Sweet Pittosporum
Tree with dense foliage, often invading sheltered urban bushland. Leaves ovate-elliptic, to 12 x 4cm long, with undulate margins, glabrous, discolorous. Flowers several in terminal clusters, scented; petals creamy white, 10–12mm long. Fruit globose, 10mm diam., pale brownish orange. **HABITAT:** Shaded hillsides, creeks, rainforest. **DIST:** Coast and ranges. Also NC, SC, NT, ST, WS, Qld, Vic, Tas, SA. **FL:** Aug–Sept.

FLOWERING PLANTS

Rhytidosporum
Erect or prostrate shrubs. Leaves alternate, entire or toothed at apex. Flowers solitary or few in corymb, with 5 free spreading petals; stamens free. Fruit a dry, ovoid, compressed capsule, 2–3 locular.

Rhytidosporum procumbens
Dwarf to prostrate shrub. Leaves crowded, narrow, to 12 x 2mm. Flowers 5-petalled white stars, 1–2 together on short stalks; buds often pink. Ovary surrounded by 5 conspicuous stamens. Fruit a leathery capsule. **HABITAT:** Dry open forest, heath. **DIST:** Coast and ranges. Most of NSW. Also Qld, Vic, Tas, SA. **FL:** Sept–Dec.

Rhytidosporum procumbens

CRASSULACEAE

Crassula
Herbs, with fleshy stems and opposite succulent leaves. Inflorescence terminal, branching. Flowers tiny, with 4–5 free or fused sepals and petals; stamens as many or twice as many as stamens. Follicles clustered.

Crassula sieberiana
Australian Stonecrop
Herb with spreading stems to 20cm long, often reddish in colour. Leaves narrow-lanceolate, to 10 x 3mm. Inflorescence spike-like. Flowers clustered, with 4 sepals and petals about 2mm long, yellow to red. **HABITAT:** Mossy sites, esp. on rocks. **DIST:** Coast and ranges. Most of NSW. All other States, NZ, LHI. **FL:** Aug–Dec.

Crassula sieberiana

ROSACEAE

Acaena
Perennial herbs with pinnate leaves. Flowers in spikes or globose heads; sessile, small; sepals usually 4; petals absent. Fruits tiny, with barbed spines, often clustered into burr-like heads.

Acaena novae-zelandiae
Bidgee-widgee
Creeping mat-forming herb with erect hairy stems. Leaves with 7–9 toothed oblong leaflets about 10mm long. Flowers in globular heads about 2cm across. Fruiting heads red-brown with spines 10mm long. **HABITAT:** Sheltered clearings in forests. **DIST:** Coast and ranges. Also NC, SC, NT, ST, WS, Qld, Vic, Tas, SA, NZ. **FL:** Nov–Feb.

Acaena ovina
Sheep's Burr
Non-stoloniferous herb to 50cm tall; stems softly hairy. Leaves pinnate, with 17–23 toothed ovate leaflets 10–20mm long, softly hairy on both surfaces. Flowers and burrs in long interrupted spikes. **HABITAT:** Grassland, roadsides, on clay soils. **DIST:** CT (e.g. Berrima, Mt Victoria, Coxs R.). Also NT, ST, WS, Qld, Vic, Tas, SA, WA. **FL:** Oct–Jan.

Acaena echinata. Similar to *A. ovina*, but leaves glabrous above. It occurs on clay soils in Western Sydney (e.g. Narellan) and Blue Mtns (e.g. Leura).

Rubus moluccanus var. *trilobus*

Rubus
Prickly shrubs and climbers. Leaves compound or simple, usually with toothed margins. Flowers with 5 persistent sepals and 5 broad petals; stamens numerous. 'Fruit' an aggregate of succulent or dryish segments (drupelets).

Rubus moluccanus var. *trilobus*
Broad-leaf Bramble
Scrambling shrub or climber with branches 2–3m long. Leaves simple, ± ovate in outline, shallowly and broadly 3- or 5-lobed, to 15 x 10cm, finely toothed, softly hairy underneath; stalk 2–6cm long. Flowers pink (sometimes white). Fruit globose, 12mm diam., red. **HABITAT:** Sheltered forest, rainforest. **DIST:** Coast and ranges. Also NC, Qld, Vic. **FL:** Nov–Feb.

Rubus nebulosus
Bush Lawyer
Climber. Branches and leaf stalks with hooked prickles. Leaves ± glabrous with 5 spreading oblong-ovate leaflets, each 4–10 x 1–4cm

Acaena novae-zelandiae

Acaena ovina

Rubus nebulosus

with finely toothed margins. Flowers with white petals 10mm long. Fruit globose, 10mm diam., red. **HABITAT:** Sheltered forest, rainforest. **DIST:** Coast, Robertson. Also NC, SC, NT, Qld. **FL:** Oct–Jan.

Rubus parvifolius
Native Raspberry
Scrambling shrub with stems to 1m long. Leaves glabrous above, white-hairy beneath, with 3–5 ovate-rhombic leaflets, each 1–4 x 1–4cm with irregularly toothed and lobed margins. Flowers red or pink. Fruit globose, 10mm diam., red. **HABITAT:** Shaded forest slopes. **DIST:** Coast and ranges. Most of NSW. Also Qld, Vic, Tas, SA, Asia. **FL:** Oct–Dec.

Rubus parvifolius

Rubus rosifolius
Forest Bramble
Weak trailing shrub with stems to 2m long. Leaves green and glandular-hairy on both surfaces, with 5–7 lanceolate leaflets, each 3–8 x 1–2cm with irregularly toothed margins. Flowers white. Fruit ovoid, 20mm long, red. **HABITAT:** Shaded forest slopes. **DIST:** Coast and ranges. Most of NSW. Also Qld, Vic, Tas, SA, Asia. **FL:** July–Dec.

Rubus rosifolius

MIMOSACEAE

Identification of wattles is best accomplished by first grouping them on the basis of foliage and inflorescence. The tiny wattle flowers (Figure **A**) are either grouped in globular heads (Figure **B**) or arranged in spikes (Figure **C**). Thirteen species in the Sydney area have bipinnate leaves (Figure **D**) and flowers in heads. Nine species have phyllodes and flowers in spikes (Figure **E**). About 45 species have phyllodes and flowers in globular heads. These can be divided into smaller groups according to whether the phyllodes are pungent-pointed (Figure **F**) or leaf-like (Figure **G–K**) and whether the flower heads occur singly in the axils (Figure **G**), in small clusters (Figure **H**) or in racemes (Figure **I**). The largest group with flower heads in racemes can be further divided depending on the size and shape of the phyllodes and whether they have longitudinal nerves (Figure **J**) or are penniveined or reticulate (Figure **K**). The number of flowers in each head varies from 4 to about 40, depending upon the species, and is useful in identification. It is easiest to count the flowers at the bud stage. The size, shape and texture of the pods (legumes) are also important.

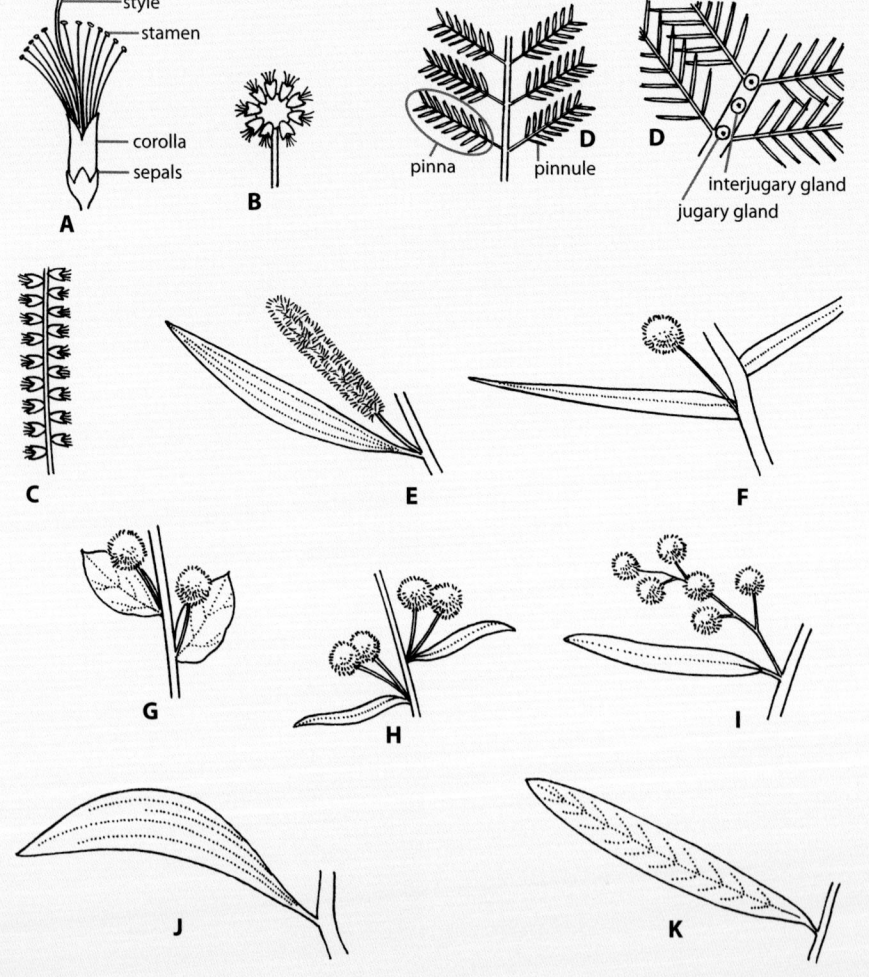

Acacia dealbata

Acacia
WATTLE

Over 60 species of wattle occur naturally in the Sydney district, with at least 12 additional species on the fringes of the area. For the purposes of identification, they are here divided into groups based on foliage and inflorescence.

GROUP 1: Wattles with bipinnate leaves.
GROUP 2: Wattles with flowers in cylindrical spikes.
GROUP 3: Wattles with flowers in globular heads solitary in axils; phyllodes pungent-pointed or branches spinescent.
GROUP 4: Wattles with globular heads solitary, paired or clustered in axils; phyllodes not pungent-pointed.
GROUP 5: Wattles with globular or ovoid heads in racemes >2cm long; phyllodes under 7cm long.
GROUP 6: Wattles with globular heads in racemes >2cm long; phyllodes more than 7cm long.

GROUP 1

Wattles with bipinnate leaves. Flowers in globular heads, the heads in axillary racemes. The number and placement of glands along the rachis is an important identifying feature for some species.

Acacia dealbata
Silver Wattle

Shrub or tree to 15m with dark grey trunk. Branchlets angled, hairy. Branches and foliage with waxy grey-green bloom. Jugary glands present. Racemes to 10cm long; heads golden

Acacia decurrens

yellow. Pod flat, to 8cm long, glaucous. **HABITAT:** Open forest, creek banks. **DIST:** Upper Blue Mtns, Sthn H'lands. Also SC, NT, ST, WS, SWP, Vic, Tas. **FL:** Aug–Nov.

Acacia decurrens
Early Green Wattle

Tree 5–15m tall, with dark grey to black bark. Branches with conspicuous wing-like ridges. Pinnules pale green, linear, 6–14mm long, widely spaced. Jugary glands present. Racemes to 10cm long; heads bright yellow. **HABITAT:** Open forest, esp. on shales. **DIST:** Coast and ranges. Also NC, SC. Naturalised elsewhere. **FL:** July–Sept.

Acacia filicifolia

Acacia elata
Cedar Wattle
Tree to 18m tall. New growth cream, contrasting with dark green older foliage. Pinnae widely spaced, with 8–14 pairs of pinnules; pinnules lanceolate, acuminate, to 50 x 10mm. Inflorescence paniculate; heads large, creamy yellow. **HABITAT:** Shaded gullies, rainforest margins. **DIST:** Hornsby Plat., Blue Mtns. Also NC, NT. **FL:** Dec–Feb.

Acacia filicifolia
Fern-leaved Wattle
Tree 3–6m tall, with dark trunk and blue-grey branchlets. New foliage creamy yellow. Rachis with 2–4 interjugary glands between each pair of pinnae. Pinnules narrow, crowded, with 35–55 pairs. Flower heads in axillary racemes 8–12cm long, bright yellow. Pod linear, 4–12cm long, blue-grey. **HABITAT:** Open forest, esp. alluvial flats. **DIST:** Hornsby Plat., Blue Mtns.

FLOWERING PLANTS

Acacia elata

Acacia irrorata subsp. *irrorata*

Also Bomaderry Ck, NC, SC, NT, WS, Qld.
FL: Aug–Sept.
   Acacia fulva. Tree. Branchlets with velvety silver-grey hairs. Rare. Restricted in Sydney area to Mt Wareng, e. of Howes Valley.

Acacia irrorata subsp. *irrorata*
Green Wattle
Tree 4–12m tall. Branchlets ridged, scabrous; new growth yellow. Pinnules sparsely hairy. Jugary glands present between upper 1–4 pairs of pinnae. Pod black, scabrous, with short pale hairs. **HABITAT:** Forest. **DIST:** Coast. Also NC, SC, NT, CT, NWS, Qld. **FL:** Nov–Jan.

Acacia jonesii

Acacia jonesii
Straggly open shrub to 3m tall. Stems and rachis often with grey hairs. Pinnae usually 5–6 pairs; pinnules 12–18 pairs. Jugary glands present. Flower heads in racemes exceeding the leaves, deep yellow, 8–15-flowered, golden yellow. **HABITAT:** Forested slopes and ridges. **DIST:** Sthn H'lands, sthn Blue Mtns. Also SC, ST. **FL:** Aug–Oct.

Acacia mearnsii
Black Wattle
Tree to 15m tall with black trunk. New growth downy yellow. Rachis hairy; pinnules minutely pubescent. Jugary glands present with an interjugary gland between pairs of pinnae. Racemes to 12cm long; heads pale yellow. Pods glabrous, leathery, to 12cm long. **HABITAT:** Open forest, river flats. **DIST:** Illawarra, Sthn H'lands. Naturalised elsewhere. Also SC, ST, Vic, Tas, SA. **FL:** Oct–Nov.

Acacia mearnsii

Acacia parramattensis

Acacia parvipinnula

FLOWERING PLANTS

Acacia parramattensis
Sydney Green Wattle
Tree to 12m tall with brown-black trunk. New growth greenish yellow. Stems and branchlets minutely ridged. Branchlets, rachis and glands with short grey hairs. Rachis with usually 1 interjugary gland between each jugary gland. Pinnae with 20–35 pairs of pinnules 3–5mm long. Flower heads pale yellow. **HABITAT:** Open forest, woodland. **DIST:** Coast and ranges, esp. Cumberland Pl. Also NC, SC, ST, WS. **FL:** Oct–Feb.

Acacia pubescens

Acacia parvipinnula
Silver-stemmed Wattle
Tree 4–10m tall with smooth silvery trunk and branches. Interjugary glands 1–3 between successive pinnae; jugary glands sometimes absent. Pinnules 14–30 pairs per pinna, crowded, 3–4mm long. Racemes 4–8cm long; heads pale yellow, with 14–18 flowers per head. **HABITAT:** Open forest, esp. on shale and lateritic sandstone. **DIST:** Hornsby Plat., Blue Mtns, Menai, Bargo. Also NC, ST, CWS. **FL:** Sept–Dec.

Acacia pubescens
Downy Wattle
Bushy shrub to 3m tall with dense white hairs on branchlets, flower stalks and rachises. Leaves pale green, glabrous, <7cm long, with 7–10 pairs of pinnae; pinnules 6–20 pairs per pinna, 2–4mm long. Racemes as long as leaves; heads golden yellow. **HABITAT:** Open forest, on shale and gravelly clay. **DIST:** Cumberland Pl.; Menai to Liverpool. Also NC, CT. **FL:** Aug–Sept.

Acacia schinoides
Shrub or tree 2–5m tall. New growth drooping, bronze-coloured. Pinnae 2–4 pairs, each with 12–20 pairs of pinnules to 18 x 3mm. Racemes exceeding leaves; heads pale yellow with up to 50 flowers. **HABITAT:** Sheltered forests, creek gullies. **DIST:** Nth of Sydney, from Lane Cove R. to Putty, Watagan Mtns. **FL:** Dec–Mar.

Acacia schinoides

Acacia terminalis

Acacia longifolia

Acacia terminalis
Sunshine Wattle
Variable spreading shrub to 5m tall; branchlets often reddish. Leaves 3–8cm long, with 2–6 pairs of pinnae. Pinnules 8–20 pairs, 10 x 2mm, dark green above, paler below. Racemes <9cm long, heads with 5–15 flowers, pale cream to golden. **HABITAT:** Open forest, woodland. **DIST:** Common. Coast and ranges. Also NC, SC, NT, ST, CWS, Vic, Tas. **FL:** Mar–July.

GROUP 2
Wattles with flowers in cylindrical spikes.

Acacia binervia
Coast Myall
Tree to 15m tall with greyish green foliage. Phyllodes elliptic, falcate, tapering to both ends, to 15 x 2cm, with 3–5 longitudinal nerves. Spikes 6cm long, deep yellow. **HABITAT:** Open forest, river banks. **DIST:** Widespread, but chiefly coast. Also NC, SC, NT, ST, CWS. **FL:** Sept–Oct.

Acacia floribunda
Sally Wattle
Shrub or tree 3–8m tall with spreading branches. Phyllodes linear with 2–4 prominent longitudinal nerves, 5–15cm x 2–12mm; gland absent or inconspicuous. Spikes loose, open, pale yellow, to 8cm long. **HABITAT:** Forested slopes, esp. along creek banks. **DIST:** Coast and ranges. Also NC, SC, NT, ST, Qld, Vic. **FL:** Aug–Sept.

Acacia longifolia
Sydney Golden Wattle
Shrub or tree 2–5m tall. Phyllodes linear to narrow-elliptic, straight, with 2–4 prominent nerves, to 20 x 2cm. Spikes dense, golden yellow, 2–5cm long. **HABITAT:** Open forest. **DIST:** Common. Coast and ranges. Also NC, SC, NT, ST, Vic. **FL:** June–Oct.

Acacia floribunda

Acacia binervia

Acacia longissima
Long-leaf Wattle
Spreading shrub to 5m tall with drooping branches. Phyllodes long, narrow, to 300 x 3mm, with one clear central nerve (unlike *A. floribunda* which has several nerves). Spikes interrupted, pale yellow, 2–5cm long. **HABITAT:** Sheltered forests. **DIST:** Chiefly nth of Sydney, less common elsewhere (e.g. Royal NP, Bangor). Also NC, SC, ST, Qld. **FL:** Nov–Feb.

Acacia maidenii
Maiden's Wattle
Tree 5–16m tall with raised spots prominent on branchlets. Phyllodes to 200 x 12mm, straight or slightly falcate, with parallel nerves. Peduncles golden hairy. Spikes pale yellow, 4–6cm long, 1–2 in axils of phyllodes. Pods narrow, cylindrical and very twisted. **HABITAT:** Moist forests, rainforest margins. **DIST:** Coast (Ourimbah to Minnamurra); Yerranderie. Also NC, SC, NT, CWS, Qld, Vic. **FL:** Jan–May.

Acacia obtusifolia

Acacia obtusifolia
Blunt-leaf Wattle
Erect shrub to small tree. New growth often reddish. Phyllodes to 20 x 2cm, with several parallel nerves; apex obtuse. Spikes to 4cm long, pale yellow, less dense than those of *A. longifolia*. **HABITAT:** Woodland, open forest, often along creeks. **DIST:** Coast and ranges. Also NC, SC, NT, ST, CWS, Qld, Vic. **FL:** Dec–Feb.

Acacia oxycedrus
Spike Wattle
Rigid shrub 2–4m tall. Phyllodes flat, sharp, 2–3cm long, with 4 or more parallel nerves. Spikes 2–3cm long, bright yellow. Pod woody, cylindrical, to 10cm long. **HABITAT:** Dry rocky soils in forest, woodland and heath. **DIST:** Hornsby Plat., lower Blue Mtns, Avon and Cordeaux dams. Also SC, Vic, SA. **FL:** July–Oct.

Acacia longissima

Acacia maidenii

Acacia oxycedrus

Acacia sophorae
Coast Wattle

Decumbent shrub often forming dense clumps on sand dunes. Similar to *A. longifolia*, but with shorter, broader phyllodes (intermediates occur). **HABITAT:** Coastal sand dunes, hind-dune heath. **DIST:** Common. Coast (e.g. Bouddi NP, Kurnell, Marley Bch, Bass Point). Also NC, SC, Qld, Vic, Tas, SA. **FL:** July–Oct.

GROUP 3
Wattles with flowers in globular heads which are solitary in axils; phyllodes with a pungent point or bearing spines.

Acacia amblygona
Fan Wattle

Sprawling shrub to 1m tall with densely hairy branches. Phyllodes sessile, asymmetrically elliptic, acuminate, 8–20 x 2–4mm, 2–3-nerved, midrib near lower margin. Heads golden

Acacia sophorae

yellow; peduncles 5–13mm long. Pods constricted between seeds, strongly curved. **HABITAT:** Open forest. **DIST:** Wollemi NP, nw. edge of Hornsby Plat. (e.g. Putty Rd). Also NC, WS, WP, Qld. **FL:** July–Oct.

Acacia amblygona

Acacia asparagoides

Acacia asparagoides
Rigid glabrous shrub to 1m tall. Phyllodes ± sessile, tapering to pungent apex, 10–20 x 1–2mm. Heads golden yellow; peduncles 2–3mm long. Pods strongly curved. **HABITAT:** Open forest, on rocky sites. **DIST:** Endemic. Upper Blue Mtns (e.g. Hassans Walls, Newnes, Radiata Plat.). **FL:** Aug–Oct.

Acacia brownii
Golden Prickly Wattle
Diffuse shrub, sparsely hairy. Phyllodes widely spaced, rigid, terete to 4-angled with raised midrib, 10–20 x 1mm. Heads deep yellow; peduncle 10–13mm long, slender. **HABITAT:** Woodland, scrub. **DIST:** Coast and ranges. Also NC, SC, NT, ST, WS, Qld, Vic. **FL:** July–Sept.

Acacia brownii

Acacia bynoeana
Bynoe's Wattle
Decumbent shrub with hairy ribbed branchlets. Phyllodes ascending, rigid, linear, 1–5 x 1–3mm, slightly curved; nerves 3. Heads pale yellow; peduncle 2–6mm long. **HABITAT:** Woodland, heath. **DIST:** Endemic. Scattered, uncommon (e.g. Maroota, Agnes Banks, Hill Top). **FL:** Dec–Apr.

Acacia echinula
Hedgehog Wattle
Shrub to 2m tall with hairy branchlets. Phyllodes crowded, ± terete, with fine recurved point, 10–13mm long. Heads 5–10mm diam., pale or golden yellow; peduncle stout, 5–15mm long. **HABITAT:** Open forest, woodland, heath. **DIST:** Coast and ranges. Also NC, CWS. **FL:** July–Aug.

FLOWERING PLANTS

Acacia bynoeana

Acacia echinula

Acacia genistifolia

Acacia gunnii

Acacia genistifolia
Spreading Wattle
Glabrous shrub about 2m tall with angular or flattened branchlets. Phyllodes well-spaced, rigid, terete or 4-angled, straight or slightly curved, 20–40 x 1–3mm, finely pointed. Heads 2–4 in axils, lemon yellow; peduncles slender, to 20mm long. **HABITAT:** Open forest, on rocky hillsides. **DIST:** Sthn Blue Mtns, Sthn H'lands. Also SC, ST, WS, SWP, Vic. **FL:** July–Oct.

Acacia gunnii
Ploughshare Wattle
Decumbent undershrub with hairy spinose branchlets. Phyllodes sessile, unequally triangular, acuminate, to 15 x 5mm, midrib prominent, green, ± glabrous. Heads cream to mid-yellow; peduncles slender, hairy, 5–15mm long. **HABITAT:** Open forest, in gravelly shale or rocky sites. **DIST:** Upper Blue Mtns, Sthn H'lands. Also NC, SC, NT, ST, WS, Qld, Vic, Tas, SA. **FL:** Aug–Sept.

Acacia paradoxa

Acacia paradoxa
Kangaroo Thorn
Shrub to 4m tall. Branches angular, often hairy. Phyllodes to 30 x 7mm, apex acute. Stipules rigid, spinescent, 10mm long. Heads deep golden yellow; peduncles to 15mm long. **HABITAT:** Open forest, woodland, scrub. **DIST:** Uncommon. Scattered (e.g. Colo Heights, Grose R., Yerranderie). Most of NSW. Also Qld, Vic, SA, WA. **FL:** Aug–Oct.

Acacia quadrilateralis
Glabrous shrub to 2m tall. Phyllodes frequently in fascicles of 2–3, ± erect, ± square in cross-section, to 60 x 1mm, glands absent. Heads cream to golden yellow; peduncles slender, 5–15mm long. **HABITAT:** Coastal heath, in deep sand. **DIST:** Rare. Bouddi NP, Anna Bay. Also Qld. **FL:** Aug–Oct.

Acacia trinervata
Three-veined Wattle
Glabrous shrub to 3m tall with angular branchlets. Phyllodes sessile, rigid, linear to linear-lanceolate, 10–40 x 1–3mm, with 2–3 prominent nerves. Heads bright yellow; peduncles 10–18mm long, slender. **HABITAT:** Open forest. **DIST:** Endemic. Blue Mtns (Glenbrook–Colo). **FL:** Mar–June.

Acacia ulicifolia
Prickly Moses
Common shrub to 2m tall. Phyllodes rigid, tapering to apex from slightly swollen base, ± angular, midrib prominent on each face, 5–15 x 1–2mm. Heads pale cream; peduncles 7–15mm long, slender. **HABITAT:** Open forest, woodland, heath. **DIST:** Coast and ranges. Also NC, SC, NT, ST, WS, Qld, Vic, Tas. **FL:** May–Oct.

The common name Prickly Moses is a corruption of 'prickly mimosa'; it is applied to a number of wattles with sharply pointed phyllodes.

GROUP 4
Wattles with globular flower heads single, paired or clustered in axils; phyllodes not pungent-pointed.

Acacia baueri subsp. *aspera*
Tiny Wattle
Decumbent shrub to about 30cm tall. Phyllodes in irregular whorls, ± terete with recurved tip, to 15 x 1mm, rough. Heads solitary in axils, bright yellow; peduncles 5–15mm long, slender, decurved in fruit. **HABITAT:** Dry heath. **DIST:** Endemic. Rare. Coast and ranges (e.g. Royal NP, Kings T'lands, Dharawal NR). **FL:** Dec–Feb.

Subsp. *baueri*. Only old records exist; probably extinct in Sydney district. Phyllodes regularly whorled, smooth.

Acacia quadrilateralis

Acacia trinervata

Acacia ulicifolia

Acacia baueri subsp. *aspera*

Acacia elongata var. *elongata*

Acacia gordonii

Acacia ixiophylla

Acacia elongata var. *elongata*
Swamp Wattle
Erect shrub 2–3m tall; branchlets ribbed, angled or flattened. Phyllodes ascending, linear, often curved, to 12cm x 4mm, apex acute, with 3–5 prominent nerves. Heads 1–3 in axils, 5mm diam., bright yellow; peduncle or raceme axis to 14mm long. **HABITAT:** Woodland, heath, esp. along creeks. **DIST:** Coast and ranges. Also NC, SC, NT, ST, WS, Vic. **FL:** July–Sept.

Acacia gordonii
Spreading shrub to 1.5m tall; branchlets terete, densely hairy. Phyllodes crowded, linear, alternate or whorled, 10–15mm long. Heads solitary in axils, golden yellow; peduncles 8–15mm long. **HABITAT:** Open woodland, heath. **DIST:** Central Blue Mtns; Glenorie (Fern Ck). **FL:** Aug–Sept.

Acacia hispidula
Rough Hairy Wattle
Shrub to 1m tall. Branchlets terete, rough, with short stiff hairs. Phyllodes asymmetric, elliptic, curved, to 22 x 5mm, rough; apex pointed, midrib prominent. Heads solitary in axils, pale

Acacia hispidula

cream to yellow; peduncles 5mm long. **HABITAT:** Dry eucalypt forest and woodland. **DIST:** Coast, lower Blue Mtns. Also NC, SC, NT, NWS, Qld. **FL:** Jan–May.

Acacia ixiophylla
Spreading shrub 1–4m tall. Branchlets flattened or angled, viscid, hairy. Phyllodes lanceolate, 1–3cm long, resinous, with several parallel veins. Heads 2–3 in racemes 2–6mm long, deep yellow. **HABITAT:** Open forest, woodland. **DIST:** Glen Davis, Capertee. Also NC, NT, WS, WP, Qld, WA. **FL:** Aug–Oct.

Acacia juncifolia

FLOWERING PLANTS

Acacia ptychoclada

Acacia stricta

Acacia juncifolia
Rush-leaved Wattle
Shrub to 2.5m tall with slender branchlets. Phyllodes ± terete to flat, 7–20cm x 1–3mm, midvein prominent. Heads 1–2 in axils, bright yellow; peduncles 4–12mm long. **HABITAT:** Open forest, rocky ridges. **DIST:** Uncommon. Coast and ranges. Also NC, WS, NWP, Qld. **FL:** Aug–Sept.

Acacia ptychoclada
Shrub to 2.5m tall with ribbed flattened or angled branchlets. Phyllodes ascending, subterete, 4–10cm x 1mm; nerves prominent; apex with recurved mucro. Heads 1–2 in axils, pale yellow; peduncles 3–4mm long, hairy. **HABITAT:** Swamps and drip ledges. **DIST:** Endemic. Blue Mtns (Woodford–Mt Victoria). **FL:** Sept–Apr.

Acacia stricta
Hop Wattle
Erect shrub or tree to 5m tall with smooth bark; branchlets angled, with yellow ribs. Phyllodes ascending, linear-oblanceolate or narrow-elliptic, obtuse or mucronate, 5–14cm x 4–10mm. Heads 2–4 in axils, cream to yellow; peduncles 2–6mm long, hairy or sticky. **HABITAT:** Open forest, esp. on clay and shale soils. **DIST:** Uncommon. Coast and ranges. **FL:** Aug–Sept.

Acacia undulifolia

Acacia undulifolia
Shrub about 2m tall with hairy branchlets. Phyllodes twisted on short stalk, broad-ovate, to 25 x 15mm, grey-green, glabrous or sparsely hairy, mid and lateral veins conspicuous, margins undulate, upwards-facing at about 60° from the branch. Heads single in axils of reduced upper phyllodes, pale yellow; peduncles to 10mm long. **HABITAT:** Open forest, woodland. **DIST:** Scattered. Blue Mtns (Megalong Valley–Bucketty). Also CWS. **FL:** Nov–Dec.

Previously part of *A. uncinata* complex.

Acacia verniciflua

Acacia amoena

Acacia verniciflua
Varnish Wattle
Shrub or tree to 5m tall; branchlets often angular, flexuose, resinous. Phyllodes variable, ± narrow-elliptic, straight, to 12cm x 12mm, with 2 prominent nerves and varnished appearance. Heads usually 2–3 per axil, golden yellow; peduncles 5-8mm long. **HABITAT:** Woodland. **DIST:** Upper Blue Mtns. Also ST, WS, SWP, Qld, Vic, Tas, SA. **FL:** Sept–Nov.

GROUP 5
Wattles with globular or (rarely) ovoid flower heads in racemes >2cm long; phyllodes under 7cm long.

Acacia amoena
Boomerang Wattle
Shrub to 2m tall with glabrous angular or flattened branchlets. Phyllodes spreading, oblanceolate, to 60 x 10mm, with 1–3 prominent glands on upper margin. Heads in racemes as long as phyllodes, pale to bright yellow. **HABITAT:** Open forest, often on rocky sites. **DIST:** Widespread, esp. sthn Blue Mtns. **FL:** Aug–Sept.

Acacia buxifolia
Box-leaved Wattle
Shrub to 3m tall with glabrous angular branchlets. Phyllodes elliptic, to 45 x 10mm, green to glaucous. Heads in racemes 1–4.5cm long, bright yellow, with 7–29 flowers. **HABITAT:** Open forest, woodland, heath. **DIST:** Blue Mtns (Blackheath–Capertee). Also NT, ST, WS, NWP, Qld, Vic. **FL:** July–Nov.

Acacia leucolobia (Katoomba to Glen Davis) is similar to *A. buxifolia*. It has pruinose branchlets, phyllodes 6–14mm wide and 5–10 flowers per head.

Acacia buxifolia

Acacia chalkeri
Chalker's Wattle
Shrub to 3m tall with glabrous angular branchlets. Phyllodes ascending, oblanceolate, to 55 x 8mm, with a short mucronate point; gland at base, not prominent. Heads in racemes 2–4cm long, golden yellow. **HABITAT:** Rocky woodland, on limestone. **DIST:** Endemic. Restricted to Wombeyan Caves area. **FL:** Nov–Jan.

Acacia chalkeri

Acacia clunies-rossiae

Acacia clunies-rossiae
Kowmung Wattle
Bushy shrub or tree 2–6m tall. New growth with grey-white hairs. Phyllodes narrowly elliptic-oblanceolate, to 50 x 6mm, 1-nerved, finely hairy; gland indented near base. Heads in racemes 2–5cm long, light golden yellow. **HABITAT:** Open forest, woodland, esp. on rocky sites. **DIST:** Endemic. Coxs R.–Kowmung R.–Yerranderie. **FL:** Aug–Sept.

Acacia dorothea
Dorothy's Wattle
Shrub, usually 1–2m tall; branchlets and phyllodes with small white hairs. Phyllodes oblanceolate, 4–7cm x 8–16mm, slightly curved; upper margin with prominent gland (rarely 2) below midway. Heads 3–8 in raceme 20–35mm long, ovoid, bright yellow. Pod oblong, to 6cm long, densely hairy. **HABITAT:** Open forest, woodland. **DIST:** Endemic. Upper Blue Mtns. Also Carrington Falls–Robertson. **FL:** Sept–Oct.

Acacia fimbriata
Fringed Wattle
Shrub or tree to 6m tall with flattened or angled finely hairy branchlets. Phyllodes oblong-elliptic, to 50 x 5mm, fringed with fine hairs. Heads in racemes 2–8cm long, small but showy, bright yellow. **HABITAT:** Moist shady eucalypt forest. **DIST:** Coast and ranges. Also NC, NT, ST, NWS, Qld. **FL:** Aug–Sept.

Acacia hamiltoniana
Hamilton's Wattle
Glabrous shrub to 3m tall with reddish ribbed branchlets. Phyllodes ascending, narrowly elliptic or oblanceolate, to 60 x 5mm slightly glaucous, 1-nerved; gland 1 or absent. Flower heads 2–7 in racemes to 4cm long, golden yellow, with <20 flowers per head. **HABITAT:** Woodland, heath. **DIST:** Upper Blue Mtns. Also Sassafras–Nerriga. **FL:** Aug–Sept.

Acacia linifolia
Flax Wattle
Slender open glabrous shrub 2–4m tall. Phyllodes spreading, linear, to 50 x 2mm, veins and gland obscure. Heads in racemes 2–5cm long, pale yellow, with 6–12 flowers per head. **HABITAT:** Open forest, woodland, heath. **DIST:** Common. Widespread, esp. on Hawkesbury sandstone. Also NC, SC. **FL:** Jan–May.

Acacia lunata
Lunate-leaved Wattle
Glabrous shrub 1–3m tall. Phyllodes asymmetric, narrow-elliptic, often with upper margin curved and lower straight, to 30 x 9mm, green to ± glaucous. Heads with 3–5 flowers, in racemes 2–4cm long, bright golden. **HABITAT:** Open forest. **DIST:** Castlereagh NR to nw. Sydney (e.g. Wheeny Ck). Also NC. **FL:** Aug–Sept.

Previously part of *A. buxifolia* complex.

Acacia dorothea

FLOWERING PLANTS

Acacia hamiltoniana

Acacia fimbriata

Acacia linifolia

Acacia lunata

Acacia meiantha

Acacia myrtifolia

Acacia meiantha
Shrub to 1.5m tall, forming clumps due to suckering. Branchlets with short stiff hairs. Phyllodes subterete, glabrous, 2–5cm long, about 1mm wide, apex obtuse with mucro. Heads in axillary raceme to 7cm long, 4–6 flowered; peduncles 2–4mm long, minutely hairy. **HABITAT:** Open forest, woodland. **DIST:** Endemic. Clarence. Also nth of Orange (CT). **FL:** July–Oct.

Acacia myrtifolia
Myrtle Wattle
Glabrous shrub to 3m tall, often with red, angular, ribbed branchlets. Phyllodes asymmetric, elliptic, to 6 x 2cm, gland below middle. Heads 3–8 in racemes to 5cm long, pale cream, with 2–8 flowers per head. **HABITAT:** Open forest, woodland, scrub. **DIST:** Coast and ranges. Also NC, SC, NT, ST, Qld, Vic, Tas, SA, WA. **FL:** May–Sept.

Acacia prominens
Gosford Wattle
Dense shrub or tree to 10m tall. Phyllodes narrow-lanceolate, 20–60 x 5–12mm, grey-green to glaucous, with 1 prominent gland on upper margin below middle. Heads in racemes 2–6cm long, bright lemon-yellow, small but abundant. **HABITAT:** Sheltered forest. **DIST:** Nth of Hawkesbury R. (e.g. Ourimbah); Oatley Park, Kogarah. Also NC, CWS. **FL:** July–Oct.

GROUP 6
Wattles with globular flower heads in racemes >2cm long; phyllodes >7cm long.

Acacia binervata
Two-veined Hickory
Tree to 12m tall with densely hairy angled or flattened branchlets. Phyllodes straight or curved, to 13 x 3cm, distinctly 2-veined. Heads in racemes 4–8cm long, pale yellow. **HABITAT:** Sheltered sites, rainforest margins. **DIST:** Illawarra, Sthn H'lands. Also NC, SC, NT, Qld. **FL:** Sept–Oct.

Acacia falcata
Sickle Wattle
Glabrous tree or shrub 3–5m tall with red-brown angled or flattened branchlets. Phyllodes falcate, 10–17 x 1–4cm, blue-grey to light green, with main vein off-centre. Racemes 2–6cm long; heads small, cream. **HABITAT:** Open forest, on clay-shale soils. **DIST:** Coast and ranges. Also NC, SC, NT, CWS, Qld. **FL:** Apr–July.

FLOWERING PLANTS

Acacia prominens

Acacia binervata

Acacia falcata

Acacia falciformis

Acacia falciformis
Broad-leaved Hickory
Shrub or tree 2–10m tall with grey-black bark; branchlets angled. Phyllodes ± pendulous, falcate, 10–20 x 1.5–3cm, grey-green, obtuse; midvein and lateral veins conspicuous; gland at base indented, connected to midvein by a nerve. Heads 10–18 on yellow-hairy peduncles in racemes 2–10cm long, pale yellow. Pods glabrous, straight, flat, 5–13 x 15–25mm.
HABITAT: Forest, woodland. **DIST:** Blue Mtns, Sthn H'lands, Bents Basin. Most of NSW. Also Qld, Vic. **FL:** Oct–Jan.

Acacia implexa
Hickory
Tree 5–12m tall with rough greyish bark; branchlets terete. Phyllodes narrow-elliptic, falcate, to 18 x 2.5cm, with 3–5 prominent nerves and numerous minor nerves; apex acute.

Acacia implexa

Heads 4–8 in glabrous racemes to 5cm long, cream to pale yellow. Pods coiled or twisted; funicle pale, folded under seed. **HABITAT:** Open forest, woodland. **DIST:** Coast and ranges. Most of NSW. Also Qld, Vic. **FL:** Jan–Mar.

NATIVE PLANTS OF THE SYDNEY REGION

Acacia penninervis

Acacia melanoxylon

Acacia obliquinervia

Acacia melanoxylon
Blackwood
Tree 8–30m tall with furrowed dark grey bark; branchlets angular or flattened. Phyllodes lanceolate to oblanceolate, straight to falcate, to 15 x 2.5cm, with 3–5 prominent longitudinal veins. Heads 3–5 in racemes to 4cm long, cream. Pods coiled or twisted; funicle red, folded around seed. **HABITAT:** Moist sheltered forest on rich soils. **DIST:** Upper Blue Mtns, Sthn H'lands. Also NC, SC, NT, WS, Qld, Vic, Tas, SA. **FL:** Aug–Nov.

Acacia obliquinervia
Shrub or tree to 15m tall with dark brown bark; branchlets angled or flattened. Phyllodes oblanceolate, falcate to straight, to 16 x 4cm, apex obtuse; midvein off-centre. Heads in racemes 2–10cm long with flexuose axes, bright yellow. Pods oblong, to 15 x 2.5cm, thin-walled; seeds transverse. **HABITAT:** Moist forest, on granite and slate. **DIST:** Upper Blue Mtns (e.g. Mt Werong, Kanangra Walls). Also ST, CWS, Vic. **FL:** Sept–Oct.

Acacia penninervis
Mountain Hickory
Shrub or tree 3–15m tall with dark grey fissured bark; branchlets terete, glabrous. Phyllodes ± oblanceolate, ± straight, to 15 x 3cm, apex acute to obtuse; lateral veins conspicuous; glands 1–2. Heads in glabrous racemes 3–12cm long, pale yellow to almost white. Pods straight, to 20 x 1.5cm, glabrous. **HABITAT:** Forest, woodland. **DIST:** Chiefly west of Blue Mtns; rare on coast (e.g. Mooney Mooney). Also NC, SC, NT, ST, WS, NWP, Qld, Vic. **FL:** Oct–Feb.

Acacia rubida
Red-stemmed Wattle
Glabrous shrub to 3m, rarely a taller tree, often with bipinnate and transitional leaves on lower branches; upper branches reddish. Phyllodes

FLOWERING PLANTS 163

Acacia rubida

Acacia saliciformis

variable, narrow-elliptic to oblanceolate, straight or curved, to 20 x 2.5cm, finely pointed, 1-nerved, with obscure lateral veins; gland prominent. Heads in racemes to 7cm long, pale to deep yellow. **HABITAT:** Open forest, riparian scrub. **DIST:** Widespread, esp. Blue Mtns, Sthn H'lands, Little R. Also SC, NT, ST, WS, Qld, Vic. **FL:** July–Sept.

Acacia saliciformis
Glabrous shrub or tree 3–6m tall with smooth grey bark and pendulous upper branches. Phyllodes narrow-lanceolate, to 12 x 1.5cm, grey-green, pointed, with obscure lateral veins; gland prominent, sometimes connected to midrib by a nerve. Flower heads in short racemes which are often arranged in panicles 4–9cm long, pale yellow. **HABITAT:** Open forest. **DIST:** NW Hornsby Plat. (e.g. Putty–Colo). Also NC. **FL:** Mar–Aug.

Acacia suaveolens
Sweet-scented Wattle
Open, sparsely branched glabrous shrub to 3m tall with stout angled or flat branchlets. New growth pinkish. Phyllodes ascending, ± linear, to 150 x 10mm, acutely pointed, mucronate, midrib prominent. Unopened racemes enclosed by deciduous bracts; heads pale cream, 3–10 flowered. **HABITAT:** Open forest, woodland, heath. **DIST:** Common. Coast and ranges. Also NC, SC, ST, Qld, Vic, Tas, SA. **FL:** Mar–Aug.

Acacia suaveolens

Pararchidendron
Genus with one species. Distinguished from *Acacia* by its united stamens.

Pararchidendron pruinosum var. *pruinosum*
Snow Wood
Shrub or tree to 15m tall. Leaves to 40cm long, bipinnate, with usually 2–3 pairs of pinnae bearing 5–11 oblanceolate pinnules to 8 x 2.5cm. Flower heads paired in axils on peduncles 3–9cm long, globular, 1–3cm diam., greenish white, yellowish with age. Pods twisted, splitting to reveal orange-red inner surface; seeds shiny, black. **HABITAT:** Coastal rainforest. **DIST:** Rare on coast (e.g. Jamberoo, Bass Point, Hardys Bay). Also NC, CWS, Qld. **FL:** Oct–Dec.

Pararchidendron pruinosum var. *pruinosum*

CAESALPINIACEAE

Senna (previously *Cassia*)
Leaves alternate, usually pinnate with a terminal leaflet. Flowers in axillary racemes, with 5 sepals, 5 petals, up to 10 fertile stamens (with anthers) and a large curved stigma which develops into a pod.

Senna aciphylla
Australian Senna
Shrub to 1.5m tall. Leaves 3–5cm long with 8–12 pairs of leaflets; leaflets linear-elliptic, 20–25 x 1–4mm, revolute, with acute tips. Flowers with all 10 stamens fertile. Pod slightly flattened, 7cm x 6mm. **HABITAT:** Wooded hillsides. **DIST:** Uncommon near coast; chiefly w. Blue Mtns (e.g. Glen Davis), Sthn H'lands (e.g. Wombeyan Caves). Also NC, NT, ST, WS, Qld, Vic. **FL:** Oct–Dec.

Senna odorata
Southern Cassia, Smooth Senna
Shrub to 2m tall. Leaves 5–15cm long with 8–13 pairs of leaflets; leaflets lanceolate-elliptic, 10–25 x 5–10mm, pale beneath, flat or recurved. Flowers with all 10 stamens fertile. Pod slightly flattened, 10cm x 6mm. **HABITAT:** Sheltered forested slopes. **DIST:** Coast and ranges (e.g. Douglas Park, Milperra, Nattai, Bullio Tunnel). Also NC, NWS, Qld. **FL:** Sept–Dec.

Senna odorata

Senna aciphylla

FABACEAE
PEA FLOWERS
KEY TO COMMON SYDNEY GENERA OF FABACEAE

1 Flower colour
a) Flowers not orange or yellow.
  i) **climbing or prostrate plants**
  Kennedia — flowers large, red to burgundy.
  Hardenbergia — flowers violet.
  Glycine — flowers pale blue to mauve.
  Desmodium — flowers mauve-pink; pod beaded (constricted between seeds).

  ii) **shrubs**
  Mirbelia — leaves simple with strong reticulate venation.
  Hovea — leaves linear to oblong, without strongly marked veins.
  Indigofera — leaves pinnate with numerous leaflets.
  Gompholobium — leaves digitate with 3–5 leaflets.

b) Flowers orange to yellow, often tinged red or red-centred .... **go to 2**

2 Stamens
a) All 10 stamens joined for most of their length.
  Bossiaea — flowers small, or leaves small or undeveloped, or stems flattened.
  Platylobium — large flowers and leaves prominently veined, >2cm, stems round.
b) All 10 stamens free .... **go to 3**

3 Ovules
(Select an old faded flower, remove petals and stamens, hold the immature pod to the light. The ovules—developing seeds—can be seen through the pod).
a) Ovules 4 or more.
  Oxylobium — leaves simple. Stem hairs simple. Stipules bristly or absent. Flowers orange or yellow with red.
  Podolobium — leaves simple. Stem hairs pubescent, 2-branched. Stipules rigid or spreading. Flowers yellow with red.
  Gompholobium — leaves digitate, with 3 or 5 leaflets, or pinnate. Flowers yellow.
  Mirbelia — leaves simple; pod divided by a partition into 2 cells. Flowers yellow or yellow and red.
b) Ovules 2 .... **go to 4**

4 Leaves
a) Leaves undeveloped.
  Jacksonia — large shrub. Branchlets grey, angular. Flowers orange.
  Viminaria — large shrub. Branchlets green, round. Flowers yellow.
  Sphaerolobium — small shrub less than 30cm high. Pods round.
b) Leaves developed normally .... **go to 5**

5 Seed pods
a) Pods inflated, triangular.
  Daviesia — flowers many, small, red-centred. Leaves variable, usually with raised veins, sometimes pungent-pointed.
b) Pods not triangular .... **go to 6**

6 Bracteoles
a) Bracteoles present on or below calyx.
  Pultenaea — bracteoles 1–3mm long, brown, scale-like, at base of calyx or attached to it. Standard 1-lobed, narrow or rounded and notched at top.
  Phyllota — bracteoles large, green, leafy, enclosing or exceeding the calyx. Standard more pointed than above, 1-lobed.
  Dillwynia — bracteoles small, deciduous, on stem below calyx. Standard broad, 2-lobed.
  Almaleea — bracteoles on stem below calyx. Standard longer than broad.
b) Bracteoles absent.
  Aotus — calyx hairy. Standard 1-lobed.
  Dillwynia — calyx not hairy. Standard 2-lobed.

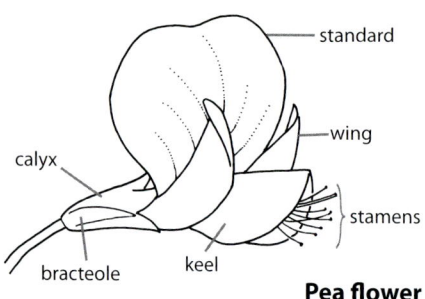

Pea flower

Almaleea

Shrubs. Inflorescence a terminal head. Differs from *Pultenaea* in having bracteoles attached to flower stalk and from *Dillwynia* in having a standard longer than broad.

Almaleea incurvata
Shrub to 1m with villous stems. Leaves distinctive, narrow-ovate, to 10 x 2mm, margins incurved. Heads subtended by hairy bracts 4–6mm long; calyx 5mm long, with lobes villous; standard petal 9–11mm long. **HABITAT:** Wet heaths. **DIST:** Endemic in Blue Mtns (Kings T'land–Clarence). **FL:** Sept–Oct.
Previously *Pultenaea incurvata*.

Aotus ericoides

Aotus

Shrubs. Leaves recurved or revolute. Flowers lacking bracteoles, distinguishing it from *Pultenaea*, *Phyllota* and *Dillwynia*, which all have bracteoles.

Aotus ericoides
Common Aotus
Shrub to 2m tall. Stems with greyish or rusty hairs. Leaves whorled, opposite or alternate, linear to narrow-lanceolate, 6–12 x 1–4mm, dark green, margins recurved. Flowers axillary, abundant along stems; standard orbicular, style not hooked, calyx hairy. Ovary with long white hairs. **HABITAT:** Heath, old dunes, scrubland, open forest. **DIST:** Coast and Sthn H'lands. Also NC, SC, NT, ST, Qld, Vic, Tas. **FL:** Aug–Nov.
   A low compact form occurs on sands at La Perouse and Kurnell.

Almaleea incurvata

Bossiaea buxifolia

Bossiaea ensata

Bossiaea
Shrubs. Stems often flattened. Leaves alternate or opposite, simple, or 1-foliolate or scale-like. Flowers yellow with red centres; stamens united into a tube. Fruit a flattened pod. Twelve species in Sydney region.

Bossiaea buxifolia
Matted Bossiaea
Decumbent shrub to 60cm tall. Stems sprinkled with soft, short white hairs. Leaves alternate, closely spaced, shortly stalked, elliptic to rounded, 2–5mm long, darker above; base asymmetric, stipules exceeding leaf stalk. Flowers single on stalks to 10mm long with a bracteole pair on upper half of stalk. **HABITAT:** Open forest, woodland, heath. **DIST:** Rare on coast, common on ranges (e.g. Lithgow, Jenolan Caves, Wingello). Also NC, SC, NT, ST, WS, Qld, Vic. **FL:** Sept–Nov.

Bossiaea ensata
Straggling shrub with weak flattened leafless stems, 30–50cm long and 7–8mm wide. Leaves reduced to scales 1–2mm long. Flowers solitary along stems; standard 7–10mm wide; stalk 2–5mm long. Pod with short distinct stalk. **HABITAT:** Sandy heath, shrubland, often in deep sands. **DIST:** Common in sandy coastal areas; less common in Blue Mtns. Also SC, NC, NT, Qld. **FL:** Aug–Nov.

Daviesia alata is similar but has flowers in clusters.

Bossiaea heterophylla

Bossiaea heterophylla
Variable shrub with flattened stems to 1m tall. Leaves alternate, in opposite rows along stems, 1-foliolate; leaflet variable, obovate to narrow-linear, 1–3cm long. Flowers solitary in axils, shortly stalked, 7–15mm long. **HABITAT:** Open forest, woodland, old dunes. **DIST:** Coast and ranges. Also NC, SC, Qld. **FL:** Apr–July.

Bossiaea kiamensis

Bossiaea kiamensis
Bushy shrub, 1–2m tall. Younger stems flattened and covered with soft white down. Leaves opposite, 1-foliolate; leaflet ovate to elliptic, 10–20mm long, shortly stalked, glabrous above, silky-hairy beneath; apex acute or obtuse, with short point. Flowers solitary on stalks less than 5mm long. **HABITAT:** Heath, shrubland. **DIST:** Jamberoo to Penrose (e.g. Belmore Falls, Carrington Falls, Barren Grounds). Also SC, ST. **FL:** Sept–Oct.

Bossiaea lenticularis
Glabrous, diffuse shrub to 1m tall with numerous tough slender stems. Leaves opposite, well-spaced on branchlets, sessile, round, 3–7mm across. Flowers shortly stalked, larger than leaves. **HABITAT:** Moist open forest. **DIST:** Scattered away from coast (e.g. Mt Tomah, Kings T'land, Maroota, Thirlmere Lakes, Colo Heights). **FL:** Aug–Oct.

Bossiaea neo-anglica
Small, sprawling or prostrate shrub with softly hairy young stems. Leaves alternate, shortly stalked, ovate, 2–7mm long, distinctly darker above; base asymmetric and cordate; apex with recurved soft point; stipules exceeding leaf stalk. Flowers axillary, single on stalks to 5mm long, with bracteole pair near mid-point. **HABITAT:** Open forest, in sheltered sites. **DIST:** Sthn T'lands (e.g. Hill Top–Oakdale, Fitzroy Falls, West Berrima). Also NC, NT, CT, NWS, Qld. **FL:** Sept–Jan.

Bossiaea obcordata
Spiny Bossiaea
Small, rigid, twiggy shrub to 1.5m high, with spiny branchlets. Leaves alternate, 1-foliolate; leaflet with broad notched or lobed tip, 2–6 x 1–3mm. Flowers axillary. **HABITAT:** Open forest, shrubland. **DIST:** Coast and ranges. Also NC, SC, NT, ST, NWS, Vic, Qld, Tas. **FL:** Aug–Oct.

Bossiaea prostrata
Small procumbent undershrub. Leaves distant, alternate, 1-foliolate with minute angled joint at base of leaflet; leaflet narrowly ovate to circular, 3–15mm long. Flowers 1–2 together on stalks to 15mm long. **HABITAT:** Heath, open forest, often on clay-shale soils. **DIST:** Coast and ranges (e.g. Menai, Munmorah SCA, Mt Gibraltar). Also NC, SC, ST, WS, Vic, Tas, SA. **FL:** Aug–Oct.

Forms from western Sydney (Marayong–Scheyville) resemble *B. buxifolia*.

Bossiaea rhombifolia
Glabrous shrub to 1m tall, with slightly flattened stems which are not spine-tipped; branchlets short, alternate, in a plane on opposite sides of stems. Leaves rhomboid, about 5–8 x 5–8mm; apex acute. Flowers single on short stalks. **HABITAT:** Open forest. **DIST:** Coast and ranges. Also NC, NT, NWS, Qld. **FL:** Aug–Oct.

Bossiaea lenticularis

Bossiaea obcordata

Bossiaea neo-anglica

Bossiaea prostrata

Bossiaea rhombifolia

Bossiaea scolopendria

Bossiaea stephensonii

Bossiaea scolopendria
Sparsely branched shrub with flat stems about 1m x 15mm. Leaves reduced to scales 1–2mm long. Flowers 10–15mm long, solitary, on stalks 1–3mm long. **HABITAT:** Open forest, woodland, heath. **DIST:** Coast (Jervis Bay–Somersby). Most common nth of Sydney. **FL:** Aug–Oct.

Bossiaea stephensonii
Erect shrub to 1m tall with flattened hairy stems. Leaves alternate, subtended by leaf-like triangular stipules 5–8mm long; 1-foliolate; leaflet 5–20mm long, softly hairy. Flowers single along stems. **HABITAT:** Open forest. **DIST:** Coast. Common in Royal NP and nth of Hawkesbury R. (e.g. Patonga, Tuggerah). Also lower NC. **FL:** Aug–Sept.

Chorizema
Shrubs. Leaves alternate. Flowers in terminal racemes; stamens free. Ovary pubescent, with 8 or more ovules. Fruit an inflated pod.

Chorizema parviflorum
Eastern Flame Pea
Shrub, 30–40cm tall with slender, ascending, angular stems. Leaves alternate, linear to obovate, 10–30mm long, with conspicuous midrib and downy undersurface. Flowers in loose clusters. Pod 5–8mm long, almost stalkless. **HABITAT:** Heath, open forest, grassland, on clay soils and coastal headlands. **DIST:** Uncommon. Cumberland Pl. (e.g. Castlereagh NR, Narellan), Lansdowne, Pleasure Point. From Albion Park to Qld. **FL:** June–Dec.

Chorizema parviflorum

Daviesia alata

Daviesia
BITTER-PEA
Shrubs, with alternate leaves, often tough, rigid and pungent-pointed, sometimes reduced to scales. Flowers yellow with red centres, with 10 free stamens. Pods triangular.

Daviesia acicularis
Shrub to 1m tall. Leaves pungent-pointed, narrow-lanceolate to linear, 2–3cm long, erect; margins very recurved. Branchlets not spine-tipped. Flowers 7mm long, usually single, on short stalks. **HABITAT:** Open forest. **DIST:** Scattered on coast and ranges (e.g. Agnes Banks NR, Sth Maroota, Mt Wilson). Also SC, NC, NT, WS, Qld. **FL:** Aug–Oct.

Daviesia alata
Low straggly shrub. Stems flat with raised wing. Leaves reduced to minute scales. Flowers clustered in small racemes, yellow or orange with red centres. Pods triangular, 9 x 7mm. **HABITAT:** Open forest, heath. **DIST:** Coast and ranges. Also lower NC, SC (near Nowra). **FL:** Aug–Nov.

This species resembles *Bossiaea ensata*, which has single flowers and oblong pods 3–4cm long.

Daviesia acicularis

Daviesia corymbosa

Daviesia corymbosa
Shrub to 2m tall. Leaves variable, usually lanceolate, 6–8cm long, with a network of fine veins. Flowers in corymbose racemes*. **HABITAT:** Open forest. **DIST:** Coast and ranges (e.g. Colo Heights, Lawson, Menai, Buxton). Also Nowra–Sassafras district. **FL:** Aug–Nov.

* The corymbose inflorescence distinguishes this species from others such as *D. latifolia*.

Daviesia latifolia
Broad-leaf Bitter-pea
Bushy shrub to 2m tall. Leaves broad, 8 x 5cm, strongly veined. Flowers numerous in dense racemes 3–6cm long. **HABITAT:** Open forest. **DIST:** Upper Blue Mtns, Sthn H'lands (e.g. Clarence, Narrow Neck, Bundanoon). Also NT, ST, WS, SC, Vic, Tas. **FL:** Sept–Nov.

Daviesia leptophylla
Slender Bitter-pea
Erect twiggy shrub to 1.5m tall. Leaves well-spaced on stems, linear, acute, 3–9cm long, rigid, wrinkled with conspicuous veins, the upper leaves smaller and more distant. Flowers abundant on slender stalks in small clusters in upper leaf axils. **HABITAT:** Open forest, often on rocks. **DIST:** Upper Blue Mtns (e.g. Wolgan Gap, Jenolan Caves), Sthn H'lands (e.g. Tallong, Wingello). Also NT, ST, CWS, Vic, SA. **FL:** Oct–Nov.

Daviesia mimosoides
Shrub to 2m tall with attractive drooping branchlets. Leaves to 8cm long, flat, narrow, without a strong network of veins; young leaves

Daviesia latifolia

Daviesia squarrosa

soft, tan-coloured, drooping. Flowers small, numerous, in racemes much shorter than leaves. **HABITAT:** Forest, on deep sands. **DIST:** Common on old dunes at Kurnell; also Yerranderie, Hill Top, Mittagong, Hunter Valley, SC, ST, WS, Qld, Vic. **FL:** July–Nov.

Daviesia squarrosa
Undershrub or bush to 1.5m tall with softly hairy stems. Leaves 6–10mm long, heart-shaped, sessile, with fine pungent point. Flowers 6mm long on stalks slightly longer than leaves. **HABITAT:** Open forest on heavier soils. **DIST:** Western Sydney and ranges (e.g. Castlereagh NR, Wedderburn, Glenbrook, Howes V.). Also NC, SC, ST. **FL:** July–Nov.

Daviesia mimosoides

Daviesia ulicifolia
Gorse Bitter-pea
Erect shrub to 1m tall with short spiny branchlets. Leaves sessile, ovate to narrow-lanceolate, 5–15 x 1–5mm, rigid, sharply pointed. Flowers on short stalks in leaf axils, singly or in groups of 2–4. **HABITAT:** Open forest, on sands or shale cappings. **DIST:** Coast and ranges. Most of NSW. Also other States (except N'Terr). **FL:** Aug–Nov.

Daviesia leptophylla

Daviesia ulicifolia

Desmodium
TICK-TREFOIL
Trailing shrubs with 3-foliolate leaves. Flowers with 9 stamens united and 1 free. Fruit a beaded pod, distinctly contracted between seeds.

Desmodium rhytidophyllum
Rusty Tick-trefoil
Prostrate trailing plant. Stems densely rusty-hairy. Leaflets elliptic, to 60 x 35mm, softly hairy on both surfaces. Flowers pink to purple in clusters at ends of branches. **HABITAT:** Open forest, on clay and fertile soils. **DIST:** Coast, lower Blue Mtns. Also NC, SC, Qld, PNG. **FL:** Dec–Mar.

Desmodium brachypodium also occurs on clay soils. Its stems have short hooked hairs and the leaflets are glabrous on the upper surface.

Desmodium varians
Slender Tick-trefoil
Prostrate trailing plant. Leaflets variable, the lower pair usually shorter (to 2cm long) and narrower than terminal leaflet (to 4cm long). Flowers pink, in small clusters on slender stalks. **HABITAT:** Grassland, woodland, open forest. **DIST:** Coast and ranges. Most of NSW. Also Qld, Vic, Tas. **FL:** Dec–Mar.

Dillwynia
EGGS AND BACON, PARROT-PEA
Shrubs, with linear or terete leaves grooved along upper surface. Flowers subtended by 2 small bracteoles on stalk well below calyx, yellow and red; standard broader than long; stamens 10, all free. Style hooked near tip. Pod ovoid, inflated.

Dillwynia acicularis
Erect shrub 1–2m high. Leaves erect or closely appressed to branches, terete, to 3cm long, sharply pointed but not pungent. Flowers in clusters at ends of branches. Calyx U-shaped. Pod 5–6mm long, usually with persistent remains of petals. **HABITAT:** Open forest. **DIST:** Hornsby Plat. (e.g. Sackville, Maroota), West Wollemi, Bargo–Avon Dam area. Also CT, CWS. **FL:** Sept–Oct.

Desmodium rhytidophyllum

Dillwynia acicularis

Desmodium varians

Dillwynia brunioides

Dillwynia brunioides
Erect shrub to 1m tall. Leaves 5–10mm long, narrow, thick, triangular in section, rough, warty. Flowers in terminal heads; buds and calyces densely covered with white hairs; ovary and fruit hairy. **HABITAT:** Heath, scrubland. **DIST:** Upper Blue Mtns (e.g. Kings T'land, Narrow Neck, Blackheath); also Barren Grounds, Budderoo NP and sth to Jervis Bay. **FL:** Oct–Nov.
   Previously *Pultenaea brunioides*.

Dillwynia elegans
Erect shrub to 1.5m tall. Young stems hairy. Leaves crowded, terete, 5–20mm long, twisted, smooth or slightly warty. Flowers paired in small clusters, often extending down the branches. **HABITAT:** Woodland, dry heath. **DIST:** Hornsby Plat., Blue Mtns to Rylstone. **FL:** Aug–Oct.
   This species was formerly called *D. floribunda* var. *teretifolia*.

Dillwynia elegans

Dillwynia floribunda

Dillwynia glaberrima

Dillwynia parvifolia

Dillwynia floribunda
Erect shrub to 1.5m tall with hairy stems. Leaves crowded, 5–15mm long, narrow, twisted, slightly rough. Flowers paired in dense terminal inflorescences 10–25cm long. **HABITAT:** Wet heaths. **DIST:** Common. Ulladulla to NC. **FL:** July–Oct.

Dillwynia glaberrima
Smooth Parrot-pea
Shrub to 80cm tall. Branches glabrous. Leaves 5–20mm long, linear, not twisted. Flowers clustered at ends of branches. **HABITAT:** Open forest, heath, on deep sand and alluvium. **DIST:** Coast, Cumberland Pl. (e.g. Mellong, Agnes Banks). Also NC, SC, Qld, Vic, Tas, SA. **FL:** Sept–Nov.

Prostrate plants occur on the seafront at La Perouse and Kurnell with at least 2 forms at the latter location: one with hairy stems, warty-hairy leaves and paired flowers on short pedicels, the other with glabrous stems, slender glabrous leaves and flowers in umbels of 1–6 on pedicels to 2cm long.

Dillwynia parvifolia
Diffuse shrub to 80cm tall. Leaves 2–3mm long, conspicuously twisted. Flowers terminal, in umbels of 1–6 on stalks to 4mm long. **HABITAT:** Woodland, on heavier clays. **DIST:** Cumberland Pl. (e.g. Menai, Salt Pan Ck, Rookwood, Villawood). Also SC. Endangered. **FL:** Sept–Oct.

Dillwynia phylicoides
Variable shrub to 1.5m tall. Branches, leaves and calyces covered with short, rigid hairs. Leaves 3–8mm long, conspicuously twisted. Flowers sessile in clusters near ends of

Dillwynia phylicoides

branches. **HABITAT:** Open forest, woodland. **DIST:** Upper Blue Mtns (e.g. Hartley, Newnes Plat., Mittagong). Also NC, NT, ST, WS, Qld, Vic. **FL:** Sept–Nov.

Dillwynia ramosissima
Much-branched shrub to 1.5m tall with short spiny side-branches. Stems glabrous. Leaves 3–5mm long, linear, not twisted. Flowers single, terminal on side-branches. **HABITAT:** Woodland, scrub, swampy areas. **DIST:** Sthn H'lands (e.g. Mittagong, Hill Top, Penrose). Also NC, SC, ST, Vic. **FL:** Aug–Nov.

Dillwynia ramosissima

Dillwynia retorta

Dillwynia rudis

Dillwynia retorta
Eggs and Bacon
Variable shrub, commonly an erect shrub to 2m tall. Leaves crowded, narrow, glabrous, 5–10mm long, conspicuously twisted. Flowers in terminal clusters of up to 9 on stalks 15–20mm long. **HABITAT:** Heath, shrubland, open forest. **DIST:** Coast and ranges. Also SC, NC, NWS, Qld. **FL:** June–Nov.

A prostrate form with leaves 3–4mm long and flower stalks 6mm long is regarded as a subspecies (unnamed). A form with sessile or shortly stalked flowers is now named *D. ericifolia*.

Dillwynia rudis
Erect shrub to 1.5m tall. Stem with dense appressed hairs. Leaves 10–18mm long, linear, not twisted, strongly warty, usually glabrous. Flowers paired and sessile in axils; keel red-brown; standard orange-red. **HABITAT:** Heath, woodland. **DIST:** Coast and ranges. Also SC, ST, CWS. **FL:** Aug–Oct.

Similar to *D. sericea* but with longer and more warty leaves.

FLOWERING PLANTS

Dillwynia sericea

Dillwynia sieberi

Dillwynia tenuifolia

Dillwynia sieberi
Erect shrub to 1.5m tall with softly hairy stems. Leaves to 15mm long, linear, rigid, pungent-pointed. Flowers in small terminal clusters. Calyx U-shaped, softly hairy. **HABITAT:** Open forest. **DIST:** Western Sydney and Sthn H'lands (e.g. Wallacia, Casula, Wingello SF). Also SC, NT, ST, WS, Qld. **FL:** Sept–Nov.

Dillwynia tenuifolia
Shrub to 80cm tall. Branchlets minutely hairy. Leaves clustered, 4–10mm long, linear, with small recurved tip. Flowers solitary on short stalks in upper axils. **HABITAT:** Open forest, on alluvium. **DIST:** Localised. Windsor to Penrith (e.g. Agnes Banks, Castlereagh NR). Also Woodford, Howes Valley, NC, ST. **FL:** June–Oct.

Glycine
Weak twining plants with pale blue-mauve flowers. Leaves divided into 3 leaflets.

Four variable species are recognised in the Sydney district. Some forms may be reclassified as new species.

Glycine clandestina
Twining Glycine
Slender trailing plant. Leaflets on short stalks of equal length, 10–80 x 2–10mm, oblong to lanceolate; secondary veins at 90° to midrib. **HABITAT:** Forest, woodland, swampy areas. **DIST:** Coast and ranges. Most of NSW. Also Qld, Vic, Tas, SA. **FL:** All year.

Dillwynia sericea
Showy Parrot-pea
Erect shrub to 1m tall. Branches with dense spreading or appressed hairs esp. on young branches. Leaves 5–10mm long, linear, grooved on upper surface, not twisted, smooth to sparsely warty, glabrous or hairy. Flowers paired in upper axils, sessile. **HABITAT:** Open forest, woodland. **DIST:** Hornsby Plat., Blue Mtns, Sthn H'lands. Also SC, NT, ST, WS, NWP, Qld, Vic, Tas, SA. **FL:** Aug–Dec.

Glycine clandestina

Glycine tabacina

Gompholobium glabratum

Glycine tabacina

Slender trailing plant similar to *G. clandestina* but with terminal leaflet on longer stalk than other leaflets and the secondary veins at an acute angle to midrib. **HABITAT:** Forest, woodland, on clay, shale, alluvium. **DIST:** Coast and ranges. Most of NSW. Also Qld, Vic, WA, NTerr, Asia, Pac. Iss. **FL:** Sept–Mar.

Glycine microphylla has leaflets with secondary veins at right angle to midrib and central leaflet on longer stalk.

Gompholobium
WEDGE-PEA

Shrubs with digitate or pinnate leaves. Flowers with 10 free stamens. Pod globular, inflated, not divided by longitudinal partition.

Gompholobium glabratum
Dainty Wedge-pea

Shrub to 30cm tall. Stems with small warts. Leaflets 4–7 (usually 5), linear, 4–10 x 1mm. Flowers to 10mm long in short terminal heads. **HABITAT:** Heath, open forest. **DIST:** Common. Coast and ranges. Also NC, SC, ST, CWS, Vic. **FL:** Aug–Nov.

Gompholobium grandiflorum

Gompholobium grandiflorum
Large Wedge-pea

Erect glabrous shrub to 1m tall. Leaves 3-foliolate; leaflets narrow-linear with recurved margins, 20–25 x 1mm; apex acute. Flowers to 25mm long; keel without white hairs. **HABITAT:** Heath, open forest. **DIST:** Coast and ranges. Also SC (Nowra). **FL:** Aug–Nov.

Gompholobium huegelii
Pale Wedge-pea

Diffuse shrub to 60cm tall. Leaves 3-foliolate; leaflets 8–20 x 1–1.5mm, margins recurved; apex acute. Flowers to 20mm long. Keel with conspicuous white fringe; calyx dark green. **HABITAT:** Woodland, open forests. **DIST:** Upper Blue Mtns (e.g. Mt Victoria, Hartley, Newnes SF). Also SC, NT, ST, WS, Vic, Tas. **FL:** Sept–Nov.

FLOWERING PLANTS

Gompholobium huegelii

Gompholobium latifolium

Gompholobium minus

Gompholobium uncinatum

Gompholobium pinnatum

Gompholobium latifolium
Broad-leaf Wedge-pea
Erect glabrous shrub to 2m tall. Leaves 3-foliolate; leaflets 30–50 x 3–5mm, flat or with recurved margins. Flowers on short stalks in upper leaf axils, to 30mm long, yellow; keel with conspicuous white fringe. **HABITAT:** Sheltered woodland, forests. **DIST:** Coast and ranges. Also NC, SC, NT, Qld, Vic. **FL:** Aug–Nov.

Gompholobium minus
Broad-leaf Wedge-pea
Low shrub, 10–30cm tall. Stems with short hairs. Leaves 3-foliolate, leaflets linear, 6–9 x <1mm. Flowers 1 or few, on stalks 5–15mm long in uppermost axils; calyx margins distinctly ridged in bud. **HABITAT:** Heath, open forest. **DIST:** Coast and ranges (e.g. Castlereagh NR, Wingello, Agnes Banks NR). Also NC, SC, ST, Vic. **FL:** Sept–Dec.

Gompholobium pinnatum
Pinnate Wedge-pea
Shrub to 30cm tall with weak, slender, smooth stems. Leaves pinnate; leaflets 20–30, linear, to 12 x 1mm. Flowers to 10mm long on short stalks near ends of branches. **HABITAT:** Heath, open forest. **DIST:** Uncommon. Coast (e.g. Agnes Banks NR, Castlereagh NR, Lane Cove). Also SC, NC, Qld. **FL:** Aug–Dec.

Gompholobium uncinatum
Red Wedge-pea
Bushy or diffuse shrub to 30cm tall. Leaves 3-foliolate; leaflets linear, 5–10 x 1mm, dark green. Flowers on stalks about 5mm long, orange-red, yellow at base. **HABITAT:** Heath, open forests, woodland. **DIST:** Upper Blue Mtns (e.g. Mt Victoria, Newnes SF). Also NC, NT, CWS, Qld. **FL:** Nov–Dec.

FLOWERING PLANTS

Goodia lotifolia

Goodia
Tall shrubs. Leaves alternate, with 3 rounded leaflets. Flowers numerous in terminal or leaf-opposed racemes; calyx 2-lipped, upper 2 teeth united into broad lip, lower 3 teeth narrow; corolla yellow with dark centre; stamens joined to form tube open on upper side. Pod with long stipe, flat.

Goodia lotifolia
Golden-tip
Shrub to 3m tall. Leaflets obovate, 10–30mm long. Flowers in loose racemes to 10cm long. Pod 2–3cm long.
HABITAT: Moist forest, sheltered slopes.
DIST: Coast, Sthn T'lands (e.g. Patonga, Kiama, Thirlmere, Fitzroy Falls). Also NC, SC, NT, ST, Qld, Vic, Tas, SA. **FL:** Sept–Nov.

Hardenbergia
Twining plants or small subshrubs with purple (rarely white) flowers. Leaves alternate, 1-foliolate. Flowers numerous in axillary racemes. Pod oblong, flat or cylindrical.

Hardenbergia violacea

Hardenbergia violacea
Purple Twining-pea
Common twiner, with stems twisting over shrubs, rocks and roadside banks. Leaflet ovate-lanceolate, strongly veined, 4–9 x 2–5cm. Corolla 8mm long. **HABITAT:** Forest, woodland, coastal scrub. **DIST:** Coast and ranges. Most of NSW. Also Qld, Vic, SA, WA. **FL:** July–Oct.

A bushy erect form occurs on clay soils in western Sydney and the Illawarra.

Hovea

Shrubs with dense hairs on most parts. Leaves usually hairy on undersurface. Flowers mauve-indigo with 10 stamens joined into sheath open on upper side. Anthers 10, alternating long and short. Pods globular or ellipsoid, inflated, hairy.

Species identification requires careful examination of nature of hairs.

Hovea linearis
Narrow-leaf Hovea
Shrub to 1m tall. Leaves narrow-linear, 5–7cm x 3mm, upper surface wrinkled and glabrous, lower surface hairy. Inflorescences 2-flowered. **HABITAT:** Heath, open forest. **DIST:** Coast and ranges. Also NC, SC, CWS, Qld. **FL:** July–Sept.

Hovea heterophylla was previously included in *H. linearis*, but differs in having wider leaves (to 17mm) hooked at the tip and 1–4-flowered inflorescences.

Hovea longifolia
Long-leaf Hovea, Rusty Pod
Erect shrub, 1–3m tall. Stems with short, brownish, closely matted hairs. Leaves narrow-linear with recurved margins, 2–8cm x 2–8mm, glossy above, with densely matted hairs below. Flowers in groups of 2–3, on stalks to 4–6mm long. **HABITAT:** Shaded moist slopes and creek banks. **DIST:** Coast and ranges sth of Newcastle. Also SC, CT. **FL:** Aug–Oct.

Hovea purpurea
Velvet Hovea
Erect shrub, 1–3m tall. Stems with short brown to white closely matted hairs. Leaves narrow-oblong, 4–7cm x 4–10mm, upper surface glabrous except for a dense line of midrib hairs; lower surface with close-matted hairs. Inflorescences sessile, 2-flowered; pedicles to 2.5mm long. **HABITAT:** Forest, woodland, stream banks. **DIST:** Rare. Blue Mtns (e.g. Jenolan Caves). Also NT, SC, Vic, SA. **FL:** Aug–Sept.

Hovea pannosa was previously included in *H. purpurea*, but differs in having longer spreading hairs on stems and branches and narrow-oblong leaves without hairs on the midrib. South from Blue Mtns to Vic.

Hovea linearis

Hovea longifolia

Hovea purpurea

Hovea speciosa
Erect shrub 1–3m tall. Stems and calyces with long brownish spreading hairs. Leaves narrow-oblong, 25–45 x 3–6mm; upper surface glabrous except for scattered midrib hairs; lower surface with both short and longer spreading hairs. Flowers in groups of 3, on stalks to 2.5mm long. **HABITAT:** Woodland, often on cliff tops. **DIST:** Cowan to Nerriga (e.g. Hill Top, Blue Mtns, Wollemi NP). **FL:** July–Sept.

Hovea speciosa

Indigofera
Herbs or shrubs. Leaves usually compound; leaflets entire, opposite. Inflorescences axillary racemes. Flowers pink-purple; stamens 9, united into tube, upper 1 free; anthers tipped with small gland. Pod globose to linear-terete, with seeds separated by pod tissue.

Indigofera australis
Native Indigo
Erect shrub to 1.5m tall. Leaves pinnate with 11–25 elliptic-obovate leaflets 10–40 x 5–10mm. Racemes as long as leaves. **HABITAT:** Sheltered forest, mostly on clay-shale soils. **DIST:** Coast and ranges (e.g. Royal NP, Thirlmere Lakes, Albion Pk). Most of NSW. All other States (except NTerr). **FL:** Aug–Sept.

Indigofera australis

Jacksonia scoparia

Kennedia rubicunda

Jacksonia
Shrubs or small trees. Leaves reduced to minute scales. Calyx tube short, divided into 5 lobes. Stamens all free. Pods oblong, swollen, hairy.

Jacksonia scoparia
Dogwood
Erect shrub to 5m tall with angular grey branches, appearing leafless. Flowers in terminal racemes and in upper axils, orange-yellow; calyx silky. Pods 6–12mm long, hairy. **HABITAT:** Woodland, on old sediments, gravelly clay soils. **DIST:** Widespread, but absent from sandstone coast. Also NC, SC, NT, WS, WP, Qld. **FL:** Oct–Nov.

Kennedia
Twining or prostrate plants. Leaves distant on stems, trifoliolate, subtended by persistent stipules. Flowers with 9 united stamens and 1 free; calyx teeth equal to tube. Pod cylindrical or compressed.

Kennedia prostrata
Running Postman
Prostrate plant with downy stems. Leaflets obovate, often with wavy margins. Flowers scarlet with yellow centre, about 18mm long. Pods cylindrical to 4cm long. **HABITAT:** Woodland, on sand dunes, gravelly clay, shale soils. **DIST:** Uncommon. Coast, Sthn H'lands (e.g. Wilton, Mittagong). Also NC, SC, ST, SWS, Vic, Tas, SA, WA. **FL:** June–Oct.

Kennedia prostrata

Kennedia rubicunda
Dusky Coral-pea
Robust climber with rusty pubescent stems. Leaflets variable, ovate to obovate, 3–10 x 2–6cm, glabrous or slightly hairy, paler beneath. Flowers in racemes on long curved peduncles, pendulous; calyx rusty pubescent; corolla deep red, to 4cm long. Pods compressed, 5–10cm long, covered with brown hairs. **HABITAT:** Forest, woodland, coastal scrub. **DIST:** Coast and ranges. Also NC, SC, NT, Qld, Vic. **FL:** July–Dec.

Lotus
Herbs. Leaves with 5 leaflets, the upper 3 leaflets palmate, the lower 2 stem-clasping at base of rachis. Flowers in umbels on stalk longer than leaves; stamens with filaments alternating long and short.

Lotus australis

Lotus australis
Australian Trefoil
Erect herb to 60cm tall. Leaflets narrow; top 3 leaflets 1–3cm long, other 2 much smaller. Flowers white to pink. **HABITAT:** Grassy woodland, on clay-shale soils. **DIST:** Uncommon. Cumberland Pl. (e.g. Prospect, Blacktown, Camden). Most of NSW; all States. **FL:** Oct–Jan.

Mirbelia
Leaves simple with strong reticulate venation. Flowers with 10 free stamens. Pod ovoid, inflated and divided into 2 by longitudinal partition.

Mirbelia baueri

Mirbelia baueri
Low or prostrate shrub with softly hairy stems. Leaves opposite or in whorls of 3, linear, 5–15mm long, pungent-pointed, margins revolute, upper surface often minutely warty. Flowers axillary, solitary, shortly stalked in leaf axils near the ends of branches. **HABITAT:** Exposed heath. **DIST:** CT (e.g. Katoomba, Fitzroy Falls, Penrose). Also SC (Nerriga). **FL:** Oct–Nov.

Mirbelia platyloboides
Prostrate shrub with softly hairy stems. Leaves ovate, 15–40 x 5–15mm, stiff, strongly veined, shiny above, silky-hairy below. Flowers axillary, sessile solitary or in small clusters. Calyx and pod with silky hairs. **HABITAT:** Heath, open woodland. **DIST:** Upper Blue Mtns, Sthn H'lands. Also SC, ST, Kandos–Rylstone. **FL:** Aug–Oct.

Mirbelia platyloboides

Mirbelia rubiifolia

Mirbelia speciosa

Mirbelia rubiifolia
Low diffuse shrub. Leaves in whorls of 3, 10–25 x 2–4mm, stiff, pungent-pointed, with a clear network of veins. Flowers small, pink-lilac, in clusters in the leaf axils. **HABITAT:** Heath, open woodland. **DIST:** Common. Coast and ranges. Also NC, SC, NT, ST, Qld. **FL:** Sept–Nov.

Mirbelia speciosa
Erect shrub to 1m tall. Leaves linear, to 25mm long, margins recurved; lower leaves in whorls of 3, upper leaves alternate. Flowers conspicuous, corolla 8–12mm, purple with 2 yellow centre spots. Calyx with white hairs. **HABITAT:** Heath, open woodland. **DIST:** Uncommon. Coast (e.g. Royal NP, Maroota, West Head, Wedderburn). Also NT, Qld. **FL:** July–Aug.

Oxylobium arborescens

Oxylobium
SHAGGY-PEA
Shrubs. Stems hairy when young; hairs simple. Leaves mostly opposite or in whorls of 3. Flowers yellow-orange with 10 free stamens. Pod inflated and not divided by a longitudinal partition.

Oxylobium arborescens
Tall Shaggy-pea
Shrub, 2–5m tall; stems appressed-pubescent. Leaves 30–50 x 5–8mm, elliptic, upper surface dark green, glabrous, lower surface silky. Flowers in short racemes. Fruit inflated, hairy, with curved pointed apex. **HABITAT:** Sheltered forests, gullies. **DIST:** Sthn H'lands (e.g. Barren Grounds, Fitzroy Falls, Penrose). Also NT, ST, Qld, Vic, Tas. **FL:** Nov–Dec.

Oxylobium cordifolium

Oxylobium cordifolium
Heart-leaved Shaggy-pea
Prostrate shrub with downy stems. Leaves opposite, alternate or in whorls of 3, 5–8 x 2–5mm, heart-shaped with recurved margins and tip. Calyx hairy; standard petal 6–8mm long. Pod almost sessile, softly hairy. **HABITAT:** Damp sandy soils in heath, rocky outcrops, coastal headlands. **DIST:** Uncommon. Chiefly coastal (e.g. Kurnell, Jibbon Hd); also Barren Grounds, Wingello SF, SC, CT, ST. **FL:** Oct–Dec.

Oxylobium ellipticum
Common Shaggy-pea
Shrub to 2m tall with pubescent stems becoming glabrous. Leaves mostly in whorls of 3–4, 5–30 x 3–10mm, elliptic, pungent-pointed, glabrous above, silver-hairy below. Flowers in short dense racemes. Calyx and pods silky-hairy. **HABITAT:** Woodland, rocky heath. **DIST:** Upper Blue Mtns (e.g. Clarence, Boyd R., Blackheath). Also SC, NT, ST, SWS, Qld, Vic, Tas. **FL:** Oct–Dec.

Oxylobium ellipticum

Phyllota

Shrubs with alternate, simple, entire leaves. Flowers yellow or yellow and red, with 10 stamens. Bracteoles green and leaf-like, attached to flower stalk below calyx. Ovules 2.

Phyllota grandiflora
Erect shrub to 1m tall, with hairy new growth. Leaves narrow-linear, 6–20mm long, paler below, margins revolute. Flowers in compact, sub-terminal leafy heads. Bracteoles long, green and larger than calyx. Calyx and ovary with long loose hairs. Ovary silky-hairy. **HABITAT:** Woodland, rocky slopes. **DIST:** Endemic. Coast nth of Sydney to Hawkesbury R.; Appin to Wedderburn. **FL:** July–Dec.

Phyllota humifusa
Dwarf Phyllota
Prostrate shrub, with hairy new growth. Leaves 3–8mm long, linear, obtuse to acute. Flowers in leafy spikes near ends of branches; bracteoles leaf-like. **HABITAT:** Woodland, swamp margins. **DIST:** Endemic. Sthn H'lands (e.g. Penrose SF, Bullio, Berrima). **FL:** Oct–Dec.

Phyllota grandiflora

Phyllota phylicoides

Phyllota humifusa

Phyllota squarrosa

FLOWERING PLANTS

Platylobium formosum subsp. *formosum*

Phyllota phylicoides
Common Phyllota
Erect shrub to 1m tall, with hairy new growth. Leaves crowded, 5–20mm long, narrow-linear, dark green. Flowers in terminal leafy clusters. Calyx with long white hairs giving fluffy appearance at bud stage. Bracteoles green, longer than calyx tube. **HABITAT:** Heath, shrubland, woodland. **DIST:** Coast and ranges. Also SC, NC, NT, CWS, Qld. **FL:** Aug–Oct.

Phyllota squarrosa
Dense Phyllota
Low shrub to 60cm tall with hairy new growth. Leaves 6–10mm long, linear, with distinct yellow recurved tips. Flowers solitary on pedicels 1–2mm long in upper axils; stamens fused to petals; ovary silky-hairy. **HABITAT:** Heath, dry open forest. **DIST:** Upper Blue Mtns (e.g. Leura, Clarence, Kanangra Walls). Also SC, ST. **FL:** Nov–Jan.

Platylobium
Leaves mostly opposite and strongly veined. Flowers relatively large; stamens united into tube. *Platylobium* means 'broad pod', a reference to flat, oblong pods.

Platylobium formosum subsp. *formosum*
Handsome Flat-pea
Straggling shrub to 1m tall. Leaves ovate to cordate, 30–40 x 15–20mm. Flowers 1–2 in leaf axils; calyx 8–10mm long, standard to 18mm long. Pods hairy on sutures. **HABITAT:** Sheltered forest. **DIST:** Coast and ranges. Also NC, SC, ST, SWS, Qld, Vic, Tas. **FL:** Aug–Nov.

A smaller-flowered variety known as subsp. *parviflorum* occurs less commonly, usually in shale soils. It has narrow-ovate leaves without basal lobes, a standard 5–8mm long and glabrous pods.

Podolobium
SHAGGY-PEA

Shrubs. Stems hairy when young; hairs 2-branched. Stipules rigid, spreading. Flowers yellow-orange with 10 free stamens.

Separated from *Oxylobium* in 1995 on basis of stipules and nature of hairs.

Podolobium ilicifolium
Native Holly

Shrub, 1–3m tall. Leaves opposite, 2–4cm long, strongly veined, with 3 or more pointed lobes. Flowers in leaf axils or in short terminal racemes. **HABITAT:** Sheltered hillsides, open forest, on sandy-clay soils. **DIST:** Coast and ranges. Also NC, SC, NT, ST, CWS, Qld, Vic. **FL:** Sept–Nov.

Podolobium scandens var. scandens
Netted Shaggy-pea

Prostrate shrub with long trailing stems. Leaves opposite, 2–5cm long, ovate, strongly veined, dark green above, paler beneath. Flowers clustered in terminal racemes. **HABITAT:** Open forest, on clay-gravel soils. **DIST:** Cumberland Pl., Woronora Plat., Illawarra. Also NC, SC, Qld. **FL:** Aug–Dec.

Pultenaea
BUSH-PEA

Flowers mostly yellow with some red, with 10 free stamens. The placement of bracteoles on the calyx or close under it is a characteristic of the genus. About 34 species in Sydney district.

Pultenaea aristata

Shrub to 1m. Leaves linear, 10–20 x 1–2mm, concave above; apex with bristle 3mm long; stipules, brown, 5–6mm long. Flowers in terminal head-like inflorescences; bracts overlapping, consisting of an enlarged stipule pair; bracteoles aristate, hairy, 7–8mm long; calyx with loose whitish hairs. **HABITAT:** Dry and wet heath and scrub. **DIST:** Endemic. Helensburgh to Mt Keira. **FL:** Sept–Oct.

Pultenaea blakelyi

Erect shrub to 2.5m tall. Leaves opposite*, 10–20mm, narrow-elliptic, flat, smooth above. Flowers ± terminal, in lax leafy inflorescences; bracts consisting of a stipule pair; calyx with silky hairs; ovary softly hairy to base. **HABITAT:**

Podolobium ilicifolium

Podolobium scandens var. *scandens*

Open forest, sheltered slopes. **DIST:** Coast, mainly sth from Royal NP; Sthn H'lands. Also NC, SC, ST. **FL:** Sept–Nov.

* Different from *P. flexilis*.

Pultenaea canescens

Erect shrub. Leaves crowded, linear-oblong, 5–15 x 1–3mm, erect, blunt, concave above, grey-hairy beneath; stipules brown, 4–5mm long. Flowers in dense terminal inflorescences with reduced leaves and enlarged stipules; bracts absent; bracteoles 7mm long, 3-lobed, on base of calyx. **HABITAT:** Damp hillsides, wet heath edges. **DIST:** Endemic. Upper Blue Mtns (e.g. Clarence, Mt Irvine). **FL:** Sept–Nov.

FLOWERING PLANTS 193

Pultenaea aristata

Pultenaea canescens

Pultenaea blakelyi

Pultenaea capitellata
Prostrate shrub. Young stems with soft grey hairs. Leaves flat, ovate, 3–12mm long, dark green above, paler below, with a reflexed point. Flowers in terminal head-like inflorescences; bracts overlapping, 3-lobed, 3–5mm long; calyx and pods with soft grey hairs. **HABITAT:** Moist heath, swamp margins. **DIST:** Upper Blue Mtns (e.g. Boyd Plat., Mt Werong). Also SC, ST, Vic. **FL:** Sept–Dec.

Pultenaea daphnoides
Large-leaf Bush-pea
Erect shrub 1–3m tall. Leaves narrow-obovate, 12–30 x 4–10mm, apex rounded or truncate. Flowers in terminal inflorescences 20–30mm across; bracts deciduous; calyx and ovary silky-hairy. **HABITAT:** Protected sites in forest, slopes and heath. **DIST:** Coast, Blue Mtns. Also NC, SC, NT, CT, Qld, Vic, Tas, SA. **FL:** Aug–Nov.

Pultenaea divaricata
Shrub to 1m tall. Leaves spreading, terete, 5–12mm long, grooved on upper surface, pimply, concolorous, acuminate. Inflorescences terminal, head-like with about 8 flowers; bracts 3-lobed, 2–3mm long; calyx with greyish hairs. Ovary hairy to base. **HABITAT:** Moist cliffs and wet sandy soils. **DIST:** Blue Mtns, Sthn H'lands, higher Woronora Plat. (e.g. Carrington Falls, Penrose). Also ST. **FL:** Sept–Nov.

Pultenaea daphnoides

FLOWERING PLANTS

Pultenaea capitellata

Pultenaea divaricata

Pultenaea echinula

Pultenaea echinula
Shrub 1–2m tall. Stems below leaves rough with old leaf bases and stipules. Leaves crowded on upper stems and branches, 10–15mm long, terete, grooved above, shortly pointed; older leaves rough and warty. Flowers in compact leafy terminal inflorescences; bracts absent; calyx with greyish hairs. **HABITAT:** Forested slopes. **DIST:** Endemic. Upper Blue Mtns and Wollemi NP (e.g. Kings T'land, Rylstone). **FL:** Aug–Sept.

Pultenaea ferruginea var. *deanei*
Erect shrub to 2m tall with softly hairy stems. Leaves obovate, 9–15 x 4–6mm, with soft white hairs. Inflorescence leafy, dense or open; calyx hairy; standard petal 12mm long. **HABITAT:** Woodland, heath. **DIST:** Terrey Hills to Ourimbah, outlier population at Bonnum Pic (Sthn H'lands). **FL:** Sep–Oct.

Pultenaea ferruginea var. *deanei*

Pultenaea ferruginea var. *ferruginea*

Pultenaea flexilis

Pultenaea ferruginea var. *ferruginea*
Shrub about 1m tall, smaller in all its parts than var. *deanei*. Leaves obovate, 3–7 x 2–3mm, with soft white hairs. Standard petal to 10mm long. **HABITAT:** Open forest. **DIST:** Colo area to Springwood. Also SC, CWS (Coxs Gap). **FL:** Oct–Nov.

Pultenaea flexilis
Erect robust shrub to 3m tall. Leaves alternate*, linear-obovate, 10–20 x 2–5mm, flat, glabrous. Flowers axillary, often crowded in upper axils; bracts absent. Calyx and ovary glabrous. **HABITAT:** Open forest, sheltered slopes. **DIST:** Coast and ranges. Also NC, SC, NT, WS, Qld. **FL:** Aug–Oct.

* Different from *P. blakelyi*.

Pultenaea hispidula

Pultenaea glabra
Erect glabrous shrub to 2m tall. Leaves alternate, spreading, with brown stipules 4–5mm long, linear, 10–20 x 1–2mm. Flowers in crowded upper axils of reduced leaves giving a head-like appearance; bracts absent. **HABITAT:** Open forest, damp heath, protected slopes. **DIST:** Endemic. Rare. Upper Blue Mtns (e.g. Wentworth Falls, Glen Davis). **FL:** Sept–Oct.

Pultenaea hispidula
Small shrub, with dense short hairs on stems and leaves. Leaves oblong with obtuse recurved apex, 6–10 x 1–2mm, concave on upper surface; stipules 2–3mm long. Flowers few in upper axils, appearing as loose heads without prominent bracts. **HABITAT:** Open forest, in shale-sandstone transition zone. **DIST:** Coast, Sthn H'lands. Also SC, Vic. **FL:** Sept–Oct.

FLOWERING PLANTS

Pultenaea glabra

Pultenaea linophylla

FLOWERING PLANTS | 199

Pultenaea microphylla

Pultenaea linophylla
Erect or prostrate shrub to 80cm tall with pubescent stems. Leaves alternate, linear to wedge-shaped, 5–25mm long; apex widening and notched with a tiny mucro. Flowers in terminal head-like inflorescences; bracts 2–3-lobed, 2–4mm long; calyx silky-hairy. **HABITAT:** Woodland, heath, on sandy soils. **DIST:** Coast and ranges. Also NC, SC, NT, ST, Vic. **FL:** Aug–Nov.

Pultenaea microphylla
Spreading Bush-pea
Much-branched shrub, often semi-prostrate; young stems with downy white hairs. Leaves alternate, linear-obovate with recurved tip, 5–8 x 1–2mm, concolorous, flat or margins slightly recurved; stipules to 2mm long. Flowers ± sessile, few in upper axils of stems and short side-branches; bracts absent; calyx hairy. **HABITAT:** Open forest, on gravelly, sandy and clayey soils. **DIST:** Upper Blue Mtns, Cumberland Pl. Also t'lands and slopes to Qld. **FL:** Sept–Nov.

Pultenaea paleacea
Shrub to 50cm tall, often semi-prostrate. Young stems with close silky hairs. Leaves ± linear with recurved tip, 15–20 x 1–2mm, margins recurved, paler beneath; stipules conspicuous, 3–10mm long. Flowers in small dense head-like clusters in upper axils with papery brown bracts 5–10mm long. Calyx and ovary softly hairy. **HABITAT:** Open forest, heath. **DIST:** Coast, chiefly nth of Wyong (e.g. Doyalson, Munmorah SCA). Also SC, CT, NC. **FL:** Aug–Oct.

Pultenaea paleacea

Pultenaea parviflora

Pultenaea parviflora
Erect shrub to 1m tall. Leaves crowded and erect on stems, ± narrow-obovate with recurved tip, 2–5 x 1–2mm, concave above. Flowers few in upper axils, 5–7mm long; pedicels 1–1.5mm long; bracts absent; bracteoles on calyx tube, 4mm long. Calyx glabrous. **HABITAT:** Woodland, on alluvium or clay soils. **DIST:** Endemic. Cumberland Pl. (e.g. Castlereagh NR). **FL:** Sept–Dec.

Pultenaea polifolia

Pultenaea polifolia
Low, often prostrate shrub to 60cm tall. Young stems, undersurface of leaves and calyx with spreading soft grey hairs. Leaves linear to elliptic, 5–25 x 2–5mm, convex with central groove and prominent reflexed point; stipules 4–7mm long. Flowers 5–10mm long on pedicels 1–2mm long in dense terminal inflorescences with persistent 3-lobed hairy bracts and simple bracteoles. **HABITAT:** Open forest, wet heath. **DIST:** Uncommon. Berowra to Royal NP; upper Blue Mtns. Also SC, NT, ST, Vic. **FL:** Aug–Oct.

Pultenaea retusa
Erect shrub to 1m tall. Leaves narrowly wedge-shaped with retuse (notched) apex, 5–15 x 2–5mm, discolorous, margins recurved; stipules 1mm long. Flowers in small terminal inflorescences 8–12mm across without bracts; calyx and ovary silky-hairy. **HABITAT:** Scrub, open forest, on sand or clay. **DIST:** Uncommon. Coast, Sthn H'lands. Also NC, SC, NT, Qld, Vic. **FL:** Aug–Nov.

Pultenaea rosmarinifolia
Shrub to 1m tall with appressed-hairy stems. Leaves linear to elliptic with recurved tip, 20–35 x 2–5mm, dark green and convex with a depressed midvein above; stipules 2–8mm long. Flowers 10mm long on pedicels 1–3mm long in terminal inflorescences with dense silky hairs and persistent 3-lobed bracts 6–8mm long. **HABITAT:** Open forest, heath. **DIST:** Coast nth of Sydney. Also SC (Jervis Bay–Milton). **FL:** Aug–Oct.

Pultenaea retusa

FLOWERING PLANTS

Pultenaea stipularis

Pultenaea subspicata

Pultenaea rosmarinifolia

Pultenaea scabra

Pultenaea scabra
Rough Bush-pea
Shrub to 1m tall; stems with dense white or brownish hairs. Leaves wedge-shaped, 3–15 x 4–10mm; apex notched, with recurved mucro; stipules 2–4mm long, brown. Flowers in small inflorescences, often at ends of short branches; bracts, bracteoles and stipules not conspicuous. **HABITAT:** Open forest. **DIST:** Chiefly Blue Mtns and nw. Hornsby Plat.; less common on coast. Also NC, SC, ST, CWS, Vic, SA. **FL:** Sept–Nov.

A form from Colo and Hornsby plateaus, with narrower leaves 3–8mm long and a 2-lobed notched apex, is known as var. *biloba*.

Pultenaea stipularis
Fine-leaf Bush-pea
Shrub with erect stems to 2m tall. Leaves crowded, linear, 15–35 x 1–2mm, with prominent brown stipules 10mm long. Flowers on pedicels 1–2mm long, in dense terminal inflorescences to 30mm across with numerous bracts composed of enlarged stipules; bracteoles aristate, 6mm long. **HABITAT:** Open forest, hillsides. **DIST:** Common. Coast from Berowra to Wollongong. Also Tomerong (SC). **FL:** Aug–Oct.

Pultenaea subspicata
Sprawling shrub with sparsely hairy stems. Leaves narrow-elliptic, 5–10 x 1–2mm; margins incurved; stipules 2–4mm long. Flowers 10mm long, in dense leafy head-like inflorescences with bracteole-stipules 5mm long. **HABITAT:** Woodland, on sand and clay soils. **DIST:** Upper Blue Mtns (e.g. Hartley, Clarence, Mt Irvine). Also NT, ST, SWS, Vic. **FL:** Oct–Dec.

Pultenaea tuberculata

Pultenaea villosa

Pultenaea tuberculata (formerly *P. elliptica*)
Erect shrub to 1m tall with hairy stems. Leaves crowded, elliptic, 7–15 x 2–5mm, flat to concave above, darker green below; stipules 2–6mm long. Inflorescence subterminal, leafy with enlarged stipules; bracteoles fused to stipules, hairy; calyx with glabrous tube and hairy lobes; flowers reddish orange to yellow. **HABITAT:** Open forest, scrub, heath. **DIST:** Coast and ranges. Also NC, SC, ST. **FL:** Sept–Feb.

Pultenaea villosa
Erect or spreading shrub with graceful drooping hairy branches. Leaves crowded, narrow-obovate, 3–8 x 1–3mm, hairy on undersurface. Flowers abundant in leaf axils; pedicels 1–4mm long; bracteoles fused to stipules, 3-lobed; calyx softly hairy. **HABITAT:** Woodland and scrub, mainly heavier soils. **DIST:** Coast to lower Blue Mtns. Also NC, SC, NT, Qld, Vic. **FL:** Aug–Oct.

Pultenaea viscosa
Erect shrub. Stems, leaves and calyx covered with weak grey hairs. Leaves linear, 10–20 x 1mm, concave, concolorous; stipules 3–4mm long. Flowers in head-like inflorescences; bracts persistent, 3–5mm long, glabrous; bracteoles 3–5mm long, hairy; calyx hairy. **HABITAT:** Open forest, on sandy clay alluvium. **DIST:** Uncommon. Nth of Sydney (e.g. Somersby, Devlins Ck, Marsfield). Also Sthn H'lands, SC, Vic. **FL:** Sept–Oct.

Pultenaea viscosa

Swainsona galegifolia

Sphaerolobium
Subshrubs with erect almost leafless branches. Pods globose with 1 or 2 seeds.

Sphaerolobium vimineum
Shrub to 80cm tall with terete stems. Leaves absent or linear and about 5mm long. Flowers in clusters of 2–3 in terminal racemes; pedicels 1–3mm long; calyx with short tube, darkly dotted; corolla yellow to orange. Pod 3–5mm diam. with filiform beak. **HABITAT:** Damp heath, woodland. **DIST:** Coast and ranges. Also NC, SC, NT, ST, Vic, Tas, SA, WA. **FL:** Sept–Nov.

Sphaerolobium minus is also widespread. It grows to 50cm tall and has calyx uniformly lead-grey.

Swainsona
Herbs or small shrubs. Leaves pinnate. Flowers in axillary racemes. Corolla usually mauve-purple; stamens 9 united into tube with upper 1 free. Chiefly a genus of inland Australia; 1 species in Sydney district.

Swainsona galegifolia
Smooth Darling Pea
Glabrous shrub with trailing or erect stems. Leaves 5–8cm long, with 15–29 obovate leaflets to 15 x 5mm. Racemes to 20cm long; corolla pink or purple. Pod swollen, to 40mm long, with beak to 15mm long. **HABITAT:** Sheltered forests, river banks. **DIST:** Upper Blue Mtns (e.g. Jenolan Caves, Hampton). Most of NSW; also Qld, Vic. **FL:** Oct–Dec.

Sphaerolobium vimineum

Viminaria
Shrubs with stems appearing leafless. Flowers in racemes, terminal or in upper axils; calyx with short equal lobes; stamens free. Pods ovoid with 2 seeds.

Viminaria juncea
Native Broom
Shrub to 4m tall with terete, glabrous, pendulous branches. Leaves reduced. Flowers abundant, yellow, near ends of branches. **HABITAT:** Swampy places, drainage of heaths. **DIST:** Coast, lower Blue Mtns. Also NC, SC, Qld, Vic, SA, WA. **FL:** Sept–Nov.

Zornia
Small trailing herbs with 2-foliolate leaves. Flowers with 9 stamens united and 1 free. Pod constricted and separating between seeds.

Zornia dyctiocarpa
Decumbent ± glabrous perennial to 20cm tall. Leaflets variable, to 30 x 7mm, on stalk to 20mm long. Flowers with 2 leafy bracts to 12mm long, often exceeding corolla, yellow to orange with red blotches. Pod 15–25mm long. **HABITAT:** Grassland, woodland, on clay-shale soils. **DIST:** Cumberland Pl. (e.g. Mt Annan, Doonside). Most of NSW. Also Qld, NTerr. **FL:** Nov–Mar.

Viminaria juncea

Zornia dyctiocarpa

PROTEACEAE

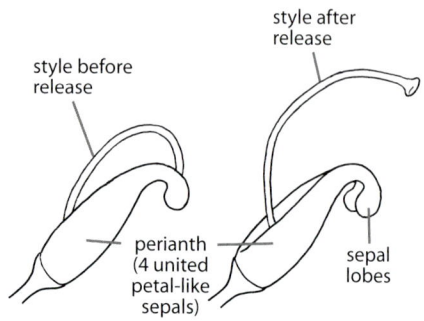

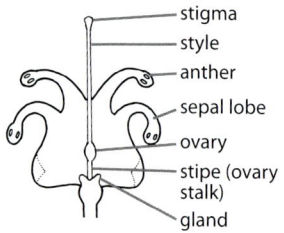

Typical flower structure of a Proteaceae (*Grevillea*)

Banksia
Woody shrubs or trees with conspicuous cylindrical flower spikes. Flowers in sessile pairs on the thick axis of the spike. Woody follicles on fruiting spikes with usually 2 winged seeds released at maturity or after fire.

Banksia aemula
Wallum Banksia
Shrub to 4m tall. Similar to *B. serrata* but leaves narrower (mostly 1–2cm wide). Spikes greenish cream or lemon. Styles straight with ovoid club-shaped stigma*. **HABITAT:** Deep sand deposits. **DIST:** Nth of Botany Bay (e.g. La Perouse, Centennial Park, North Head, Bouddi NP), Agnes Banks. Common on dunes nth to Qld. **FL:** Feb–May.

* Unlike *B. serrata*, which has cylindrical stigmas.

FLOWERING PLANTS

Banksia cunninghamii

Banksia aemula

Banksia cunninghamii
Single-stemmed shrub or tree to 5m tall, without lignotuber. Leaves narrow, to 70 x 5mm, mostly entire, but with small teeth near apex. Spikes to 15cm long, yellow, with yellow to red or deep purple hooked styles. **HABITAT:** Sheltered slopes, open forest. **DIST:** Mainly upper Blue Mtns. Also O'Hares Ck, Fitzroy Falls, SC, Qld, Vic. **FL:** Apr–July.

Banksia ericifolia var. *ericifolia*
Heath Banksia
Large shrub to small tree, with dense foliage. Leaves linear, crowded, to 15mm long, apex notched with 2 small teeth. Spikes 10–20cm long, showy, red-orange, rich in nectar. Style hooked at tip. **HABITAT:** Scrub, heath, on sand. **DIST:** Coast, CT. Also SC, ST. **FL:** Apr–Aug.

Banksia ericifolia var. *ericifolia*

Banksia integrifolia subsp. *integrifolia*

Banksia integrifolia subsp. *integrifolia*
Coast Banksia
Tree to 15m tall. Adult leaves entire or with a few teeth, narrow-obovate, 5–10 x 1.5cm, dark green above, silvery below. Spikes to 10cm long, yellow, with straight styles. **HABITAT:** Hinddunes near sea, coastal forests on sandy soils. **DIST:** Chiefly coast; inland at Thirlmere Lakes, Wingello, Elderslie (Camden). Also Vic, Tas, Qld. **FL:** Jan–June.

Subsp. *monticola* has leaves 10–20cm long. It occurs in upper Blue Mtns (e.g. Mt Wilson, Blackheath). Also NC, NT.

Banksia marginata
Silver Banksia
Spreading shrub to 2m tall in Sydney area. Leaves to 60 x 10mm, entire, cut off squarely at apex. Spikes to 10cm long, yellow, with straight styles. **HABITAT:** Woodland, heath. **DIST:** Coast to ranges. Most of NSW. Also Vic, Tas, SA. **FL:** Feb–July.

Banksia oblongifolia
Multi-stemmed shrub to 3m tall, with lignotuber. New growth rusty-hairy. Leaves elliptic, 5–10 x 1.5–2cm, toothed; underside whitish, often with rusty hairs on midvein. Spikes 10–15cm long, yellow, greyish in bud; perianth 20–25mm long; styles straight. **HABITAT:** Heath, rocky hillsides. **DIST:** Coast, Wentworth Falls. Also NC, CWS, Qld. **FL:** Mar–July.

Banksia paludosa
Swamp Banksia
Multi-stemmed shrub to 1.5m tall. Young branches glabrous. Leaves obovate, 4–10 x 1–3cm, toothed near tip, silver beneath. Spike to 13cm long, golden brown turning golden yellow. Styles small, straight. **HABITAT:** Sandy heath and woodland. **DIST:** Woronora Plat., Illawarra, Blue Mtns. Also SC, ST. **FL:** Apr–Aug.

Banksia penicillata (syn. *B. conferta* var. *penicillata*)
Erect shrub to 4m tall with tan new growth. Leaves whorled, obovate, 5–10 x 1–3cm, toothed. Spikes to 19cm long, yellow, greybrown in bud. Cones with up to 100 follicles remaining on plant until fire. **HABITAT:** Open forest, woodland, steep rocky slopes. **DIST:** Clarence to Wollemi NP (e.g. Newnes Tunnel). Also CWS. **FL:** Apr–July.

FLOWERING PLANTS

Banksia marginata

Banksia paludosa

Banksia oblongifolia

Banksia penicillata

Banksia robur
Swamp Banksia, **Large-leaf Banksia**
Shrub to 2m tall. Leaves about 25 x 8cm, leathery, toothed, with yellow midvein; undersurface covered with rusty brown hairs. Spikes to 17cm long, green in bud, becoming cream-green. Style spreading, straight. **HABITAT:** Swampy or seepage areas and drainage lines. **DIST:** Coast, from Wollongong north. Also NC, Qld. **FL:** Jan–July.

Banksia robur

Banksia spinulosa var. *collina*

Banksia spinulosa var. *spinulosa*

Banksia serrata

Banksia serrata
Old Man Banksia, Saw Banksia
Spreading tree or shrub with gnarled appearance to 10m tall, often dwarfed on coastal heaths. Bark warty. Leaves toothed, to 16 x 4cm. Spikes to 16cm long, grey in bud, greyish cream in flower. Styles straight with cylindrical stigmas*. **HABITAT:** Open forest, heath, coastal hind-dunes. **DIST:** Coast, upper Blue Mtns. Also NC, SC, Vic, Tas, Qld. **FL:** Dec–May.

* Unlike *B. aemula*, which has ovoid, club-shaped stigmas.

Banksia spinulosa var. collina
Hill Banksia
Multi-stemmed shrub to 2m tall, with lignotuber. Leaves broad-linear, 5–10cm x 3–8mm, recurved. Otherwise similar to var. *spinulosa*. **HABITAT:** Woodland, heath, coastal scrub. **DIST:** North of Hawkesbury R.; Mt Wilson. Also NC, NT, Qld. **FL:** Apr–Aug.

Banksia spinulosa var. spinulosa
Hairpin Banksia
Multi-stemmed shrub to 2m tall, with lignotuber. Leaves narrow-linear, 5–8cm x 1–2mm, with notched tip and a few teeth near apex. Spikes to 15cm long, golden yellow; styles hooked, dark red or blue-black with yellow tips. **HABITAT:** Woodland. **DIST:** Coast sth of Brisbane Water NP, Blue Mtns. Also SC, ST, Qld. **FL:** Mar–Sept.

There is a gradual change to var. *collina* between Sydney and Newcastle.

Conospermum
CONESEEDS, SMOKEBUSH
Shrubs with small tubular flowers and irregular lobes appearing as 2 lips. Fruit a hairy nut about 3mm long.

Conospermum ellipticum
Coneseeds
Erect shrub to 80cm tall, with terminal clusters of white flowers. Leaves ascending, crowded, 5–25 mm. **HABITAT:** Open forest, heath, in moist shallow sandy soils. **DIST:** Coast to ranges. **FL:** July–Oct.

Some botanists regard different leaf forms as different species, but forms tend to intergrade.

Conospermum ellipticum. Leaves flat, not twisted, 3–8mm wide. Coast from Bouddi NP to Bulli. Also Jervis Bay.

Conospermum ericifolium. Leaves almost terete, to 1mm wide. Coast, Port Jackson nth to Kariong.

Conospermum taxifolium. Leaves flat, twisted, 1–3mm wide. Coast to ranges. Also NC, SC, NT, CWS, Qld, Vic.

Conospermum ellipticum

Conospermum longifolium
Long-leaf Coneseeds
Shrubs with erect stems to 1.5m tall, long flat leaves and dense white corymbose panicles. **HABITAT:** Dry forest, heath, in sandy soils. **DIST:** Blue Mtns, Sthn H'lands (e.g. Glenbrook, Hill Top, Robertson). **FL:** July–Oct

A variable species divided into 3 intergrading subspecies, esp. in western parts of district.

Subsp. *longifolium*: leaves lanceolate to 25cm x > 8mm. Newcastle–Sydney (e.g. Ku-ring-gai Chase, Brisbane Water NP). Also Putty, Cambewarra.

Subsp. *angustifolium*: leaves 1–4mm wide. Sth of Sydney, Georges R.–Waterfall–Picton.

Subsp. *mediale*: leaves lanceolate, 4–8mm wide. Chiefly Blue Mtns.

Conospermum longifolium

Conospermum tenuifolium
Sprawling Coneseeds
Slender sprawling plant, often forming a large tangled mass. Leaves 8–16cm x 1mm, almost cylindrical but channelled above, curling from the tip. Flowers in small heads on erect stalks 18–40cm long, blue or lilac. **HABITAT:** Wet sandy soils. **DIST:** Coast and ranges (e.g. Heathcote, Mt Wilson, Katoomba, Bargo). Also SC. **FL:** Sept–Jan.

Conospermum tenuifolium

Grevillea

Shrubs, or rarely trees. Flowers in long cylindrical or 1-sided racemes or short umbel-like racemes; perianth segments 4, tubular, rolled back to 1 side; style captive in bud, released at flowering. Fruit a follicle containing 1 or 2 seeds.

Grevillea acanthifolia subsp. acanthifolia
Shrub to 1.5m tall, often with red new growth. Leaves rigid, to 9cm long, deeply divided into about 7 or 9 primary lobes which are divided into 3 triangular sharply pointed lobes. Flowers in terminal tapering 1-sided racemes to 9cm long; perianth with silver woolly hairs; styles red. **HABITAT:** Wet hillsides, swampy heath, sandy creek banks. **DIST:** Endemic. Upper Blue Mtns. **FL:** Sept–Dec.

Grevillea acanthifolia subsp. *acanthifolia*

Grevillea aspleniifolia
Spreading shrub 2–5m tall with terete tomentose branchlets. Leaves linear to narrow-lanceolate, to 25 x 2cm, irregularly toothed or lobed, frequently entire, dull and dark green above, with matted curled brownish hairs beneath. Flowers in 1-sided racemes to 6cm long, pink-red. **HABITAT:** Open forest, often in rocky sites. **DIST:** Warragamba and Kowmung catchments (e.g. Nattai NP, Kedumba V., lower Coxs R.). Also ST. **FL:** July–Nov.

Grevillea arenaria subsp. arenaria
Spreading shrub to 3m tall. Leaves ± elliptic-obovate, 20–60 x 5–15mm, mucronate, densely hairy beneath, margins revolute. Inflorescences terminal, open, few-flowered. Perianth red or orange with green and yellow; lobes pointed; style green, softly hairy. **HABITAT:** Woodland, scrub. **DIST:** Sthn Blue Mtns, Nepean–Nattai rivers. Also ST. **FL:** May–Oct.

Grevillea arenaria subsp. *arenaria*

Grevillea aspleniifolia

FLOWERING PLANTS 211

Grevillea buxifolia subsp. *buxifolia*

Grevillea caleyi

Grevillea baueri subsp. *baueri*
Compact shrub to 1m tall. Leaves crowded, sessile, elliptic, to 25 x 7mm. Inflorescence umbel-like, to 4 x 2cm; flowers pink-red with green, sometimes becoming black; style and ovary hairy. **HABITAT:** Open forest, woodland. **DIST:** Sthn H'lands (e.g. Appin, Hill Top, Berrima). Also SC. **FL:** June–Dec.

Grevillea baueri subsp. *baueri*

Grevillea buxifolia subsp. *buxifolia*
Grey Spider Flower
Spreading shrub 1–2m tall with brown hairs on new growth. Leaves crowded, ovate-elliptic, to 25 x 8mm; soon becoming glabrous above, villous beneath. Inflorescences terminal, umbel-like, to 4 x 4cm, densely hairy; perianth grey, tube 4–5mm long; style with horn-like appendage 2–3mm long; pollen presenter round. **HABITAT:** Open forest, heath. **DIST:** Coast, nearby plateaus. **FL:** July–Nov.

Subsp. *ecorniculata* (Colo–Putty) has no horn-like appendage.

Grevillea caleyi
Spreading shrub 2–4m tall. New growth tan-red with soft rusty brown hairs. Leaves to 18 x 7cm, divided to midrib, with 19–36 oblong lobes to 35 x 5mm with recurved margins. Racemes 1-sided, 4–8cm long, with hairy axis; perianth rusty-hairy; style red. Fruit to 2 x 1cm, pale-hairy with reddish stripes. **HABITAT:** Open forest. **DIST:** Endemic. Terrey Hills–Belrose. **FL:** Oct–Dec.

Grevillea capitellata
Prostrate or low shrub with rusty hairs on new growth; branchlets angular. Leaves oblong-elliptic, 20–90 x 2–8mm, mucronate, ± glabrous above, silky with longer dark hairs beneath. Inflorescence terminal, pendent on stout peduncle 5mm long, crowded, about 20 x 25mm; flowers deep crimson to dark maroon. **HABITAT:** Woodland, heath. **DIST:** Endemic. Mt Keira–Cordeaux Dam (e.g. Appin Rd). **FL:** July–Dec.

The rusty hairs on new growth help separate this species from the similar *G. diffusa*.

Grevillea diffusa subsp. diffusa
Spreading shrub to 50cm tall, with silky-hairy branchlets. Leaves lanceolate, 1–7cm x <5mm, acute, silvery-silky below. Flowers maroon to red, densely silky in compact pendent clusters on slender peduncles to 15mm long. **HABITAT:** Open forest, woodland, heath. **DIST:** Endemic. Menai–Holsworthy–Cordeaux Dam. **FL:** July–Nov.

Grevillea diffusa subsp. constablei
Erect shrub to 2m tall. Leaves 5–10cm long. Flowers pink to red in clusters on longer peduncles than subsp *diffusa*. **HABITAT:** Sheltered sites and creek lines. **DIST:** Endemic. Port Hacking–Helensburgh. **FL:** June–Sept.

Subsp. *filipendula*. Shrub to 1m tall. Flowers sparsely silky-hairy in pendent clusters on slender peduncles to 20mm or more long. Endemic. Brooklyn to Gosford.

Grevillea juniperina subsp. juniperina
Prickly Spider Flower
Prickly shrub to 1.5m tall. Leaves linear, 1–2cm long, crowded, needle-like. Flowers orange-yellow, with greenish tips. **HABITAT:** Woodland, cleared areas, on clay-loam soils. **DIST:** Western Sydney (e.g. Castlereagh NR). **FL:** June–Nov.

This subsp. is endemic in western Sydney. Other forms and subsp. occur elsewhere in NSW and Qld.

Grevillea kedumbensis
Shrub to 1m tall. Leaves narrow-lanceolate, 30 x 4mm. Flowers creamy green, with pink styles; perianth limb distinctly obtuse. **HABITAT:** Open forest. **DIST:** Endemic. Kedumba Valley–Mt Cookem. **FL:** Sept–Nov.

Grevillea laurifolia
Prostrate shrub forming mats to 3m across. Leaves elliptic, 3–12cm long, shiny above, silky below. Flowers red in 1-sided racemes to 8cm long. **HABITAT:** Open forest, woodland, swamp margins. **DIST:** Endemic. Blue Mtns. **FL:** Oct–Dec.

G. laurifolia hybridises with *G. acanthifolia* to form *G.* x *gaudichaudii* which has ovate leaf lobes.

Grevillea linearifolia
Open shrub to 3m tall. Leaves narrow-linear, 10–90 x 2–5mm, silky-hairy beneath. Flowers white, occasionally pink. **HABITAT:** Woodland, often in gullies. **DIST:** Coast nth of Sydney, Audley, Lawson, Putty. Also SC. **FL:** July–Dec.

Grevillea capitellata

Grevillea diffusa subsp. *diffusa*

Grevillea diffusa subsp. *constablei*

Grevillea juniperina subsp. *juniperina*

Grevillea kedumbensis

Grevillea laurifolia

Grevillea linearifolia

Grevillea longifolia

Grevillea mucronulata

Grevillea longifolia
Spreading shrub 2–5m tall with angular silky-hairy branches. Leaves 20 x 1.5cm, serrations less than halfway to midrib, silky grey beneath. Flowers red in 1-sided racemes to 7cm long. **HABITAT:** Banks of perennial streams. **DIST:** Endemic. Woronora Plat. (e.g. Heathcote Ck, Woronora R., O'Hares Ck). **FL:** July–Jan.

Grevillea mucronulata
Green Spider Flower
Spreading shrub, 1–2m tall. Leaves variable, elliptic to lanceolate with small point, 5–20mm long, upper surface flat or convex. Flowers green, with red to black style. **HABITAT:** Open forest, woodland. **DIST:** Coast and ranges. Also SC, CWS. **FL:** Apr–Oct.

Grevillea oldei
Decumbent shrub to 1m tall. Stems sprinkled with long hairs. Leaves clustered in groups of 2–3 on stems, elliptic with a short mucro, to 10mm long. Racemes terminal, pendent, on peduncles to 15mm long; perianth deep red. **HABITAT:** Open forest, woodland, heath. **DIST:** Endemic. Mangrove Mtn–Woy Woy. **FL:** Sept–Oct.

Grevillea oleoides
Red Spider Flower
Erect shrub to 2m tall. Leaves narrow-lanceolate, 5–10cm long, undersurface silky-hairy. Flowers red in sessile clusters of 12–16. **HABITAT:** Open forest, woodland, heath, rocky creek banks. **DIST:** Endemic. Sth of Sydney to Illawarra, Blue Mtns, Sthn H'lands. **FL:** July–Oct.

Grevillea oldei

Grevillea oleoides

Grevillea patulifolia

Grevillea parviflora subsp. *parviflora*
Shrub to 1m, often suckering. Leaves narrow-linear, 20–35 x 1mm. Flowers white with rusty hairs. **HABITAT:** Woodland, on ridges and rocky slopes. **DIST:** Endemic. Upper Georges R. to Tahmoor (e.g. Moorebank, Wedderburn, Appin, Bargo). **FL:** July–Dec.

Similar to *G. linearifolia* but with smaller narrower leaves

Grevillea patulifolia
Shrub to 1.5m tall. Leaves narrow, to 5cm long, pungent, margins recurved. Flowers pale pink to white; style <10mm. **HABITAT:** Woodland, wet heath. **DIST:** Heathcote to Sthn H'lands (e.g. Dharawal NR, Budderoo NP). Also SC, Vic. **FL:** Aug–Jan.

Grevillea parviflora subsp. *parviflora*

Grevillea phylicoides

Erect to spreading shrub. Leaves linear–lanceolate, 20–30 x 2–6mm. Flowers grey; perianth tube 2–3mm long; pollen presenter elliptic; style with horn-like appendage 1–4mm long. **HABITAT:** Open forest, heath. **DIST:** Endemic. Lower Blue Mtns to Kings T'land, sth to Oakdale. **FL:** Aug–Nov.

Sometimes regarded as a subsp. of *G. buxifolia*.

Grevillea raybrownii

Bushy shrub to 1.5m tall. Leaves divided into 3–5 narrow primary lobes, these divided into 2–5 spreading pungent-pointed secondary lobes 5–25mm long. Inflorescence a dense ovoid cluster to 2cm long, mainly terminal; perianth white with rusty-hairy limb. **HABITAT:** Open forest, scrubland. **DIST:** Endemic. Sthn H'lands (e.g. Joadja, Mittagong, Bullio), also Dapto. **FL:** Sept–Nov.

Grevillea rivularis
Carrington Falls Grevillea

Spreading shrub to 2m tall, often forming dense thickets. Leaves stiff, deeply divided into 3–9 lobes which are again divided; ultimate lobes linear, 1–3cm long, revolute, pungent. Flowers delicate pink; styles lime green at tip. Follicle hairy with reddish stripes. **HABITAT:** Moist stream banks. **DIST:** Endemic. Carrington Falls area. **FL:** Aug–Nov.

Grevillea phylicoides

Grevillea raybrownii

Grevillea rivularis

Grevillea sericea subsp. *riparia*

Grevillea sericea subsp. *sericea*

Grevillea sericea* subsp. *riparia
Shrub to 2m tall. Leaves narrow-linear, 6–12cm long, 1–3mm wide. Flowers purplish pink, silky-hairy. **HABITAT:** Rocky riparian zone of permanent streams. **DIST:** Endemic. Eastern escarpment of Blue Mtns (e.g. Grose R., Glenbrook Ck). **FL:** Aug–Nov.

Grevillea sericea* subsp. *sericea
Pink Spider Flower
Sparse shrub to 2m tall. Leaves narrow-elliptic, 10–30 x >3mm, densely silky-hairy below. Flowers whitish to deep pink, silky-hairy; style 10–15mm long. **HABITAT:** Open forest, woodland, heath. **DIST:** Common. Coast and ranges. **FL:** July–Nov.

Grevillea shiressii
Shrub to 4m tall. Leaves narrow-lanceolate, 16 x 3cm, distinctly veined. Inflorescences with 2–9 flowers; perianth green, becoming blue-grey then cream; ovary glabrous on stipe 11–13mm long. **HABITAT:** Open forest, close to streams. **DIST:** Endemic. Mullet Ck, Mooney Ck (Brisbane Water NP). **FL:** Aug–Dec.

Grevillea speciosa

Grevillea shiressii

Grevillea speciosa
Red Spider Flower
Bushy shrub to 2m tall with silky new growth. Leaves ovate, mucronate, 20–40 x 5–10mm, pale below. Inflorescences 2–4cm across, pendent on short branches; perianth red, silky-hairy. Follicle glabrous. **HABITAT:** Woodland, heath. **DIST:** Endemic. Common. Nth of Sydney to Narara and Maroota. **FL:** June–Sept.

Grevillea sphacelata
Erect to spreading shrub to 2m tall. Leaves linear–lanceolate, 10–30 x 2–4mm. Flowers grey with brown buds; perianth tube 4–5mm long; pollen presenter oblong-elliptic; style with appendage 1mm long. **HABITAT:** Open forest, heath. **DIST:** Endemic. Coast. Woronora Plat. to Dapto and Mittagong. **FL:** Aug–Nov.

Grevillea sphacelata

Hakea

Small trees or shrubs. Leaves flat or terete. Flowers in axillary clusters, white, rarely yellow or pink; perianth segments rolled back to release style, becoming free (i.e. separate). Fruit a woody follicle splitting to release 2 winged seeds.

Hakea bakeriana

Hakea bakeriana
Multi-stemmed shrub, 1–2m tall. Leaves terete, 5–9cm long; mucro 1mm long. Flowers on older branches, pink with styles 4cm long. Fruit 4–5cm across, rough, with a short beak. **HABITAT:** Woodland, heath. **DIST:** Hornsby Plat. (Glenorie–Morisset). Also NC. **FL:** May–July.

Hakea constablei
Slender shrub 2–3m tall with hairy young growth. Leaves terete, 5–10cm long; mucro 2mm long. Flowers white, abundant. Fruit to 4cm across, dark, very warty, not beaked. **HABITAT:** Rocky outcrops, cliffs. **DIST:** Upper Blue Mtns, Wanganderry T'land. **FL:** Oct–Dec.

Hakea dactyloides
Broad-leaved Hakea
Erect single-stemmed shrub to 4m tall. New foliage silky tan. Leaves linear to oblanceolate, to 10 x 1.5cm, with 3 clear longitudinal veins. Flowers white, often with pink tinge. Fruit ovate, with many small warts and short pointed apex. **HABITAT:** Open forest, heath. **DIST:** Coast to ranges. Also SC, ST, CWS, Vic. **FL:** Sept–Nov.

Hakea constablei

Hakea dactyloides

Hakea gibbosa

Hakea dohertyi
Shrub or tree to 4m tall. Leaves upwards-pointing, linear, appearing terete but triangular in cross-section, to 35cm long. Flowers white, in clusters of 4–6. Fruit ovoid, 25mm long, finely warted with short pointed beak. **HABITAT:** Open rocky forest, woodland. **DIST:** Restricted. Kanangra-Boyd NP–Lake Burragorang. **FL:** Aug–Sept.

Hakea gibbosa
Compact upright shrub 1–3m tall; new growth softly hairy. Leaves terete, 4–8cm long. Flowers white to cream. Fruit to 3cm across, deeply wrinkled, distinctly beaked. **HABITAT:** Woodland, heath. **DIST:** Coast (e.g. Bouddi NP, Royal NP). Also SC. **FL:** June–Sept.

Hakea laevipes subsp. laevipes
Lignotuberous multi-stemmed shrub to 3m tall. New growth and leaves tan. Leaves lanceolate to obovate, to 12 x 3cm, with 3–5 longitudinal veins. Flowering fire-dependent, infrequent. **HABITAT:** Woodland, heath. **DIST:** Widespread, but scattered (e.g. Curra Moors, Lucas Heights, Newnes SF, Wingello). Also NC, SC, ST. **FL:** Nov–Jan.

Differs from *H. dactyloides* in its lignotuber and dark brown tomentum on new growth.

Hakea dohertyi

Hakea laevipes subsp. *laevipes*

Hakea microcarpa
Small-fruited Hakea
Shrub to 2m tall. Leaves, terete or narrow-elliptic, pungent-pointed, 5–12cm x 1.5–4mm. Flowers profuse in axillary clusters, white. Fruit 15 x 6mm, smooth, elongated, pointed. **HABITAT:** Wet heath, damp scrub. **DIST:** Upper Blue Mtns, Sthn H'lands (e.g. Penrose SF, Lett R., Boyd R.). Also SC, NT, ST, WS, Qld, Vic, Tas. **FL:** Sept–Jan.

Hakea pachyphylla
Compact shrub to 2m tall. Leaves terete, needle-pointed, rigid, 1–4cm long, soon becoming glabrous. Flowers yellow. Fruit 30–35 x 25mm, warty, with small beak. **HABITAT:** Wet heath. **DIST:** Endemic. Upper Blue Mtns (e.g. Mt Victoria, Leura, Newnes SF). **FL:** Aug–Oct.

This species was formerly regarded as an upper Blue Mtns form of *H. propinqua*. The two differ in habit, leaf, flower, fruit, habitat and flowering time.

Hakea propinqua
Bushy shrub, 1–4m tall. Leaves terete, pointed, somewhat flexible, 3–7cm long. Flowers white. Fruit 35–45 x 25–35mm, with large irregular warts; apex bluntly pointed or with tiny mucro.

Hakea pachyphylla

HABITAT: Woodland, heath. **DIST:** Coast and ranges (e.g. Royal NP, Gosford, Leura, Hill Top). **FL:** May–July.

Hakea salicifolia subsp. salicifolia
Willow-leaved Hakea
Shrub or tree, 2–5m tall. Leaves narrow-elliptic to lanceolate, 5–12cm x 8–18mm, pale green, centre vein prominent. Flowers white. Fruit ovoid, 2–3 x 1–2cm, with blunt or beaked tip and prominently raised warts. **HABITAT:** Protected forest, creek banks. **DIST:** Coast and ranges. Also NC, SC, NT. **FL:** Sept–Nov.

Subsp. *angustifolia* has narrower leaves (4–7mm wide). It occurs in creek beds between Hornsby and upper Georges R.

Hakea sericea
Silky Needle Bush
Bushy shrub to 3m tall with white-pubescent branchlets. Leaves spreading, terete, sharply pointed, 2–6cm long, silky-hairy when young. Flowers white. Fruit globular, wrinkled, about 2cm across, with short smooth beak. **HABITAT:** Open forest, heath, scrub. **DIST:** Common. Coast and ranges. Also NC, SC, ST, Qld. **FL:** June–Sept.

Hakea teretifolia
Dagger Hakea
Rigid shrub to 3m tall with silky hairy new growth. Leaves spreading, terete, rigid, sharply pointed, 2–7cm long. Flowers white. Fruit to 30 x 8mm with long tapering beak. **HABITAT:** Wet heath, woodland. **DIST:** Common. Coast and ranges. Also NC, SC, ST, Vic, Tas. **FL:** Jan–Apr. Often forming dense thickets after fire.

Hakea microcarpa

Hakea propinqua

Hakea salicifolia subsp. *salicifolia*

Hakea teretifolia

Hakea sericea

Isopogon
DRUMSTICKS

Shrubs. Leaves opposite, deeply divided 2–3 times into flat or terete segments (rarely entire and simple). Flowers in terminal globose cone-like structures. Fruit a tiny nut with long hairs, numerous, in globular cones.

Isopogon anemonifolius
Broad-leaf Drumsticks

Erect shrub 1–2m tall. Leaves divided in upper part into 3 segments which are again divided; segments flat, 3–5mm wide, pointed but not pungent. Flowers yellow. Cones 10–16mm diam. **HABITAT:** Open forest, heath. **DIST:** Common. Coast and ranges. Also NC, SC, NT, ST, Qld. **FL:** Aug–Dec.

Isopogon anethifolius
Narrow-leaf Drumsticks

Erect shrub 1–2m tall. Stems often reddish. Leaves divided into narrow terete non-pungent segments which point upwards. Flowers yellow. **HABITAT:** Woodland, heath. **DIST:** Coast to ranges. Also SC, ST. **FL:** Sept–Dec.

Isopogon anethifolius

Isopogon dawsonii
Dawson's Drumsticks

Erect shrub to 3m tall. Leaves divided, flat, pointing upwards. Flowers silky-hairy, white, tipped with yellow and pink. Cone with silver-downy hairs. **HABITAT:** Dry forest, rocky sites. **DIST:** Blue Mtns (e.g. Glen Davis, Glenbrook, Nortons Basin, Grose Vale). Also CWS. **FL:** Aug–Nov.

Isopogon anemonifolius

Isopogon dawsonii

FLOWERING PLANTS 225

Isopogon fletcheri

Isopogon prostratus

Lambertia formosa

Isopogon fletcheri
Fletcher's Drumsticks
Erect shrub to 1.5m tall. Leaves entire, to 12cm long, 2cm wide. Flowers cream-green. **HABITAT:** Damp ledges of cliff faces. **DIST:** Endemic. Blackheath. **FL:** Oct–Nov.

Isopogon prostratus
Prostrate Drumsticks
Ground-hugging shrub with softly hairy new growth. Leaves divided, leaf segments narrow (2mm wide). Flowers yellow in cone 12–16mm across. **HABITAT:** Open forest, heath, on sandy soil. **DIST:** Newnes SF, Tallong–Penrose. Also SC, ST, Vic. **FL:** Nov–Jan.

Lambertia
Shrubs with leaves in whorls of 3. Flowers in erect terminal clusters exceeded by involucral bracts; perianth tubular; style filiform, projecting beyond perianth. Fruit woody, with a short beak and a horn on each valve.

Lambertia formosa
Honey Flower, Mountain Devil
Medium-sized shrub. Leaves rigid, pungent, 3–5cm long. Flowers in clusters of 7, erect, red, tubular to 5cm long, enclosed by overlapping bracts. **HABITAT:** Open forest, heath, scrub. **DIST:** Common. Coast and ranges. Also NC, SC, ST. **FL:** Chiefly Sept–May.

Lomatia ilicifolia

Lomatia myricoides

Lomatia
Shrubs or small trees. Leaves alternate, simple or bipinnate. Inflorescences raceme-like, 10–20-flowered. Flowers mostly white or yellowish. Fruit a thin-walled follicle, differing from *Grevillea* in having more than 2 winged seeds in 2 rows.

Lomatia ilicifolia
Holly-leaved Lomatia
Erect stiff shrub to 2m tall. Leaves lanceolate to elliptic, to 18 x 5cm long, coarsely serrated, leathery, strongly veined. Inflorescences terminal, to 30cm long, exceeding the leaves. Perianth creamy white. **HABITAT:** Open forest. **DIST:** Sthn H'lands (e.g. Moss Vale, Fitzroy Falls, Penrose). **FL:** Nov–Jan.

Lomatia myricoides
River Lomatia
Large shrub to 5m tall. Leaves narrow-oblong, entire or coarsely toothed, to 20 x 2cm. Inflorescences axillary, raceme-like, shorter than leaves. Perianth greenish yellow. Follicle 2–3cm long. **HABITAT:** Gullies, creek margins. **DIST:** Coast and ranges. Also SC, ST, CWS, Vic. **FL:** Dec–Jan.

Lomatia silaifolia

Lomatia silaifolia
Crinkle Bush
Erect stiff shrub to 1m tall. Leaves 10–35cm long, divided 1–4 times, very variable; segments narrow with lobed or toothed margins. Inflorescences terminal, raceme-like, exceeding the leaves. Perianth white. **HABITAT:** Open forest. **DIST:** Coast and ranges. Also NC, NT, NWS, CWS, Qld. **FL:** Nov–Jan.

A large-leaf form occurs between Sackville, Maroota and Colo.

Persoonia
GEEBUNG
Flowers axillary, yellow, 10–15mm long, tubular with 4 spreading segments. Fruit an ovoid drupe with a persistent style, edible though astringent.

Persoonia acerosa
Erect shrub. Leaves crowded, upwards-pointing, 10–20 x 0.5mm, subterete, channelled above. Flowers in upper leaf axils; tepals 8–10mm long with subulate tips 1.5–2mm long. Leaves between flowers same as stem leaves. Drupe glabrous. **HABITAT:** Open forest, heath. **DIST:** Blue Mtns to Hill Top. **FL:** Dec–Apr.

Persoonia chamaepitys
Prostrate Geebung
Trailing shrub forming mats of light green foliage 1–2m across. Leaves linear to terete, glabrous, crowded, 10–15mm long. Flowers in small clusters. **HABITAT:** Open forest, heath. **DIST:** Upper Blue Mtns, Colo, Wingello. Also CWS. **FL:** Dec–Jan.

Persoonia hirsuta subsp. hirsuta
Hairy Geebung
Spreading shrub to 1m tall with hairy new growth. Leaves narrow-oblong, 6–12 x 1.5mm, with recurved margins. Flowers, ovary and fruit distinctly hairy. **HABITAT:** Woodland, heath. **DIST:** Endemic. Coast (Gosford–Royal NP). **FL:** Nov–Jan.

Subsp. *evoluta* (Putty–Hill Top) has elliptic leaves to 5mm wide. Intergrades between the 2 subspecies occur between western Sydney and the lower Blue Mtns.

Persoonia acerosa

Persoonia chamaepitys

Persoonia hirsuta subsp. *hirsuta*

Persoonia isophylla
Erect shrub to 1.5m tall with sparsely hairy branchlets. Leaves terete with recurved tip, to 30 x 0.5mm, glabrous, grooved below. Flowers in upper leaf axils; tepals 7–8mm long with apiculate tips <1mm long. Drupe glabrous. **HABITAT:** Woodland, heath. **DIST:** Endemic. Ourimbah to Terrey Hills. **FL:** Dec–June.

Persoonia lanceolata
Lance-leaf Geebung
Erect shrub to 3m tall with hairy young growth. Leaves oblanceolate, 3–9 x 1–3cm, flat, with distinct mucro. Flowers erect on hairy stalks 1–5mm long; tepals 10–12mm long, hairy. Drupe glabrous. **HABITAT:** Open forest, woodland, heath, coastal dunes. **DIST:** Coast, Sthn H'lands (e.g. Hill Top), Warrimoo. Also NC, SC, ST. **FL:** Jan–May.

Persoonia laurina subsp. laurina
Laurel Geebung
Small erect shrub with dense short rusty hairs on new growth. Leaves opposite, decussate, oblong, smooth, discolorous, 3–10 x 1–5cm. Flowers in stalked clusters; tepals 12–15mm long, densely rusty-hairy outside. Ovary densely hairy. Drupe to 20 x 15mm, yellow-green or purplish at maturity; style bent back. **HABITAT:** Forest, heath. **DIST:** Coast and ranges. Also NC, ST. **FL:** Oct–Dec, later in mtns.
   Subsp. *intermedia* (Georges R.–Wombeyan Caves) has scabrous mature leaves and a hairy ovary.
   Subsp. *leiogyna* (Kanangra Walls–Budawangs) has scabrous leaves and a glabrous ovary.

Persoonia levis
Broad-leaf Geebung
Tall shrub with loose flaky bark, deep red-brown under black outer surface. Leaves variable, elliptic to obovate, 8–16 x 2–6cm, bright green, smooth. Flowers on stalks 3–8mm long; tepals 10–14mm long. **HABITAT:** Forest, woodland, heath. **DIST:** Coast and ranges. Also NC, SC, ST, Vic. **FL:** Sept–Jan.

Persoonia isophylla

Persoonia linearis
Narrow-leaf Geebung
Tall shrub with loose flaky bark, red under dark outer surface; new growth with downy hairs. Leaves linear-oblong, 2–8cm x 2–5mm. Flowers near ends of branches; tepals 10–14mm long, hairy. **HABITAT:** Open forest, woodland. **DIST:** Coast and ranges. Also NC, SC, NT, ST, CWS, Vic. **FL:** Dec–July.

Persoonia mollis
Soft Geebung
Erect to prostrate shrub. Branchlets, new leaves and buds with silky white hairs. Leaves variable, linear to lanceolate, 2–12cm x 1–17mm, recurved or revolute. Flowers on erect hairy pedicels 1–3mm long; tepals to 10mm long, hairy. Ovary glabrous. **HABITAT:** Open forest, heath. **DIST:** Coast, Blue Mtns. Also SC, ST. **FL:** Dec–Mar.
   Six subsp. recognised in Sydney district: *ledifolia* (Sthn H'lands); *livens* (Penrose–Tallong); *maxima* (Cowan–Turramurra); *mollis* (Blue Mtns); *nectans* (Illawarra–Hill Top); *revoluta* (Mittagong–Bullio).

Persoonia mollis

Persoonia linearis

Persoonia lanceolata

Persoonia levis

Persoonia laurina subsp. *laurina*

Persoonia nutans

Persoonia myrtilloides subsp. *myrtilloides*

Persoonia oblongata

Persoonia myrtilloides subsp. *myrtilloides*
Myrtle Geebung
Spreading shrub to 2m tall with hairy new growth. Leaves lanceolate-elliptic, 20–40 x 5–30mm, glabrous when mature. Flowers nodding on short hairy stalks; tepals with terminal appendage 2mm long. **HABITAT:** Dry forest, scrub. **DIST:** Endemic. Upper Blue Mtns. **FL:** Oct–Feb.

Persoonia nutans
Nodding Geebung
Spreading shrub to 1.5m tall. Leaves linear, 15–30 x 1–2mm. Flowers glabrous. Fruit pendulous on fine stalks to 10mm long; tepals with terminal appendage 2mm long. **HABITAT:** Dry forest, scrub. **DIST:** Endemic. Cumberland Pl. (e.g. Agnes Banks, Castlereagh NR). **FL:** Dec–Apr.

Persoonia oblongata
Spreading shrub to 2m tall with hairy new growth. Leaves ovate to lanceolate, 20–50 x 5–25mm, light green, smooth. Flowers and fruit pendulous on slender stalks to 2cm long; tepals without terminal appendage. **HABITAT:** Open forest, woodland. **DIST:** Blue Mtns to Howes V. (e.g. Winmalee, Cattai, Glen Davis). **FL:** Feb–June.

Persoonia oxycoccoides
Prostrate or leafy spreading shrub to 90cm tall. New stems reddish, moderately hairy. Leaves elliptic to ovate, 5–11 x 2–6mm. Flowers erect to spreading; pedicels, tepals, ovary and fruit glabrous. **HABITAT:** Open forest, sandy heath. **DIST:** Endemic. Sthn H'lands (e.g. Budderoo NP, Tallong). **FL:** Dec–Jan.

FLOWERING PLANTS

Petrophile pedunculata

Persoonia oxycoccoides

Persoonia pinifolia
Pine-leaf Geebung
Attractive shrub to 4m tall with hairy new growth. Leaves crowded, terete, pine-like, 30–50 x 0.5mm. Flowers in dense terminal racemes in axils of reduced leaves; tepals 8–9mm long, hairy. Fruit glabrous. **HABITAT:** Open forest. **DIST:** Endemic. Coast, lower Blue Mtns. **FL:** Jan–July.

Petrophile
CONESTICKS
Erect shrubs with stiff divided leaves and flowers in dense terminal or axillary spike-like inflorescences. Fruit a nut, numerous, in woody ovoid cones.

Petrophile canescens
Shrub to 2m tall with downy young shoots. Leaves ascending with terete non-pungent leaf segments, undivided part longer than divided part. Cones sessile or on stalks <10mm long. Cones ovoid-oblong to 25 x 15mm. **HABITAT:** Open forest, on deep sands. **DIST:** Upper Blue Mtns (e.g. Newnes SF), Yerrinbool, Castlereagh NR. Also NC, NT, ST, WS, Qld. **FL:** Dec–Feb.

Petrophile pedunculata
Stalked Conesticks
Glabrous shrub to 2.5m. Leaves spreading, divided, to 16cm long and non-pungent. Flower cones stalked, 15–25mm long in leaf axils. **HABITAT:** Dry forest, heath. **DIST:** Nepean, Blue Mtns, Sthn H'lands (e.g. Nortons Basin, Lawson, Wingello, Budderoo NP). Also SC. **FL:** Nov–Jan.

Persoonia pinifolia

Petrophile canescens

Petrophile sessilis

Petrophile pulchella
Common Conesticks
Erect shrub to 3m tall. Leaves divided to halfway, segments ascending but not pungent-pointed. Cones 1 to few, sessile, 3–5cm long. **HABITAT:** Open forest, heath. **DIST:** Common. Coast and ranges. Jervis Bay to NC, Qld. **FL:** Dec–Jan.

Intermediate forms exist between *P. pulchella*, *P. sessilis* and *P. canescens*.

Petrophile sessilis
Prickly Conesticks
Stiff erect shrub to 3m tall with pubescent new growth. Leaves at right angles to stems; segments spreading, rigid, pungent, undivided part shorter than divided part. Inflorescences terminal or axillary. Fruiting cones 20–30mm long, usually sessile. **HABITAT:** Open forest, heath. **DIST:** Coast sth of Sydney to Budawangs (e.g. Royal NP, Agnes Banks NR, Bargo, Bundanoon). **FL:** Dec–Jan.

Petrophile pulchella

Stenocarpus salignus

Symphionema montanum

Stenocarpus
Rainforest trees and shrubs with alternate leaves. Flowers in umbels. Fruit a narrow follicle containing winged seeds.

Stenocarpus salignus
Scrub Beefwood
Large shrub or small tree. Leaves entire, lanceolate, 4–10 x 1–4cm, glossy above, paler below, with 3 longitudinal veins. Inflorescences terminal, with 1 or more umbels, each with 10–20 flowers. Perianth creamy green. Follicle 4–6cm long, with 5–8 flat seeds. **HABITAT:** Rainforest gullies. **DIST:** Widespread (e.g. Gosford, Royal NP, Albion Park, Maroota, Colo, Glen Davis). **FL:** Oct–Jan.

Symphionema
Erect subshrub. Leaves alternate or opposite, divided into narrow segments. Inflorescence spike-like, exceeding leaves. Tepals 4, equal, spreading; stamens 4. Fruit an achene with 1 seed.

Symphionema paludosum

Symphionema montanum
Mountain Symphionema
Many-stemmed shrub to 60cm tall. Leaves 2–4cm long, divided into narrow flat segments. Flowers 4–5mm long, with 4 pale yellow spreading lobes. **HABITAT:** Wet heath, open forest. **DIST:** Upper Blue Mtns (e.g. Woodford, Narrow Neck). **FL:** Sept–Nov.

Symphionema paludosum
Swamp Symphionema
Shrub 30–50cm tall, with few stems. Leaves 1–3cm long, divided into 3–9 segments, each segment terete. Flowers 4–5mm long, with 4 pale yellow spreading lobes. **HABITAT:** In and near wet heath. **DIST:** Coast, Sthn H'lands. Also NC, SC, ST. **FL:** Aug–Oct.

Telopea speciosissima

Telopea
WARATAH
Erect shrubs with terminal red flower heads subtended by red-pink bracts. Flowers in pairs; perianth incurved in bud. Fruit a woody follicle with numerous winged seeds in 2 rows.

Telopea mongaensis
Monga Waratah
Multi-stemmed shrub 2–4m tall. Leaves narrow-obovate, 5–16cm x 5–20mm, entire, leathery. Flowers in loose terminal heads, pink or greenish. Follicle to 6cm long. **HABITAT:** Creek banks in open forest. **DIST:** Sthn H'lands (Meryla SF, Lamonds Ck, Barren Grounds) to Monga near Braidwood. **FL:** Oct–Nov.

Telopea speciosissima
Waratah
Multi-stemmed shrub to 3m tall. Leaves toothed, strongly veined, to 15 x 4cm. Flowers in terminal heads to 15cm diam. Follicle boat-shaped, remaining on plant after release of seeds. **HABITAT:** Woodland, open forest. **DIST:** Coast and ranges. Also SC, ST. **FL:** Sept–Oct; to Dec in mtns.

Telopea mongaensis

FLOWERING PLANTS | 235

Xylomelum pyriforme

Xylomelum
WOODY PEAR
Shrubs or trees. Leaves opposite, leathery. Flowers in dense spikes. Fruit large, woody, pear-shaped, splitting to release 2 large winged seeds.

Xylomelum pyriforme
Woody Pear
Shrub or small tree with rusty hairs on new shoots. Leaves lanceolate, to 20cm long, strongly veined; margins entire, young leaves toothed and reddish. Flowers small, tightly crowded in tan spikes, 5–8cm long. Fruit to 9cm long. **HABITAT:** Open forest, on sandy soils, coastal heaths. **DIST:** Coast to ranges. Also NC. **FL:** Oct–Nov.

MYRTACEAE

Acmena
Trees or shrubs with opposite leaves. Flowers with 4 petals, 4 deciduous sepals and numerous free conspicuous stamens. Fruit a purple-white globular berry.

Acmena smithii
Lillypilly
Shrub from 1m tall on sea-cliffs to tree 20m tall in protected gullies. Leaves opposite, 4–10 x 1–4cm, tapering to long blunt point, glossy above. Flowers small, numerous in branched clusters, creamy white or pinkish. Fruit globular, 10–20mm diam, whitish to pale purple, showy in early winter. **HABITAT:** Coastal rainforest and moist forests. **DIST:** Coast and ranges. From Gippsland to n. Qld. **FL:** Nov–Feb.

Forms with smaller leaves occur north of Sydney.

Acmena smithii

Angophora, Corymbia, Eucalyptus

Subgroups in the Sydney district

1. Inflorescence terminal, irregularly branched and clustered
    2. Flowers in broad, flat corymbose clusters. Fruit usually >10mm across.
        3. Operculum absent (bud), petals present (flower). Fruit thin, ribbed
        .... ***Angophora***
        3*. Operculum present (bud), petals absent (flower). Fruit woody with insert valves, not ribbed ....***Corymbia***
    2*. Flowers in narrow panicles. Fruit usually <10mm across, not ribbed .... ***Eucalyptus*** (subgenus ***Symphyomyrtus***)

1.* Inflorescence axillary, simple (very rarely compound).
    4. Scar present on mature buds due to early detachment of outer operculum
    ....…................................. ***Eucalyptus*** (subgenus ***Symphyomyrtus***)
    4*. Bud scar absent .... ***Eucalyptus*** (subgenus ***Monocalyptus***)

ANGOPHORA

Flowers with 5 or 4 broad-based petals and free (i.e. separate) sepals persistent as teeth on rim of fruit. Stamens creamy white, showy. Adult leaves in opposite pairs. Fruit papery or thinly woody, often ribbed.

Angophora bakeri
Narrow-leaved Apple

Compact rough-barked tree to 12m tall, sometimes much smaller. Leaves shortly stalked, narrowly lanceolate, to 9 x 1cm. Buds to 5mm x 6mm. Fruit ovoid, often narrowing towards rim, to 10 x 10mm. **HABITAT:** Dry sandy soils. **DIST:** Patchy occurrence. Coast and CT, centred on Sydney. Large examples common in Agnes Banks NR. **FL:** Dec–Jan.

A variant with larger leathery leaves (to 11 x 1.5cm) and larger fruit (to 14 x 14mm) is now called *A. crassifolia*. It is restricted but locally common on Ku-ring-gai Plat. (e.g. Mona Vale Rd).

Angophora inopina grows in woodland on deep white sandy soils between Wyee and Charmhaven, nth of Wyong. It has broader

Angophora bakeri

Angophora costata

lanceolate leaves (to 11 x 2.6cm), narrow, distinctly ribbed fruit (to 15 x 7mm) and hispid inflorescences.

Angophora costata
Smooth-barked Apple, Sydney Red Gum, Rusty Gum
Short-trunked tree with twisted branches and trunk stained by dark kino exudate on shallow dry sandstone sites or up to 25m tall with straight trunk for half its height on better sites. Bark shed in patches in Nov. leaving smooth dimpled pinkish brown surface. Leaves lanceolate, acutely pointed, 10–15 x 2–3cm on stalks 10–25mm long. Flowers whitish, in large terminal clusters, flowering much more prolifically some years than others. Fruit slightly woody, 12–14 mm across. **HABITAT:** Open forest with shrub understorey. **DIST:** Coast, lower Blue Mtns. Also NC, SC. **FL:** Chiefly Nov–Dec.

    A. euryphylla, a variant with broader leaves and larger fruit, grows in the Putty–Howes V.–Wollombi district.

Angophora floribunda
Rough-barked Apple
Medium to large spreading tree with grey-green canopy and short furrowed fibrous bark, frequently with large crooked branches

Angophora floribunda

descending close to ground. Leaves ovate to lanceolate, acutely pointed, to 12 x 3cm. **HABITAT:** Watercourses, alluvial flats, undulating country in open forest and woodland. **DIST:** Eastern Vic. to Qld on coast, t'lands and slopes. **FL:** Oct–Dec.

    The leaf stalks (about 10mm long) and tapering leaf bases help to distinguish this species from *A. subvelutina*. Flowers and fruit are similar.

Angophora hispida
Shrub or straggly tree with loose flaky rough bark, usually 1.5–2.5m tall. Leaves stiff, broad, sessile, dull yellowish green above, greyish beneath, obtuse at tip, heart-shaped at base. Young branches, leaves and inflorescence with stiff reddish hairs. Flowers and fruit larger than other species. Fruit 15–19mm across. **HABITAT:** Dry sandstone sites. **DIST:** Common. Coast. **FL:** Oct–Dec.

Angophora subvelutina
Broad-leaved Apple
Large tree with rough bark and dense crown to 20m. Leaves with pointed tip, often velvety, otherwise similar to those of *A. hispida*. Buds 4–7mm across. Flowers small. Fruit 7–9mm across. **HABITAT:** Woodland, open forest, on loamy soils near streams. **DIST:** Mainly Camden to Richmond. Also SC, CT to se. Qld. **FL:** Oct–Dec.

Resembles *A. floribunda* and has a similar preference for heavy alluvial soils, but differs in its almost sessile leaves with heart-shaped bases.

Angophora hispida

CORYMBIA
Consisting of bloodwoods and ghost gums, *Corymbia* is closely related to *Angophora* and shares many characteristics. Flowers are in broad terminal inflorescences similar to *Angophora* but flower buds have sepals and/or petals fused into a small cap called an operculum or calyptra. The fruit is a medium-large rounded woody 'gumnut' with deeply enclosed valves.

Corymbia gummifera
Red Bloodwood
Small mallee in coastal heaths, single- or multi-trunked tree 10–15m high in dry open forest, or medium to large tree in gullies and valleys. Leaves pale on undersurface with close parallel veins at large angle (about 65°) to midrib. Inflorescences whitish with mostly 7-flowered clusters. Buds without scar (inner and outer opercula fall together at flowering). **HABITAT:** Mostly poor soils with other eucalypts in open forest. **DIST:** Coast and CT below 900m. From Vic to Qld. **FL:** Mainly Feb–March, to May at higher elevations.

Angophora subvelutina

Corymbia gummifera

Corymbia eximia

Corymbia eximia
Yellow Bloodwood

Spreading tree to 20m tall with rough tessellated flaky yellow-brown bark extending to smaller branches. Adult leaves stalked, lanceolate or falcate, to 20cm x 25mm, grey-green both sides. Inflorescences cream with mostly 7-flowered clusters. Flowers and fruit sessile. Buds with scar due to early detachment of outer operculum. **HABITAT:** Poor sandstone soils. **DIST:** Nowra to Hunter and Howes V., patchy, mostly upper slopes, escarpments and plateaus of Nepean and Hawkesbury districts and lower Blue Mtns. **FL:** Mainly Oct.

Corymbia gummifera and *C. eximia* do not grow intermixed and do not appear to hybridise.

Corymbia maculata
Spotted Gum

Tall smooth-barked forest tree with characteristic dimpling and grey and cream mottling of trunk. Inflorescences terminal with 3-flowered clusters. Fruit to 14 x 12mm, woody, ovoid to urn-shaped. **HABITAT:** Open forest on well-drained moderately heavy shale soils. **DIST:** Coast (e.g. Appin, Liverpool region, West Pittwater, Gosford). From Bundaberg to Bega; isolated occurrence in Vic. **FL:** Apr–June.

Corymbia maculata

EUCALYPTUS (SUBGENUS *MONOCALYPTUS*)

Operculum on flower bud formed from calyx (petals absent). Includes stringybarks, bastard mahoganies, ashes, mallee ashes, snowgums, peppermints and scribbly gums. Usually found on low-nutrient acidic soils, frequently occupying distinctive sites, sometimes in nearly pure stands. Most or all species dependent on mycorrhizal associations in the soil. Hybridisation and intergrades common.

Note: The following species are arranged according to relationships within this subgenus.

Eucalyptus umbra
Bastard or Broad-leaved White Mahogany
Small to medium tree with fine fibrous fissured bark persisting to small branches. Leaves firm, dark green, concolorous, 10–14 x 2–3.5cm, tapering to acute point. Flowers and fruit shortly stalked, in clusters of 7–11, not tightly packed. Fruit 6–10mm across, hemispherical with narrow to broad flat disc and valves at rim level. **HABITAT:** Dry forests, often on rocky ridges near salt water. **DIST:** Coast. Middle Harbour to n. Qld. **FL:** Aug–Sept.

Eucalyptus acmenoides
White Mahogany
Large tree with fibrous grey-brown bark on trunk and branches. Leaves thin, distinctly paler on undersurface, 10–14cm long, lanceolate, with fine reticulate veins between parallel lateral veins, frequently with irregular or wavy margins. Clusters axillary with 7 or more flowers. Fruit 7mm across, stalked, with

Eucalyptus acmenoides

narrow obscure disc, thin rim and 4 valves at rim level. **HABITAT:** Fertile soils on hillsides in open and tall open forest, often near rainforest. **DIST:** Uncommon. Coast. Epping–Nth Rocks, Gosford, Wyong, to n. Qld. **FL:** Nov–Jan.

Eucalyptus muelleriana
Yellow Stringybark
Large tree. Bark long-fibred on trunk and branches, grey-brown outside, yellow near cambium when freshly cut. Leaves paler on undersurface, 10–15 x 2–3cm, lanceolate-falcate, with fine tip and oblique base. Flowers in clusters of 7–11 on stalks 1–3mm long. Fruit not tightly packed, 10mm across, truncate-globose with flat disc; valves enclosed. **HABITAT:** Slopes of sheltered valleys on coast and ranges. **DIST:** Wollongong to Wilsons Prom. (e.g. Macquarie Pass, Mt Kembla). **FL:** Nov–Mar.

Eucalyptus umbra

Eucalyptus muelleriana

Eucalyptus blaxlandii
Blaxland's Stringybark
Medium-sized, short-boled spreading tree on dry shallow soils to columnar, densely crowned tree 30m tall on deep fertile soils at Mt Wilson. Bark furrowed and fibrous on trunk and larger branches, smooth on upper branches. Leaves thick, glossy, 6–12 x 2.5cm, oblique at base; intramarginal vein distant. Buds sessile in clusters of 7 or more; operculum rounded or obtusely conical. Fruit in tight clusters, 9–12mm across, domed with convex ascending disc and usually 4 exserted valves. **HABITAT:** Forests on sandy, granitic or basaltic soils. **DIST:** Upper Blue Mtns (e.g. Hassans Walls, Shipley Plat., Wingello). Also ST. **FL:** Oct–Nov.

Eucalyptus blaxlandii

Eucalyptus camfieldii
Heart-leaved Stringybark, Camfield's Stringybark
Mallee or small tree with fissured fibrous bark throughout, 1–4m high. Juvenile leaves 2–5cm long, round or heart-shaped, bristly. Adult leaves 7–10cm long, dark green, thick, leathery, broadly lanceolate or with rounded mucronate tip. Buds sessile, >10 on thick flattened peduncle, angular, 6–7mm long. Fruit crowded, compressed-spherical, 4–6 x 6–9mm with broad convex disc and 3 or 4 slightly exserted valves. **HABITAT:** Shallow sandstone soils on plateaus bordering coastal heath. **DIST:** Bulli Pass to Gosford. **FL:** Irregular. Chiefly Apr–Dec. An endangered species largely represented by a few stands in Ku-ring-gai Chase and Royal NPs. It resembles stunted forms of E. capitellata but can be distinguished by the juvenile leaves.

Eucalyptus camfieldii

Eucalyptus capitellata
Brown Stringybark
Stunted tree 2m high to a thin tree 15 m high, often misshapen. Juvenile leaves to 14cm long, ovate to broad-lanceolate. Adult leaves broad-lanceolate, about 10 x 3–4cm, thick, dark green, glossy in dwarfed coastal forms or about 16 x 2–3.5cm in taller trees. Buds sessile on thick flattened peduncle in clusters of 7 or more, angular, with short conical operculum. Fruit compressed-spherical, broader than tall, often misshapen due to crowding; disc broad, reddish, flat or convex; valves enclosed or tips exsert. **HABITAT:** Shallow infertile sandstone soils, rarely on shales. **DIST:** Nowra to Port

Eucalyptus capitellata

Stephens, mostly nth of Sydney, e.g. St. Ives (tall form), Munmorah SCA (stunted form). **FL:** Dec–Feb.

Eucalyptus agglomerata
Blue-leaved Stringybark
Tall tree with straight trunk and thick, furrowed, long-fibred bark to small branches. Leaves about 14 x 3cm, broad-lanceolate; intramarginal vein 3–5mm from margin. Buds sessile in clusters of 11 or more. Fruit tightly packed, 8–9mm across, broader than long with convex disc and 4 enclosed valves in narrow opening. **HABITAT:** Valley slopes. **DIST:** Coast and t'lands from Wauchope to se. Vic, inland to Mudgee and Rylstone. **FL:** Apr–July.

A useful timber tree recognisable from a distance by its emergent blue-green crown above the greyish or brownish forest canopy.

Eucalyptus eugenioides
Thin-leaved Stringybark
Forest tree with stringy bark. Adult leaves thin, lanceolate with oblique base, to 12 x 2.5cm. Operculum conical, pointed. Fruit 8mm across with rounded base, not compressed or misshapen; disc reddish, broad, flat.
HABITAT: Open forest on moderately fertile soils. **DIST:** Lower rainfall areas of coastal plain, lower e. slopes of Blue Mtns, Kanimbla, Megalong and Hartley Valleys, coast and t'lands of NSW and se. Qld. **FL:** Sept–Nov.

Resembles, and sometimes intergrades with, *E. globoidea* but differs in buds and fruit: shortly stalked in *E. eugenioides*, sessile in *E. globoidea*.

Eucalyptus agglomerata

Eucalyptus eugenioides

Eucalyptus globoidea
White Stringybark

Tree with thick fibrous bark throughout. Juvenile leaves (on seedlings and coppice growth) disjunct, >2cm wide with rounded base, crinkled irregular margins and fine stellate hairs, soon replaced by intermediate then glossy lanceolate adult leaves to 12.5 x 2.5cm. Buds 5–7mm long in sessile clusters of 11–15, broadly spindle-shaped, pointed. Fruit 6–7 x 8–10mm, hemispherical to almost globular; disc broad, flat; valves enclosed or at rim level. **HABITAT:** Sandy and gravelly loams and on shale capping. **DIST:** Coast and t'lands of NSW. Also Vic. **FL:** Jan–Aug.

Eucalyptus globoidea

Eucalyptus oblonga
Narrow-leaved Stringybark, Sandstone Stringybark

Slender, irregularly branched small tree with 1 or several trunks 20–30cm diam. Juvenile and adult leaves and flower clusters similar to *E. globoidea*. Fruit 5–7 x 4–7mm, hemispherical to nearly globular with small orifice; disc ascending; valves at rim level or slightly exserted. **HABITAT:** Dry open forest on poor sandstone soils. **DIST:** Yerrinbool to Gosford and lower Blue Mtns. **FL:** Jan–Apr. Hybrids often flower winter to spring.

Eucalyptus sparsifolia is regarded by some authorities as a separate species, by others as included in, or synonymous with, *E. oblonga*. It

Eucalyptus oblonga

has linear juvenile leaves, adult leaves often <15mm wide, narrow buds with conical operculum and pedicels <1.5mm long. It intergrades with about 10 species.

Eucalyptus imitans is closely related. It is smaller (to 10m tall) with leaves up to 3.8cm wide (to 2.8cm for *E. oblonga*), buds with rounded operculum and fruit to 9mm across, but generally very similar. It is restricted to plateaus in the Tallong–Nerriga–Nowra district and intergrades with *E. sparsifolia*.

Eucalyptus ligustrina
Privet-leaved Stringybark
Mallee or dense tree to 6m high, often 2–3m, with matted criss-crossed fibrous bark on trunk and main branches and flaky or smooth bark on upper branches. Leaves thick, obscurely veined, dark green both sides, 4–7 x 0.7–1.5cm. Buds small, sessile, in clusters of 7 or more; buds and stamens often yellowish. Fruit crowded, hemispherical to truncate-globular, 4–5 x 4–5mm; disc descending; valves to rim level. **HABITAT:** Exposed shallow sites on sandstone. **DIST:** Coast and t'lands of NSW, locally on coastal escarpment near Wollongong (Sublime Point–Brokers Nose), Barren Grounds, West Dapto, above Wentworth Falls. **FL:** May–June, occasionally spring.

Eucalyptus pilularis
Blackbutt
Medium to tall tree with dark finely fibrous bark at base or on most of trunk shedding in strips and leaving upper trunk and limbs smooth greenish white. Adult leaves stalked, lanceolate, concolorous, to 16 x 3cm. Buds and flowers in axillary clusters of 7–15 on flattened peduncles 1–2cm long; operculum pointed or beaked. Fruit hemispherical or truncate-globular, about 10 x 10mm, on pedicels to 5mm long; disc level or descending; valves enclosed. **HABITAT:** Tall open forest in valleys, often on deep sandy soils on slopes above creeks or rainforest. **DIST:** Coast and foothills of ranges in NSW and se. Qld. **FL:** Sept–Mar.

A common and conspicuous tree in coastal ranges of NSW and probably the most important timber tree in coastal forests.

Eucalyptus obliqua
Messmate
Large tree with thick, fibrous, rough brown-grey bark extending to small branches. Juvenile leaves very broad, discolorous with oblique base and abruptly pointed tip, narrower and concolorous in intermediate and adult stages. Adult leaves 10–15 x 2–3cm, lanceolate, glossy dark green. Buds in axillary clusters of 11–15; operculum rounded with tiny raised point. Fruit ovoid to globular-truncate, 8–9mm across with descending disc and 3 or 4 enclosed valves. **HABITAT:** Deep heavy soils. **DIST:** Near Robertson and Mittagong–Wingello; from Tas to se. Qld; also SA. **FL:** Dec–Mar.

Eucalyptus ligustrina

Eucalyptus obliqua

FLOWERING PLANTS | 245

Eucalyptus pilularis

Eucalyptus fastigata

Eucalyptus fastigata
Brown Barrel
Large tree with rough fibrous brown bark peeling from upper trunk and branches in fine ribbons. Leaves lanceolate, to 16 x 2.5cm, with fine lateral veins at small angle to midrib. Flower clusters with 11–15 flowers, often paired in leaf axils; peduncles to 14mm long. Buds 4mm across with conical or beaked operculum. Fruit 6–8 x 5–7mm; disc flat or slightly convex; valves 3, slightly exserted. **HABITAT:** Deep fertile soils in cool moist t'lands. **DIST:** Bowral–Mittagong district, Macquarie Pass, Mt Wilson, Mt Tomah, Mt Irvine, Browns Gap, Hassans Walls, near Jenolan Caves. From NT to se. Vic. **FL:** Dec–Feb.

Pairing of flower clusters is unusual and distinctive.

Eucalyptus oreades
Blue Mountains Ash
Tall straight-shafted tree in valleys and hillsides, shorter and spreading on plateaus. Old bark deciduous in large strips, leaving smooth white upper trunk and branches, and small stocking of dark rough bark at base with strips hanging from it. Adult leaves to 17 x 2.5cm, lanceolate, concolorous with acute lateral veins characteristic of ash group. Buds curved inwards, in clusters of 7 on long flattened peduncles. Fruit nearly sessile, about 10 x 10mm, barrel-shaped (ovoid-truncate) with descending disc and 4 or 5 enclosed valves. **HABITAT:** Open forest on deep Hawkesbury sandstone soils, slopes at edge of basalt caps, tall open forest bordering rainforest. **DIST:** Blue Mtns above Springwood, Nattai V. to Qld border. **FL:** Jan–Feb.

Readily distinguished from *E. cypellocarpa* by absence of bud scar below the operculum.

Eucalyptus luehmanniana
Yellow-top Ash
Mallee with numerous stems 3–6m high and <10cm diam; lower stems smooth and whitish; young branchlets yellow, 4-sided, narrowly

Eucalyptus oreades

Eucalyptus luehmanniana

Eucalyptus consideniana

winged. Adult leaves 15–20 x 3–5cm, falcate, thick, grey-green with obscure venation; petioles 3cm long, yellow, flattened, twisted. Buds and new growth often glaucous. Flowers in clusters of 11–15 on broad flattened striate peduncles; operculum pointed. Fruit about 12 x 10mm, ribbed with wide rim and 5 enclosed valves. **HABITAT:** Low fertility sandy soils. **DIST:** Scattered. Bulli to Gosford (e.g. Royal NP, Ku-ring-gai Chase NP). **FL:** July–Dec.

Eucalyptus consideniana
Yertchuk
Small to medium crooked tree with rough, finely furrowed grey over yellow bark persistent to small branches. Adult leaves to 18 x 3cm, leathery, with irregular lateral veins at small angle to midrib. Flowers in clusters of 11–15 on slightly flattened peduncles 12mm long; operculum short, rounded with small point. Fruit funnel-shaped, 8mm across with broad disc, usually 4 slightly exserted valves. **HABITAT:** Open forest on skeletal sandy or clayey soils. **DIST:** Uncommon. Coast and t'lands from Gippsland to Blue Mtns (e.g. Clarence, Warrimoo–Hazelbrook, Wedderburn, Audley, Wingello). **FL:** Nov–Dec.

Eucalyptus sieberi

Eucalyptus multicaulis

Eucalyptus pauciflora subsp. *pauciflora*

Eucalyptus sieberi
Black Ash, Silvertop Ash
Tree of variable form depending upon location. Bark thin and flaky on young trees, dark brown to black, thick, hard and furrowed on trunk and main branches on mature trees; upper branches white and smooth; smallest branchlets shiny red. Leaves lanceolate, to 15 x 2.8cm, concolorous, fairly thick, glossy and smooth with obscure irregular veins at small angle to midrib. Buds, flowers and fruit tapering to pedicel. Operculum short, rounded with small point. Flowers white. Fruit 8mm across with broad flat red disc and usually 3 slightly enclosed valves. **HABITAT:** Coast and ranges in open forest and woodland on variety of sites and soils. **DIST:** Melbourne to Watagan Mtns; also ne. Tas. **FL:** Sept–Jan.

E. sieberi is related to *E. consideniana* and has a similar range but is much more common in the Sydney district.

Eucalyptus multicaulis
Whipstick Mallee Ash
Uncommon multi-stemmed tree to 6m high closely related to *E. sieberi*, differing mainly in its mallee habit, short-fibred bark at the base and slightly smaller buds and fruit. **HABITAT:** Dry sandstone ridges and hillsides. **DIST:** Scattered. Blue Mtns (Faulconbridge–Woodford), coast (Bulli–Gosford), sth to Budawangs. **FL:** Variable, mainly June–July.

Eucalyptus pauciflora subsp. *pauciflora*
Snow Gum, White Sally
Small to medium tree, often of poor form; trunk smooth, white mottled with cream and grey, often with insect scribbles, sometimes with persistent rough bark at base. Juvenile leaves to 20 x 7cm, thick and greyish. Adult leaves 7–15 x 2–3cm, glossy green with lateral veins diverging only slightly from midrib. Fruit tapering to short pedicel, 10mm across with broad disc and 3 valves. **HABITAT:** Cold woodland on a variety of soils. **DIST:** Upper Blue Mtns in Clarence–Wolgan area, Wallerawang to Hampton and Boyd Plateau, a few occurrences near Bell and in Hartley and Kanimbla Valleys, fairly common in Berrima–Marulan and Moss Vale–Bundanoon districts. NSW, Vic and Tas. **FL:** Oct–Apr.

Contrary to the specific name, which means 'few-flowered', this species often flowers profusely in late spring and summer.

Eucalyptus gregsoniana

Eucalyptus gregsoniana
Mallee Snow Gum, Wolgan Snow Gum
Mallee with several thin stems from a lignotuber or rootstock, usually 2–3m tall. Related to *E. pauciflora* but differing in its mallee habit, narrower leaves 12–25mm wide and short peduncles <8mm long. **HABITAT:** Scattered in low scrub in skeletal sandstone soils at high altitudes, often in association with *E. stricta*. **DIST:** Mt Wilson to Clarence and Wolgan V., also Budawangs. **FL:** Nov–Dec.

Eucalyptus dendromorpha
Giant Mallee Ash, Budawang Ash
Medium to tall tree with dark grey fibrous bark on lower or entire trunk shed in ribbons, smooth greyish or greenish above. Leaves lanceolate-falcate, 12 x 2.5cm, glossy green with many oil dots. Flowers in clusters of 7–11 on flattened peduncles 10–15mm long. Operculum conical with small pointed tip. Fruit globular-truncate or urn-shaped, about 10mm across with thin rim, descending disc and 4 enclosed valves. **HABITAT:** Moist slopes above cliffs on sandstone. **DIST:** Wentworth Falls–Blackheath, Wynne's Rocks at Mt Wilson, escarpment from Mt Keira to Barren Grounds, Fitzroy and Carrington Falls, Sugarloaf Mtn and Budawangs. **FL:** July–Dec.

This, possibly relict, species is closely related to *E. stricta* and *E. obstans* but differs in its tree form.

Eucalyptus obstans
Port Jackson Mallee
E. obstans is distinguished from other mallees on the coast by its thin bark shed in long narrow strips leaving smooth greenish white stems, its short, broad, thick, glossy leaves which are often held erect, its 7-flowered axillary clusters on broad flattened peduncles, its club-shaped buds with a short rounded or obtusely conical operculum, and its rather large barrel-shaped or urn-shaped fruit 10–11mm across with a narrow rim, descending disc and 4 deeply enclosed valves. **HABITAT:** Shallow sandstone soils in heath and shrubland. **DIST:** Coast. Patchy occurrence from Jervis Bay to Broken Bay. **FL:** June–Sept.

FLOWERING PLANTS | 251

Eucalyptus dendromorpha

Eucalyptus obstans

Eucalyptus burgessiana

Eucalyptus stricta

Eucalyptus burgessiana
Faulconbridge Mallee Ash
Thin-stemmed mallee to 5m tall with smooth grey bark peeling in strips to a tan-coloured surface. Juvenile leaves alternate, grey-green, thick, to 20 x 7cm on twisted stalks to 25mm long. Adult leaves similar but smaller, 14 x 2.5cm, with prominent midvein and obscure lateral veins. Fruit in clusters of 7–11 on flattened peduncles 15–20mm long; fruit similar to *E. obstans* but larger (to 13 x 13mm). **HABITAT:** Sandstone ridges and plateaus. **DIST:** Springwood, Faulconbridge, Linden, e. slopes of Mt Tomah. **FL:** Sept–Jan.

Eucalyptus stricta
Blue Mountains Mallee Ash
Mallee 1–4m tall, related to *E. obstans* but with narrower leaves and smaller buds and fruit. Adult leaves lanceolate, about 9 x 1cm, thick, dark green, glossy. Buds in mostly 7-flowered clusters, club-shaped, with numerous oil glands, 7 x 3mm. Fruit slightly constricted below rim, to 10 x 8mm; rim thin; disc descending; valves deeply enclosed. **HABITAT:** In small colonies on exposed rocky, sandy sites. **DIST:** Common from Newnes Plat. to Budawangs. Rare on coast (e.g. Terrey Hills, Stanwell Tops). **FL:** Dec–May.

Eucalyptus apiculata
Narrow-leaved Mallee Ash
Mallee to 4m tall, occasionally to 6m. Leaves narrow-lanceolate with recurved point at apex, 8–12cm x 5–7mm on flat or channelled stalks 3–5mm long, glossy, leathery, without visible veins. Clusters 7-flowered. Buds and fruit marginally smaller than *E. stricta* (to 9 x 8mm) but otherwise similar. **HABITAT:** Open forest on sandstone plateaus. **DIST:** Sporadic between Linden and Berrima (e.g. West Berrima, Wanganderry T'land, Hill Top, Mt Keira). Fl: Nov–Feb.

Eucalyptus laophila is closely related to *E. apiculata*. It grows to 7m, has adult leaves with a bluish sheen, 5–11cm x 4–10mm on flat stalks 6–10mm long, 7-flowered clusters on peduncles up to 15mm long and fruit up to 11 x 10mm. It is restricted to the Newnes Plat. (e.g. Wolgan Gap), usually near pagoda rock formations. It hybridises and intergrades with *E. stricta*.

Eucalyptus apiculata

Eucalyptus cunninghamii
Cliff Mallee Ash
Mallee 1–3m tall with erect leaves near top of thin grey-brown stems; new leaves often pinkish; older leaves greyish green, narrow-lanceolate, 60–70 x 4–7mm; apex sometimes hooked. Clusters 7-flowered, partly hidden among leaves. Fruit mostly on bare stems below leaves, rounded, slightly constricted below rim, about 7 x 7mm; rim thin; disc descending; valves enclosed. **HABITAT:** Confined to rocky slopes above cliffs of Jamison, Kedumba, Megalong and Grose Valleys in Blue Mtns. **DIST:** Sites include Kings T'land (Sunset Rock), Sublime Point, Gordon Falls, Leura, Pulpit Rock. **FL:** May–July.

Related to *E. stricta*, *E. apiculata* and *E. obstans*, but smaller with smaller buds and fruit.

Eucalyptus stellulata
Black Sally
Spreading woodland tree; trunk short with dark, rough, hard bark; branches smooth, shiny, olive green to greyish. Juvenile leaves opposite, sessile, rounded and very broad; adult leaves 6–10 x 1.5–2cm, alternate or disjunct, stalked, lanceolate or elliptic with several longitudinal veins. Buds sessile in small stellulate (star-shaped) axillary clusters of up to 20 or more, narrowly conical at each end, sometimes curiously expanded at top of floral tube. Fruit sessile, about 5 x 5mm, with descending disc and 3 enclosed valves. **HABITAT:** Cold wet places at high altitudes, often near streams and swampy flats. **DIST:** Gt Div. Ra. from Vic Alps to Qld border. **FL:** Feb–May.

Eucalyptus copulans, one of Australia's rarest plants with just 2 living examples from the head of a creek at Wentworth Falls, provides a link between *E. stellulata* and *E. moorei*.

Eucalyptus moorei
Little Sally, Narrow-leaved Sally
Mallee 2–3m tall; stems smooth, pale, with strips of old bark at base. Juvenile leaves narrow-lanceolate, <1cm wide. Adult leaves to 7 x 1cm, usually held erect, thick, dark green, glossy, dotted with oil glands; veins faint except midrib. Buds and fruit similar to *E. stellulata*; flowers in clusters of 11–15. **HABITAT:** Thickets on exposed damp sandstone slopes. **DIST:**

Eucalyptus cunninghamii

Eucalyptus stellulata

Wentworth Falls and Mt Wilson to Clarence (e.g. Narrow Neck, Minnehaha Falls). Also ST, NT. **FL:** Jan–Apr.

E. moorei differs from *E. stellulata* in its mallee habit, narrow juvenile leaves, smaller adult leaves and smaller flower clusters.

Eucalyptus moorei

Eucalyptus radiata subsp. *radiata*
Narrow-leaved Peppermint

Forest tree with pendulous branches or short-trunked spreading woodland tree with large dense canopy. Leaves thin, narrow-lanceolate, to 15 x 1.5cm, with typical peppermint venation (i.e. with fine, sparse lateral veins at very acute angle to midrib, and intramarginal vein distant from margin). Inflorescences also typical of the group (i.e. with 11–20 or more flowers radiating from centre of cluster). **HABITAT:** Mostly on sandy soils with heavier base. **DIST:** Blue Mtns west of Woodford, Sthn H'lands to Vic. **FL:** Oct–Jan.

Readily distinguished from *E. elata* by its rough grey bark on upper trunk and large branches.

Eucalyptus radiata subsp. *radiata*

Eucalyptus elata
River Peppermint
Tree with hard dark fissured bark on lower trunk, shed from upper trunk and branches in long hanging ribbons leaving smooth white surface. Leaves similar to *E. radiata*. Clusters with up to 40 flowers. Buds club-shaped, 2–5mm long. Fruit globular, 4–6mm across; disc usually descending (flat in *E. radiata*). **HABITAT:** Fertile alluvial soils on banks of rivers and larger creeks, also sheltered slopes. **DIST:** Coast and ranges from Gippsland to Putty; locally Bowral, Kangaroo R., Mt Kembla, Thirlmere Lakes, Bents Basin, Cattai Ck, Upper Colo, Macdonald V. **FL:** Aug–Nov.

Eucalyptus piperita subsp. piperita
Sydney Peppermint
Common tree, straight-trunked to 25m tall at its best, short-boled with spreading branches and 15m tall on shallow exposed sites. Trunk and larger branches with shortly fibrous grey bark; upper branches smooth or with ribbony bark. Canopy with juvenile, intermediate and adult leaves; adult leaves to 12 x 2cm with oblique base and fine pointed apex. Buds in stellate clusters of 11 or more; operculum conical, as long as floral tube. Fruit to 7 x 7mm, ovoid (barrel-shaped); disc descending; valves enclosed. **HABITAT:** Open forest on sandstone soils, preferring valley slopes. **DIST:** Coast and t'lands from Bullahdelah to Illawarra, where it intergrades with subsp. *urceolaris*. **FL:** Nov–Jan.

Eucalyptus oil, one of the first exports from the Sydney colony, was distilled from leaves of this species by Considen, Surgeon of the First Fleet. A tree of this species, known as the Explorers Tree, was marked by Blaxland, Wentworth and Lawson in 1813 during the first crossing of the Blue Mtns.

Eucalyptus piperita subsp. urceolaris
Urn-fruited Peppermint
Similar to subsp. *piperita* except tree often larger (possibly due in part to better growing conditions), leaves longer (to 18cm), operculum narrowly pointed or beaked, much longer than floral tube, and fruit to 9 x 7mm, urceolate (contracted below rim, necked). **HABITAT:** Open forest on sandy soils. **DIST:** Batemans Bay to Sthn H'lands. **FL:** Jan–Feb.

This subspecies intergrades with subsp. *piperita* south of Macquarie Pass.

Eucalyptus haemastoma
Scribbly Gum
Mallee or tree with short deformed trunk or medium-sized tree with light crown. Trunk and branches with smooth patchy white to yellowish or greyish bark marked by insect scribbles. Leaves thick, broad, leathery, up to 15 x 3cm. Peduncles 12–25mm long. Buds 7 x 4–5mm, club-shaped with short rounded or conical operculum. Fruit 8 x 8mm, hemispherical or pear-shaped with thick rim and broad reddish disc. **HABITAT:** Open forest, scrub and heath. **DIST:** Eastern part of Sydney district. **FL:** Chiefly Nov–Mar.

The scribbly gums form a complex in which differences between species are not always clear-cut. This is one of 3 closely related intergrading species on the central coast of NSW.

Eucalyptus rossii
Scribbly Gum, White Gum
Tree, usually 10–20m tall with short trunk and large irregularly branched crown; trunk smooth to base, usually mottled cream to grey due to irregular shedding of old bark and marked by insect scribbles. Juvenile and adult leaves alternate, stalked, narrowly lanceolate, 7–15 x 1–1.5cm, greyish green, concolorous. Operculum short, rounded. Fruit hemispherical to globose, to 5 x 6mm, on slender peduncles and pedicels. **HABITAT:** Woodland in seasonally cold low-rainfall sandy plateaus, ridges and slopes. **DIST:** Capertee and Wolgan valleys, tops and w. slopes of Gt Div. Ra. **FL:** Dec–Feb.

Eucalyptus piperita subsp. *piperita*

FLOWERING PLANTS | 255

Eucalyptus elata

Eucalyptus piperita subsp. *urceolaris*

Eucalyptus haemastoma

Eucalyptus rossii

Eucalyptus racemosa

Eucalyptus sclerophylla
Hard-leaved Scribbly Gum

Medium to tall tree with open crown. Bark smooth, white to grey with insect scribbles. Leaves similar to *E. haemastoma*. Peduncles 8–12mm long. Buds 5 x 3mm; operculum rounded; bud scar absent. Fruit 5 x 5mm, hemispherical, disc flat; valves at or below rim level. **HABITAT:** Open forest on shallow to deep sandy soils. **DIST:** Blue Mtns and t'lands from Howes V. to Sassafras, also coast (e.g. Holsworthy, East Hills, Agnes Banks).
FL: Dec–Mar.

This is the common scribbly gum throughout the Blue Mtns where it often resembles *E. mannifera* with which it grows. The latter species has flower buds with a scar, and fruit with an ascending disc and exserted valves.

Eucalyptus racemosa
Scribbly Gum, Snappy Gum

Tree, usually taller and less misshapen than *E. haemastoma*. It has thinner, narrower juvenile and adult leaves (up to 15 x 1.5cm) and smaller buds (to 5 x 5mm) and fruit (to 5 x 7mm).
HABITAT: Deep sandy and silty soils. **DIST:** Coast, Pokolbin to Sydney Harbour. 'Pure' forms occur n. and nw. of Sydney. Intergrades occur sth of Sydney Harbour. **FL:** Aug–Nov.

Eucalyptus racemosa and *E. haemastoma* occupy different habitats and normally flower at different times, hence opportunities for hybridisation are limited. Trees regarded as intergrades are common in some areas from which one or both species may be absent.

Eucalyptus sclerophylla

EUCALYPTUS (SUBGENUS *SYMPHYOMYRTUS*)

Symphyomyrtus occurs in all parts of Australia in a wide range of environments, usually on deeper, more fertile soils than the other groups. Hybridisation occurs frequently. In all sections except one, the outer operculum is shed at an early stage of bud development leaving a scar around the bud. In the Sydney area, 3 box species (*E. moluccana, E. bosistoana, E. melliodora*) and one ironbark (*E. sideroxylon*) shed the double operculum (calyx and corolla) in one piece at flowering and therefore lack a bud scar. *Symphyomyrtus* includes a large section of medium to large rough- and smooth-barked trees occurring throughout the highlands of se. Australia in which juvenile leaves in opposite pairs are retained from sapling to mature tree. Another section includes the ironbarks and boxes with characteristic terminal paniculate inflorescences.

Note: The following species are arranged according to relationships within this subgenus.

Eucalyptus deanei

Eucalyptus deanei
Mountain Blue Gum, Round-leaved Gum
Tall shaft-like tree with smooth white bark showing patches of cream and pale grey-blue, occasionally with a little rough flaky bark at the base. Juvenile leaves large (6–15 x 4–8cm), ovate to orbicular. Intermediate leaves large, ovate to broad-lanceolate, common on mature trees among adult leaves which are narrower (8–14 x 2–3.5cm). Flowers in axillary clusters of 7–11 on angular or flattened peduncles. Buds pedicellate; operculum hemispherical with pointed apex. Fruit bell-shaped, 6 x 6mm with 3 or 4 short broad valves. **HABITAT:** Tall open forest on deeper loams in e. valleys of ranges and on soils derived from volcanic intrusions (e.g. Murphy's Glen). **DIST:** Picton Lakes to Hunter R., also McPherson Ra. on NSW/Qld border. **FL:** Apr–Aug.

Eucalyptus saligna

Eucalyptus saligna
Sydney Blue Gum
Tall tree resembling *E. deanei* and *E. grandis*, but with more coastal distribution than former and more southerly distribution than latter. In gullies near Gosford, where their ranges overlap, *E. saligna* is distinguished from *E. deanei* by its rough persistent bark at base of trunk, sessile buds with conical operculum and fruit with 3–4 narrowly pointed, spreading exserted valves. Further north, *E. grandis* can be recognised by its fruit with 4–5 broad blunt incurved valves. From Wollongong to Batemans Bay, *E. saligna* intergrades or hybridises with rough-barked *E. botryoides*, producing trees in a variety of habitats with intermediate characteristics. **HABITAT:** Tall open forest on deep, moist but well-drained alluvial loams. **DIST:** Port Jackson nth. Locally common on Nth Shore (e.g. Hornsby V., Lane Cove). Also NC, NT, Qld. **FL:** Jan–Apr.

Eucalyptus botryoides

Eucalyptus botryoides
Bangalay, Southern Mahogany
Spreading tree with dense crown and brownish, flaky-fibrous, thick, fissured bark on trunk and large branches. Leaves thick, glossy, dark green above, paler below, with close regular fine veins at large angle to midrib. Buds and fruit sessile in clusters of 7 on flattened peduncles. Buds 9 x 5mm; operculum rounded-conical. Fruit to 12 x 9mm. **HABITAT:** Sandy soils near salt water, coastal scrub on or behind sand dunes. **DIST:** Bairnsdale (Vic) to nth of Newcastle (NSW). **FL:** Mainly Dec–Apr.

Eucalyptus robusta
Swamp Mahogany
Rough-barked tree. Leaves similar to *E. botryoides*. Buds and fruit stalked, in clusters of 9–15 on flattened peduncles. Buds 20 x 7mm; operculum beaked. Fruit to 18 x 12mm, valves 4, joined at tips across mouth of fruit. **HABITAT:** Saline swampy ground near estuaries and lagoons. **DIST:** Moruya to Rockhampton, locally at Kurnell, La Perouse, Warriewood Wetlands, Avalon. **FL:** Apr–Sept.

Eucalyptus scias
Large-fruited Red Mahogany
Medium-sized tree; bark thick, rough, fibrous on trunk and branches. Flowers in axillary clusters of 3 or 7 on flattened peduncles. Buds >10mm across; operculum rounded with short point to conical with long beak, often broader than calyx tube. Fruit 10–17mm across with prominent concave rim, flat disc and 4 strongly exserted valves. **HABITAT:** Open forest on

Eucalyptus robusta

Eucalyptus scias

Eucalyptus notabilis

sandstone ridges and hillsides near the sea. **DIST:** Sporadic. Bega to Cessnock, mostly nth of Sydney (e.g. McCarrs Ck, Warrah Trig, Bouddi NP), also sth of Helensburgh. **FL:** Nov–Feb.

Eucalyptus notabilis
Blue Mountains Mahogany

Medium to fairly large tree with rough, soft, fibrous, fissured light grey to red-brown bark throughout. Leaves on flattened stalks, lanceolate, 10–16 x 2–3cm, dark green, glossy, discolorous. Flowers in clusters of 7–11 on flattened peduncles. Buds 5–6mm across; operculum obtusely conical, broader but not longer than calyx tube. Fruit hemispherical, 9–10mm across with 3–4 exserted valves. **HABITAT:** Sandy or clayey soils of medium fertility and depth. **DIST:** Lower slopes of Blue Mtns and nearby Cumberland Pl., Colo and Putty districts. Also NT, se. Qld. **FL:** Jan.

This species intergrades with *E. resinifera*.

Eucalyptus resinifera subsp. *resinifera*
Red Mahogany

Medium to tall densely crowned tree with persistent soft, fibrous, reddish brown fissured bark on trunk and branches. Leaves lanceolate to 16 x 3cm on channelled stalks 15–30mm

Eucalyptus resinifera subsp. *resinifera*

long, discolorous, dark green and glossy above with wide-angled transverse veins. Buds 5–7mm across; operculum conical, often beaked, longer but not broader than floral tube. Fruit similar to *E. notabilis* but valves more exserted and out-turned. **HABITAT:** Open to tall open forest on sheltered slopes and valley flats. **DIST:** Coast. From Jervis Bay to n. Qld. **FL:** Nov–Jan.

The epithet resinifera ('resin-bearing'), appears to be a mistaken reference to kino, which exudes profusely from *Angophora costata* and the bloodwoods but not from the mahoganies.

Eucalyptus punctata

Eucalyptus squamosa

Eucalyptus longifolia

Eucalyptus punctata
Grey Gum
Mallee-like tree on poor sites to medium-sized woodland or forest tree. Bark thin, mottled, matt or granular, shedding in large irregular flakes to expose orange new bark which weathers to light and dark grey. Leaves lanceolate, 8–15 x 1.5–3cm, glossy dark green above, pale below with numerous fine transverse veins. Flowers in clusters mainly of 7 on flattened or angular peduncles; buds ovoid or obtusely pointed; scar at or below mid-height. Fruit 10mm across with thick rim, raised disc and 3–4 short exserted valves.
HABITAT: Open forest on sandy, occasionally rocky soils or on sandy soils with a clay base. **DIST:** Coast from Jervis Bay to upper Hunter V.; Blue Mtns below Springwood and Bilpin. **FL:** Dec–Mar.

Eucalyptus longifolia
Woollybutt
Spreading tree with low crooked branches to straight-trunked tree 30m tall with rough grey flaky-fibrous bark and pendulous dull greyish green lanceolate leaves to 20cm x 25mm. Clusters 3-flowered, large, loose, pendulous on slender peduncles 15–30mm long with pedicels 5–20mm long. Buds about 20 x 10mm; operculum conical, as long as calyx tube. Fruit 10–15mm across; rim formed by broad bud scar sloping upwards to narrow staminal ring; disc descending; valves at or below rim level.
HABITAT: Open forest on moist alluvial soils derived from shales, often on flats. **DIST:** Coast (e.g. Dapto, Milperra, Castlereagh NR), Vic to Newcastle. **FL:** Feb–May.

Eucalyptus squamosa
Scaly-bark
Straggly tree with short trunk and crooked, often pendulous branches. Bark grey, rough and scaly or flaky on trunk and branches. Leaves lanceolate with fine tapered point, to 12 x 1.5cm, greyish green, dull, thin. Clusters of 7–13 flowers, often paired in leaf axils. Operculum conical. Fruit 5–6 x 6–7mm, bell-shaped with 3 exserted valves. Buds and fruit frequently disfigured by warty outgrowths.
HABITAT: Sandstone ridges and plateaus. **DIST:** Coast (e.g. Engadine–Bulli, Lucas Heights, Sackville, Winmalee, Cowan). **FL:** Irregular, mainly Aug–Oct.

Eucalyptus parramattensis subsp. *parramattensis*

Eucalyptus parramattensis subsp. *parramattensis*
Drooping Red Gum
Tree usually <8m tall with slender trunk of poor form and drooping branches; bark smooth, mottled white, grey and dark grey. Leaves lanceolate, dull, to 20 x 3cm. Flowers in clusters of 7. Buds 4–8 x 4–5mm; operculum conical or rounded, longer than calyx tube. Fruit hemispherical, about 6 x 8mm. **HABITAT:** Low-lying flat clayey soils, alluvial sandy soils with clay subsoil. **DIST:** Scattered. Coast (e.g. Agnes Banks, Hammondville, East Hills, Thirlmere, Bargo, Glenbrook–Lapstone, Wallaby Swamp on Putty Rd). Also NC, CWS. **FL:** Nov–Jan.

Eucalyptus amplifolia subsp. *amplifolia*
Cabbage Gum
Tree to 30m tall, sometimes forking close to ground. Juvenile leaves on saplings ovate to orbicular, to 20 x 15cm, dull green. Intermediate and adult leaves 15–20 x 2.5–3.5cm. Flowers in clusters of mostly 11–15, pedicellate. Buds with long conical operculum, to 20 x 5mm. Fruit globose, to 6 x 8mm; disc ascending; valves exserted. **HABITAT:** Low-lying clay soils with high watertable, often in small stands. **DIST:** Common w. of Liverpool (e.g. Cobbitty, Badgerys Ck, South Ck, Eastern Ck). Coast and t'lands. **FL:** Nov–Jan.

Eucalyptus amplifolia subsp. *amplifolia*

Eucalyptus tereticornis
Forest Red Gum
Medium to tall tree with straight trunk, ascending major branches and large open crown. Bark matt, mottled, white, grey and bluish grey due to shedding of old bark in large sheets and flakes, frequently with a little rough dark bark at base. Juvenile leaves on saplings ovate to broad-lanceolate, to 20 x 10cm, glossy blue-green. Adult leaves lanceolate, to 20 x 2.5cm, glossy green both sides. Flowers in clusters of 7–11, frequently 7. Operculum conical or flared at open end, much longer than calyx tube. Fruit ovoid to globose, to 6 x 8mm. **HABITAT:** Well-drained moist alluviums, often with clay subsoil, mostly on slopes and hillsides. **DIST:** Common w. of Liverpool. Coast and t'lands from Vic to PNG. **FL:** Aug–Nov.

Eucalyptus ovata
Swamp Gum
Medium to tall tree with straight trunk and dense crown or <15m tall with spreading branches in exposed parts of Sthn H'lands. Bark shed in strips from branches and most of trunk leaving smooth whitish surface and short stocking of rough grey bark. Leaves ovate or lanceolate, to 15 x 3cm, glossy green, often with undulate margins and acuminate tip. Inflorescences with 7-flowered clusters. Operculum conical or beaked. Fruit tapering or bell-shaped with flat disc and valves at or slightly below rim level. **HABITAT:** Poorly drained flats and cold swampy ground at high elevations. **DIST:** Wingello, Tallong and Marulan districts. South from Oberon to Vic, Tas and SA. **FL:** Mar–Sept.

Eucalyptus aggregata
Black Gum
Tree with curly-flaky dark bark on trunk and branches; branches twisted. Adult leaves lanceolate, about 10cm x 18mm, medium green, dullish and concolorous with faint, widely spaced, very oblique lateral veins; intramarginal vein distant from edge. Flowers in crowded clusters of 7 on short peduncles and pedicels; buds tapering to both ends. Fruit hemispherical, 4–5 x 4–5mm, with thick rim, flat disc and 3–4 slightly exserted valves. **HABITAT:** Cold damp flats and hollows on t'lands. **DIST:** Wallerawang, Hartley V., Berrima, Wingello. Also CT, ST. **FL:** Dec–Feb.

Eucalyptus tereticornis

Eucalyptus ovata

Eucalyptus aggregata

Eucalyptus mannifera
Brittle Gum, Powder Bark, Mountain Spotted Gum
Slender tree with smooth powdery white or grey bark, sometimes with patchy older bark. Juvenile leaves alternate, dull green, linear to broad-lanceolate on short stalk. Adult leaves dull grey-green, lanceolate, 10–20 x 1–3cm. Flowers on short pedicels in clusters of 7; buds 3–6mm long. Fruit rounded or tapering, 4–7 x 5–7mm, with ascending disc and 3 short exserted valves. **HABITAT:** Woodland on shallow sandstone and granite soils. **DIST:** Blue Mtns above Lawson, Sthn T'lands, higher ranges. Also CT, ST, Vic. **FL:** Oct–Mar.

A polymorphic species including forms previously regarded as subsp. *mannifera*, *maculosa* and *gullickii*.

Eucalyptus mannifera

Eucalyptus bridgesiana
Apple Box
Woodland tree with short bole and large spreading crown with crooked branches. Bark rough, pale grey, soft, friable, tessellated, persistent on trunk and branches. Juvenile leaves retained on saplings, opposite, glaucous, broadly heart-shaped with finely toothed margins. Adult leaves on long stalks, narrowly lanceolate, to 25 x 3.5cm, dark green, slightly glossy. Flowers in clusters of 7 on slightly flattened peduncles; fruit hemispherical, about 6 x 6mm, with fine concave ring surrounding the broad disc; valves 3–4, exserted. **HABITAT:** Open forest and woodland in deep moist loamy soils on slopes. **DIST:** T'lands and w. slopes from Qld to Vic, extending to coast in places. **FL:** Jan–May.

Eucalyptus bridgesiana

Eucalyptus cypellocarpa

Eucalyptus maidenii

Eucalyptus quadrangulata

FLOWERING PLANTS | 265

Eucalyptus macarthurii

Eucalyptus cypellocarpa
Mountain Grey Gum
Tall tree with smooth yellow-white to greyish bark shed in ribbons. Intermediate and adult leaves stalked, dark glossy green both sides, frequently very large (15–35 x 2–5cm). Flowers in clusters of 7 on flattened peduncles. Buds elongated, to 12 x 5mm, with ribs or angles continuous from pedicel to floral tube, scar present; operculum much shorter than floral tube. Fruit goblet-shaped, to 10 x 9mm, valves mostly 3, usually below thin rim. **HABITAT:** Tall open forest on sheltered slopes and valleys with deep fertile soils. **DIST:** Mountain Lagoon, Mt Tomah, Mt Wilson, Megalong and Kanimbla valleys, Bundanoon, Penrose. Mainly on ranges from Tamworth to Vic, coastal in sth. **FL:** Dec–Feb.

Eucalyptus maidenii
Maiden's Gum
Tall smooth-barked tree with rough bark peeling in strips at base; branchlets and inflorescences sometimes glaucous. Adult leaves 12–25 x 1.5–2.5cm, on stalks 1.5–3cm long. Flowers in clusters of 7 on broad flattened peduncles 1–2.5cm long; buds 10 x 7mm, sessile, angular or ribbed. Fruit tapering to base, about 10 x 10mm with broad raised disc and 3 or 4 exserted valves. **HABITAT:** Mountain valleys, slopes and tops. **DIST:** Wingello to Vic. **FL:** Mar–Sept.

Eucalyptus bicostata is closely related to *E. maidenii* and has sessile, glaucous, 3-flowered inflorescences with larger buds and fruit. It occurs at Jenolan Caves, Nullo Mtn near Rylstone and in other isolated stands on the ranges of NSW and Vic.

Eucalyptus quadrangulata
Coast White Box, **White-topped Box**
Tall forest tree with fine grey box-like bark. Juvenile leaves opposite, discolorous, stem-clasping for many pairs on square winged stems. Adult leaves narrow-lanceolate with scalloped margins. Flowers in clusters of 7 on slightly flattened peduncles; pedicels absent or very short. Fruit tapering to base; disc level; valves 3–4, exserted. **HABITAT:** Tall open forest on fertile heavy soils. **DIST:** Coast and e. slopes of ranges from Qld border to Bundanoon (e.g. Stanwell Park, Mt Kembla, Macquarie Pass, Robertson, Colo Vale). **FL:** Jan–Mar.

Eucalyptus macarthurii
Paddy's River Box, **Camden Woollybutt**
Woodland tree. Bark thick, coarsely fibrous, shed imperfectly from upper branches. Leaves on saplings and sucker growth opposite, sessile, heart-shaped, becoming alternate, stalked, broad-lanceolate, finally 8–16cm x 8–16mm at adult stage. Leaves and freshly cut bark with strong geranium scent. Flowers nearly sessile in clusters of 7. Fruit hemispherical to bell-shaped, to 5 x 6mm; disc broad, ascending; valves 3, exserted. **HABITAT:** Heavy alluvial moist soils in cold parts of t'lands. **DIST:** Jenolan Caves, Boyd Plat., Mittagong to Goulburn. Common in Berrima–Burradoo–Paddy's R. area. **FL:** Nov–Feb.

Eucalyptus smithii
Gully Gum

Tall shaft-like tree with light crown in coastal gullies, shorter with denser crown in Sthn H'lands. Bark hard, dark, furrowed, smooth white on upper trunk and branches. Adult leaves narrow-lanceolate, acuminate, 10–18 x 0.8–1.5cm, green, concolorous. Flowers in clusters of 7 on flattened peduncles 10mm long; pedicels slender, 2–6mm long. Operculum conical. Fruit ovoid, 5–7 x 4–6mm; valves 3, exserted. **HABITAT:** Moist clay loams on slopes and flats, also poorer sandstone sites. **DIST:** Mt Kembla, West Dapto, Robertson, Hill Top, Mittagong. Also CT, SC, ST, Vic. **FL:** Jan–Mar.

Eucalyptus viminalis
Ribbon Gum, Manna Gum

Tall forest tree or spreading shady tree in open country. Bark rough at base, shed in ribbons leaving some old grey bark over part or most of trunk. Juvenile leaves opposite, sessile, lanceolate, 6–12 x 1.5–3cm, pale green. Adult leaves narrow-lanceolate, 10–20 x 0.8–2.5cm, concolorous. Flowers in clusters of 3 on flat or angular peduncles 6mm long; pedicels short. Buds ovoid, 5–8mm long. Fruit 5–8 x 5–9mm; disc raised; valves 3–4, exserted. **HABITAT:** Forest and woodland in valleys and cool highlands. **DIST:** Coast and t'lands of se. Aust; locally at Mt Wilson and Megalong, Hartley and Kanimbla Valleys. **FL:** Jan–May.

E. viminalis is sometimes difficult to distinguish from *E. dalrympleana* and *E. rubida*, particularly the former, but can be recognised by the narrow juvenile leaves on seedlings and saplings near the parent tree.

Eucalyptus benthamii
Camden White Gum, Nepean River Gum

Tree to 40m tall with ascending main branches. Bark smooth, white to cream or greyish, often with peeling brown old bark near base. Juvenile leaves sessile, broadly heart-shaped, greyish or subglaucous. Adult leaves lanceolate, 10–13 x 1.5–2cm, concolorous, green, dull, thin. Flowers and fruit nearly sessile in clusters of 7. Buds club-shaped. Fruit 5 x 5mm, disc raised; valves 3 or 4, slightly exserted. **HABITAT:** Sandy flats or ridges near streams. **DIST:** Lower Kedumba Ck and Bents Basin; isolated trees Wallacia–Camden and in Reedy and Cedar Cks. **FL:** Apr–May.

Eucalyptus dalrympleana subsp. dalrympleana
Mountain Gum

Tall, straight-trunked tree. Bark smooth, white, blotched due to patchy shedding of old bark,

Eucalyptus smithii

Eucalyptus viminalis

FLOWERING PLANTS

Eucalyptus benthamii

Eucalyptus dalrympleana subsp. *dalrympleana*

rough at base. Juvenile leaves opposite, orbicular to broadly ovate, sessile, pale green, dull. Adult leaves lanceolate, 10–20 x 1.5–2.5cm, stalked, undulate, concolorous, glossy green. Flowers in clusters of 3 on peduncles 3–8mm long. Buds ovoid, to 8 x 3mm. Fruit sessile, hemispherical, to 8 x 9mm; disc ascending; valves 3 or 4, exserted. **HABITAT:** Forests on deep fertile soils. **DIST:** Mt Wilson, Boyd Plateau, Wombeyan Caves, Jenolan Caves; uncommon Blackheath to Clarence. Along Gt Div. Ra. in NSW and Vic, also Tas. **FL:** Mar–June.

This sp. can be distinguished from *E. viminalis* by differences in juvenile leaves and by its preference for slopes and plateau tops rather than valleys.

Eucalyptus rubida subsp. *rubida*
Candlebark

Tree to 40m but usually much smaller, low-branched, spreading and irregular. Bark decorticating unevenly leaving patches of reddish brown old bark on trunk and branches and dark bark at the base. Juvenile leaves sessile, opposite, orbicular on seedlings and saplings. Adult leaves lanceolate, 9–15 x 1–2.4cm, glaucous to dull grey or bluish green. Flowers sessile in clusters of 3 on short peduncles. Fruit 5 x 6mm. **HABITAT:** Woodland and open forest on variety of soil types. **DIST:** Plateaus and

Eucalyptus rubida subsp. *rubida*

valleys of upper Blue Mtns and Sthn H'lands. Also ST, SWS, Vic, Tas. **FL:** Jan–Apr.

E. rubida occurs on drier, shallower soils at lower elevations than *E. dalrympleana* and *E. viminalis*, often in frost hollows of open mountain valleys.

Eucalyptus cinerea subsp. cinerea
Argyle Apple
Straggly tree to 15m. Bark rough, fibrous, red-brown weathering to grey on trunk and larger branches, smooth and red-brown to grey above. Leaves opposite, thick, glaucous, sessile, orbicular, about 5 x 5cm, becoming broadly heart-shaped to ovate in intermediate foliage which is retained. Flowers in clusters of 3 in upper leaf axils; buds sessile, glaucous. Fruit cone-shaped, to 9 x 9mm, with broad convex disc and 3–4 exserted valves. **HABITAT:** Woodland in undulating country on infertile soils. **DIST:** Locally common. Berrima to Marulan, Tallong, Wingello. Also CT, ST. **FL:** Oct–Dec.

Eucalyptus moluccana
Grey Box
Tree to 25m tall with fibrous tessellated bark on lower trunk; upper trunk and branches smooth, pale grey. Juvenile leaves on saplings to 1m tall alternate, stalked, ovate, to 15 x 10cm, leathery, dull green. Adult leaves broadly lanceolate, about 14 x 3cm, glossy green, concolorous. Inflorescence paniculate with 7-flowered clusters. Fruit goblet-shaped, 6–9 x 5–6mm, disc depressed, valves 3–4, enclosed. **HABITAT:** Woodland in moist well-drained undulating country with clay soil or subsoil. **DIST:** Common from Parramatta and Campbelltown to Nepean R., absent from silts, gravels and sands from Penrith to Windsor. Coast and t'lands of NSW and Qld. **FL:** Feb–Jul.

Eucalyptus fibrosa subsp. fibrosa
Broad-leaved Ironbark
Medium to tall tree with trunk to half tree height. Bark grey-black, furrowed, soft and flaky on trunk and branches. Adult leaves lanceolate, 10–18 x 2–2.5cm, grey-green; intermediate leaves broader. Inflorescence terminal with clusters of 7–11 flowers. Buds 12–16 x 4–5mm, tapering to each end. Fruit 6–12 x 5–10mm; valves 4–5, usually exserted. **HABITAT:** Open forest on ridge tops and flat, poorly drained clay, silt or sandy soils. **DIST:** Kemps Creek–Mulgoa district, Penrith–Richmond, Castlereagh NR, East Hills, Menai on shale cap. From Bodalla to Rockhampton. **FL:** Nov–Feb.

Eucalyptus crebra
Narrow-leaved Ironbark
Medium-sized tree with slender trunk to two-thirds tree height and narrow pendulous branchlets and leaves. Bark hard, black, coarsely furrowed, persistent to small branches. Adult leaves lanceolate, acuminate, 7–15 x 1–1.5cm, dull, concolorous. Inflorescence paniculate with clusters of 7–11 flowers on slender peduncles to 12mm long. Operculum rounded-obtuse. Fruit 4–6mm across with thin rim and 3–4 enclosed valves. **HABITAT:** Woodland in undulating country, confined to Wianamatta shale soils in Sydney area, also Kurrajong, Glenbrook. **DIST:** Picton to n. Qld on coast and t'lands. **FL:** Aug–Dec.

Eucalyptus siderophloia is similar to *E. crebra* but has a conical-acute operculum. It occurs between Milperra and Padstow.

Eucalyptus cinerea subsp. *cinerea*

Eucalyptus moluccana

Eucalyptus fibrosa subsp. *fibrosa*

Eucalyptus crebra

Eucalyptus baueriana
Blue Box
Woodland tree with light to dark grey fibrous-flaky bark on trunk and main branches and a large bluish green crown. Juvenile and intermediate leaves opposite and orbicular, later alternate and broadly ovate, to 10 x 7.5cm, dull grey-green, thin, concolorous, often persisting on mature trees. Adult leaves ovate to broad-lanceolate, 6–10 x 3–4cm, green to glaucous, semi-glossy, with wavy margins. Flowers in clusters of 7 in terminal panicles. Buds club-shaped. Fruit tapering to short stalk 6mm across, thin, flared; disc broad, descending; valves 3–4, enclosed. **HABITAT:** River flats on heavy or loamy soils, sometimes in small stands. **DIST:** Putty district (e.g. Cobbitty, w. of Picton, Moorebank, Liverpool Cemetery) to Bairnsdale (Vic). **FL:** Aug–Sept.

Note: The outer stamens of *E. baueriana*, *E. paniculata*, *E. beyeriana*, *E. melliodora* and *E. sideroxylon* are reduced to staminodes (i.e. without anthers).

Eucalyptus paniculata subsp. *paniculata*

Eucalyptus paniculata subsp. paniculata
Grey Ironbark
Forest tree with hard deeply furrowed grey bark on trunk and branches. Adult leaves lanceolate, to 18 x 3cm, thin, discolorous, finely but densely veined. Flowers in clusters of 7. Buds angular; operculum conical, shorter and narrower than floral tube. Fruit 7–9 x 6–8mm, with thick rim, descending disc and usually 5 slightly enclosed valves. **HABITAT:** Open forest on fertile heavy soils, often as dominant species. **DIST:** Coast. Bega to Bulahdelah. **FL:** May–Nov.

Easily recognised by the light grey bark, pale leaf undersurface and the 5-valved fruit.

Eucalyptus beyeriana
Beyer's Ironbark
Woodland tree with large crown and hard dark furrowed bark on trunk and branches. Juvenile and adult leaves stalked, lanceolate, 6–12 x 0.8–2cm, dull grey-green. Buds, flowers and fruit similar to *E. crebra* except operculum shorter and narrower than floral tube and stamens inflexed in bud. **HABITAT:** Undulating country on sandy or clayey soils. **DIST:** Bargo, Kingswood, Windsor, Nortons Basin, Wheeny Ck. Scattered from Nowra to NWS. **FL:** Aug–Oct.

An uncommon species resembling *E. crebra*, but floral characters indicate membership of the *paniculata* group.

Eucalyptus baueriana

FLOWERING PLANTS | 271

Eucalyptus beyeriana

Eucalyptus melliodora
Yellow Box

Medium to tall woodland tree with large spreading crown of fine pendulous grey-green leaves and trunk less than half tree height. Bark very variable, yellow-brown or red-brown or nearly black, fibrous or flaky, thick, coarse; inner bark usually yellow. Juvenile and adult leaves concolorous, with pleasant cineol odour when crushed. Adult leaves lanceolate, 7–14cm x 8–18mm, faintly veined; intramarginal vein distinct, distant from margin. Clusters axillary, 7-flowered, crowded towards ends of branchlets. Buds small, pointed, lacking scar. Fruit hemispherical, 6 x 6mm, on slender pedicels, often with remains of staminal ring inside mouth; valves 5, enclosed. **HABITAT:** Woodland on undulating country. **DIST:** Wstn Vic to se. Qld, mainly on t'lands and w. slopes (e.g. w. of Lithgow, Hartley, Jenolan Gorge; also Kanimbla V.), absent from Blue Mtns. **FL:** Sept–Feb.

Eucalyptus sideroxylon
Mugga Ironbark

Tree with short trunk and large irregular crown with pendulous branches bearing dull grey-green lanceolate leaves to 14cm x 18mm. Bark thick, hard, deeply furrowed, dark due to kino impregnation. Clusters axillary, 7-flowered, on long slender peduncles. Flowers on long slender pedicels, white to creamy, rarely pink; stamens inflexed in bud. Fruit to 10 x 9mm; disc descending; valves 5, enclosed.
HABITAT: Open forest on clayey or gravelly low ridges. **DIST:** Holsworthy, Moorebank, Parramatta, Kemps Creek, Londonderry, Richmond. From Vic to Qld on coast, ranges, slopes and plains. **FL:** Apr–Nov.

Closely related to *E. melliodora*.

Eucalyptus microcorys
Tallowwood

Tall tree with soft, fibrous, light brown bark. Adult leaves to 12 x 2.5cm, lanceolate, glossy dark green above, dull and paler below, with fine lateral veins and minutely irregular margins. Inflorescence compound; clusters

Eucalyptus melliodora

Eucalyptus sideroxylon

Eucalyptus microcorys

7–11-flowered on flattened peduncles 6–18mm long. Stamens in 4 indistinct groups. Operculum hemispherical, marked by cross sutures. Fruit tapering, to 8 x 5mm; valves 3–4, ± at rim level. **HABITAT:** Tall open forest bordering rainforest on deep moist fertile soils. **DIST:** Cooranbong, Dora Creek and Watagan Mtns. Coastal ranges from Cooranbong (NSW) to Maryborough (Qld). **FL:** Aug–Oct.

An isolated species which does not hybridise with any other eucalypt.

Austromyrtus
Shrubs or small trees with opposite leaves. Flowers with 5 petals and 5 sepals (Sydney region); stamens numerous, free, in several rows. Fruit a berry.

Austromyrtus tenuifolia

Austromyrtus tenuifolia
Narrow-leaf Myrtle
Shrub 1–2m high with flaky bark; young growth softly hairy. Leaves to 40 x 3mm, linear with recurved margins; apex acute with small point. Flowers solitary in upper axils on pedicels 5–10mm long, white, pinkish in bud. Fruit an indigo berry. **HABITAT:** Damp sheltered places in forests, often beside streams. **DIST:** Endemic in Sydney district on coast and ranges (e.g. Girrakool, Pennant Hills Pk, Glenbrook Ck, The Needles on Woronora R., Grose R.). **FL:** Nov–Dec.

Austromyrtus acmenoides is a shrub or tree with large lanceolate leaves and short axillary inflorescences with 2–4 flowers. It occurs in dry rainforest in the Illawarra.

Babingtonia, Baeckea, Euryomyrtus, Ochrosperma

BABINGTONIA
HEATH-MYRTLE
Similar to *Baeckea*. Flowers in axillary clusters of 1–7 or more on short peduncles and pedicels; petals and sepals 5; anthers fused to staminal filament, opening by terminal pores or short divergent slits. Ovary and fruit 3-locular; seeds disc-shaped.

Babingtonia densifolia
Shrub to 1.5m tall with grey scaly bark. Leaves crowded, often appressed to stem, 3–6mm long, linear to terete with 2 rows of oil glands. Flowers solitary on peduncle to 3mm long, white, 6mm across; stamens 8–12. **HABITAT:** Rocky heath, woodland, open forest. **DIST:** Coast to ranges. Also NC, NT, CWS, NWS, NWP, Qld. **FL:** Dec–May.

Babingtonia virgata
Twiggy Heath-myrtle, Tall Baeckea
Spreading shrub to 3m tall. Leaves lanceolate to elliptic, often pointed, to 25 x 6mm, with distinct midvein. Flowers in cymose clusters of 3–4 or more on peduncle to 10mm long; white, 6–8mm across; stamens 10–20. **HABITAT:** Sheltered forests, creek banks. **DIST:** Scattered on coast, Nepean R., lower Blue Mtns and Boyd Plat. Also coast from Qld to Vic, NCal. **FL:** Chiefly Dec–Jan.

A new name may be required for this taxon since it appears that the name, as published, applies only to plants from New Caledonia.

BAECKEA
HEATH-MYRTLE
Small shrubs with opposite leaves. Flowers solitary in axils, nearly sessile; petals and sepals 5; stamens 5–20 (if 5, not always opposite a sepal); anthers versatile, opening by long parallel slits. Capsule 2-locular; seeds disc-shaped, without aril.

Baeckea brevifolia
Short-leaved Heath-myrtle
Small undershrub usually <1m tall. Leaves narrow-oblong, triangular in cross-section, <2mm long. Flowers white, occasionally pink, 4–5mm across, stamens 8–15. **HABITAT:** Rocky heath, flat sandstone outcrops. **DIST:** Coast and ranges (e.g. Hill Top, Mt Banks, Kanangra Walls, West Head). Also SC, ST. **FL:** Irregular; flowers observed July–Aug and Feb–Apr.

B. brevifolia typically grows in clumps in shallow depressions on bare sandstone. When protected from bushfires on such sites it becomes gnarled and woody with age.

Baeckea diosmifolia
Shrub to 1m tall. Leaves narrow-ovate to oblong, with recurved tip, 2–6 x 1–2mm, concave, minutely hairy and fringed when young. Flowers white, 4–5 mm across; sepals minutely fringed; stamens usually 7–8. **HABITAT:** Wet heath, often in rocky ground. **DIST:** Coast to ranges (e.g. Brisbane Water NP, Canoe Grounds, Agnes Banks, Macquarie Pass). Also SC, ST, NC, Qld. **FL:** Sept–Feb.

Baeckea imbricata
Shrub to 1m tall. Leaves crowded, overlapping, 4-ranked along stems, broadly ovate, about 4 x 3mm, concave with recurved tip. Flowers white 4–5mm across; stamens 5–7. **HABITAT:** Damp sandy sites in heath, close to sea. **DIST:** Coast to ranges. Also NC, SC, ST, Qld. **FL:** Dec–May.

Babingtonia densifolia

Baeckea diosmifolia

Baeckea imbricata

Babingtonia virgata

Baeckea brevifolia

Baeckea linifolia

Baeckea linifolia
Swamp Baeckea, Flax-leaf Heath-myrtle
Spreading shrub to 1.5m tall with long drooping branches. Leaves linear with fine point, 5–15mm long. Flowers white, 5–6mm across, often abundant on leafy branches. **HABITAT:** Damp gullies, stream banks, near waterfalls. **DIST:** Coast and ranges. Also NC, SC, ST, Qld, Vic. **FL:** Mainly Jan–Feb.

EURYOMYRTUS HEATH-MYRTLE
Similar to *Baeckea*. Flowers single or paired in axils, stalk (peduncle + pedicel) exceeding leaves. Stamens 3–29, some opposite petals; anthers dorsifixed or versatile, opening by long slits. Ovary and fruit 3-locular. Seeds kidney-shaped without aril.

Euryomyrtus ramosissima
Rosy Heath-myrtle
Small decumbent shrub with rosy pink flowers. Leaves narrow-linear to linear-lanceolate, 4–10 x 1–2mm. Flowers solitary, 6–10mm across; stamens mostly 5–10, occasionally fewer. **HABITAT:** Heath and open forest in both damp and exposed dry sandy sites. **DIST:** Common on coast from Coffs Harbour to Vic. Also Tas, SA. **FL:** Jun–Jan.

OCHROSPERMA HEATH-MYRTLE
Similar to *Baeckea*. Flowers axillary, single, or paired on common peduncle. Stamens 5, each opposite a sepal. Ovary and fruit 3-locular. Seeds kidney-shaped, with whitish aril.

Ochrosperma oligomerum
Dense spreading shrub to 50cm high. Leaves opposite, elliptic to obovate, 2–6 x 1–3mm, glabrous, apex recurved. Flowers 3mm across; petals white, obovate to round. **HABITAT:** Forest and scrub on ridges and in gullies near rock outcrops. **DIST:** Blackheath to Rylstone (e.g. Cape Horn, Wolgan Pinnacle, Newnes Glow Worm Tunnel). **FL:** Sept–Nov.

Ochrosperma oligomerum

FLOWERING PLANTS 277

Euryomyrtus ramosissima

Backhousia

Trees or shrubs. Leaves opposite, with prominent veins. Flowers with 4–5 persistent sepals and petals and numerous free stamens in several whorls. Fruit dry.

Backhousia myrtifolia
Grey Myrtle
Commonly a shrub 3–5m tall, rarely a tree to 10m with brown fissured flaky bark; young stems and leaves softly hairy. Leaves ovate, 4–7 x 1–3cm, on stalks 5mm long. Flowers creamy or greenish white, in axillary inflorescences on peduncles 20–35mm long; sepals 6mm long; petals 3mm long. Capsule enclosed within enlarged calyx. **HABITAT:** Rainforest in sheltered gullies, stream banks, hillsides. **DIST:** Coast, ranges, Bermagui to Qld. Also CWS. **FL:** Nov–Dec.

Backhousia myrtifolia

Callistemon, Melaleuca

CALLISTEMON
BOTTLEBRUSH
Shrubs or small bushy trees with alternate leaves and flowers in large cylindrical spikes. Stamens conspicuous, usually brightly coloured, free (i.e. separate), rarely shortly united at base. Capsule sessile, woody, with 3–4 valves, remaining on old wood for some years. The genus may in future be included in *Melaleuca*.

Callistemon citrinus
Red Bottlebrush, Lemon Bottlebrush, Crimson Bottlebrush
Erect shrub 1–2m tall. New growth flesh-coloured. Leaves to 7 x 1cm, lanceolate with short rigid tip. Flower spikes to 12 x 6cm; filaments red; anthers dark. Capsule 5–7mm diam. **HABITAT:** Rocky creek banks, damp places. **DIST:** Common. Coast and ranges. Also NC, SC, ST, WS, Qld, Vic. **FL:** Peak Oct–Dec.

Callistemon linearifolius
Shrub 3–4m tall with pilose new growth. Leaves linear-lanceolate, 8–12cm x 5–10mm, pointed; lateral veins raised. Flower spikes 9–12cm long; rachis pubescent; filaments red. **HABITAT:** Open forest in sheltered places. **DIST:** Uncommon. Endemic (e.g. Brooklyn–Patonga, Munmorah SCA). **FL:** Oct–Nov.

Callistemon linearis
Narrow-leaved Bottlebrush
Spreading shrub to 2m tall, with silky new growth. Leaves to 12cm x 2mm, channelled above or flat, stiff, pungent. Flower spikes to 10cm long; filaments pale red; anthers dark. **HABITAT:** Creek banks, damp places. **DIST:** Coast and ranges. Also NC, SC, NT, WS, NWP, Qld. **FL:** Oct–Nov.

Callistemon pinifolius
Pine-leaved Bottlebrush, Green Bottlebrush
Shrub to 2m tall. Leaves to 8cm x <2mm, channelled above or almost terete, pungent. Flower spikes to 7cm long, greenish yellow (rarely red). **HABITAT:** Damp sites, esp. on clay-shale soils. **DIST:** Coast (e.g. Casula, Rookwood, Agnes Banks). Also WS. **FL:** Sept–Nov.

FLOWERING PLANTS

Callistemon citrinus

Callistemon linearifolius

Callistemon linearis

Callistemon pinifolius

Callistemon pityoides

Callistemon pityoides
Alpine Bottlebrush
Compact shrub 2–3m tall, with silvery grey new growth. Leaves upwards-pointing, to 20 x 2mm, linear to subterete, pungent. Flower spikes 3–4cm long, creamy, sometimes tinged pink. **HABITAT:** Seepage areas, damp to wet boggy ground. **DIST:** Upper Blue Mtns (e.g. Clarence, Boyd Plat.). Also t'lands from Qld to Vic. **FL:** Dec–Jan.

Callistemon rigidus
Stiff Bottlebrush
Stiff erect shrub, 2–3m tall; new growth with silky appressed hairs. Leaves to 7cm x 4mm, stiff, pointed; margins thickened; veins obscure; lower surface dotted with oil glands. Flowers spikes to 10cm long; filaments red; anthers dark. **HABITAT:** Heath, shrubland, on clay or damp sandy soils. **DIST:** Coast (e.g. Rookwood, La Perouse). Also SC, NC, NT. **FL:** Oct–Nov.

Callistemon salignus
Willow Bottlebrush, White Bottlebrush, Pink Tip
Tree to 6m with papery bark and bronze-pink new growth. Leaves narrow-elliptic, to 10cm x 12mm, apex acute, base finely tapered, main veins distinct. Flower spikes to 5cm long; filaments creamy yellow. **HABITAT:** Swamps, creek banks and damp places. **DIST:** Uncommon. Coast and lower Blue Mtns (e.g. Bents Basin, Lane Cove R., Kangaroo R.). Also SC, NC, NT, Qld. **FL:** Sept–Oct.

Callistemon shiressii
Shrub 3–10m tall with papery bark and softly hairy new growth. Leaves lanceolate, to 4cm x 8mm, dark, acute to acuminate, pungent; midvein distinct. Flower spike to 4cm long; rachis with dense woolly hairs; calyces hairy; filaments white to cream. **HABITAT:** Rainforest margins, forest on clay-shale soils. **DIST:** Rare. Recorded from Colo, Gosford, Cattai Ck, Ourimbah, Yengo NP. **FL:** Sept–Oct.

Callistemon subulatus
Shrub to about 1m tall. Leaves dense, upwards-pointing, linear with long subulate point, 35 x 2mm. Flower spikes to 6cm long; filaments red. **HABITAT:** Rocky stream banks. **DIST:** Woronora R., Upper Kangaroo R. Also SC, ST, Vic. **FL:** Dec–Mar.

FLOWERING PLANTS

Callistemon rigidus

Callistemon salignus

Callistemon subulatus

Callistemon shiressii

MELALEUCA
PAPERBARK

Shrubs or trees with hard corky or papery bark. Leaves opposite, alternate or irregular. Flowers in spikes or clusters, the terminal bud usually growing on and forming a leafy shoot. Stamens conspicuous, joined at the base into 5 bundles or claws opposite the petals.

Melaleuca armillaris
Bracelet Honey-myrtle

Large spreading shrub with rough hard bark. Leaves alternate, 12–25 x <1mm, linear, acutely pointed with recurved tip, dark green. Flower spikes cylindrical, 3–7 x 2.5cm, below new leafy growth, dense, white; staminal claw 5–6mm long. Capsule globular, 3–5mm across with persistent sepals on rim. **HABITAT:** Damp heath and coastal headlands. **DIST:** Coast, from Qld to Vic and Tas. Also NT and CWS. **FL:** Sept–Nov.

Melaleuca biconvexa. Small tree with white to pale yellow flowers. Leaves to 18mm long, channelled above with 3 longitudinal veins. Common Ourimbah–Wyong.

Melaleuca armillaris

Melaleuca capitata

Shrub to 2 x 2m, with softly hairy new growth. Leaves erect, alternate, 10–25 x 1–2mm, subulate, pungent, stiff. Flowers in terminal pale yellow globular heads 25mm across; staminal claw 2mm long. Capsules in loose irregular clusters; 4–6mm across, rim smooth, valves deep in orifice. **HABITAT:** Heath, often on damp sandy or rocky sites. **DIST:** Sthn Blue Mtns and Bundanoon sth (e.g. Beecroft Peninsula, Budawangs, Wingello SF). **FL:** Sept–Dec.

The larger flower heads and capsules distinguish this species from *M. nodosa*.

Melaleuca deanei
Deane's Paperbark

Shrub 1–2m tall with villous new growth and grey stringy-papery bark. Leaves alternate, elliptic to oblanceolate, to 2.5cm long. Flowers creamy white in terminal spikes up to 6cm long; petals silky-hairy outside. Capsules 6–7mm across with smooth narrow orifice. **HABITAT:** Lateritic, rocky or sandy ridges and slopes. **DIST:** Endemic. Near Sydney (e.g. Brisbane Water NP, Royal NP, Heathcote NP, Wedderburn, Holsworthy, Lucas Heights). **FL:** Sept–Nov.

Regarded as rare and endangered owing to its small scattered populations close to development. Growth and flowering are fire-stimulated and sporadic.

Melaleuca capitata

FLOWERING PLANTS 283

Melaleuca deanei

Melaleuca decora
Ironwood Myrtle, Ridge Myrtle
Shrub or tree with papery bark to 7m tall. Leaves alternate, linear to narrow-lanceolate, 10–24 x 1–2.5mm, with conspicuous midvein. Flower spikes 3–5cm long with creamy white sweet-smelling flowers loosely arranged; staminal claw 3–4mm long. Capsules 2–4mm across; rim smooth; sepals not persistent. **HABITAT:** Damp or swampy ground and heavy soils subject to flooding. **DIST:** Clayey shale soils of Cumberland Pl. Coast from Nowra to Qld. **FL:** Nov–Jan.
   A common, sometimes dominant, tree on low-lying land in western Sydney.

Melaleuca ericifolia
Swamp Paperbark
Erect shrub usually <3m tall, sometimes a tree >6m; bark grey, flaky or corky. Leaves alternate, narrow-linear, 10–15 x <1mm, crowded. Flower spikes to 20 x 15mm, the axis growing on; stamens creamy white with staminal claw 2mm long. Capsules 3–4mm across; rim undulate due to persistent sepals. **HABITAT:** Wet places, near coastal lagoons and swamps in heath and low open forest. **DIST:** Coast. From Hastings R. to Vic and Tas. **FL:** Sept–Nov.
   Could be confused with *M. armillaris* or *M. parvistaminea* but has shorter leaves and flower spikes than the former, and flowers with longer stamens than the latter.

Melaleuca decora

Melaleuca ericifolia

Melaleuca erubescens
Rosy Paperbark, Blush Honey-myrtle
Shrub 1–2m tall with thin hard bark. Leaves alternate, about 10 x 1mm, terete with tip curved down. Flower spikes 2–4cm long below the leafy shoot tip; staminal claw 4–6mm long. Capsules in crowded oblong clusters. **HABITAT:** Open forest near watercourses or periodically waterlogged land, on clay soils. **DIST:** Coast (Picnic Point, Rookwood, Wilton, Castlereagh NR, Newnes). Also slopes and plains, Qld. **FL:** Nov–Apr.

Melaleuca hypericifolia
Red Honey-myrtle, Hillock Bush
Spreading shrub 2–4m tall. Leaves evenly spaced, opposite, nearly sessile, decussate, narrow-elliptic, to 4 x 1cm; midvein prominent. Flower spikes orange-red to red, 3–5cm long, on older wood within foliage; staminal claw 12–16mm long. Capsules 8–10mm across with prominent erect calyx teeth on rim. **HABITAT:** Damp sites in sandy heath and woodland, often on coastal headlands and tops. **DIST:** Blue Mtns, coast sth to Bermagui. **FL:** Oct–Feb.

Melaleuca erubescens

Melaleuca hypericifolia

Melaleuca linariifolia

Melaleuca linariifolia
Flax-leaf Paperbark, Snow in Summer
Tree to 8m tall with papery bark. Leaves opposite, narrow-elliptic, 20–40 x 2–3.5mm, acutely pointed. Flowers white in fluffy feathery spikes 3.5–4cm long; stamens appearing pinnate on long narrow claw. Capsules 3–4mm across. **HABITAT:** Heavier soils, near swamps, sheltered creek banks. **DIST:** Coast and ranges nth from Ulladulla. Also Qld. **FL:** Oct–Jan.

Melaleuca nodosa

Melaleuca megalongensis
Megalong Bottlebrush
Shrub to 5m tall with greyish new growth. Leaves narrow-elliptic, to 45 x 5mm; apex acute; oil glands moderately dense. Flowers in spikes to 8cm long, magenta pink, with dark red anthers. **HABITAT:** Swampy flats. **DIST:** Rare. Few sites near Megalong Ck in Blue Mtns. **FL:** Nov–Dec.

This species was first recognised in 2002.

Melaleuca nodosa
Shrub 1–4m tall with brown corky to papery bark. Leaves terete or linear, 10–30 x 1–3mm, dark green, rigid, pungent. Flower heads 10–15mm across, yellow to white; staminal claw 1–2mm long. Capsules 2–3mm across in tight globular clusters 6–10mm diam.; rim smooth; orifice nearly closed. **HABITAT:** Coastal heath, open forest, hind-dunes, headlands. **DIST:** Coast. Also NC, NT, NWS, Qld, SA. **FL:** Sept–Nov.

Melaleuca parvistaminea
Shrub or tree to 4m with rough corky bark. Leaves irregular or in groups of 3, narrow-linear, crowded, erect on stems, about 9 x <1mm. Flower spikes longer and narrower than *M. ericifolia* with styles protruding conspicuously from spike, pale lemon. Capsules 3mm across in tightly packed oblong clusters about 22mm long. **HABITAT:** Wet areas in open forest. **DIST:** Sthn fringe of Sydney district (near Tallong), intermittent on SC and ST to Vic. **FL:** Oct–Nov.

Melaleuca megalongensis

FLOWERING PLANTS

Melaleuca parvistaminea

Melaleuca quinquenervia

Melaleuca quinquenervia
Broad-leaved Paperbark
Common tall or short tree with pale papery bark. Leaves alternate, elliptic, 3–7 x 1–2.5cm, stiff, leathery, with 5 longitudinal veins, 3 prominent. Flowers white or cream, in bottlebrush-like spikes 2–5cm long. **HABITAT:** Coastal swamps and brackish lagoons. **DIST:** Coast from Kurnell to Qld, also PNG, NCal. **FL:** Feb–May, some flowers through winter.

Melaleuca sieberi
Tree or shrub to 5m with papery bark. Leaves narrow-lanceolate, 4–14 x 1–4mm. Flowers white, in fluffy loose terminal spikes about

Melaleuca sieberi

2cm long; calyx and rachis hairy; claw 2–3mm long. Capsules barrel-shaped, 2–4mm across; not closely packed. **HABITAT:** Deep, wet sandy ground, swampy areas. **DIST:** Limited on coast (e.g. Wyong, Morisset), chiefly NC. **FL:** Sept–Oct.

Melaleuca squamea

Melaleuca styphelioides

Melaleuca squamea
Swamp Honey-myrtle, Mealy Honey-myrtle
Uncommon shrub to 2m tall, often <1m; bark whitish, corky. Leaves crowded, upwards-pointing, narrow-lanceolate, 10 x 2.5mm long, 3–5-veined, concave on upper surface. Flowers in terminal clusters, bright purplish pink; staminal claw <1mm long. Capsules 5–7mm across, in short crowded clusters. **HABITAT:** Wet coastal heath. **DIST:** Chiefly coastal (e.g. Royal NP, O'Hares Ck, Darkes Forest); upper Blue Mtns (Centennial Glen–Blackheath). Also NC, Vic, Tas, SA. **FL:** July–Nov.

Melaleuca squarrosa
Scented Paperbark
Tree to >6m tall, more often a dense shrub to 3m; bark corky to papery. Leaves opposite, decussate, ovate, 5–12 x 4–7mm, acute to acuminate, 5–7-veined; pubescent on new growth. Flowers in terminal spikes 15–40 x 15mm, yellow, pleasantly scented; staminal claw <2mm long. Capsules 3–4mm across, rim undulate, sepals not persistent. **HABITAT:** Damp heath and woodland. **DIST:** Mainly sth of Sydney (e.g. Illawarra Esc., Budderoo NP, Macquarie Pass) to Vic, Tas, SA. **FL:** Sept–Dec.

Melaleuca styphelioides
Prickly-leaved Paperbark
Shrub or tree to 20m high (usually <6m) with papery bark. Leaves alternate, ovate-lanceolate, twisted, tapering to acute pungent point, 10–15 x 4–6mm, sessile, with numerous fine longitudinal nerves; new growth softly hairy. Flowers creamy white, in spikes 2–3cm long;

Melaleuca thymifolia

staminal claw 3–4mm long. Capsules ovoid, 3–4mm across with prominent acuminate sepals. **HABITAT:** Sheltered moist situations, near streams, on periodically inundated land. **DIST:** Coast nth from Nowra, gullies of Blue Mtns. Also NC, NWS, Qld. **FL:** Nov–Dec.

Melaleuca thymifolia
Thyme Honey-myrtle
Small spreading shrub with corky bark. Leaves opposite, lanceolate to elliptic, 5–12 x 2–3mm, erect on stems. Flowers deep mauve-pink, occasionally white, with inwards-curled stamens in irregular clusters on older wood. Capsules 3–5 x 3–5mm, with prominent calyx teeth on rim. **HABITAT:** Damp sandy soils in forest and heath. **DIST:** Coast and ranges, from SC to Qld, also CWS. **FL:** Oct–Jan, some flowers in autumn.

Melaleuca squarrosa

Calytrix tetragona

Choricarpia leptopetala

Calytrix
Shrubs with linear leaves spirally arranged. Petals 5, spreading, pink or white; sepals 5. Long bristle-like appendages (awns) develop on sepals as they age.

Calytrix tetragona
Common Fringe Myrtle
Bushy shrub to 2m tall. Leaves linear, 4–8mm long, finely toothed, triangular in section, upwards-pointing. Flowers axillary; petals usually pink, rarely white or yellow (Newnes Plat.), deciduous; sepals broad-ovate, 2.5mm long with awn 15–17mm long, turning tan-red in fruit. **HABITAT:** Heath, shrubland. **DIST:** All States except NTerr. **FL:** July–Dec.

Choricarpia
An endemic genus of 2 species with 4-angled young stems. Leaves opposite, with oil glands. Flowers in pedunculate globular heads; corolla tiny; stamens numerous, free. Fruit a capsule.

Choricarpia leptopetala
Brown Myrtle, Brush Turpentine
Shrub or tree to 10m tall; new growth and inflorescences rusty tomentose. Leaves ovate to elliptic, 8–12 x 4–5cm, smooth green above, greyish beneath. Flowers with rusty tomentose petals and cream stamens; peduncles to 3cm long. **HABITAT:** Rainforest margins, often along watercourses. **DIST:** Stanwell Park, Colo R., Yarramalong; NC to Qld. **FL:** Aug–Nov.

Darwinia biflora

Darwinia

Shrubs. Leaves opposite, usually laterally compressed and triangular in section. Flowers subtended by paired bracts and bracteoles, tubular with 5 small petals and sepals and long protruding style.

Darwinia biflora

Erect shrub to 80cm tall. Leaves decussate, 6–10mm long. Flowers paired, greenish white, 4–8mm long, surrounded by scaly bracts and red bracteoles as long as tube; bracteoles deciduous when flower opens; style green, straight, 10–14mm long. **HABITAT:** Shale soils and laterite overlying sandstone. **DIST:** Ridges on Hornsby Plat. (e.g. Mt Cowan, Mt Colah, Hornsby, Maroota, Pennant Hills Park). **FL:** July–Nov.

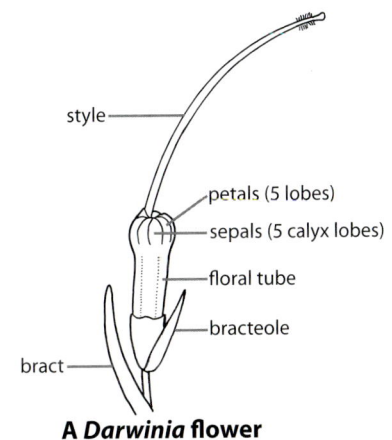

A *Darwinia* flower

Darwinia camptostylis

Erect shrub about 30cm tall. Leaves decussate, 6–15 x <1mm. Flowers paired in clusters of 2–4, cream-white when young, 3–6mm long; bracteoles shorter than floral tube, yellow-green or brownish, often persistent; sepals >two-thirds length of petals; style slightly exceeding tube, sharply bent. **HABITAT:** Heath, on sandy soils. **DIST:** Little R., Barren Grounds, Budderoo NP. Mainly sth of Shoalhaven R., SC, Vic. **FL:** Aug–Nov.

Darwinia camptostylis

Darwinia diminuta

Darwinia diminuta
Shrub to 1m tall. Leaves decussate, 6–11mm long. Flowers in clusters of 2–5 on short side-branches; bracteoles 3–4mm long; floral tube 3–5mm long, white when young, pink-red with age; sepals and petals <1.5mm long. **HABITAT:** Damp or dry sandy heath. **DIST:** Rare and restricted to 2 areas: Sutherland to Barren Grounds (e.g. Lucas Heights, Helensburgh, O'Hares Ck) and Terrey Hills to Manly. **FL:** Aug–Jan, also June.

Darwinia fascicularis subsp. *fascicularis*
Shrub 1m or more tall. Leaves subterete, about 15mm long. Flowers in clusters of up to 20, white, turning red with age; bracteoles deciduous, 3–5mm long; floral tube 5–7mm long; style straight, 12–18mm long. **HABITAT:** Sandy shrubland and heath. **DIST:** Gosford to Bulli within 30km of coast. **FL:** June–Sept, some flowers at other times.

The number of flowers per cluster varies, even on the one plant. Nth of the Hawkesbury R. the average is 6, in Ku-ring-gai Chase NP 11, sth of Sydney 14. Hybrids with *D. procera* and *D. glaucophylla* occur at Piles Creek near Kariong, and with *D. grandiflo*ra and *D. diminuta* at Helensburgh.

Darwinia fascicularis subsp. *oligantha*
Decumbent shrub <30cm high, sometimes rooting along branches. Leaves similar to subsp. *fascicularis*. Flowers mostly in clusters of 4, otherwise similar to subsp. *fascicularis*. **HABITAT:** Heath on shallow sandy soils, often at edge of flat sandstone outcrops. **DIST:** Blue Mtns (Mt Banks to Kings T'land and Woodford), Maroota–Wisemans Ferry, Nattai–Burragorang. **FL:** Oct–Feb.

Darwinia glaucophylla
Prostrate plant with short erect branchlets, often forming extensive mats. Leaves glaucous with reddish tinge, decussate, 8–17mm long. Flowers in clusters of 2–4, white when young, pink-red with age, 7–8mm long; bracteoles red-brown, deciduous, 4–6mm long; style

Darwinia fascicularis subsp. *fascicularis*

Darwinia leptantha

Darwinia fascicularis subsp. *oligantha*

straight, to 16mm long. **HABITAT:** Heath and open forest on hillsides near sandstone platforms. **DIST:** Restricted to small areas around Calga, Kariong, Mt Kariong, Piles Ck, Somersby Falls. **FL:** July–Nov.

Rare species, vulnerable because of its limited distribution.

Darwinia grandiflora
Prostrate shrub with ascending branchlets to 50cm high. Leaves 8–18 x <1mm. Flowers in clusters of 4–6; floral tube white to pink when young, turning red with age, 7–12mm long; bracteoles yellowish, 4–8mm long; sepals and petals 1.5–2mm long. **HABITAT:** Wet heath on sandy soils. **DIST:** Waterfall to Dapto. **FL:** July–Aug. with some flowers at other times.

Darwinia leptantha
Small shrub about 80cm tall. Leaves decussate, 7–11 x 0.5mm. Flowers small, in clusters of 2–5, almost hidden by leaves. Floral tube narrow, 4–6mm long, white ageing to pink, sepals <two-thirds length of petals. Bracteoles deciduous, 2–3.5mm long, yellow-brown or reddish. **HABITAT:** Coastal heath on sandy soils. **DIST:** Coast from Laurieton to Clyde River. Rare near Sydney (e.g. North Head, Long Bay, Little Bay, La Perouse); more common in Jervis Bay area, Myall Lakes NP. **FL:** Apr–Nov.

Darwinia glaucophylla

Darwinia grandiflora

Darwinia peduncularis

Darwinia procera

Darwinia taxifolia subsp. *macrolaena*

Darwinia taxifolia subsp. *taxifolia*

Darwinia peduncularis
Spreading shrub to 1.5m tall. Leaves decussate, not crowded, 7–12mm long. Flowers in clusters of 2, rarely 4, creamy white, often with pink petals; pedicels 4–7mm long; bracteoles red, 4–8mm long; floral tube 9–12mm long. **HABITAT:** Open forest, on or near rocky outcrops. **DIST:** Rare endemic. Elouera Reserve, Brooklyn, Mt Ku-ring-gai, Kings T'land (Waratah Trig), Glen Davis. **FL:** June–Nov.

Darwinia procera
Erect shrub to 2–3m tall. Leaves widely spaced on branchlets, deciduous on lower stems, 10–25mm long. Flowers in groups of 4–8, cream with red lobes; pedicels 1–2mm long; bracteoles 8–11mm long, red-brown; floral tube 5–8mm long; style 14–20mm long, red. **HABITAT:** Sheltered slopes in open forest on sandy soils. **DIST:** Restricted to Gosford–Manly area (e.g. Brisbane Water NP, Cowan Ck, Upper Middle Harbour). **FL:** July–Oct.

Darwinia taxifolia subsp. *macrolaena*
Decumbent shrub (occasionally erect) <1m high. Leaves keeled beneath, 5–12mm long. Flowers in clusters of 3–4, red, 6–9mm long; bracteoles 7–14mm long; style 15–24mm long. **HABITAT:** Heath on sandy soils. **DIST:** Sthn H'lands (e.g. Moss Vale, West Berrima, Sutton Forest, Sassafras-Nerriga Rd). Also SC, ST. **FL:** Oct–Nov.

Darwinia taxifolia subsp. *taxifolia*
Similar to subsp. *macrolaena* but with smaller floral parts. Flowers in clusters of 2–4 (usually 3, rarely 6), red, 5–6mm long; bracteoles persistent, 5–8mm long; style curved, 6–12mm long. **HABITAT:** Heath on sandy soils. **DIST:** Endemic in Blue Mtns above 915m (e.g. Blackheath, Narrow Neck, Echo Point, Leura, Wentworth Falls). **FL:** Sept–Dec.

Kunzea
Shrubs with (usually) alternate leaves. Flowers in terminal heads, short leafy clusters or solitary in upper axils; petals and sepals 5, exceeded by stamens. Fruit a capsule, thin-walled with persistent sepals.

Kunzea ambigua
Tick Bush
Dark green bushy shrub to 3m tall with hairy new stems. Leaves crowded on short side-branches, linear, about 10 x 1mm, concave, finely hairy to glabrous. Flowers abundant in upper axils, sessile, white, heavily scented. **HABITAT:** Common in heath, woodland, dry open forest and regenerated urban bushland. **DIST:** Coast and ranges. Also SC, ST, Vic, Tas. **FL:** Mid-Oct–Dec.

Kunzea ambigua

Kunzea cambagei

Kunzea cambagei
Decumbent or prostrate plant <50cm high. Leaves elliptic to obovate, 4–9 x 2–3mm, thick, with appressed white hairs on undersurface. Flowers creamy white in terminal clusters on short side-branches; bracts and floral tube villous; stamens 2–3mm long. **HABITAT:** Heath and scrub on damp sandstone soils. **DIST:** Endemic in sthn Blue Mtns (near Mt Werong) and Sthn H'lands (e.g. Coleby, West Berrima, Wanganderry Plat.). **FL:** Sept–Nov.

Rare and vulnerable species because of its restricted range.

Kunzea capitata
Pink Kunzea, Pink Buttons
Bushy shrub to 1.5m tall. Leaves 3-veined, concave, to 9 x 4mm, broadening towards apex; tip pointed, mucronate, recurved. Flowers pink, in terminal heads; calyx with 5 triangular sepals, villous; stamens 5mm long, with yellow anthers. Capsule 4mm long, narrow. **HABITAT:** Heath, open forest. **DIST:** Coast and ranges. Also NC, SC, ST. **FL:** Sept–Nov.

Easily distinguished from pink-flowered *K. parvifolia* by its larger 3-veined leaves. White-flowered forms occur at Barren Grounds, Budderoo NP and Bonnum Pic.

Kunzea parvifolia
Small-leaved Kunzea, Crimson Kunzea
Spreading shrub to 1.5m tall. Leaves appressed to stems, oblanceolate, to 4 x 1mm, midvein prominent. Flowers in terminal heads; rosy pink; calyx usually glabrous; stamens 2–3mm long with yellow anthers. Capsule rounded, 2 x 2mm. **HABITAT:** Heath and woodland, in shallow soils. **DIST:** Mainly Sthn H'lands (e.g. Penrose, Medway); Boyd Plat., coast. Also ST, WS, Vic. **FL:** Sept–Jan.

Kunzea rupestris
Rock Kunzea
Spreading suckering shrub to 1.5m tall with softly hairy young stems. Leaves crowded, oblanceolate, 8–14 x 1.5–3mm, hairy to glabrous. Flowers creamy white in terminal clusters; calyx with long white hairs; stamens 3–5mm long. Capsule villous, cone-shaped, 4–5mm long. **HABITAT:** Depressions on large, flat sandstone outcrops in open scrub. **DIST:** Marramarra NP (Sth Maroota), Yeomans Bay (Ku-ring-gai Chase NP). **FL:** Sept–Oct.

Endangered endemic shrub occurring in small populations on Hornsby Plat. It resprouts after fire and some plants may be quite old.

Kunzea sp. E
Mt Cookem Kunzea
Shrub 1–2m tall with hairy young stems. Leaves ovate, 3–5 x 1.5–2mm, concave, sharply pointed, shortly stalked. Flowers in terminal heads on small branches, aromatic, greenish cream; calyx and new growth softly hairy; stamens 1.5–2mm long. Capsule 4–5mm long, glabrous. **HABITAT:** Rocky outcrops. **DIST:** Cliffs above Kowmung River (e.g. Sombre Dome, Mt Cookem); Axehead Mtn near Yerranderie. **FL:** Sept–Nov.

Kunzea sp. E

Kunzea capitata

Kunzea parvifolia

Kunzea rupestris

Leptospermum
TEA-TREE

Shrubs with fibrous or papery bark and small alternate leaves. Flowers with 5 small round white to pink or red petals and 5 smaller greenish sepals attached to rim of floral tube (hypanthium); stamens erect, usually white, in 1 ring. Capsule with 3–12 compartments (locules), usually 5, opening by valves.

Leptospermum arachnoides
Spider Tea-tree

Low stiff spreading shrub with short leafy side-branches on horizontal stems, sometimes erect; bark rough, peeling in flakes. Leaves lanceolate, 10–20 x 1–3mm, stiff, thick, grooved, pungent. Flowers solitary, white, about 8mm across with woolly floral tube and sepals. Fruit 8–10mm across, grey-hairy. **HABITAT:** Heath and woodland in exposed places. **DIST:** Coast and t'lands, throughout NSW except drier plains. Also Qld. **FL:** Nov–Jan.

Leptospermum continentale

Slender straggly shrub 1–4m tall; young stems pubescent. Leaves spreading to deflexed, incurved, 5–15 x 1–3mm, narrow-lanceolate, tapering to long pungent point. Flowers white, 10mm across, solitary, nearly sessile. Fruit 6–8mm across, domed with 5 exserted valves. **HABITAT:** Shrubby swamp, open forest. **DIST:** Coast (e.g. Maddens Plains) and ranges (e.g. Newnes Plat., Mittagong). Also SC, ST, WS, Vic, SA. **FL:** Oct–Jan.

Leptospermum emarginatum

Tall shrub with long arching stems and rough hard bark. Leaves narrow-obovate, 15–35 x 3–6 mm wide, flat, glabrous, apex notched. Flowers white, often paired, in clusters of up to 5, 6–7mm across; floral tube glabrous, tapering to stalk. Fruit 4–5mm diam. on stalk 3–4mm long; valves exserted. **HABITAT:** Creek margins and rocky creek beds. **DIST:** Common in Nattai and Nepean catchments and lower Blue Mtns (e.g. Grose R., Erskine Ck, Nattai R.). Also SC to Vic. **FL:** Oct–Jan.

Leptospermum grandifolium
Woolly Tea-tree

Erect shrub to 4m tall; bark partially shed in patches and strips; young stems hairy. Leaves oblong with firm pointed tip, 10–30 x 3–7mm, often hairy on undersurface. Flowers solitary, white, 15mm across, floral tube and sepals hairy. Fruit domed, 10–12mm across, flaking to almost glabrous. **HABITAT:** Creek banks, wet or waterlogged areas, swamp margins, damp rocky sites. **DIST:** Coast and t'lands, common in Blue Mtns, Hawkesbury R. to Vic. **FL:** Oct–Jan.

Leptospermum juniperinum
Prickly Tea-tree

Slender shrub 2–3m tall with smooth hard bark; young stems with appressed hairs. Leaves clustered, ± sessile, erect, 5–15 x 1–2mm, narrow-lanceolate with pungent point, slightly incurved towards apex. Flowers white, 8–10mm across, solitary, nearly sessile. Fruit 5–7mm across, domed, with 5 exserted valves. **HABITAT:** Swamps, wet heath, escarpments, sandy peaty soils. **DIST:** Coast and ranges, from Ulladulla to Qld. **FL:** Oct–Dec.

Leptospermum arachnoides

Leptospermum continentale

Leptospermum emarginatum

Leptospermum grandifolium

Leptospermum juniperinum

Leptospermum laevigatum
Coast Tea-tree
Tall coastal shrub. Bark thin, hard, shedding in strips. Leaves narrow-obovate, 15–30 x 5–7mm, grey-green, often with short mucro. Flowers white, 14–18mm across; usually with red ring at base of stamens. Fruit woody, 7–8mm diam., with 8–12 valves around the top. **HABITAT:** Hind-dunes near sea, often forming thickets; on headlands. **DIST:** Coast. NC to Vic, Tas, SA. **FL:** July–Oct.

Leptospermum macrocarpum
Shrub 1–2m tall with hard bark; young stems densely pubescent. Leaves elliptic, 10–20 x 5–10mm, dull, flat, pointed. Flowers to 30mm across, pink, red or lemon yellow. Buds, floral tube and young capsules softly hairy. Capsules >15mm across, domed, glabrous with age. **HABITAT:** Open forest on dry, sandy, rocky slopes. **DIST:** Mostly nth of Bells Line of Road in Blue Mtns (yellow form at Mt Wilson and Zig-zag Railway, pink form along Mountain Lagoon Rd, red form on tops of Colo Gorge). **FL:** Oct–Dec.

Leptospermum morrisonii
Shrub 2–5m tall with firm bark becoming coarse; new growth with silky appressed hairs. Leaves narrow-elliptic with 3 longitudinal veins, 15–35 x 2–8mm, aromatic, with many small oil glands; new leaves reddish, softly hairy; older leaves green, glabrous. Flowers 15mm across, white to greenish cream. Fruit 6–10mm across. **HABITAT:** Woodland and scrub, often in rocky places. **DIST:** Woronora Plat. to upper Blue Mtns and Sthn H'lands. **FL:** Dec–Jan.

Closely related to *L. polygalifolium*.

Leptospermum myrtifolium
Grey Tea-tree, Swamp Tea-tree
Shrub to 3m tall, with hard bark. Leaves greyish, elliptic to oblanceolate, to 12 x 5mm, oil glands visible. Flowers white, about 10mm across. Buds, floral tubes and young capsules with short grey hairs. Fruit domed, 4–6mm across, 5-locular, glabrous with age. **HABITAT:** Heath and waterlogged sandy soils. **DIST:** Western Blue Mtns (e.g. Hartley V., Black Ra., Newnes Tunnel); a t'lands species from Orange to Vic. **FL:** Sept–Jan.

Leptospermum obovatum
Creek Tea-tree
Shrub to 3m tall with hard rough bark and abundant flowers along branches; young stems pubescent. Leaves obovate 6–20 x 3–8mm, rounded or notched at tip; oil glands large. Flowers single in axils, white with greenish tinge, 10mm across. Capsule woody, 7mm across, with 5 valves exceeding rim after opening. **HABITAT:** Locally common in thickets on creek banks and nearby flats. **DIST:** Upper Blue Mtns, Sthn H'lands. Also SC, ST, Vic. **FL:** Nov–Dec.

Leptospermum laevigatum

Leptospermum macrocarpum

FLOWERING PLANTS | 301

Leptospermum morrisonii

Leptospermum myrtifolium

Leptospermum obovatum

Leptospermum parvifolium

Leptospermum polygalifolium

Leptospermum polyanthum

Leptospermum parvifolium
Small-leaf Tea-tree
Thin shrub 1–2m tall with flaky bark. Leaves narrow-obovate, 3–8 x 1–2mm. Flowers white or pinkish, to 10mm across; floral tube with long weak hairs. Capsule woody, 4mm across, domed, with 3–5 compartments. **HABITAT:** Open forest, on sandy soils. **DIST:** Coast and ranges (e.g. Moorebank, Maroota, Narrow Neck). Also NWP, NWS, CWS, NT, SC. **FL:** Sept–Nov.

Leptospermum polyanthum
Shrub or small tree, often pendulous. Bark firm, furrowed; new branchlets silky-pubescent. Leaves elliptic-oblong, 10–25 x 2–4mm, glabrous. Flowers solitary at many adjacent leafless nodes, white, 5–6mm across; hypanthium glabrous. Fruit not persisting, 3mm across, glossy; valves thin, wider than hypanthium, shorter than sepals. **HABITAT:** Rocky ground near streams and on escarpments. **DIST:** On coast, but mainly Wentworth Falls to higher ranges, Wombeyan Caves to WS. **FL:** Oct–Jan.

Leptospermum polygalifolium
Yellow Tea-tree
Shrub 3–4m tall, sometimes taller, with smooth firm bark. Leaves narrow-elliptic, 8–20 x 1–5mm, green, flat, glabrous, with conspicuous oil glands. Flowers about 12mm across, creamy white with green centre, sometimes distinctly yellow. Fruit 8mm across, glabrous, domed with 5 exserted valves. **HABITAT:** Open forest, woodland, heath on damp sandy soils. **DIST:** Coast and ranges. Also NWS, NWP, Qld. **FL:** Aug–Feb.

Leptospermum rotundifolium

Leptospermum rotundifolium
Round-leaf Tea-tree
Spreading shrub to 2m with thin hard bark becoming rough; young stems pubescent. Leaves round with short recurved tip, 5–6 x 5–6mm, folded along midvein. Flowers about 30mm across, white to deep pink with dark green centre and pale stamens. Fruit 10–12mm across, domed with exserted valves larger than hypanthium. **HABITAT:** Open forest, woodland, scrub. **DIST:** Woronora Plat., Sthn H'lands. Also Jervis Bay–Ulladulla. **FL:** Oct–Dec.

Leptospermum spectabile
Shrub to 3m tall; young stems hairy. Leaves narrow-elliptic, 20–30 x 3–5mm, flat, apex acute and stiff. Flowers solitary, dark red, 20mm across; sepals silky. Fruit 9–12mm diam. **HABITAT:** Sandy alluvium and rocky sites in sandstone gorge. **DIST:** Endemic. Colo R. above Upper Colo. **FL:** Nov–Jan.
    Vulnerable because of its restricted range.

Leptospermum spectabile

Leptospermum sphaerocarpum

Leptospermum sphaerocarpum
Round-fruited Tea-tree
Spreading shrub to 2m tall with smooth hard bark. Leaves narrow-lanceolate, 10–20 x 2–5mm, concave with edges incurved near acute tip. Flowers pink to white, 15–25mm across; floral tube silky-hairy. Fruit very domed, ovoid, with rim midway, 8–10mm across. **HABITAT:** Heath, open forest on shallow sandy soil, often on ridges and escarpments. **DIST:** Blue Mtns (e.g. Blackheath, Clarence, Newnes Plat.) and nw. Hornsby Plat. to Rylstone. **FL:** Sept–Dec.

Leptospermum squarrosum
Pink Tea-tree, Peach Blossom Tea-tree
Erect shrub 1–3m tall with firm bark; young stems silky-hairy. Leaves crowded, stiff, lanceolate with long acute pungent point, 8–15 x 1–2mm. Flowers solitary on older branches, about 16mm across, pink in bud, white to deep pink in flower; sepals deep pink, clearly visible between petals. Fruit 12mm across, domed, glabrous, shiny. **HABITAT:** Woodland, shrubland, heath. **DIST:** Coast; also SC and ST. **FL:** Chiefly Jan–Aug.

Leptospermum trinervium
Flaky-barked Tea-tree, Slender Tea-tree
Variable shrub 3–4m tall, with ascending branches and rough flaky bark. Leaves broad-obovate and blunt, or narrow-elliptic and pointed, 10–20 x 1–6mm, silky. Flowers 12–15mm across, white, often with red ring; floral tube with long weak grey hairs. Capsule thin-walled, 4–6mm across, pubescent, flat, valves not exserted. **HABITAT:** Dry open forest, scrub. **DIST:** Coast and ranges from Qld to Vic. Also CWS. **FL:** Aug–Oct (coast), to Jan (mountains).

The Blue Mtns form is robust with narrow leaves.

Leptospermum squarrosum

Leptospermum trinervium

Micromyrtus ciliata

Micromyrtus
Low shrubs with small opposite leaves crowded along stems. Flowers 1–3 on common peduncle in upper axils with 5 petals and sepals and 5 or 10 stamens. Fruit a tiny nut.

Micromyrtus ciliata
Fringed Heath-myrtle
Spreading shrub to 1m. Leaves decussate, crowded, linear with fringed margins, 2–3 x 1mm, triangular in section, dotted with oil glands. Flowers abundant, solitary, white or pinkish; petals 1.5–4mm long; floral tube 2–5mm long; stamens 5, opposite petals. **HABITAT:** Sandy coastal heaths, rocky slopes and rock platforms. **DIST:** Coast and ranges sth from Hunter region. Also Vic, SA. **FL:** Aug–Oct.

Micromyrtus blakelyi is similar but is fringed, with long hairs on the sepals and short hairs on the leaf keel and margins. It is rare and restricted to Cowan–Canoelands–Maroota.

Micromyrtus minutiflora
Shrub <1m tall. Leaves linear with fringed margins, 3 x 1mm, appressed to stem on older branches. Flowers similar to *M. ciliata* but smaller; petals 1mm long; floral tube 1mm long. **HABITAT:** Shrubby woodland on consolidated river sediments. **DIST:** Restricted to Castlereagh vicinity. **FL:** Sept–Dec, June–Mar.

Micromyrtus minutiflora

Rhodamnia
Tall shrub or small tree. Leaves opposite, distinctly 3-veined from the base. Flowers with 4 petals and sepals and numerous stamens in several whorls. Fruit a berry.

Rhodamnia rubescens
Brush Turpentine, Scrub Stringybark
Shrub or tree to 25m but seldom >6m in Sydney district. Bark reddish brown, flaky, stringy on trunk; densely tomentose on young stems. Leaves ovate, 5–12 x 2–4.5cm, 3-veined with minor reticulate veins, sparsely to densely hairy, pale beneath. Flowers in several 3-flowered clusters in leaf axils, white, 10mm across. Fruit 6mm diam., crowned by calyx remains, green turning red then glossy black. **HABITAT:** Rainforest and rainforest margins. **DIST:** Chiefly coast but also t'lands (e.g. Minnamurra Spit, upper Hacking R., Macquarie Pass, Lithgow). From Milton to NT and Qld. **FL:** Sept–Oct.

Syncarpia glomulifera

Syncarpia
Tree with fibrous bark. Leaves opposite, crowded or whorled at ends of branches. Floral tubes (hypanthia) of flowers and fruit fused into irregular clusters of 7 on long peduncles.

Syncarpia glomulifera
Turpentine
Common medium to tall tree with red-brown (grey outside) fibrous stringy bark on trunk and branches. Leaves ovate, 7–12 x 2.5–5cm, thick, greyish-hairy on undersurface. Flowers with prominent cream stamens; petals 5–8mm long. Fruit clusters woody, 1–2cm across; sepals persistent. **HABITAT:** Wianamatta shale clay soils or shale-sandstone transition, rainforest margins, emergent in rainforest, alone or in small stands in eucalypt forest. **DIST:** Coast and t'lands, from Ulladulla to Qld. **FL:** Aug–Dec.

Syzygium
Shrubs or trees with hard grey bark and opposite leaves. Flowers with 4 petals and sepals and many conspicuous free stamens. Fruit a succulent berry with sepal remains at the top.

Syzygium australe
Brush Cherry
Tree usually 5–8m tall with firm finely fissured bark on trunk and branches; branchlets green, 4-angled with small expansion above leaf insertions. Leaves elliptic to obovate with abrupt acute tip, 4–10 x 1.5–3cm, glossy dark green above, paler and duller beneath. Petals white, 4–6mm long; sepals rounded; stamens

Rhodamnia rubescens

Syzygium australe

Syzygium oleosum

Tristania neriifolia

white 15–20mm long. Fruit red, 20 x 15mm; seed with 1 embryo. **HABITAT:** In or near rainforest in sheltered gullies and on littoral sands, often near water. **DIST:** Coast. From Batemans Bay to n. Qld. **FL:** Jan–April, also spring.

Syzygium paniculatum is similar. Its branchlets are terete and its fruit is magenta and up to 25mm across. The seed contains more than 1 embryo (usually 3–4). It occurs close to the sea and is uncommon in the Sydney area.

Syzygium oleosum
Blue Cherry
Small rainforest tree similar to *S. australe*, but much less common. New branchlets terete. Leaves lanceolate to elliptic with long tapering narrow apex, 3–10 x 1–4cm; oil dots numerous and conspicuous. Flowers similar to *S. australe*. Fruit a purplish blue berry with mauve-pink seed containing 1 embryo. **HABITAT:** In or near rainforest and in sheltered moist gullies. **DIST:** Usually close to coast (e.g. Burraneer Point, Patonga, Barrenjoey Head, Strickland SF). From Mt Kembla to n. Qld. **FL:** Dec–Mar.

Tristania, Tristaniopsis
Trees or shrubs with cymose inflorescences in upper axils. Flowers yellow, 5-merous with stamens fused into bundles opposite the petals. Fruit a 3-locular capsule fully or partly exserted from the hypanthium. Former *Tristania* species with alternate leaves and exserted fruits with flat winged seeds were reclassified as *Tristaniopsis* in 1982.

Tristania neriifolia
Water Gum
Shrub to 3m tall; bark smooth to slightly flaky. Leaves opposite, narrow-lanceolate, 40–80 x 5–15mm, paler beneath; veins obscure. Petals 6–10 x 3–6mm; stamens fused at base into bundles of 3–4. Fruit 4–5mm diam. **HABITAT:** Creek beds and banks among rocks. **DIST:** Endemic. Common on coast and t'lands. **FL:** Nov–Jan.

Tristaniopsis collina

Tristaniopsis laurina

ONAGRACEAE

Epilobium
WILLOW-HERB
Perennial herbs, with 4 sepals and petals, 8 stamens and 4-locular ovary. Fruit a terete capsule splitting into 4 from the tip; seeds numerous, with hair tufts.

Epilobium billardierianum
Smooth Willow-herb
Erect herb 30–60cm tall. Leaves linear to narrow-ovate, to 30 x 5mm, greyish, toothed, softly hairy. Flowers in upper axils, mauve-pink to white. Capsule to 7cm long. **HABITAT:** Open forest, woodland, on moist clay-shale and basalt soils. **DIST:** Coast and ranges. Most of NSW. All other States (except NTerr), NZ, LHI. **FL:** Nov–Mar.

Ludwigia
Shrubs or herbs, erect or with prostrate stems rooting at nodes in mud or shallow water. Flowers solitary, yellow.

Ludwigia peploides subsp. *montevidensis*
Water Primrose
Aquatic herb with spongy stems to 3m long. Leaves alternate, glossy green, ovate, 8 x 3cm. Flowers with 5 bright yellow petals, to 25mm across. Fruit a 5-celled cylindrical capsule, to 25mm long. **HABITAT:** Shallow streams, ponds, lagoons. **DIST:** Coast, Cumberland Pl. Also NC, SC, WS, WP, Qld, Vic, SA, NZ, S. Amer. **FL:** Dec–Apr.

Tristaniopsis collina
Mountain Water Gum
Shrub to 3m tall, a mallee-like tree on dry ridges or a much taller tree on rainforest margins. Bark fibrous on older trees. Leaves elliptic, 4–9 x 1–2.5cm, acuminate, glabrous, green above, yellowish beneath; oil glands numerous. Petals about 3 x 2.5mm; stamens as long as petals. Capsule 7 x 4mm, valves exserted. **HABITAT:** Forested slopes and ridges, rainforest margins. **DIST:** Coast and Blue Mtns, common on Illawarra esc. Coast and t'lands from Narooma to Qld. **FL:** Dec–Feb.

Tristaniopsis laurina
Kanooka, Water Gum
Spreading tree to 15m tall along streams in sheltered gullies and rainforest, shrubby and much smaller on open sites. Bark light grey, smooth on young trees, flaky and blotched on older trees. Leaves alternate or irregular, oblanceolate, 9–12 x 2–3cm, dark green and semi-glossy above, pale yellow-green and dull beneath. Petals about 5 x 4mm; stamens shorter than petals. Capsule about 8 x 5mm with exserted valves. **HABITAT:** Creek banks. **DIST:** Coast and t'lands. From Vic to Qld. **FL:** Dec–Feb.

HALORAGACEAE

Gonocarpus
RASPWORT

Herbs or subshrubs, often scabrous, with small toothed leaves and tiny green or brown flowers. Flowers bisexual, single in spike-like or branched inflorescences. Fruit a nut <2mm long.

Gonocarpus micranthus subsp. *micranthus*
Creeping Raspwort

Prostrate herb to 10cm tall with creeping glabrous stems. Leaves and stems green or reddish. Leaves opposite, round to heart-shaped, to 8mm long. Flowers in terminal spike-like racemes, minute, reddish. **HABITAT:** Wet ground, margins of swamps. **DIST:** Coast and ranges. Also NC, SC, NT, ST, SWS, Qld, Vic, Tas, SA, NZ, Malesia. **FL:** Sept–Feb.

Subsp. *ramosissimus* is erect (to 60cm) and has diffuse branched inflorescences.

Gonocarpus teucrioides
Common Raspwort

Multi-stemmed plant to 30cm tall with hairy 4-angled stems. Leaves opposite, decussate, ovate, to 15mm long, toothed, with rough scabrous hairs; upper leaves reduced to bracts. Flowers in axillary racemes; petals 3mm long, green to red; bracts 3–4mm long; bracteoles 2.5mm long, green, entire. Nut scabrous. **HABITAT:** Open forest, on sandy soils. **DIST:** Chiefly coast. Also Blue Mtns and Sthn H'lands, NC, SC, NT, ST, Qld, Vic, Tas. **FL:** Sept–Jan.

Gonocarpus micranthus subsp. *micranthus*

Epilobium billardierianum

Ludwigia peploides subsp. *montevidensis*

Gonocarpus teucrioides

Haloragis heterophylla

Haloragis
Subshrubs or herbs with usually toothed leaves and tiny green or brown flowers mostly in cymose clusters of 3–7 in long inflorescences in axils of alternate leaves or bracts. Fruit a nut.

Haloragis heterophylla
Rough Raspwort
Herb to 50cm tall with scabrous stems and leaves. Leaves alternate (lower ones opposite), to 3cm long, deeply divided into 3 or more linear lobes. Inflorescences in upper axils, of 1–3-flowered cymes, exceeding foliage; bracts linear, 3–5mm long; petals reddish, 3mm long. **HABITAT:** Swamp margins, drainage channels, on clay-shale soils. **DIST:** Cumberland Pl. (e.g. Agnes Banks, Doonside); Illawarra, Sthn H'lands. Most of NSW. Also Qld, Vic, Tas, SA. **FL:** Dec–Mar.

Haloragodendron
Shrubs or small trees. Inflorescence narrow, spike-like. Flowers with 4 petals, 8 stamens.

Haloragodendron lucasii
Shrub to 1.5m tall with glabrous 4-angled stems. Leaves opposite, sessile, lanceolate, to 30 x 5mm, toothed. Petals 4, white, 9–12mm long, twisted in bud. Fruit with 4 well-developed wings. Seed does not appear to be produced; suckers from roots. **HABITAT:** Sheltered slopes, among ferns. **DIST:** Endemic. Restricted to Garigal NP (St Ives). **FL:** Oct–Nov.

H. lucasii was believed to be extinct until rediscovered in 1986.

Haloragodendron gibsonii. Occurs in Bungleboori Ck (Blue Mtns). Fruit not strongly winged and seeds produced. *H. gibsonii* was named in 2006 after discovery in 1982.

Haloragodendron lucasii

Myriophyllum variifolium

Myriophyllum
WATER-MILFOIL
Aquatic monoecious plants, rooting in mud of shallow water. Leaves variously arranged, often whorled; submerged leaves usually different from emergent leaves. Flowers solitary or 2–3 in axils of emergent leaves. Male flowers with 4 sepals and petals; female flowers usually lacking both. Fruit with 4 carpels, splitting into 1-seeded mericarps.

Myriophyllum variifolium
Common Water-milfoil
Perennial with bright green leafy stems to 5mm diam. Leaves in whorls of 5, linear-terete, to 15 x 1mm, entire; submerged leaves toothed. Flowers solitary; petals 2–4mm long, yellow or reddish (males); stigmas 4, white (females). **HABITAT:** Wetlands, margins of creeks, swamps. **DIST:** Coast and ranges. Also NC, SC, NT, ST, CWS, Vic, Tas, SA. **FL:** Oct–Feb.

Alectryon subcinereus

Cupaniopsis anacardioides

SAPINDACEAE

Alectryon
Monoecious trees with alternate simple or pinnate leaves. Flowers in axillary racemes or panicles; calyx 4–6-lobed; petals absent or 4–6. Fruit with several 1-seeded globose lobes.

Alectryon subcinereus
Native Quince
Tree to 8m tall, finely hairy on branchlets and inflorescences. Leaves with 4–6 leaflets; leaflets elliptic, to 15 x 5cm, margins toothed in upper part, often acuminate, veins distinct. Flowers in loose panicles to 12cm long, tiny. Fruit with 2 (rarely 3) lobes, each with shiny black seed and red aril. **HABITAT:** In and near rainforest. **DIST:** Coast, Kanangra, Robertson NR. Also NC, SC, CWS, Qld, Vic. **FL:** Nov–Dec.

Cupaniopsis
Trees with small corky outgrowths (lenticels) along branches. Leaves alternate, pinnate. Flowers unisexual, solitary or in small cymes or clusters in axillary inflorescences; calyx 5-lobed; petals 5. Fruit a 3-lobed capsule.

Cupaniopsis anacardioides
Tuckeroo
Tree to 10m tall. Leaves with 4–11 leaflets; leaflets ovate-elliptic, 6–12cm long, entire, dark green above, paler green beneath, blunt or notched at apex, with a short swollen stalk. Flowers small, yellowish, in panicles to 30cm long. Fruit orange, to 16mm long, each lobe containing 1 black seed enclosed in orange aril. **HABITAT:** Coastal dunes, estuarine woodland. **DIST:** Coast (e.g. Jibbon Bch, Towra Point, Tuggerah). Also NC, Qld. **FL:** Feb–July.

Diploglottis australis

Dodonaea boroniifolia

Dodonaea camfieldii

Diploglottis
Trees with large pinnate leaves. Flowers small in axillary panicles. Fruit a large capsule with 2–3 lobes and a fleshy acidic aril.

Diploglottis australis
Native Tamarind
Rainforest tree, to 30m tall, with rusty-hairy new growth. Leaves pinnate, 40–100cm long, with 6–12 leaflets (no terminal leaflet); leaflets to 30 x 8cm; lower surface rusty-hairy. Flowers 3mm long, creamy brown. Capsule brown-hairy; seeds enclosed by orange-yellow aril. **HABITAT:** Rainforest. **DIST:** Illawarra, Gosford district. Also NC, SC, Qld. **FL:** Sept–Nov (usually every second year).

Dodonaea
HOP BUSH
Shrubs with alternate, simple or pinnate leaves. Flowers unisexual (plants dioecious), rarely bisexual, insignificant; sepals 3–7 (often 4); petals absent. Fruit a capsule with 2–6 papery wings.

Dodonaea boroniifolia
Fern-leaf Hop Bush
Shrub 1–2m tall, with sparsely hairy branches and viscid leaves. Leaves to 3cm long with 7–15 leaflets; leaflets 3–7mm long with sunken glands near upper margin and 3–6 lobes at apex. Flowers 2–3 in axils on stalks 4–8mm long. Capsule 4-winged with wings 2.5–5mm wide. **HABITAT:** Open forest, on sandy soils. **DIST:** Uncommon on coast and ranges (e.g. Kurrajong, Capertee, Culoul Ra.). Most of NSW. Also Qld, Vic. **FL:** Sept–Nov.

Dodonaea camfieldii
Camfield's Hop Bush
Shrub to 30cm tall, often prostrate. Leaves sessile, 10–30 x 3–7mm long, oblong, entire or irregularly lobed. Flowers solitary or paired on short stalks. Capsule 4-winged, 10–15mm long, turning red or indigo with age. **HABITAT:** Open forest, woodland, on sandy soils. **DIST:** Mainly Hornsby Plat. (e.g. Mt White, Maroota). Also Royal NP, Lucas Hts, Cataract Dam, Glenbrook and Jervis Bay–Nowra district. **FL:** Sept–Nov.

Dodonaea falcata
Thread-leaf Hop Bush
Erect, sticky shrub to 2m tall; branches minutely, hairy. Leaves sessile, linear, to 50 x 1mm, channelled above. Flowers in terminal clusters of 3–4. Capsule 4-winged, to 12 x 14mm, with sparse minute hairs, turning red with age. **HABITAT:** Dry open forest. **DIST:** Cumberland Pl. (e.g. Castlereagh NR, Kenthurst), Hornsby Plat. (e.g. Cattai, Putty). Also NC, NT, NWS, NWP, Qld. **FL:** Sept–Dec.

Dodonaea multijuga
Erect shrub to 1.5m tall with short hairs. Leaves pinnate, to 6cm long with 16–28 leaflets 4–7mm long. Flowers in axillary clusters, on stalks 7–14mm long. Capsule 3-winged, wings 3–4mm wide. **HABITAT:** Open forest, often near streams. **DIST:** Uncommon on coast and ranges (e.g. Deepwater Park, Erskine Ck, Clarence, Wingello). Also NC, SC, ST, Qld. **FL:** Sept–Mar.

Dodonaea pinnata
Pinnate Hop Bush
Spreading shrub to 1.5m tall with dense long hairs. Leaves pinnate, to 35mm long, with 10–16 leaflets 4–9mm long. Flowers solitary in axils of upper leaves or bracts; stalks 3–7mm long. Capsule 4-winged, wings 3.5–6mm wide. **HABITAT:** Open forest, on sandy soils. **DIST:** Endemic. Uncommon. Lower Blue Mtns and nth of Sydney (e.g. Springwood, Maroota, Bobbin Head, Patonga). **FL:** Sept–Dec.

Dodonaea falcata

Dodonaea multijuga

Dodonaea pinnata

Dodonaea triquetra

Dodonaea truncatiales

Dodonaea triquetra
Common Hop Bush
Glabrous shrub to 3m tall. Leaves elliptic, to 10 x 4cm, with pointed apex. Capsule 3-winged, about 1cm long, green turning purple-brown.
HABITAT: Open forest. **DIST:** Common on coast, less so in ranges (e.g. Blackheath, Hill Top, Bundanoon). Also NC, SC, NT, NWS, Qld, Vic.
FL: Chiefly July–Oct.

A short-lived species, regenerating from seed and often abundant after bushfires.

Dodonaea truncatiales
Erect shrub to 3m tall with angular branches. Leaves narrow-elliptic, 5–10cm long, entire or with toothed margins, glabrous, often viscid. Flowers and fruit in axillary clusters, on stalks 3–8mm long. Capsule with 3 or 4 wings; capsule and wings broader than long.
HABITAT: Open forest, on slopes. **DIST:** Scattered, away from coast (e.g. Nattai R., Putty, Nortons Basin, Glen Davis). Also NC, SC, CWS, Vic.
FL: July–Aug.

Dodonaea viscosa subsp. *angustifolia*

Dodonaea viscosa subsp. *angustifolia*
Sticky Hop Bush
Erect shrub 1–4m tall. Leaves narrow-lanceolate, to 10 x 1cm, acutely pointed, tapering to stalk, often appearing varnished. Fruit with 3 (rarely 4) wings, green turning red or brown. **HABITAT:** Open forest, often in rocky areas. **DIST:** Scattered. Coast and ranges (e.g. Dapto, Lansdowne, Colo Hts, Kanangra). Also NC, SC, NT, ST, WS, NWP, Qld, Vic, Amer., Afr., Asia, Pac. Iss. **FL:** Aug–Oct.

Dodonaea viscosa subsp. *angustissima*
Narrow-leaf Hop Bush
Erect shrub 1–4m tall. Leaves sessile, linear, to 90 x 1–5mm, with acute or blunt apex; margins recurved, minutely toothed. Fruit with 3 (rarely 4) wings, green, turning red to purplish. **HABITAT:** Open woodland, often near rocky areas. **DIST:** Valleys of Blue Mtns (e.g. Goodmans Ford, Coxs R.) and Mt Gibraltar near Bowral. Most of NSW (except coast). All mainland States. **FL:** Sept–Nov.

This subspecies intergrades with subsp. *spatulata*. Subsp. *spatulata* has obovate leaves to 7cm long. It occurs at St Marys, Wallacia, Clarence and w. of Blue Mtns.

Subsp. *cuneata* has wedge-shaped leaves to 3cm long. It occurs Cumberland Pl. and Capertee. Common w. of Blue Mtns.

Dodonaea viscosa subsp. *angustissima*

Guioa
Monoecious trees with pinnate leaves. Flowers tiny in large axillary inflorescences; calyx lobes 5; petals 5. Fruit a capsule with 3 compressed wing-like lobes.

Guioa semiglauca
Guioa
Tree to 6m tall with smooth grey bark. Leaves with 2–6 leaflets; leaflets often alternate on rachis, obovate-elliptic, to 10 x 4cm, glabrous and dark green above, glaucous and finely hairy beneath. Flowers in axillary panicles to 10cm long, yellowish green or white. Capsule to 10 x 15mm, reddish. **HABITAT:** Littoral rainforest, protected gullies. **DIST:** Coast (e.g. Bass Point, Minnamurra Spit, Burning Palms, Calga, Razorback Mtn). Also NC, SC, Qld. **FL:** Oct–Nov.

Guioa semiglauca

Melia azedarach

MELIACEAE

Melia
WHITE CEDAR
Deciduous trees. Leaves alternate, 2–3-pinnate. Flowers bisexual or unisexual, in axillary panicles; calyx lobes and petals 5; stamens 8–10, united. Fruit a drupe.

Melia azedarach
White Cedar
Spreading tree to 20m tall. Leaves to 45cm long, bipinnate with 3–5 pairs of pinnae, each with 30–70 leaflets (pinnules); leaflets ovate, 2–5cm long, entire or toothed. Panicles shorter than leaves; flowers lilac, strongly scented. Drupe turning yellowish with age. **HABITAT:** Rainforest margins, river banks. **DIST:** Nepean R. (e.g. Cattai, Kurrajong, Picton), Illawarra (e.g. Macquarie Pass, Budderoo NP). Also NC, SC, CWS, Qld, WA, Asia. **FL:** Oct–Dec.

Synoum
Small tree. Leaves alternate, pinnate. Flowers bisexual, in short panicles; sepals and petals usually 4, rarely 5; stamens usually 8, united. Fruit a 2–3-valved capsule. Monotypic genus.

Synoum glandulosum
Scentless Rosewood, **Bastard Rosewood**
Tree to 5m tall. Leaves with usually 5–9 leaflets, terminal leaflet largest; leaflets oblanceolate,

Synoum glandulosum

3–10cm long, dark green above, paler beneath. Panicles 2–4cm long. Flowers white to pink. Capsule 2–3cm diam., red, ripening Sept–Dec. **HABITAT:** Rainforest margins, sheltered gullies. **DIST:** Common. Coast. Also Robertson. From Bega (SC) to Qld. **FL:** Mar–July.

Toona
RED CEDAR
Deciduous trees. Leaves alternate, pinnate. Flowers unisexual, in pendent panicles; sepals and petals 5; stamens 5, free. Fruit a 5-valved capsule.

Toona ciliata
Red Cedar

Partly deciduous tree reaching 40m but usually much smaller. New shoots pink. Leaves pinnate, to 45cm long, with 4–8 pairs of leaflets; leaflets ovate, to 14 x 4cm, shortly stalked, asymmetric at base. Panicles to 40cm long; flowers numerous, 5mm long, white or pinkish. Capsule ellipsoid, to 25 x 10mm. **HABITAT:** Rainforest, esp. along gullies and streams. **DIST:** Coast, Blue Mtns (e.g. Kowmung R.). Also NC, SC, NT, ST, CWS, Qld, PNG, Indon. **FL:** Oct–Nov.

T. ciliata may be confused with *Polyscias murrayi* but leaves lack the terminal leaflet characteristic of that species.

Toona ciliata

RUTACEAE

Acronychia
Trees or shrubs with opposite 1- or 3-foliolate leaves. Flowers with 4 sepals, 4 petals, 8 stamens and 4-locular ovary. Petals touching in bud. Fruit succulent with a central stone.

Acronychia oblongifolia
Common Acronychia, Yellowwood

Small glabrous rainforest tree. Leaves 1-foliolate; leaflets with rounded tip, usually notched, 5–10 x 1–4cm on stalk about 25mm long with distinct elbow at the top, shiny, dotted with oil glands. Flowers creamy white in loose axillary clusters. Fruit obscurely 4-lobed, white to green, 10–12mm across. **HABITAT:** Warm coastal rainforest. **DIST:** Royal NP, Illawarra, Gosford district. Coast from Qld to Vic. **FL:** Jan–Apr.

Acronychia oblongifolia

Asterolasia

Shrubs with dense stellate hairs on most parts. Leaves alternate, simple, entire. Flowers with 5 sepals, 5 petals and 10 free stamens surrounding superior 5-lobed ovary with 5 fused styles. Fruit with 1–5 beaked segments (cocci).

Asterolasia buckinghamii

Slender branching shrub to 1.5m tall. Branches, buds and midvein of leaf on underside with matted brown stellate hairs. Leaves obovate-elliptic, 10–17 x 7–10mm, green and rough above, paler beneath with silvery brown hairs. Flowers solitary on short stalks in axils, bright yellow; petals 4–9mm long; stamens with orange anthers. **HABITAT:** Sheltered forested gullies. **DIST:** Restricted to small areas near Penrose, Wingello, West Berrima. **FL:** Oct.

Asterolasia correifolia

Asterolasia correifolia
Star-bush

Shrub to 2m with matted brown stellate hairs on branches. Leaves elliptic to lanceolate, 3–8 x 1–2cm, green and glabrous above, densely stellate-hairy beneath. Flowers in small axillary clusters; petals white, 5–6mm long, spreading or reflexed; stamens yellow. **HABITAT:** Moist sheltered forest on hillsides, gullies on Narrabeen shale. **DIST:** Hornsby Plat. (e.g. Cowan Ck, Girrakool, Olney SF). Also NC, ST. **FL:** Chiefly Sept–Oct.

Boronia

Shrubs with opposite simple or compound leaves. Inflorescence axillary or terminal. Flowers with 4-lobed calyx, 4 petals and 8 stamens. Capsule with 4 distinct compartments (cocci).

Boronia algida

Small, often scrambling shrub with symmetrical branching; young stems red, softly hairy. Leaflets 5–9 (usually 5), obovate-spathulate, 4–6 x 2–3mm with terminal leaflet shortest, glabrous, thick. Flowers usually solitary; petals valvate, light pink, 5–6mm long, pointed; staminal filaments glabrous. **HABITAT:** Open forest on sandy soils. **DIST:** Upper Blue Mtns, mainly t'lands and w. slopes in NSW. Also Vic. **FL:** Sept–Dec.

Boronia anemonifolia var. *anemonifolia*
Sticky Boronia

Odorous sticky shrub to 1m tall; branchlets with sparse hairs in 2 lines. Leaves with 3 (sometimes 5) leaflets on stalk 5–15mm long;

Asterolasia buckinghamii

FLOWERING PLANTS

Boronia anemonifolia var. *anemonifolia*

Boronia algida

leaflets narrow-cuneate with 3-toothed apex, 3–8mm long, incurved or folded, glandular-warty. Flowers 1 or 2 in short axillary clusters; petals pink and white, 3–4mm long. **HABITAT:** Open forest on sandy soils. **DIST:** Upper Blue Mtns (e.g. Narrow Neck), Sthn H'lands (e.g. Belmore Falls, Bundanoon), coast (e.g. Girrakool, Thirlmere Lakes). Coast and t'lands of NSW, Vic, Tas. **FL:** Aug–Nov.

Boronia anethifolia
Narrow-leaved Boronia

Sparse shrub about 50cm tall with angular or grooved glandular-warty branchlets. Leaves bipinnate with 3–7 narrow acute leaflets mostly 5–8 x 1.5mm, some further divided into 3. Flowers axillary in tight clusters; petals pale pink to white, 3–4mm long. **HABITAT:** Open forest, woodland on rocky, sandy soils. **DIST:** Uncommon. Coast and Blue Mtns (e.g. Nortons Basin, Erskine Ck). Coast to w. slopes in NSW. Also Qld. **FL:** July–Sept.

Boronia anethifolia

Boronia barkeriana

Boronia barkeriana
Barker's Boronia
Shrub <1m high with glabrous reddish branchlets. Leaves 1-foliolate, elliptic-oblanceolate, 15–30mm long, acute, erect, minutely serrated. Flowers in clusters of 2–8 on pedicels 15–18mm long in upper axils; petals pale to deep pink, 6–8mm long. **HABITAT:** Swampy ground in heath and scrub. **DIST:** Mainly upper Blue Mtns, Illawarra (e.g. Barren Grounds, Carrington Falls). Also SC, ST, Qld. **FL:** Sept–Dec.

Boronia deanei
Dean's Boronia
Branching shrub to about 1m with glabrous warty branchlets. Leaves strongly aromatic, 1-foliolate, semi-terete to linear, 3–10 x 1–2mm, obtuse, smooth above, warty beneath due to raised oil glands. Flowers in terminal clusters of 1–3, rarely also axillary (Fitzroy Falls); petals pale to deep pink, rarely white, 5mm long. **HABITAT:** Swamps, wet ground in heath and scrub. **DIST:** Upper Blue Mtns (Clarence, Kanangra-Boyd NP), Sthn H'lands (Moreton NP, Budderoo NP). **FL:** Sept–Dec.

Boronia floribunda
Pale Pink Boronia
Aromatic shrub to 1m tall with scented flowers; branchlets glabrous. Leaves pinnate with usually 5 leaflets on winged axes, terminal leaflet shortest; leaflets 5–25 x 2–3mm, acute, mucronate, glandular but glabrous, paler beneath. Inflorescences axillary, several-flowered. Petals pale pink, 10–15mm long, imbricate; anthers hairy, yellow-white; stigma enlarged. **HABITAT:** Open forest and heath, usually in very well-drained positions. **DIST:** Coast and Blue Mtns. **FL:** Sept–Nov (as late as Jan in mtns).

Boronia fraseri
Fraser's Boronia
Shrub to 2m tall, similar to *B. mollis* but with 4-angled glabrous or sparsely hairy branchlets. Leaves pinnate with 3–7 leaflets, terminal leaflet much the largest. Inflorescences axillary, 2–6-flowered, much shorter than leaves. Sepals broadly ovate. Petals pink, 6–8mm long, valvate, hairy outside. **HABITAT:** Moist open forest, sheltered slopes and gullies. **DIST:** Endemic. Hornsby Plat. to lower Blue Mtns (e.g. Glenbrook). **FL:** Sept–Nov.

This species occasionally intergrades with *B. mollis*.

Boronia deanei

FLOWERING PLANTS | 321

Boronia floribunda

Boronia fraseri

Boronia ledifolia

FLOWERING PLANTS

Boronia microphylla

Boronia mollis

Boronia ledifolia
Sydney Boronia, Ledum Boronia
Dainty shrub usually 30–100cm tall with brown stellate-hairy branchlets. Leaves 1- or 3-foliolate; leaflets linear-elliptic, 10–30 x 2–4mm, green with few hairs above, paler and densely hairy beneath, margins recurved. Petals bright pink, 5–10mm long, valvate. **HABITAT:** Open forest, woodland and heath. **DIST:** Widespread but mainly coast. Also SC, NT, ST, CWS, Vic. **FL:** July–Sept.

Boronia microphylla
Small-leaved Boronia
Shrub 60–100cm tall; branchlets glandular-warty. Leaves pinnate with 9–11 leaflets on winged axis, terminal leaflet shortest; leaflets obovate with mucronate or notched tip, mostly 5–6 x 2–3mm. Inflorescences axillary, 1–5 flowered. Petals deep rosy pink, 5–6mm long. **HABITAT:** Woodland and heath on rocky, sandy soils. **DIST:** Upper Blue Mtns (e.g. Katoomba, Bell, Mt Wilson), Nattai T'land (e.g. Hill Top, West Berrima). Also SC, NC, NT, CWS, Qld. **FL:** Sept–Dec.

Boronia mollis
Soft Boronia
Robust shrub to 2m tall with dense soft brown stellate hairs on branches and leaf axes. Leaves with unpleasant odour, pinnate with 5–9 moderately hairy oblong-elliptic leaflets, terminal leaflet longest (to 40mm). Inflorescences axillary, 2–6-flowered, shorter than leaves. Sepals narrow-triangular. Petals

Boronia nana var. *hyssopifolia*

greyish pink, 8–10mm long, valvate, hairy outside on midrib. **HABITAT:** Sheltered moist forested slopes and gullies. **DIST:** Coast to Nepean R. but usually not far from sea. Also NC. **FL:** Aug–Nov.

Boronia nana var. hyssopifolia
Dwarf Boronia, Waxy Boronia
Small trailing shrub, often only 10–15cm tall with stem hairs in 2 lines separated by decurrent leaf bases. Leaves 1-foliolate, narrow-elliptic, 6–12 x 1–2mm, acute, concolorous, thick, glabrous to sparsely hairy. Flowers solitary in axils; petals white or pinkish, 3–4mm long; anthers with minute appendage. **HABITAT:** Grassy understorey of woodland and open forest. **DIST:** Upper Blue Mtns (e.g. Newnes SF, Kanimbla V.). Also t'lands and slopes to Vic, Tas. **FL:** Nov–Jan.

Boronia parviflora

Boronia pinnata

Boronia polygalifolia

Boronia parviflora
Swamp Boronia
Herb with erect glabrous stems seldom >30cm high. Leaves 1-foliolate, 10–15 x 2–5mm, entire or finely serrated near tip. Flowers solitary or in axillary clusters of 2–3; sepals pointed, visible between petals; petals pink, 3–4mm long.
HABITAT: Peaty and swampy ground in heath and scrub. **DIST:** Mainly coast. Also CT, NC, SC, ST, Qld, Vic, Tas, SA. **FL:** Some flowers all year.

Boronia pinnata
Pinnate Boronia
Aromatic glabrous shrub about 1m tall with smooth angular branchlets. Leaves pinnate with 5–9 leaflets on winged axis; leaflets oblong-elliptic, about 15 x 2mm, entire, acute, thick. Flowers numerous in small clusters in upper axils; petals pink, imbricate; stigma pointed.
HABITAT: Low open eucalypt forest and heath.
DIST: Coast (mainly nthn part) and lower Blue Mtns (uncommon). Also NC, SC. **FL:** July–Nov.

Boronia polygalifolia
Milkwort Boronia
Weak shrub or herb, similar to *B. nana* but with glabrous branchlets and leaves paler on undersurface. Leaves 1-foliolate, 6–15 x 1–2mm, acute, discolorous. Flowers solitary in axils, pale to bright pink; anthers with minute appendage. **HABITAT:** Grassy understorey of woodland and open forest on heavier soils and shales. **DIST:** Cumberland Pl. (Castlereagh NR, Shanes Park), Sthn H'lands (Joadja Ck), Hornsby Plat. (Warrah Sanct.). Coast and t'lands of NSW and Qld. **FL:** Sept–Jun.

Boronia rubiginosa

Boronia rubiginosa
Bushy shrub to 1m or more tall with stellate-hairy branchlets. Leaves with strong odour when crushed, usually with 3 or 5 leaflets; leaflets elliptic, 7–17 x 3–8mm with terminal leaflet longest, glabrous or nearly so. Inflorescences axillary, 1–3-flowered. Petals pink, 6–10mm long, valvate. Stamens alternating long and short; filaments with white hairs. **HABITAT:** Open forest, sandy soils on rocky sites. **DIST:** nw. Blue Mtns (e.g. Glen Davis–Newnes, Wolgan Pinnacle), Colo, Berrima. **FL:** Aug–Nov.

Boronia serrulata
Native Rose, Rose Boronia
Aromatic shrub usually <1m tall with erect spreading stems. Leaves crowded, overlapping and flattened on stems, broadly ovate, acute, minutely serrated, 10–15 x 4–8mm. Flowers bunched at end of stems, bright rose pink, cupped with overlapping (imbricate) petals; anthers yellow. **HABITAT:** Heath and woodland in moist or periodically dry sandy soils. **DIST:** Urban outskirts from Brisbane Water NP to Darkes Forest (O'Hares Ck). **FL:** Aug–Nov.

One of the gems of the Sydney sandstone flora, preserved in NPs and Holsworthy Military Reserve but much rarer today than in the past.

Boronia serrulata

Boronia thujona

Boronia thujona

Glabrous shrub to 3 or 4m with peppermint smell. Leaves pinnate with 9–13 leaflets; leaflets narrow-elliptic, about 15 x 3–4mm, smaller towards apex, minutely serrated with conspicuous marginal oil glands, acute, thin. Flowers in axillary clusters, numerous; petals bright pink, imbricate, often cupped rather than spreading; stigma flat. **HABITAT:** Sheltered shady forest. **DIST:** Coast, Sthn H'lands, Blue Mtns (Kings T'land) to Budawangs. **FL:** Aug–Nov.

Correa

Shrubs with stellate hairs and opposite simple leaves. Calyx cup-shaped with smooth or slightly 4-lobed rim. Corolla tubular with 4 short reflexed lobes; 1 sp. (*C. alba*) with short tube and 4 large spreading lobes. Stamens 8, strongly exserted.

Correa alba var. alba
White Correa, Coastal Correa

Low dense spreading coastal shrub with rusty-tomentose stems. Leaves broadly obovate to almost round, 2–4 x 1.5–3cm, thick, dark grey-green and sparsely hairy above, pale with short, closely matted white or rusty hairs beneath. Flowers white, 10–15mm long with short deeply divided tube and 4 spreading lobes. **HABITAT:** Exposed coastal cliff tops, headlands and sandy heath near sea. **DIST:** Coast. From NC to Vic, also Tas. **FL:** Apr–Sept.

Correa lawrenciana var. macrocalyx
Mountain Correa

Uncommon shrub about 1.5m tall in Sydney area, much taller elsewhere. Stems rusty-floccose. Leaves ovate, 4–8 x 2–5cm, dark green above, paler with dense yellow-brown stellate

Correa alba var. *alba*

Correa lawrenciana var. *macrocalyx*

tomentum beneath. Flowers with green lobed calyx 5–10mm long and greenish yellow floral tube 20–25mm long with dense velvety hairs. **HABITAT:** Sheltered forest. **DIST:** Recorded only from Patonga Ck, Hill Top and Minnamurra Falls. **FL:** May–June.

Var. *lawrenciana* is a common plant in the high country of the ST and Vic, but var. *macrocalyx* is much less common.

Correa reflexa var. *reflexa*
Common Correa, Native Fuchsia
Variable shrub with several forms. Local form 50–80cm tall with rusty floccose stems and stem-clasping leaves to 3cm long with sparse to dense covering of stellate hairs. Flowers pendent, terminal, often subtended by small bract pair; corolla tubular, 2–3cm long, yellow-green or red with green lobes. **HABITAT:** Eucalypt forest and heath on sandy soils, also clay soils. **DIST:** Coast to w. slopes in NSW. Also Qld, Vic, Tas, SA. **FL:** March–Sept.

The 2 colour forms rarely, if ever, grow intermixed.

Correa reflexa var. *reflexa* (red)

Correa reflexa var. *reflexa* (green)

Crowea exalata

Crowea saligna

Crowea
Shrubs with alternate simple leaves. Flowers solitary in axils with 5 free sepals, 5 overlapping petals, and 10 free stamens with broad-based tapering filaments which are contiguous or overlap to form an erect structure enclosing ovary. The genus is close to *Philotheca* but differs in having anthers with conspicuous white-bearded appendages.

Crowea exalata
Small Crowea
Shrub <1m tall. Leaves linear to narrow-obovate, 20–40 x 2–6mm, apex rounded or shortly pointed, paler beneath. Flowers mostly solitary in upper axils; sepals softly hairy to almost glabrous; petals pale to deep pink, 6–10mm long; stamens erect with filaments overlapping. **HABITAT:** Open forest and woodland on rocky sandstone slopes and gullies. **DIST:** Coast (mainly); t'lands and w. slopes in NSW. Also Vic. **FL:** Dec–July.

Crowea saligna
Lance-leaf Crowea
Soft green shrub to 1.5m tall with glabrous angled or narrowly winged stems. Leaves sessile, elliptic, 3–6cm x 6–14mm, acute or obtuse, with short point. Flowers axillary; petals bright pink, 12–18mm long; stamens erect with filaments cohering by prominent marginal hairs. **HABITAT:** Sheltered open forest. **DIST:** Restricted to coastal zone around Sydney. **FL:** Feb–June.

Eriostemon
Shrubs with stellate and longer simple hairs on new stems and leaves. Leaves alternate, simple, 3-veined (Sydney sp.) or 5-veined (n. Qld sp.). Flowers with 5 free sepals, 5 imbricate 5-veined petals and 10 free stamens incurved over the ovary, also pedicels with several prominent bracteoles, staminal filaments with woolly-ciliate margins, anthers without appendages, a capitate stigma and carpels without a beak.

FLOWERING PLANTS

Eriostemon australasius

Eriostemon australasius
Pink Wax-flower
Erect bushy or straggly thin shrub to 2m tall; branchlets angular, dull green, stellate-hairy. Leaves linear to lanceolate, 30–80 x 8–14mm, light green, obscurely veined. Flowers mostly solitary in upper axils; petals waxy, pink, 5-veined from base, 13–18mm long, stellate-scaly on outside. **HABITAT:** Open forest, heath and coastal dunes. **DIST:** Common on NSW coast, rare on CT. Also Qld. **FL:** July–Oct.

Geijera
Trees or shrubs with alternate 1-foliolate or simple leaves. Inflorescences axillary or terminal. Flowers with 5 free sepals, 5 petals, 5 stamens opposite the sepals and ovary with 5 distinct carpels which are not beaked. Fruit with 1–5 cocci, each with 2 glossy black seeds.

Geijera salicifolia var. *latifolia*

Geijera salicifolia var. *latifolia*
(also *G. latifolia*)
Brush Wilga, Scrub Wilga, Green Satinheart
Uncommon tree 10–20m tall with grey scaly bark; branchlets and inflorescences minutely hairy. Leaves ovate to lanceolate, 7–11 x 3–6cm on stalks 1–1.5cm long, glossy green above, paler beneath. Flowers in large paniculate clusters; petals 2–3mm long, white, valvate, glabrous. Fruit brown, ovoid, 6mm across. **HABITAT:** Rainforest remnants on coast. **DIST:** Camden district (e.g. Cobbity), Illawarra district (e.g. Mt Keira, Berkeley Hill) sth to Minnamurra. Also NC, Qld, PNG. **FL:** Sept–Nov.

Leionema, Nematolepis, Phebalium

Most *Phebalium* species in the Sydney area were reclassified in 1998 as *Leionema* or (1 species) *Nematolepis*, largely on differences in indumentum, anthers and bracteoles.

LEIONEMA
Stems with non-scaly indumentum. Leaves alternate, simple. Flowers with cup-like calyx, 5 petals and 10 spreading stamens; anthers lacking a tiny sterile point (apiculum).

Leionema dentatum
Toothed Phebalium
Slender shrub to 5m tall with minute stellate hairs on young stems. Leaves 60–80 x 3–7mm with conspicuous oil dots, dark green above, greenish white with raised midrib beneath; margins recurved, sometimes minutely toothed. Flowers abundant in axillary clusters of about 10; petals and stamens pale cream.
HABITAT: Forest in sheltered sandstone gullies. **DIST:** Widespread on coast and ranges. Also NC, NT. **FL:** Aug–Oct.

Leionema diosmeum
Bushy shrub 1m or more tall with long loose hairs on stems. Leaves 8–15mm long, appearing terete due to strongly revolute margins, glabrous or with short or long hairs above and with simple or stellate hairs beneath, apex obtuse. Flowers in compact terminal clusters on short stout pedicels; calyx pubescent with long

Leionema lamprophyllum subsp. *orbiculare*

narrow lobes; petals yellow, 6mm long.
HABITAT: Heath and eucalypt forest. **DIST:** Coast sth of Sutherland and Sthn H'lands (e.g. Royal NP, Carrington Falls, Barren Grounds, Wingello). Also SC, ST, Vic. **FL:** Aug–Nov.

Leionema lamprophyllum subsp. orbiculare
Shining Phebalium
Shrub to 1m tall with downy and very warty young stems. Leaves obovate to almost circular, about 6 x 5mm, convex and glossy above (*lampro*, 'like a lamp'). Flowers solitary in upper axils or in clusters of 1–3; petals and stamens creamy white.
HABITAT: Heath and woodland, and near pagoda rock formations. **DIST:** Upper Blue Mtns between Lithgow–Glen Davis–Rylstone. Also ST, Vic. **FL:** Sept–Jan.

Leionema diosmeum

Leionema dentatum

Nematolepis squamea subsp. *squamea*

Phebalium squamulosum subsp. *lineare*

NEMATOLEPIS

Shrubs with stellate-scaly indumentum, especially on new growth. Leaves alternate, simple, dotted with oil glands. Flowers in terminal inflorescences or solitary, with 2 bracteoles at or near top of pedicel; sepals 5; petals 5; stamens 10, spreading, with slightly flattened filaments and versatile anthers without an apical gland. Seeds smooth and shiny.

Nematolepis squamea subsp. *squamea*
Satinwood
Shrub or tree to 12m, usually much less. Leaves lanceolate-elliptic, flat, 8–10 x 1–2cm, light green and glabrous above, silvery white beneath. Flowers in axillary clusters; petals white, 4–5mm long, gland-dotted. **HABITAT:** Sheltered gullies. **DIST:** Widespread on coast and ranges. Also NC, SC, Qld, Vic, Tas. **FL:** Sept–Oct.

PHEBALIUM

Shrubs with stellate-scaly indumentum, especially on new growth. Leaves alternate, simple, with dotted or raised warty oil glands. Flowers in terminal inflorescences, or solitary; calyx 5-lobed, scaly outside; petals 5, yellow to white in Sydney district; stamens 10, with slender filaments and basifixed anthers with prominent spherical gland at apex. Seeds longitudinally ridged.

Phebalium squamulosum subsp. *argenteum*
Scaly Phebalium
Leaves oblong-elliptic with rounded apex, dull green above, with silvery stellate scales beneath. Petals creamy white with silvery scales outside. Stamens creamy. **HABITAT:** Forest, woodland and heath, usually on sands, also on Narrabeen shales at Mooney Mooney Ck and clayey sea-cliffs at Garie Bch. **DIST:** Coast from Port Stephens to Vic; uncommon on CT (e.g. Bundanoon). **FL:** Aug–Nov.

In the Georges R.–Hacking R. area this subsp. grades into a thin-leaved form of subsp. *squamulosum*.

Phebalium squamulosum subsp. *lineare*
Leaves linear with obtuse or flat apex, 10–25 x 1mm, grooved above. Flowers clustered, on slender stalks; calyx with reddish brown scales; petals pale yellow. **HABITAT:** Forest on rocky slopes. **DIST:** Shoalhaven R. (Badgerys Crossing). Also upper Hunter V., CWS. **FL:** Aug–Oct.

Phebalium squamulosum subsp. *squamulosum*

Phebalium squamulosum subsp. *ozothamnoides*

Phebalium squamulosum subsp. *ozothamnoides*
Everlasting Phebalium, Alpine Phebalium

Compact shrub to 1m tall (locally) with rusty scales on stems. Leaves obovate to nearly circular with obscure midrib, about 10 x 6mm, glossy dark green above, stellate-hairy when young, whitish and smooth beneath, margins recurved. Flowers bright yellow in terminal clusters on short side-branches; petals with rusty or silvery scales outside. **HABITAT:** Heath and exposed rocky sites on Narrabeen Sandstone. **DIST:** Upper Blue Mtns (e.g. Hassans Walls, Mt York, Mt Victoria). Also NT, ST, Vic. **FL:** Sept–Dec.

Phebalium squamulosum subsp. *argenteum*

Phebalium squamulosum subsp. *squamulosum*
Scaly Phebalium

Shrub 1–3m tall with brownish and silvery scales. Leaves oblong with rounded, notched or pointed tip, 20–50 x 2–10mm, green above, silver grey-hairy and scaly beneath. Flowers in terminal clusters; calyx unlobed; petals, 2–4mm long, pale to mid-yellow on inside. **HABITAT:** Forest, woodland and heath. **DIST:** Coast to slopes and plains in NSW. Also Qld, Vic. **FL:** Mainly July–Nov.

Subsp. *squamulosum* is the most widely distributed and common of the 4 subspecies in the Sydney region.

Philotheca
WAX-FLOWER
Endemic genus including (since 1998) a number of species formerly within *Eriostemon* but differing in having leaves and petals with 1 longitudinal vein, peduncles without bracts or bracteoles, indistinct stigma and beaked carpels.

Philotheca buxifolia subsp. buxifolia
Box-leaf Wax-flower
Compact shrub <1m tall with ribbed finely hairy stems. Leaves sessile, broadly elliptic with acutely pointed tip and cordate base, 6–15 x 4–10mm, concave above, strongly recurved lengthwise, thick, keeled beneath. Flowers solitary in axils on short stalks, pink in bud; petals white, 8–15mm long. **HABITAT:** Coastal heath on sand or sandy soils. **DIST:** Dee Why to Port Hacking and Point Perpendicular (Jervis Bay). **FL:** Aug–Nov.

This subspecies intergrades with *P. scabra* in Royal NP. Intergrades have narrower leaves and petals about 10mm long.

Philotheca buxifolia subsp. obovata
Box-leaf Wax-flower
Similar to subsp. *buxifolia* except leaves obovate, narrower, with narrow truncate base and warty margins. **HABITAT:** Shrubby eucalypt woodland on sandy soils, usually near the sea. **DIST:** Gosford to Manly, Wondabyne, Patonga, Calga, Peats Ridge. Also SC near Ulladulla. **FL:** Aug–Nov.

This subsp. hybridises with *P. myoporoides* in the lower Hawkesbury area.

Philotheca buxifolia subsp. *buxifolia*

Philotheca buxifolia subsp. *obovata*

Philotheca hispidula
Hairy Wax-flower, Spreading Wax-flower
Shrub <1m high; stems with short hairs, not warty. Leaves sessile, linear or narrowly obovate, mucronate with straight or recurved point, 10–30 x 3–4mm, margins revolute and irregular due to prominent oil glands. Flowers axillary, mostly solitary on short pedicel; petals white to pale pink, 6mm long. **HABITAT:** Open forest on sandy soils. **DIST:** Endemic. Nattai T'land (e.g. Colo Vale), through lower Blue Mtns (e.g. Linden, Glenbrook) to Colo Heights. **FL:** May–Oct.

A longer-leaf form in the lower Blue Mtns is believed to be a hybrid with *P. myoporoides*.

Philotheca myoporoides subsp. myoporoides
Native Daphne, Long-leaf Wax-flower
Glabrous bushy shrub to 2m tall; stems terete, green, with raised oil glands. Leaves flat, narrowly oblong-elliptic with pointed apex, 4–10cm x 5–12mm, dark green, dotted with oil glands. Flowers in axillary clusters of 3–8 on slender pedicels from a longer stout peduncle to 20mm long, pink in bud; petals white, 8mm long. **HABITAT:** Forest on sheltered rocky slopes, near streams and on shaded escarpments. **DIST:** Scattered in Blue Mtns and Nepean R.–Tahmoor district. Also SC, ST, CWS, Qld, Vic. **FL:** Aug–Nov.

Of the 6 subsp., only the type subsp. occurs in the area. In parts of the Blue Mtns (e.g. Lapstone, Linden) and at Tahmoor, it appears to hybridise with *P. hispidula*. In the lower Hawkesbury it hybridises with *P. buxifolia*.

FLOWERING PLANTS

Philotheca hispidula

Philotheca myoporoides subsp. *myoporoides*

Philotheca obovalis

Philotheca obovalis
Silky Wax-flower, Fairy Wax-flower
Small compact shrub with downy non-warty stems. Leaves thick, rounded at apex, narrowing to base, 6–10 x 4–6mm, concave and smooth above, somewhat warty beneath, midvein faint. Flowers solitary on short stalk in axils; petals white, often tinged pink, 6–8mm long. **HABITAT:** Rocky outcrops on exposed sandstone heaths. **DIST:** Endemic. Upper Blue Mtns (e.g. Wolgan Gap, Mt Banks, Mt Wilson, Blackheath). **FL:** Mainly Oct–Jan.

Philotheca salsolifolia subsp. *salsolifolia*

Philotheca scabra subsp. *scabra*

Philotheca salsolifolia subsp. *salsolifolia*
Philotheca
Shrub, variously ground-hugging on exposed heaths, low dense shrub on old sand dunes or open shrub to 1m tall in open eucalypt forest. Leaves alternate, crowded, narrow-linear to terete, 4–15mm long, thick, glabrous, with numerous oil glands. Flowers mostly solitary at ends of branches; petals mauve-purple, 6–12mm long; stamens hairy, united at base, free above. **HABITAT:** Sandy and rocky heath, dunes, woodland and open forest. **DIST:** Coast and ranges (e.g. West Head, La Perouse, Jibbon Head, Agnes Banks, Narrow Neck). Also NC, SC, ST, WS, NWP. **FL:** July–Dec.

Philotheca reichenbachii is similar to *P. salsolifolia*. Its anthers are concealed by upwards-pointing hairs on filaments and have an apical tuft of long hairs. It occurs in rocky and sandy coastal heath north and south of Sydney, also CWS.

Philotheca scabra subsp. *scabra*
Finely branched shrub usually <60cm high with terete stems; lower stems with brown hairy indumentum, not warty; upper stems glossy green with short stiff hairs, warty. Leaves terete, pointed, concave above, convex and warty below, 10–15 x 1–2mm. Flowers solitary on short stalks in axils, pink in bud; petals white to pale pink, 7–8mm long. **HABITAT:** Open forest, heath, shaded sandstone escarpments. **DIST:** Mainly in Georges R.–Woronora R. area, extending to Nowra. **FL:** May–Nov.

A broad-leaf form known as subsp. *latifolia* occurs at Bundanoon and Wingello SF and from Nowra to Nerriga. It has very warty stems and smooth or slightly warty oblong-elliptic leaves to 30 x 2–5mm. Subsp. *scabra* hybridises with *P. buxifolia* sth of Botany Bay. Subsp. *latifolia* hybridises with *P. myoporoides* subsp. *myoporoides*.

Philotheca trachyphylla
Rock Wax-flower
Usually a tall shrub with warty glabrous stems. Leaves sessile, oblong-elliptic, mucronate, 30–45 x 5–8mm, glabrous, dark green, thin, with prominent oil glands; margins recurved. Flowers axillary, solitary or in clusters of 3 on pedicels 6–20mm long; petals white, 5–8mm long; carpels green, united; stamens free,

FLOWERING PLANTS

Philotheca trachyphylla

Melicope micrococca

Sarcomelicope simplicifolia

spreading; anthers orange. **HABITAT:** Forest, on rocky hillsides. **DIST:** Scattered and uncommon (e.g. west of Dapto, West Berrima, Howes V., Mt Savage–Sombre Dome). Also SC, ST, CWS, Vic. **FL:** Sept–Oct.

Melicope
Trees or tall shrubs with opposite 3-foliolate leaves. Inflorescences axillary or terminal. Flowers with 4 sepals, 4 petals, and 4 stamens opposite the sepals. Carpels not beaked. Fruit with 1–4 cocci, each with 2 glossy black seeds.

Melicope micrococca
Hairy-leaved Doughwood, White Euodia
Tree to 10m tall. Leaves clustered near ends of branchlets, opposite, with 3 leaflets digitate, on stalk 5–11cm long; leaflets obovate with bluntly acuminate tip, 4–10 x 2–4cm, dotted with oil glands. Flowers whitish in short clusters; petals 3–4mm long, stamens slightly longer. Cocci 4–6mm long, dark, bristly. **HABITAT:** Rainforest, sheltered gullies, hind-dunes and talus slopes. **DIST:** Illawarra (e.g. Minnamurra Spit), Royal NP and Gosford–Wyong district (e.g. Ourimbah Ck Rd). Also NC, Qld. **FL:** Jan–Mar.

Sarcomelicope
A small genus originally called *Bauerella*. Later included in *Acronychia*, from which it differs in being dioecious and in having petals overlapping in bud and persisting at base of fruit.

Sarcomelicope simplicifolia
Yellowwood, Yellow Acronychia
Shrub or small tree. Leaves 1-foliolate; leaflet on jointed stalk 1–5cm long, elliptic to obovate, 7–12 x 4–5cm, shiny above, paler beneath, dotted with oil glands. Flowers in small axillary panicles to 4cm long; calyx minutely hairy; petals cream, fringed; staminal filaments fringed; stigmas 4-lobed. Fruit a slightly fleshy yellow-brown or whitish berry 10–15mm across. **HABITAT:** Coastal rainforest and margins. **DIST:** Coast (e.g. Saddleback Mtn, Minnamurra Falls, Royal NP). From SC (Mt Dromedary) to Qld. Also LHI. **FL:** May–Sept.

Zieria

Shrubs with opposite 3-foliolate leaves, often with oil glands. Flowers in axillary cymes; calyx 4-lobed; petals 4, usually white or pink, stamens 4; fruit with 4 separate carpels.

Zieria arborescens subsp. arborescens
Stinkwood, Tree Zieria
Shrub or tree 3–5m tall resembling *Z. smithii* and with similar unpleasant odour. Leaflets narrow-elliptic, to 10cm x 25mm on stalk 15–30mm long, recurved, not warty, lower surface hairy. Flowers white, 7–13mm across in loose many-flowered cymes shorter than leaves. **HABITAT:** Sheltered forested gullies and near rainforest, mostly on fertile heavier soils. **DIST:** Coast and ranges (e.g. Jamberoo Pass, Olney SF, Mt Tomah). Also SC, NC, NT, Qld, Vic, Tas. **FL:** Aug–Nov.

Zieria covenyi
Stellate-pubescent shrub to 2m tall. Leaflets elliptic-ovate, 25–35 x 7.5–10mm, recurved, upper surface dark green, lower surface grey-green; apex obtuse. Inflorescence as long as leaves, 3–20-flowered. Flowers white to pale pink, 12mm across. **HABITAT:** Ridgetop open forest. **DIST:** Restricted. Narrow Neck, Katoomba. **FL:** Oct–Nov.

Zieria arborescens **subsp.** *arborescens*

Zieria covenyi

Zieria cytisoides

FLOWERING PLANTS

Zieria fraseri subsp. *compacta*

Zieria cytisoides
Downy Zieria
Compact shrub 1–3m tall; leaves and stems velvety with soft stellate hairs, often greyish. Leaflets obovate, to 3cm long, the central leaflet largest, paler and more densely hairy on undersurface. Cymes with 3 or more flowers, shorter than leaves; sepals 3mm long, tomentose; petals white or pink, 3–5mm long; carpels and fruit hairy. **HABITAT:** Open forest on rocky slopes and ridges, near pagoda formations, also woodland and heath. **DIST:** Nepean area, Blue Mtns (e.g. Bents Basin, Narrow Neck, Newnes). Most of NSW except drier w. plains. Also Qld, Vic. **FL:** Mainly Aug–Oct.

Zieria fraseri subsp. *compacta*
Bushy shrub 0.6–1.5m tall with dense short soft hairs; older stems and leaf upper surface glabrous. Leaflets narrow-elliptic, about 25 x 8mm on petiole 2–8mm long. Flowers in dense axillary cymes, usually pale pink; carpels glabrous. **HABITAT:** Open forest on rocky hillsides and ridges. **DIST:** Coast to w. slopes (e.g. Nortons Basin, Erskine Ck, Wolgan Pinnacle, Bonnum Pic). Also Qld. **FL:** Aug–Nov.

Zieria granulata
Aromatic shrub 1–3m tall with prominent raised glands on stems and leaves. Upper stems glabrous and rough. Leaflets linear, 15–30 x 1mm, with revolute irregular margins. Cymes many-flowered, shorter than leaves; flowers white, 7mm across. **HABITAT:** Basalt hillsides near rainforest. **DIST:** Very restricted. Sth of Sydney (e.g. Saddleback Mtn, Albion Park). **FL:** Aug–Oct.

Zieria granulata

Zieria laevigata

Zieria murphyi

Zieria pilosa

Zieria smithii

Zieria laevigata
Shrub to 1m tall with glabrous 4-angled upper stems. Leaflets linear, 20–40 x 1–3mm, acutely pointed, margins revolute. Cymes usually 3-flowered, ± as long as leaves; sepals short, broad; petals pale pink or white, 5mm long; anthers orange with tiny white reflexed point (apiculum). **HABITAT:** Eucalypt forest and heath. **DIST:** Coast and ranges. Also SC, NC, NT, Qld. **FL:** June–Oct.

Zieria murphyi
Murphy's Zieria
Softly hairy shrub to 1.5m tall. Leaflets 1 or 3, narrow-elliptic, 15–50 x 5–8mm, recurved, stellate-hairy beneath. Cymes with 3 or more flowers, shorter than leaves; sepals triangular, 2mm long, hairy; petals 3mm long, pale pink to white. Fruit hairy. **HABITAT:** Open forest in sheltered sites, often below cliff lines. **DIST:** Endemic. CT (e.g. Mt Tomah, Penrose, Bundanoon). **FL:** Aug–Oct.

Zieria pilosa
Hairy Zieria
Shrub to 1m tall, usually about 30cm; stems round with simple hairs. Leaflets narrow-elliptic, 7–25 x 2–7mm, acute or obtuse, with slightly recurved margins. Cymes 1–3-flowered, often 1, distinctly shorter than leaves; sepals hairy, narrowly triangular, nearly as long as petals; petals white, 2–3mm long; anther appendage minute. **HABITAT:** Eucalypt forest and heath on sandy soils. **DIST:** Coast and ranges. Also NC, SC. **FL:** Mainly May–Oct.
   The specific epithet *pilosa* refers to the long simple hairs on the stems, a feature which separates it from *Z. laevigata*. The long, finely pointed sepals are also characteristic. Flowering is usually sparse.

Zieria smithii
Sandfly Zieria
Glabrous shrub to 2m tall with glandular stems. Leaves with strong unpleasant odour; leaflets narrowly elliptic-ovate, 25–50 x 4–10mm on stalk 10–20mm long, recurved, slightly warty, lower surface glabrous. Flowers white, rarely pale pink, 5–9mm across in loose many-flowered axillary clusters. **HABITAT:** Sheltered eucalypt forests. **DIST:** Coast and ranges. Also SC, NC, NT, Qld, Vic. **FL:** Aug–Oct.

GERANIACEAE

Erodium
CROWFOOT
Herbs or shrubs with lobed or deeply divided or compound leaves. Flowers solitary or in umbels of 2–7, with 5 sepals and petals, and 5 stamens alternating with 5 staminodes. Fruit a schizocarp splitting into 5 hairy, awned mericarps, each with 1 seed.

Erodium crinitum
Blue Crowfoot, Blue Heronsbill
Annual herb to 50cm tall with hairy stems. Leaves at base to 6 x 3cm, deeply dissected into 3 or more lobes, which are in turn jaggedly lobed. Stem leaves opposite, smaller. Flowers in umbels of 2–6, deep blue. **HABITAT:** Grassland, woodland, on clay-shale soils. **DIST:** Cumberland Pl., Blue Mtns (e.g. Kurrajong, Wombeyan Caves). Most of NSW. All other States. **FL:** Aug–Nov.

Erodium crinitum

Geranium
CRANESBILL
Herbs and shrubs with deeply divided or lobed leaves. Inflorescences 1- or 2-flowered. Flowers regular, with 5 sepals and petals and 10 stamens. Fruit a schizocarp separating into 5 awned mericarps each with 1 seed.

Geranium homeanum
Decumbent perennial with stems to 70cm long, sparsely covered with coarse appressed hairs on stem. Leaves to 4 x 5cm, with 3–5 lobes, each divided into 3 secondary lobes. Flowers paired, pink to white; petals 2–4mm long. Mericarp to 14mm long; seed dark brown. **HABITAT:** Moist forest, woodland, grassland, esp. on clay and basaltic soils. **DIST:** Coast and ranges. Also NC, SC, ST, SWS, Qld, Vic. **FL:** Nov–Mar.

Geranium neglectum
Decumbent herb. Flowering stems to 1m long, much-branched, often reddish. Leaves to 3 x 6cm deeply divided into 5–7 lobes, each with smaller apical lobes, with scattered appressed hairs on both surfaces. Flowers solitary; petals pink with darker veins, 15mm long; stalks to 10cm long with two bracteoles mid-way.

Geranium homeanum

HABITAT: Swamps, creek banks. **DIST:** Upper Blue Mtns (e.g. Boyd Plat.), Sthn H'lands (e.g. Wingecarribee). Also SC, NT, ST, Qld, Vic. **FL:** Dec–Mar.

Geranium neglectum

Geranium potentilloides
Hairy perennial with decumbent stems often rooting at the nodes. Stem leaves opposite on stalks 2–4cm long; lamina 2–4 x 2–5cm, with 5–7 deeply divided lobes, each lobed or toothed at the end, pale or purplish beneath. Flowers solitary on stalk 2–4cm long; petals pink or white, 5–6mm long. Mericarps with stiff hairs; seed brown or rarely black. **HABITAT:** Woodlands, damp grasslands, swamp margins. **DIST:** Upper Blue Mtns. Also SC, NT, ST, CWS, Vic, Tas, SA, NZ, PNG. **FL:** Nov–May.

Geranium solanderi var. *solanderi*
Australian Cranesbill
Decumbent or ascending perennial herb with coarse hairs. Leaves opposite, on stalks to 5cm long; lamina to 3 x 5cm, with 5–10 obovate lobes, each with several apical lobes. Flowers paired or rarely solitary on peduncles to 4cm long and pedicels to 5cm long; petals pink, rarely white, 7–10mm long. Mericarps hairy; seeds black. **HABITAT:** Grassland, open woodland, on clay-shale soils. **DIST:** Coast, upper Blue Mtns. Most areas of NSW. All other States (except NTerr), NZ. **FL:** Sept–Mar.

Geranium graniticola grows in swamps of Boyd Plat. It has solitary flowers with stalks about 3cm long and white petals 5mm long.

Pelargonium
STORKSBILL
Annual or perennial herbs or shrubs. Inflorescences with >2 flowers. Flowers with unequal sepals and petals; upper sepal spurred; petals white to deep pink, the upper pair longer than lower 3 and with dark markings; stamens 10. Mericarps separating from column (style), bristly with curved long-hairy awn.

Pelargonium australe
Native Storksbill, Wild Geranium
Softly hairy perennial with sprawling or erect stems to 50cm long, resembling the cultivated *Geranium*. Leaves on stalks to 12cm long; lamina ± ovate, to 9 x 8cm, with 5–7 shallow lobes having toothed margins. Umbels with 4–12 pink flowers. Mericarps with weak hairs. **HABITAT:** Sea-cliffs, sand-dune scrub, rocky outcrops. **DIST:** Coast and upper Blue Mtns. Also NC, SC, ST, WS, WP, all other States (except NTerr), LHI. **FL:** Nov–Feb.

Geranium potentilloides

Geranium solanderi var. *solanderi*

Pelargonium australe

Comesperma ericinum

POLYGALACEAE

Comesperma
MATCHHEADS, MILKWORT

Shrubs and climbers. Leaves alternate, often reduced to scales or deciduous. Flowers in short racemes; sepals 5, 2 petal-like, spreading and larger than others; petals 3 (or 5 with 2 much reduced), the lower 2 forming a keel around the stamens. Fruit a 2-lobed narrowly winged capsule.

Comesperma ericinum
Pink Matchheads, Heath Milkwort

Slender erect shrub to 1m tall with reddish stems. Leaves linear-oblong to 20 x 4mm, margins recurved, paler beneath. Flowers lilac-pink, rarely white, with yellow anthers; wing sepals 6–8mm long, exceeding keel.
HABITAT: Woodland, heath. **DIST:** Common. Coast and ranges. Also NC, SC, NT, ST, WS, Qld, Vic, Tas, SA. **FL:** Sept–Dec.

Comesperma volubile
Love Creeper

Delicate plant with glabrous twining stems to 1m long. Leaves few, narrow-elliptic, 50 x 5mm, recurved, often reduced or deciduous. Flowers blue to mauve with a darker centre. **HABITAT:**

Comesperma volubile

Heath, open forest, esp. in shaded areas.
DIST: Coast and ranges. Also NC, SC, NT, ST, Qld, Vic, Tas. **FL:** Aug–Nov.

Two other *Comesperma* sp. in the Sydney area have reduced leaves, non-twining stems and blue flowers. *C. defoliatum* has outer sepals 1mm long, wing sepals 2–3mm long. *C sphaerocarpum* has outer sepals 3mm long, wing sepals 6mm long.

TREMANDRACEAE

Tetratheca
BLACK-EYED SUSAN
Small heath-like shrubs with simple leaves. Flowers axillary, usually solitary, with 4 (rarely 5) dark sepals, deep pink to white petals, and 8 black stamens. Fruit a flattened capsule.

Tetratheca bauerifolia
Compact shrub to 30cm tall. Stems with short curled hairs. Leaves in whorls of 4–6, narrow-elliptic, to 10 x 3mm. Flowers deep lilac-pink; petals 6–10mm long; pedicels 10–20mm long, slender, glabrous, hooked beneath flower. **HABITAT:** Woodland, heath, often on rocky sites. **DIST:** T'lands (e.g. Kanangra Walls, Budderoo NP). Also ST, Vic. **FL:** Oct–Dec.

Tetratheca decora
Shrub to 40cm tall with densely hairy stems. Leaves alternate, rarely opposite, ± sessile, linear-elliptic, to 12 x 1.5mm, revolute. Flowers deep lilac-pink; petals 6–15mm long; pedicels 3–6mm long, with pale hairs. **HABITAT:** Sandy heath, dry scrub. **DIST:** Scattered. Recorded from 3 areas in Sydney district: Howes Valley–Putty; Ingleside–Mona Vale; Narrow Neck–Burragorang. Also SC, NWS. **FL:** Sept–Oct.

Tetratheca decora

Tetratheca ericifolia

Tetratheca ericifolia
Spreading shrub 15–40cm tall. Stems with short, pale brown, upwards-pointing (antrorse) hairs. Leaves in whorls of 4–6 with short but distinct stalk, to 10 x 1mm, flat to recurved, with tubercles on upper surface. Flowers pale to deep pink; sepals with reddish glandular hairs; petals 6–12mm long; ovary glabrous; pedicels glabrous, 10–20mm long. **HABITAT:** Sandy heath, open forest. **DIST:** Scattered. Port Jackson–Bateau Bay, Engadine–Appin Rd; t'lands (e.g. Mt Tomah, Springwood, Mittagong). Also NC, ST. **FL:** July–Nov.

Tetratheca bauerifolia

Tetratheca glandulosa
Spreading shrub to 50cm tall. Stems with downwards-pointing hairs and small warts. Leaves opposite, alternate or rarely whorled, to 20 x 2mm, revolute, with small stiff hairs near margins. Pedicels and sepals glandular hairy; petals deep pink, 5–10mm long; pedicels 3–10mm long. **HABITAT:** Sandy heath, scrub. **DIST:** Endemic. Hornsby Plat. (e.g. Glenorie, Colo Heights, Cowan, Duffys Forest). **FL:** July–Jan.

Tetratheca juncea
Weak sprawling shrub with winged stems to 60cm long. Leaves alternate, reduced to scales or narrow-elliptic to 20 x 5mm. Flowers 1–2 in axils on glabrous stalks 5–10mm long; petals lilac-pink, 7–10mm long. **HABITAT:** Heath, open forest. **DIST:** Lake Macquarie–Munmorah to Bulahdelah. **FL:** Aug–Oct.

Tetratheca neglecta
Compact or diffuse shrub to 50cm tall. Stems with loose long and short hairs. Leaves in sessile whorls of 4–6, linear, to 15 x 1mm, revolute. Flowers deep pink on glabrous stalks to 15mm long; sepals glabrous; petals 5–10mm long. **HABITAT:** Sandy heath, woodland. **DIST:** Endemic. South of Port Jackson; also near Yerranderie. **FL:** July–Nov.

Tetratheca rubioides
Compact shrub with erect stems to 60cm tall with short white hairs bent downwards. Leaves in whorls of 5–7, crowded along the stem, linear, to 15 x 1mm, revolute. Flower stalks and sepals ± glabrous; petals deep pink, 5–10mm long. **HABITAT:** Sandy heath, open forest. **DIST:** Blue Mtns (e.g. Newnes Plat., Blackheath), St Albans, Bucketty. Also NC, ST. **FL:** Sept–Jan.

Tetratheca rupicola
Compact or diffuse shrub to 30cm tall. Stems terete or 4-angled with fine spreading greyish hairs. Leaves crowded, usually in sessile whorls of 4–6, linear, to 15 x 1mm, revolute; apex acute. Flowers deep pink, petals 5–15mm long; stalks 7–13mm long. **HABITAT:** Open forest, in sandy soils. **DIST:** Blue Mtns (e.g. Lawson, Kings T'land, Govetts Leap, Mt Tomah). **FL:** Oct–Dec.

Tetratheca glandulosa

Tetratheca juncea

Tetratheca neglecta

Tetratheca rubioides

Tetratheca rupicola

Tetratheca shiressii

FLOWERING PLANTS | 349

Astrotricha floccosa

Tetratheca shiressii
Sprawling shrub with wiry stems of 2 kinds: (1) glabrous with opposite linear leaves to 20 x 1mm and (2) with stiff pale brown hairs and ovate leaves to 12 x 6mm in whorls of mostly 3–4. Flowers deep pink; sepals glabrous; petals 10–20mm long; stalks glabrous, 6–12mm long, dark coloured. **HABITAT:** Heath, open forest. **DIST:** Endemic. Coast in 2 separate areas: Hawkesbury R.–Watagan SF (e.g. Somersby, Girrakool) and Royal NP–Bulli Pass. **FL:** July–Oct.

Tetratheca thymifolia
Straggly shrub with dense spreading fawn hairs on stems. Leaves usually in whorls of 3–5, elliptic, to 20 x 8mm, ± recurved, glabrous or hairy; stalk 1mm long. Flowers deep pink; sepals with white hairs or red bristles; petals 6–15mm long; stalks to 20mm long. **HABITAT:** Sandy heath, open forest. **DIST:** Coast and ranges. Also NC, SC, NT, ST, Qld, Vic. **FL:** Aug–Nov.

ARALIACEAE

Astrotricha
STAR-HAIR
Shrubs with stellate hairs (*astro*, 'star'; *tricha*, 'hair') on branchlets, leaf stalks and leaf undersurface. Leaves alternate, simple, entire. Flowers in umbels arranged in large terminal panicles; calyx 5-lobed; petals 5; stamens 5. Fruit a schizocarp separating into 2 mericarps.

Astrotricha floccosa
Woolly Star-hair
Erect shrub to 3m tall with dense soft woolly hairs. Leaves elliptic-ovate, to 20 x 5cm, acuminate, dull pale green above; stalk 1–2cm long. Flowers small, cream to white. **HABITAT:** Sheltered forested slopes, esp. beside estuaries. **DIST:** Endemic. Coast between Port Jackson and Gosford. Also Casula, lower Blue Mtns (e.g. Hawkesbury L'out). **FL:** Oct–Jan.

Astrotricha latifolia
Broad-leaf Star-hair
Shrub to 3m tall with spreading branches. Young branches with loose indumentum. Leaves ovate-elliptic, to 20 x 6cm, glossy green above with thin indumentum beneath; stalk 3–8cm long. Flowers greenish yellow. **HABITAT:** Sheltered forest, gullies. **DIST:** Coast and ranges (e.g. Patonga, Kentlyn, Hawkesbury L'out, Mt Wilson). Also NC, SC, NT, Qld. **FL:** Oct–Jan.

Astrotricha latifolia

Tetratheca thymifolia

Astrotricha ledifolia

Astrotricha longifolia

Astrotricha ledifolia
Shrub to 1.5m tall with hairy branchlets. Leaves narrow-oblong, to 50 x 7mm, obtuse; upper surface rough, lower surface with close indumentum. Panicle loose, 10–15cm long; flowers creamy green to white. **HABITAT:** Open forest, in rocky sites. **DIST:** Upper Blue Mtns (e.g. Clarence), Sthn H'lands (e.g. Wombeyan Caves, Wingello). Also ST, Vic. **FL:** Oct–Dec.

Astrotricha longifolia
Long-leaf Star-hair
Spreading or erect shrub 1–2m tall; axes of branchlets and inflorescence often purplish with few hairs in coastal form. Leaves narrow-elliptic with acute apex, to 12cm x 10mm (coast form) or to 25mm wide (inland form), dull green above, downy white below. Panicle narrow, to 60cm long (coast form) or to 25cm long (inland form); flowers white to cream. **HABITAT:** Sheltered forest, on sandy soils. **DIST:** Inland form: upper Blue Mtns (e.g. Bell, Newnes Plat., Hill Top); coast form: Galston Gorge, lower Blue Mtns. Also NT, ST, NWS, WP, Qld. **FL:** Oct–Dec.

Cephalaralia
Climber or scrambling shrub with 3-foliolate leaves. Flowers sessile in small heads in racemes or panicles; petals and sepals 5. Fruit a drupe. Monotypic genus.

Cephalaralia cephalobotrys
Climbing Panax
Climber to 5m. New growth and inflorescences with scattered brown hairs. Leaflets oblong to lanceolate, to 12 x 6cm, finely toothed. Flowers deep reddish purple. Drupe 5mm diam., black. **HABITAT:** Rainforest, on talus slopes and fertile soils. **DIST:** Rare. Coast (e.g. Burning Palms, Minnamurra Falls, Gap Ck FR) and near Mountain Lagoon. Also NC, SC, Qld. **FL:** Jan–Mar.

Polyscias
Shrubs or trees with compound leaves. Flowers in compound terminal inflorescences, usually with 5 petals and 5 stamens. Fruit a drupe.

Polyscias elegans
Celerywood
Tree to 30m tall, usually much smaller. Leaves usually bipinnate, to 1m long; secondary

Cephalaralia cephalobotrys

Polyscias elegans

Polyscias murrayi

Polyscias sambucifolia

rachises to 10cm long with 3 or more ovate leaflets, 5–10 x 3–6cm, in opposite pairs with a large terminal leaflet. Flowers solitary in large compound inflorescences, purple. Drupe dull purplish black, 5–7mm diam. **HABITAT:** Littoral rainforest. **DIST:** Uncommon. Coast. Royal NP to Beecroft Pen. on SC (e.g. Werrong, Kellys Falls, Bass Pt). Also NC, Qld, PNG. **FL:** Mar–July.

Polyscias murrayi
Pencil Cedar, Umbrella Tree
Columnar tree to 20m tall, usually much smaller, usually unbranched, with large pinnate leaves radiating from trunk. Leaves 1m or more long, with about 15 pairs of minutely toothed elliptic leaflets to 16 x 9cm and a slightly smaller terminal leaflet. Flowers numerous, in small dense umbels in large paniculate inflorescences, creamy green. Drupe pale blue, 5mm long. **HABITAT:** Rainforest gullies. **DIST:** Coast, lower Blue Mtns (e.g. Jamberoo, Mt Kembla, Bola Creek, Strickland SF, Woodford). Also NC, SC, Qld, Vic. **FL:** Mar–Apr.

Polyscias sambucifolia
Elderberry Panax
Diffuse shrub to 3m tall. Leaves variable, pinnate with 5–23 elliptic or oblong leaflets or bipinnate with narrow, lobed, ultimate leaflets. Flowers in small dense umbels in large panicles, cream to greenish yellow. Drupe bluish*, 4mm diam. **HABITAT:** Forest, on sheltered slopes and gullies. **DIST:** Coast and ranges. Also NC, SC, NT, ST, Qld, Vic. **FL:** Nov–Jan.

* The similar but unrelated *Sambucus australasica* has yellow drupes.

APIACEAE

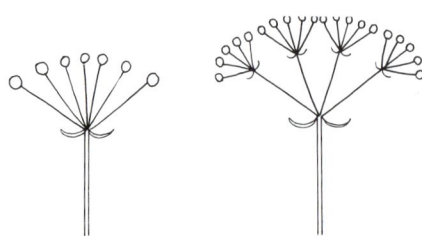

Simple umbel

Compound umbel

Actinotus
FLANNEL FLOWER

Herbs, with silky narrow-tubular flowers in crowded umbel resembling a head surrounded by larger radiating petal-like bracts (involucre) with dense soft hairs. Leaves divided into narrow lobes.

Actinotus forsythii

Sprawling branched herb to 30cm tall. Leaves divided to base into narrow segments. Flowers pink, surrounded by white involucre to 20mm across; peduncles 4–10cm long. **HABITAT:** Exposed heath, on rocky or sandy sites. **DIST:** Upper Blue Mtns (e.g. Narrow Neck, Hargraves L'out). Also SC, ST, Vic. **FL:** Dec–May.

Rarely seen because of its restricted distribution and ephemeral nature. Germinating after fire and suitable rain, it can be locally common, then die off to be absent for years.

Actinotus gibbonsii

Short-lived tufted herb. Leaf blades to 18mm long, mostly divided into 3 acute lobes. Leaf stalk to 20mm long. Umbels pinkish, 10mm diam., surrounded by 7–9 pointed green bracts to 5mm long; peduncles 2–6mm long. **HABITAT:** Sandstone rock platforms, often mossy areas. **DIST:** Rare. Upper Blue Mtns (e.g. Wingello, Newnes Plat., Boyd R.). Also SC, NT, ST, WS, WP, Qld. **FL:** Dec–May.

Actinotus gibbonsii

Actinotus helianthi
Flannel Flower

Short-lived perennial*. Leaves soft, green-grey, deeply divided into a number of narrow lobes. Umbels creamy white, 12–20mm diam., surrounded by involucre of flannel-like green-tipped bracts 5–8cm diam. **HABITAT:** Old dunes, coastal heath; open forest. **DIST:** Coast and ranges. Also NC, SC, NT, ST, WS, NWP, Qld. **FL:** Chiefly Oct–Apr (depending on fire).

 * Dune form is more robust, to 90cm tall, and may last for 15 years.

Actinotus forsythii

Actinotus helianthi

Actinotus minor
Lesser Flannel Flower
Thin sprawling plant with wiry branches. Leaves grey-green, 7mm long, with 3 lobed segments, glabrous above, silky-hairy beneath. Umbels white, 5–8mm diam., surrounded by white bracts 6mm long in starry involucre; peduncles 6–10cm long. **HABITAT:** Open forest, heath, on sandy soils. **DIST:** Coast (Milton to Ourimbah) and ranges. Also ST. **FL:** Some flowers all year.

Actinotus minor

Apium
CELERY
Herbs with hollow grooved stems and leaves with 3 lobes or leaflets. Flowers white, in compound umbels. Fruit dry, separating into 2 flattened mericarps.

Apium prostratum
Sea Celery
Prostrate herb with thick rootstock and fleshy spreading branches to 70cm long. Leaves with linear leaflets (var. *prostratum*) or elliptic and lobed (var. *filiforme*). Umbels 2cm across. Mericarps with 5 close thick ridges. **HABITAT:** Salt marsh, lake foreshores, coastal cliffs. **DIST:** Coast. Also NC, SC, CWS, Qld, Vic, Tas, SA, WA, NZ. **FL:** Sept–Mar.

Apium prostratum

Centella

Herb with erect stems or leaves on long creeping runners. Flowers in small axillary umbels. Fruit with 2 compressed mericarps on undivided axis.

Centella asiatica

Centella asiatica
Swamp Pennywort

Leaves grouped at nodes; lamina cordate to ± circular, 1–3cm diam., crenate; stalk to 20cm long. Umbels ± sessile or with short peduncle, 2–3-flowered; flowers tiny, white or pinkish. Fruit about 3 x 4mm; mericarps with 3–5 ribs on each side. **HABITAT:** Swamp margins, damp sites in woodland, scrub. **DIST:** Coast. Also NC, SC, NT, SWS, all other States (except NTerr), Asia, Pac. Iss. **FL:** Nov–Dec.

Eryngium

Prickly plants, with rigid spiny leaves and sessile flowers in heads surrounded by longer spiny bracts.

Eryngium vesiculosum
Pricklefoot

Prostrate spiny plant with stems to 30cm long spreading from a rootstock. Leaves basal, with a few opposite on stems, narrow-oblanceolate with spiny teeth, 4–15cm long, flat, tough. Heads ovoid, 8mm diam., pale blue. **HABITAT:** Woodland, in waterlogged soils and damp areas. **DIST:** Upper Blue Mtns (e.g. Hartley Vale, Kanangra, Long Swamp). Also NT, ST, WS, Vic, Tas, NZ. **FL:** Dec–Mar.

Eryngium vesiculosum

Hydrocotyle
PENNYWORT

Herbs with stems rooting at nodes or with underground stolons. Leaves stipulate, with long erect stalks; lamina ± circular, cordate or peltate. Flowers usually in simple umbels. Mericarps flattened, ribbed. *H. bonariensis* is a common invasive weed on coastal sand dunes. It has circular leaves to 10cm diam., and flowers in tiny umbels on branched inflorescences.

Hydrocotyle geraniifolia
Forest Pennywort

Lax trailing herb. Stems 50cm long or more, rooting at nodes. Leaves deeply divided into 3–5 narrow toothed lobes to 6cm long. Umbels with 15–20 tiny white flowers. **HABITAT:** Sheltered sites in forest, rainforest margins. **DIST:** Coast and ranges. Also NC, SC, NT, Vic, SA. **FL:** Nov–Mar.

Hydrocotyle tripartita is a widespread herb from damp places. It has leaves divided to the base into 3–5 segments and umbels of 3–6 flowers.

Hydrocotyle geraniifolia

Platysace clelandii

Oreomyrrhis
Tufted herbs, often with fleshy taproot. Leaves in basal rosette, pinnately dissected. Inflorescence a simple umbel. Mericarps 5-ribbed.

Oreomyrrhis eriopoda
Australian Carraway
Perennial herb to 50cm tall. Leaves to 15cm long, divided into 15–25 acute leaflets, villous. Leaf sheaths ciliate, often hirsute. Umbels with 15–35 flowers. Petals white, often with pink tinge, hirsute on outside. **HABITAT:** Grassy woodland. **DIST:** Upper Blue Mtns. Also NC, SC, NT, ST, WS, Vic, Tas, SA. **FL:** Dec–Feb.

Platysace
Herbs or subshrubs with carrot-like smell. Leaves alternate, entire or lobed. Flowers small, in terminal compound umbels. Mericarps 2, flattened, often rough or pimply.

Platysace clelandii
Shrub to 50cm tall with hairy stems. Leaves fan-shaped, hairy, to 7 x 10mm, with 3–5 lobes. Flowers white. **HABITAT:** Open forest, on hillsides. **DIST:** Endemic. Hornsby Plat., Colo–Wollemi (e.g. Putty, Colo R., Marramarra NP). **FL:** Dec–Apr.

Oreomyrrhis eriopoda

Platysace ericoides
Heath Platysace
Diffuse shrub to 50cm tall with scabrous stems. Leaves linear or oblong with acute apex, to 15 x 2mm, sometimes bristly; stalk <1mm long. Flowers in umbels to 2cm across, white. **HABITAT:** Heath, scrub, open forest, on sandy soils. **DIST:** Coast and ranges. Also NC, NT, WS, NWP, Qld. **FL:** Nov–Apr.

Platysace lanceolata
Lance-leaf Platysace
Erect shrub to about 1m tall. Leaves variable, linear to elliptic, to 50 x 15mm, glabrous, shortly stalked. Flowers white, in umbels to 4cm across. **HABITAT:** Heath, scrub, open forest, on sandy soils. **DIST:** Common. Coast and ranges. Also NC, SC, NT, ST, WS, SWP, Qld, Vic. **FL:** Dec–Apr.

Platysace linearifolia

Platysace stephensonii

Platysace linearifolia
Narrow-leaf Platysace
Open spreading shrub to 1m tall with glabrous stems. Leaves sessile, linear, to 25 x 1mm, flexible, acute. Flowers white, in umbels to 20mm across. **HABITAT:** Open forest, on sandy soils. **DIST:** Common. Coast and ranges. Also SC, Qld. **FL:** Dec–May.

Platysace stephensonii
Shrub to 50cm tall with slightly hairy stems. Leaves crowded, rigid, to 15mm long, deeply divided into 3 pointed lobes or the side lobes again divided giving 5 lobes. Umbels ± hidden in upper leaves, to 30mm across; flowers white. **HABITAT:** Heath, esp. on and near rocky areas. **DIST:** Coast. Bundeena–Broken Bay (e.g. La Perouse, Ku-ring-gai Chase NP). Also SC. **FL:** Nov–Feb.

Platysace ericoides

Platysace lanceolata

FLOWERING PLANTS

Trachymene incisa subsp. *incisa*

Trachymene
Herbs with deeply dissected leaves with long stalks. Umbels simple. Mericarps 2, hairy, smooth or pimply.

Trachymene incisa subsp. incisa
Erect glabrous perennial herb to 50 cm tall with thick rootstock. Leaves chiefly basal; lamina ± circular in outline, about 4cm diam., divided into 3–5 lobes, the lobes again deeply divided; stalk 5–15cm long. Umbels 1–2cm across, with 50 or more white or pinkish flowers; peduncles to 15cm long. **HABITAT:** Woodland, in sandy soils. **DIST:** Coast, Cumberland Pl. Also NC, SC, NT, NWS, Qld. **FL:** Aug–Jan.

Xanthosia
Herbs and small shrubs with 3-lobed leaves. Umbels usually compound, sometimes simple or reduced to 1 or a few flowers. Flowers sessile, white, pinkish or greenish; sepals conspicuous, ovate. Mericarps 2, with 7–9 ribs.

Xanthosia atkinsoniana
Perennial with wiry stems to 60cm long. Leaves mostly basal; lamina 2–4cm diam. in outline, divided into 3 lobed segments which are again divided once or twice; stalk to 10cm long. Flowers white, often pinkish in bud, in umbels of 6–10 surrounded by 3 white petal-like bracteoles. **HABITAT:** Open forest, on sandy soils. **DIST:** Blue Mtns (e.g. Mt Wilson, Colo Hts), Sthn T'lands (e.g. Penrose, Wombeyan Caves). Also NT, ST, WA. **FL:** Nov–Feb.

Xanthosia atkinsoniana

Xanthosia pilosa

Xanthosia pilosa
Variable shrub, 30–50cm tall with long greyish hairs and matted silver-brown stellate hairs. Leaf laminae 10–50 x 5–35mm with 3–7 broad lobes. Umbels simple, paired in axils, 1–3-flowered; flowers white to creamy green. **HABITAT:** Open forest, hillsides and gullies. **DIST:** Common. Coast and ranges. Also NC, SC, NT, ST, Qld, Vic, Tas. **FL:** Chiefly July–Feb.

Xanthosia scopulicola and *X. stellata* are both hairy subshrubs from Katoomba. *X. scopulicola* has umbels with 3 flowers; *X. stellata* has umbels with 1–2 flowers.

Xanthosia dissecta is a subshrub with basal dissected leaves to 15mm long. It occurs in upper Blue Mtns swamp margins.

Xanthosia tridentata

Xanthosia tridentata
Rock Xanthosia
Variable shrub to 25cm tall, often with loose hairs on stems and leaves. Leaves irregular on stems, wedge-shaped, to 12 x 5mm, with 3 teeth at apex. Umbels with peduncles to 10mm long, terminal or adjacent to axils, with 1–3 greenish yellow flowers. **HABITAT:** Woodland, on rocky hillsides. **DIST:** Coast, lower Blue Mtns. Also SC, ST, Vic, Tas, WA. **FL:** Sept–Mar.

OLACACEAE

Olax
Climbers, shrubs or trees parasitic on roots of other plants. Leaves alternate, simple, entire, leathery; stipules absent. Flowers bisexual or unisexual. Fruit a drupe.

Olax stricta
Olax
Erect shrub to 1.5m tall with terete branchlets and yellowish green appearance. Leaves distichous, linear-oblong, 5–10 x 1–2mm; apex mucronate or recurved. Flowers solitary; petals 5, white or yellowish, 4–5mm long; staminodes 5, opposite petals; stamens 3. Drupe ellipsoid, 6–8mm long, enclosed in enlarged calyx. **HABITAT:** Woodland, heath. **DIST:** Coast and ranges. Also NC, SC, NT, ST, WS, Qld, Vic. **FL:** Most of year.

Olax stricta

Choretrum candollei

SANTALACEAE

Choretrum
SOURBUSH
Glabrous hemiparasitic shrubs, appearing leafless. Flowers bisexual, axillary, solitary or clustered; tepals 5, incurved, thick. Fruit drupaceous, crowned by persistent tepals.

Choretrum candollei
White Sourbush
Shrub to 3m tall; branches angular, ridged or lined below leaves. Leaves ± erect, subulate, to 3mm long. Flowers solitary, on pedicels <2mm long with 4 or more leaf-like bracteoles; tepals white or cream, 1mm long. Fruit globular, dark green, 4–5mm long. **HABITAT:** Open forest, on sandy soils. **DIST:** Coast and ranges, more common in higher areas. Also NC, SC, NT, ST, WS, Qld. **FL:** June–Sept.

Choretrum pauciflorum
Dwarf Sourbush
Decumbent shrub to 1.5m tall, with rigid green terete branches, and scale-like subulate leaves to

Choretrum pauciflorum

1mm long. Flowers ± sessile with about 10 bracteoles; tepals white or cream, 1mm long. Fruit globular, 4–5mm long. **HABITAT:** Open forest, on sandy soils. **DIST:** Upper Blue Mtns, Sthn T'lands. Also SC, NT, ST, Vic. **FL:** Sept–Jan.

Exocarpos cupressiformis

Exocarpos
BALLART
Hemiparasitic shrubs or small trees with leaves reduced to scales, rarely well-developed. Flowers in axillary spikes or clusters. Fruit a globose drupe or nut on a colourful swollen pedicel.

Exocarpos cupressiformis
Native Cherry, Cherry Ballart
Shrub or tree to 8m tall with many light green furrowed pendulous branchlets. Leaves reduced to minute scales. Flowers tiny in short spikes, only 1 developing a swollen stalk. Mature fruit dark, 4mm diam., on a succulent red pedicel 4–6mm long. **HABITAT:** Woodland, esp. hillsides. **DIST:** Coast and ranges. Most of NSW. Also Qld, Vic, Tas, SA. **FRUIT:** Ripe July–Dec.

Exocarpos strictus
Pale Ballart
Shrub to 3m tall, with erect, angular, striate, bronze-green branchlets. Leaves 1–3mm long, pointed, falling early leaving a tooth-like base. Flowers 2–4 in ± sessile axillary clusters, greenish yellow or tinged red. Fruit a black drupe 2.5–4mm diam. on a swollen red, mauve or white pedicel 4–5mm long. **HABITAT:** Open forest. **DIST:** Coast and ranges. Most of NSW. Also Vic. **FRUIT:** Ripe Nov–Jan.

Exocarpos strictus

Leptomeria
CURRANT-BUSH
Hemiparasitic shrubs with slender, glabrous, striate, ridged branches. Leaves scale-like or well-developed. Flowers bisexual, solitary or in spikes or racemes. Fruit a dry or fleshy drupe.

Leptomeria acida
Native Currant
Shrub to 2m tall, with slender green semi-terete striate branches. Leaves 1–2mm long, deciduous. Flowers 1mm long, 20 or more clustered in spikes 1–4cm long, green-brown. Drupe 5mm diam., succulent, with acidic taste. **HABITAT:** Open forest, in sheltered sites. **DIST:** Common. Coast and ranges. Also NC, SC, ST, Qld, Vic. **FL:** Feb–Apr.

Leptomeria acida

Omphacomeria
SOURBUSH
Dioecious shrubs with alternate scale-like deciduous leaves. Male flowers in short axillary spikes, female flowers solitary. Fruit a drupe with thin flesh. Monotypic genus.

Omphacomeria acerba
Leafless Sourbush
Erect hemiparasitic shrub to 1m tall with striate terete branchlets. Drupe ovoid, 6–9mm long, green or purplish, slightly rough; flesh acidic.
HABITAT: Open forest, often on rocky sites.
DIST: Coast and ranges. Also SC, ST, CWS, Vic.
FL: Aug–Nov.

Omphacomeria acerba

Santalum
SANDALWOOD
Hemiparasitic shrubs or small trees. Leaves usually opposite. Tepals 4, fleshy; stamens 4. Fruit a globose drupe.

Santalum obtusifolium
Sandalwood
Glaucous shrub 1–2m tall with ridged or angular branchlets. Leaves lanceolate, to 60 x 15mm, dark green above, pale to glaucous below. Flowers greenish white, about 4mm long, in axillary clusters. Drupe blue-black, succulent, 8–10mm long, with circular scar on summit. **HABITAT:** Sheltered forest, on hillsides and in gullies. **DIST:** Uncommon. Coast and ranges (e.g. Ingleburn, Royal NP, Springwood). Also NC, SC, NT, ST, Qld, Vic. **FL:** Oct–Dec.

Santalum obtusifolium

LORANTHACEAE

Amyema
MISTLETOE
Hemiparasites. Leaves opposite or whorled. Flowers in triads in umbel-like inflorescences; petals usually 5, free to base; anthers attached to filament by their base. Fruit berry-like, with 1 seed.

Amyema cambagei
She-oak Mistletoe
Spreading to pendulous plant with grey-tomentose stems. Leaves terete, to 12cm x 2mm. Flowers 15–20mm long, pink with white hairs. Fruit globular, pink-red, 5–6mm diam. **HABITAT:** Mainly on branches of she-oaks (esp. *Casuarina cunninghamii*), mimicking leaves of host. **DIST:** Coast (e.g. Nepean R., Wisemans Ferry) and ranges (e.g. Wombeyan Caves). Most of NSW. Also Qld. **FL:** June–Dec.

Amyema congener subsp. *congener*

Amyema gaudichaudii

Amyema congener subsp. *congener*
Erect or spreading plant. Leaves broadly lanceolate, to 10 x 4cm, obscurely veined; apex rounded; stalk short. Flowers to 35mm long, green or yellow with pink stamens and style. Fruit elliptic to globular, green, 8mm long. **HABITAT:** Parasitic on a number of species, including exotics (often on *Allocasuarina littoralis*, rarely on *Eucalyptus*). **DIST:** Coast. Also NC, SC, Qld. **FL:** June–Nov.

Amyema gaudichaudii
Paper-bark Mistletoe
Erect, compact plant. Leaves lanceolate, to 35 x 5mm; apex rounded. Flowers slender, 7–10mm long, deep red. Fruit red, globular, about 4mm diam. **HABITAT:** Parasitic on *Melaleuca* spp., esp. *M. decora*. **DIST:** Clay areas of western Sydney (e.g. Milperra, Castlereagh, Rookwood) and Illawarra. Also NC, SC, Qld. **FL:** Nov–Dec.

Amyema cambagei

Amyema miquelii

Amyema miquelii
Pendulous plant with stems to 2m long. Leaves elliptic, to 35 x 3cm, often falcate, narrowing to stalk 1–4cm long, sometimes pinkish or yellowish. Flowers all stalked, 15–28mm long, red, with 5–7 petals. Fruit ovoid, yellowish red, 8–12mm long. **HABITAT:** Clay-shale woodland, on eucalypts (esp. *E. moluccana*) and acacias. **DIST:** Western Sydney (e.g. Wallacia, Berkshire Park), upper Blue Mtns. Most of NSW. All mainland States. **FL:** Chiefly Dec–Jan.

Amyema pendulum subsp. *pendulum*

Amyema pendulum subsp. pendulum
Drooping Mistletoe
Pendulous plant to 1.5m long. Leaves lanceolate, to 40 x 2cm. Flowers to 4cm long, red, the central flower of each triad sessile. Fruit ovoid, brown, 10mm long. **HABITAT:** On *Eucalyptus* and *Acacia* in open forest and woodland. **DIST:** Coast and ranges. Also NC, SC, NT, ST, WS, Vic, SA. **FL:** Apr–Nov.

Amylotheca
Aerial hemiparasitic shrubs. Leaves opposite, prominently veined. Flowers in triads in axillary racemes, the central flower sessile; petals 6, united for about half their length; anthers attached by their back.

Amylotheca dictyophleba

Amylotheca dictyophleba
Spreading plant. Leaves elliptic, 6–12 x 2–6cm, glossy above; stalk 2–8mm long. Corolla to 4cm long, slightly inflated, red and green with yellow apex. Fruit globular, red or purple, 10–12mm diam. **HABITAT:** Parasitic on rainforest trees. **DIST:** Coast. Illawarra (e.g. Mt Keira, Kangaroo V.) and Wyong area. Also NC, Qld, PNG, NCal. **FL:** Nov–Jan.

Atkinsonia ligustrina

FLOWERING PLANTS

Atkinsonia
Shrub, parasitic on roots of nearby plants. Leaves opposite, penniveined. Flowers in axillary racemes, shorter than leaves; petals 6, free; anthers versatile. Fruit drupaceous. Monotypic genus.

Atkinsonia ligustrina
Glabrous shrub to 1m tall. Leaves lanceolate, to 50 x 14mm, apex obtuse. Corolla yellow and red, 6–8mm long. Fruit ovoid, scarlet, fleshy, about 12mm long. **HABITAT:** Open forest, woodland. **DIST:** Endemic. Rare. Blue Mtns–Colo Plat. (e.g. Linden, Mt Wilson, Colo Gorge). **FL:** Oct–Dec.

Dendrophthoe
Aerial parasitic shrubs. Leaves variously arranged, penniveined. Inflorescence axillary, racemose, sometimes reduced to 1 or 2 flowers; petals 5, united for half their length; anthers attached by their back, not versatile.

Dendrophthoe vitellina
Spreading to pendulous plant. Leaves irregular or alternate, elliptic, to 16 x 3cm, leathery,

Dendrophthoe vitellina

midvein prominent; apex rounded; stalk to 15mm long. Inflorescence axis to 5cm long, with up to 20 flowers. Petals yellow with red tip, 4cm long. Fruit ovoid, to 15mm long, yellow to red. **HABITAT:** Parasitic on trees, mainly Myrtaceae, in open forest and woodland. **DIST:** Coast and lower Blue Mtns. Also NC, SC, NWS, Qld, Vic. **FL:** July–Dec.

Muellerina
MISTLETOE
Aerial parasitic shrubs. Leaves opposite. Flowers single or in triads with the middle flower sessile, arranged into terminal racemes; petals 5, free; anthers versatile.

Muellerina celastroides
Erect to spreading plant. Leaves oblong to nearly circular, 3–7 x 2–4cm, discolorous. Raceme axis about 1cm long. Flowers with green tips, pink bases and red styles, to 35mm long. Fruit green to pink, 7–11mm long. **HABITAT:** Parasitic on many hosts but not eucalypts, in tall forest and rainforest. **DIST:** Coast (e.g. Mt Keira, Berowra). Also Wingello, NC, SC, CWS, Qld, Vic. **FL:** Nov–Feb.

Muellerina celastroides

Muellerina eucalyptoides

Muellerina eucalyptoides
Spreading or pendent plant with stems 1–2m long. Leaves lanceolate, 8–20 x 1–3cm, pale green, with indistinct veins. Raceme axis 1–4cm long. Flowers greenish yellow with red centres, 30–45mm long. Fruit green to yellow, 8–15mm long. **HABITAT:** Parasitic mainly on *Eucalyptus* and *Angophora* spp. in open forest. **DIST:** Coast and ranges. Most of NSW. Also Qld, Vic, SA. **FL:** Dec–Feb.

VISCACEAE

Korthalsella
Stems flattened, with constricted nodes bearing rudimentary leaves and many tiny flowers. Fruit <2mm long, with 1 seed.

Korthalsella rubra subsp. *rubra*
Jointed Mistletoe
Erect plant with green flattened stems to 20cm x 3–9mm with 3-veined obovate internodes 1–2cm long. Flowers unisexual, green or yellowish, in 2–5 rows. **HABITAT:** Parasitic on rainforest trees, esp. *Doryphora*. **DIST:** Illawarra (e.g. Mt Keira, Cambewarra Mtn). Also Colong Caves (CT), NC, SC, NT, ST, Qld, Vic, LHI. **FL:** Sept–Dec.

Notothixos
Monoecious plants with densely matted branched or stellate hairs. Leaves developed, petiolate. Flowers 4-merous, in cymes of 3–13 in compound terminal inflorescences.

Notothixos subaureus
Golden Mistletoe
Spreading parasitic plant with bright golden hairs. Leaves opposite, ± ovate, to 3 x 2cm; stalks 3–5mm long. Cymes in 3s, each with 5–11 flowers 1–2mm long. Fruit elliptic, 7mm long. **HABITAT:** On other mistletoes in rainforest and open forest. **DIST:** Mainly coast; Blaxland. Also NC, SC, WS, Qld, Vic. **FL:** Most of the year.

Korthalsella rubra subsp. *rubra*

FLOWERING PLANTS

Notothixos subaureus

Citronella moorei

Pennantia cunninghamii

ICACINACEAE

Citronella
Trees with bisexual flowers, or dioecious or polygamous. Leaves alternate, entire, with domatia at junctions of mid and lateral veins. Flowers 5-merous, in narrow thyrsoid inflorescences. Fruit a 1-seeded drupe.

Citronella moorei
Silky Beech, Churnwood
Dioecious tree to 40m tall with grey finely fissured bark. Leaves elliptic, 5–12 x 3–6cm, glossy green above, paler below. Inflorescences longer or shorter than leaves; flowers creamy green, 2–3mm long. Drupe globose or ovoid, black, about 20mm long. **HABITAT:** Rainforest, sheltered valleys and slopes. **DIST:** Coast. Chiefly Illawarra (e.g. Cambewarra, Mt Keira); rare Ourimbah–Wyong. Also NC, SC, Qld. **FL:** May–Sept.

Pennantia
Trees with alternate leaves bearing domatia in forks of secondary veins. Inflorescences terminal or axillary corymbose panicles. Flowers 5-merous, bisexual and unisexual in various combinations on same or separate trees. Fruit a 1-seeded drupe.

Pennantia cunninghamii
Brown Beech
Glabrous tree with fluted dark bark, rarely exceeding 10m tall; branchlets zigzagged between consecutive leaves. Leaves broadly elliptic, 10–15 x 5–8cm, shiny green above; domatia prominent. Panicles 5–12cm long; flowers white, 3–4mm long. Drupe ellipsoid, 12–14 x 7–9mm long. **HABITAT:** Rainforests, on rocky hillsides and along streams. **DIST:** Illawarra (e.g. Minnamurra Falls, Belmore Falls). Also NC, SC, NT, ST, CWS, Qld. **FL:** Nov–Jan.

Cassine australis var. *australis*

Celastrus australis

Celastrus subspicata

CELASTRACEAE

Cassine
OLIVE PLUM
Trees and shrubs, usually dioecious and glabrous. Leaves usually opposite, simple, with tiny stipules. Flowers in axillary simple or compound cymes. Fruit a drupe with 1–3 seeds.

Cassine australis var. *australis*
Red Olive Plum
Tree to 8m tall. Leaves elliptic with crenate margins (sometimes only in upper half), to 12 x 5cm, thick, leathery; stalk to 1cm long. Flowers in cymes shorter than leaves, pale green, with 4-lobed calyx and 4 petals. Drupe ovoid to ellipsoid, 12–24 x 10–15mm, orange-red with 1 seed. **HABITAT:** Littoral and dry rainforests. **DIST:** Coast (e.g. Bouddi, Royal NP, Bass Point). Also NC, SC, NT, CWS, Qld. **FL:** Aug–Nov.

Celastrus
STAFF VINE
Scrambling dioecious shrubs or climbers; branchlets with lenticels. Leaves alternate. Flowers in racemes or panicles, 5-merous. Fruit a 3-lobed capsule with 1–6 seeds enveloped in colourful fleshy aril.

Celastrus australis
Staff Vine
Woody climber. Leaves lanceolate to elliptic, 3–9 x 1–4cm. Flowers in dense terminal panicles shorter than leaves, yellow-green, 2–3mm long. Capsule 3–6 x 3–6mm, orange or light brown with red spots inside, aril orange. **HABITAT:** Rainforest. **DIST:** Coast and ranges (e.g. Razorback Mtn, Mt Keira, Mt Wilson, Robertson). Also NC, SC, ST, WS, Qld, Vic. **FL:** Oct–June.

Celastrus subspicata
Large-leaf Staff Vine
Woody climber. Leaves lanceolate to elliptic, 5–14 x 2–7cm, discolorous. Flowers in loose terminal panicles shorter than leaves, yellow-green, 2–3mm long. Capsule 5–9 x 5–9mm, orange, without red spots inside; aril orange or yellow-green. **HABITAT:** In or near dry vine scrub. **DIST:** Rarer and more restricted than *C. australis*. Illawarra (e.g. Cambewarra, Mt Keira). Also NC, NT, Qld. **FL:** Nov–Dec.

Maytenus
Shrubs or small trees; branchlets without lenticels. Leaves alternate. Flowers mostly bisexual in axillary inflorescences, 4- or 5-merous. Fruit a 2-valved 2-lobed capsule.

Maytenus silvestris
Narrow-leaved Orangebark
Shrub to 3m. Leaves narrow-elliptic, 2–8cm x 2–15mm, toothed or entire, paler below. Flowers solitary, clustered or in short racemes, 5-merous, 2mm long, pale green. Capsule orange, 5–8mm long; seeds with an aril. **HABITAT:** Moist forests, often near rainforest. **DIST:** Coast (e.g. Wyong, Ingleburn, Razorback Mtn); Wingello. Also NC, NT, WS, Qld. **FL:** Oct–Jan.

Maytenus silvestris

STACKHOUSIACEAE

Stackhousia
Perennial and annual herbs with alternate simple entire leaves. Flowers with petals fused into a tube for half their length below 5 spreading star-like lobes.

Stackhousia monogyna
Creamy Candles
Erect perennial to 50cm tall. Leaves linear to lanceolate, to 30 x 4mm. Flowers white to yellow in a dense spike, 3–10cm long; floral tube 5–8mm long, lobes to 5mm long. **HABITAT:** Open forest, woodland, heath. **DIST:** Upper Blue Mtns, Sthn T'lands (e.g. Kanangra-Boyd NP, Fitzroy Falls). Most of NSW. All other States (except NTerr). **FL:** Aug–Jan.

Stackhousia monogyna

Stackhousia muricata has branched stems and yellow-green flowers in clusters of 1–5. It occurs on clay shales from Glenfield to Camden.

Stackhousia nuda
Leafless Stackhousia
Erect perennial with branched stems to 80cm tall. Leaves reduced to scales. Flowers in clusters of 1–4 in a loose 1-sided spike, yellow to yellow-green; floral tube 2–3mm long, lobes to 2mm long. **HABITAT:** Wet heath, moist sand, swamp margins. **DIST:** Rare. Coast (e.g. Garigal NP, Maddens Plains); Carrington Falls. Also NC, SC, Qld, Vic. **FL:** Nov–Feb.

Stackhousia nuda

Stackhousia spathulata

Stackhousia spathulata
Coast Stackhousia
Erect perennial with stems to 50cm tall. Leaves spathulate, to 30 x 15mm, thick, obtuse. Flowers white in dense spikes; floral tube 6–7mm long, lobes 4–5mm long. **HABITAT:** Beach sand dunes, margins of coastal lagoons. **DIST:** Coast (e.g. Wamberal Bch). Also NC, SC, Qld, Vic, Tas, SA. **FL:** Aug–Nov.

Stackhousia viminea
Slender Stackhousia
Slender erect perennial with unbranched stems to 60cm tall. Leaves sparse but well-developed, elliptic, to 40 x 8mm. Flowers green to yellow (sometimes reddish), in scattered clusters of 1–5 in spike-like inflorescences; floral tube 2–4mm long, lobes 2–3mm long. **HABITAT:** Woodland, esp. on clay soils. **DIST:** Coast and ranges. Also NC, SC, NT, ST, NWS, all other States (except SA). **FL:** Oct–Jan.

RHAMNACEAE

Alphitonia
Trees. Buds, young stems and undersurface of leaves with white or rusty hairs. Leaves alternate. Flowers with 5 spreading petals. Fruit a drupe.

Alphitonia excelsa
Red Ash
Tree 6–10m tall, with rusty-hairy branchlets and leaf stalks. Leaves broad-elliptic, to 15 x 5cm, dark green and glossy above, silvery white with brown veins beneath. Flowers in branched axillary inflorescences shorter than leaves, numerous, small, creamy. Drupe 5–10mm diam., black. **HABITAT:** Rainforest, protected drier gullies. **DIST:** Coast, Blue Mtns. Also NC, SC, WS, NWP, Qld, WA, NTerr. **FL:** Dec–Mar.

Cryptandra
Shrubs with alternate leaves. Flowers sessile, surrounded by persistent brown bracts, with short hypanthium or calyx tube, 5 sepals on rim of tube and 5 petals hooded over anthers; anthers sessile on rim of tube. Fruit a capsule enclosed in the persistent calyx tube.

Cryptandra amara
Shrub to 60cm tall with rigid intertwined branches often ending in a spine. Leaves variable, clustered, 2–5mm long, flat or recurved. Flowers in upper axils of branchlets, 2–6mm long, white or pinkish; sepals longer than hypanthium. **HABITAT:** Heath, woodland. **DIST:** Coast and ranges. Most of NSW. Also Qld, Vic, Tas, SA. **FL:** June–Sept.

Stackhousia viminea

Alphitonia excelsa

Three varieties occur in Sydney district. Var. *amara* (widespread) has elliptic leaves and flowers 2–4mm long. Var. *longiflora* (upper Blue Mtns) has oblong leaves and flowers 4–6mm long. Var. *floribunda* (CT) has terete leaves and flowers crowded at end of branches.

Cryptandra ericoides
Shrub to 60cm tall. Leaves clustered, linear, 4–8mm long, revolute. Flowers white, in terminal head-like clusters; floral tube 4mm long, silky-hairy. **HABITAT:** Heath, on sandy soils. **DIST:** Coast sth of Gosford (e.g. Kurnell, Royal NP), t'lands (e.g. Mt Banks, Robertson). Also SC. **FL:** Mar–June.

Cryptandra ericoides

Cryptandra amara

Cryptandra propinqua

Cryptandra spinescens

Cryptandra propinqua
Silky Cryptandra
Rigid shrub to 1m tall with many small stellate-hairy branches. Leaves linear, 3–8mm long, glabrous above, hairy beneath. Flowers white, bell-shaped, 5–7mm long, clustered at ends of branches; floral tube silky-hairy, ± hidden by bracts. **HABITAT:** Open forest, woodland. **DIST:** Chiefly coast (e.g. Maroota, Brooklyn); rare in Blue Mtns. Most of NSW. Also Qld, Vic, SA, WA. **FL:** Apr–Sept.

Cryptandra spinescens
Spiny Cryptandra
Spreading shrub with many small spiny-tipped side-branches. Leaves obovate, 3–4 x 1–2mm, glabrous. Flowers white, crowded on short side-branches; floral tube 3–4mm long, glabrous at base, silky-hairy above; bracts on stalk below flower. **HABITAT:** Woodland, on clay-shale soils. **DIST:** Cumberland Pl. (e.g. Doonside, Douglas Park); rare in t'lands. Also upper Hunter V. **FL:** June–Oct.

Emmenosperma alphitonioides

Emmenosperma
Rainforest trees. Flowers in short cymes or panicles; hypanthium short; sepals 5; petals 5, hooded over stamens. Fruit a drupe.

Emmenosperma alphitonioides
Yellow Ash, Bonewood
Tree to 20m tall. Leaves opposite, ovate, to 10 x 5cm, glabrous, shiny green above, paler beneath. Flowers numerous, white, 5mm across; petals tiny. Fruit 5–9mm diam., splitting into 2, conspicuous when mature from May–July. **HABITAT:** Rainforest. **DIST:** Mainly coastal (e.g. Ourimbah, Hacking R., Whispering Gully); Kurrajong, Colo R., Gap Ck. Also NC, Qld. **FL:** Sept–Oct.

Pomaderris

Shrubs with simple hairs and close-matted stellate hairs. Leaves alternate, simple, stalked, with stipules. Flowers in small cymes, usually arranged in terminal panicles or corymbs; sepals 5; petals 5 or absent; stamens 5. Fruit a capsule separating into 3 carpels.

Identification is assisted by noting whether petals are present or absent. However, petals are often deciduous and a number of newly opened flowers should be examined. Petal colour and shape are often useful diagnostically. The nature of hairs is also important.

GROUP 1: FLOWERS WITH PETALS

Pomaderris adnata
Shrub to 2m tall with close grey stellate hairs and longer simple hairs. Leaves mostly narrow-elliptic, to 30 x 8mm, glabrous above, greyish stellate below; secondary veins clear with rusty or greyish hairs; minor veins not visible; margins strongly recurved. Flowers cream-yellow, in short dense axillary cymes; petals ± ovate, hooded. **HABITAT:** Woodland near top of escarpment. **DIST:** Restricted. Sublime Point (above Bulli). **FL:** Sept.

Pomaderris adnata

Pomaderris andromedifolia
Shrub to 1.5 m tall, with dense simple hairs on branchlets. Leaves ± lanceolate, to 50 x 15mm, dark green above, pale or rusty with silky hairs and prominent rusty veins beneath. Flowers cream-yellow, in short dense terminal panicles to 4cm across. **HABITAT:** Open forest. **DIST:** Coast sth of Sydney (e.g. Menai, Heathcote Ck, Kentlyn), t'lands (e.g. Clarence, Mt Banks, Penrose). Also NC, SC, NT, ST, CWS, NWP, Qld, Vic. **FL:** Sept–Oct.

Pomaderris andromedifolia

Pomaderris elliptica
Shrub 2–3m tall with stellate hairs on branches. Leaves elliptic, to 12 x 4cm, dull green and glabrous above, with greyish mat of short stellate hairs beneath. Flowers yellow, in large terminal branched panicles. **HABITAT:** Open forest, in gullies. **DIST:** Coast and ranges. Also NC, SC, ST, CWS, Vic, Tas. **FL:** Sept–Oct.

Pomaderris elliptica

Pomaderris ferruginea
Rusty Pomaderris
Shrub 2–4m tall with loose rusty simple hairs. Leaves lanceolate, to 10 x 3cm, glabrous and dull dark green above, with pale tomentum and longer curly rusty hairs beneath. Flowers in open terminal panicles to 10cm diam., floral tube densely hairy; petals whitish or yellow. **HABITAT:** Open forest, woodland. **DIST:** Coast and ranges. Also NC, SC, ST, CWS, Qld, Vic. **FL:** Sept–Oct.

Pomaderris intermedia
Shrub to 3m tall with white stellate tomentum and longer sparse tan woolly hairs. Leaves elliptic-ovate, to 12 x 4cm, glabrous and dark green above, undersurface with hairs as on branches. Flowers yellow, in large branched terminal panicles; buds and base of floral tube with simple hairs. **HABITAT:** Open forest, woodland. **DIST:** Coast and ranges. Also NC, SC, NT, ST, CWS, Vic, Tas. **FL:** Sept–Oct.

Pomaderris lanigera
Woolly Pomaderris
Shrub to 3m tall with soft rusty curly hairs. Leaves lanceolate to ovate, to 8 x 3cm, dull green and sparsely hairy above, with short white and longer rusty hairs beneath. Flowers golden yellow, in showy terminal panicles; floral tube densely grey-hairy. **HABITAT:** Open forest, in gullies. **DIST:** Coast and ranges. Also NC, SC, NT, ST, WS, Qld, Vic. **FL:** Aug–Sept.

Pomaderris ledifolia
Shrub to 1.5m tall with short white simple hairs on branches. Leaves narrow-elliptic, to 20 x 4mm, glabrous above, with short greyish hairs and obscure veins beneath. Flowers yellow, in short dense panicles 2–3cm across; floral tube and sepals with white hairs. **HABITAT:** Open forest, heath. **DIST:** Upper Blue Mtns, Sthn H'lands, Woronora R.–Waterfall–Appin. Also NC, SC, ST, Vic, Qld. **FL:** Sept–Oct.

Pomaderris velutina
Velvet Pomaderris
Shrub to 2m tall. Branchlets and leaf stalks with long rusty simple hairs. Leaves ovate-elliptic, to 30 x 20mm, velvety and dull green above, pale greyish beneath with scattered long hairs on veins and stalk. Flowers yellow, in loose terminal panicles; floral tube with dense long grey hairs. **HABITAT:** Open forest. **DIST:** Restricted to Nattai R. and Burragorang V. Also ST, SWS, Vic, Tas. **FL:** Sept–Oct.

Pomaderris ferruginea

FLOWERING PLANTS | 375

Pomaderris intermedia

Pomaderris lanigera

Pomaderris ledifolia

Pomaderris velutina

Pomaderris aspera

GROUP 2: FLOWERS WITHOUT PETALS

Pomaderris aspera
Hazel Pomaderris
Shrub to 5m tall with dense brown stellate or branched hairs on branchlets and leaf stalks. Leaves lanceolate with obscurely toothed margins, to 18 x 7cm, dark green and slightly glossy with impressed veins above, whitish with stellate hairs and raised lateral veins below. Flowers greenish yellow in large terminal panicles. **HABITAT:** Moist forest gullies. **DIST:** Illawarra, Sthn H'lands, Blue Mtns (e.g. Macquarie Pass, Tallong, Berrima, Mt Wilson). Also NC, SC, NT, ST, WS, Vic, Tas. **FL:** Oct–Nov.

Pomaderris brunnea
Shrub to 3m tall with brownish simple hairs above a pale tomentum. Leaves elliptic, 2–3 x 1–1.5cm, glabrous and dark green above, with hairs as on branches beneath; margins with hair tufts at vein ends. Flowers cream, in dense panicles; floral tube and sepals hairy. **HABITAT:** Woodland, on clayey alluvial soils. **DIST:** Rare. Endemic. Upper Nepean R. (e.g. Menangle, Wirrimbirra, Colo R.). **FL:** Sept–Oct.

Pomaderris discolor
Shrub to 3m tall with pale stellate tomentum and longer simple hairs. Leaves elliptic, to 8 x 3cm, recurved, glabrous and green above, with greyish stellate tomentum and a few longer hairs on veins beneath. Flowers in terminal panicles, pale yellow, petals absent or falling early. **HABITAT:** Open forest, in gullies. **DIST:** Mainly coast (e.g. Lane Cove NP, Royal NP, Kentlyn); also CT (e.g. Bowral, Yerranderie), NC, SC, SWS, Qld, Vic. **FL:** Aug–Sept.

Pomaderris eriocephala
Shrub 2–3m tall with woolly tomentum and longer brown simple hairs. Leaves ovate to nearly round, to 20 x 15mm, velvety, with deeply impressed lateral veins above, with hairs as on branches beneath; margins with hair tufts at vein ends. Flowers yellow, in compact clusters. Capsule with long rusty hairs. **HABITAT:** Woodland and forested gullies. **DIST:** Upper Blue Mtns (e.g. Jenolan Caves, Newnes). Also NC, SC, NT, ST, NWS, Vic. **FL:** Sept–Nov.

Pomaderris ligustrina
Shrub to 2m tall with stellate tomentum and longer rusty hairs on branches. Leaves lanceolate, 2–6 x 1–2.5cm, recurved, glabrous above, with dense silky-rusty simple hairs beneath. Flowers creamy, in loose terminal panicles; floral tube with long grey hairs. **HABITAT:** Open and sheltered forest. **DIST:** Coast and ranges. Also NC, SC, NT, Qld, Vic. **FL:** Aug–Sept.

Pomaderris brunnea

Pomaderris discolor

Pomaderris ligustrina

Pomaderris eriocephala

Pomaderris mediora

Pomaderris phylicifolia subsp. *ericoides*

Pomaderris mediora
Shrub to 2m tall with grey stellate tomentum and grey or rusty longer simple hairs on branches and inflorescence. Leaves narrow-elliptic, 1–2cm x 2–5mm, recurved; glabrous above, with dense yellow-rusty hairs obscuring secondary veins beneath. Flowers creamy, in small axillary clusters. Floral tube white silky-hairy. **HABITAT:** Heath, scrub, on cliff tops. **DIST:** Rare. Endemic. Coast. Palm Bch–Bulli (e.g. Turrimetta Hd, Sublime Point). **FL:** Sept–Oct.

Pomaderris phylicifolia subsp. *ericoides*
Erect shrub to 2m tall with short branched hairs and longer simple hairs on branchlets. Leaves revolute, appearing terete, to 10 x 1.5mm wide. Flowers pale yellow, in axillary clusters of about 6; floral tube hairy; style deeply 3-cleft. **HABITAT:** Wooded slopes. **DIST:** Upper Blue Mtns, Sthn H'lands, Sublime Point (Bulli). Also SC, ST, WS, Vic. **FL:** Sept–Nov.

Subsp. *phylicifolia* has flat recurved leaves >1.5mm wide. Recorded from Hill Top and Joadja Ck.

Pomaderris prunifolia
Shrub to 2m tall with rusty stellate hairs on branches. Leaves elliptic, to 45 x 20mm, with short stiff simple hairs and impressed veins above and rusty stellate hairs beneath. Flowers yellow, in short panicles; floral tube hairy. **HABITAT:** Sheltered forest. **DIST:** Endangered population Dundas–Parramatta, t'lands (e.g. Mt Caley, Bundanoon). Also SC, NT, ST, WS, Qld, Vic. **FL:** Oct–Nov.

Pomaderris prunifolia

Cayratia clematidea

Cissus hypoglauca

VITACEAE

Cayratia
Climbers with branched tendrils. Leaves compound with 3–12 leaflets. Inflorescences leaf-opposed cymes with long peduncles. Flowers tiny, 4-merous. Fruit a berry.

Cayratia clematidea
Slender Grape
Slender climber with stems to 2m long. Leaflets 5, the terminal leaflet largest, ovate, to 8 x 4cm, acuminate, irregularly toothed, glabrous, dull, wilting easily. Flowers greenish. Berry black, 6mm diam. **HABITAT:** Moist forest, rainforest margins. **DIST:** Coast, Kowmung R., Wollondilly R. Also NC, SC, NT, NWS, Qld. **FL:** Jan–Feb.

Cissus
NATIVE GRAPE, WATER VINE
Woody climbers with simple or 2-branched tendrils. Leaves simple, or palmately compound with terminal leaflet largest. Inflorescences cymose, extra-axillary on long peduncles. Flowers tiny, usually 4-merous. Fruit a berry.

Cissus antarctica
Water Vine
Large climber, with rusty matted woolly hairs on young shoots and leaf stalks. Leaves alternate, simple; lamina ovate, 7–10 x 2–5cm, toothed or entire; glabrous above, rusty-hairy beneath; stalk 1–3cm long. Flowers yellowish. Berry purplish black, 15mm diam. **HABITAT:** Rainforest. **DIST:** Coast, incl. Kurrajong, Razorback Mtn. Also NC, SC, NT, ST, CWS, Qld. **FL:** Dec–Jan.

Cissus antarctica

Cissus hypoglauca
Five-leaf Water Vine
Large climber with soft rusty-hairy new growth and 2-branched tendrils. Leaves with 5 elliptic leaflets to 12 x 3.5cm, glabrous and green above, glaucous beneath; domatia absent; petiole 2–4cm long; leaflet stalks 1–2cm long. Flowers yellow, in terminal umbels. Berry purple-black, to 10mm diam. **HABITAT:** Rainforest, sheltered gullies. **DIST:** Chiefly coast; less common CT. Also NC, SC, NT, CWS, Qld, Vic, PNG. **FL:** Nov–Dec.

Cissus opaca. Climber with 5 narrow leaflets to 60 x 15mm. Rare in Sydney region. Occurs in dry rainforest scrub at Razorback Mtn.

Cissus sterculiifolia. Climber similar to *C. hypoglauca*. Leaflets not grey underneath, with prominent domatia. Occurs in rainforest, e.g. Ourimbah, Burning Palms.

OLEACEAE

Notelaea
NATIVE OLIVE, MOCK OLIVE
Shrubs or small trees with opposite leaves. Flowers bisexual, in axillary racemes, sometimes reduced; calyx lobes and petals 4; stamens 2. Fruit an olive-like drupe.

Notelaea longifolia
Large Mock Olive
Shrub or tree to 6m tall with softly hairy branchlets. Leaves narrow-lanceolate to ovate, to 15 x 4cm, entire, ± glabrous to velvety, leathery with age, discolorous; veins uneven in thickness and spacing. Racemes <2cm long, lengthening in fruit. Petals cream-green, 2mm long. Drupe globular to ovoid, 10–15mm long, bluish black. **HABITAT:** Open forest, rainforest. **DIST:** Chiefly coast; less common CT. Also NC, SC, ST, CWS. **FL:** Apr–July.

Notelaea venosa
Smooth Mock Olive
Shrub to 4m tall with ± glabrous branchlets. Leaves lanceolate to ovate, to 14 x 4cm, glabrous or softly hairy when young, discolorous; veins distinct, fine and evenly reticulate. Racemes to 3cm long, lengthening in fruit. Petals pale yellow or whitish, 2mm long. Drupe ellipsoid to ovoid, 15–20mm long, dark purplish. **HABITAT:** Moist forest, rainforest margins. **DIST:** Coast and ranges. Also NC, SC, ST, CWS, Qld, Vic. **FL:** Oct–Feb.

Notelaea ovata. Small shrub. Leaves ovate and glabrous. Occurs along coast in open forest in sandy soils.

Notelaea neglecta. Small shrub. Leaves narrow (5–7mm). Occurs upper Blue Mtns, Tallong–Wombeyan Caves.

Notelaea longifolia

Notelaea venosa

CAPRIFOLIACEAE

Sambucus
Trees or small shrubs with prominent lenticels on branchlets. Leaves opposite, pinnate. Inflorescence umbellate or corymbose with flowers ultimately in cymes. Flowers bisexual, 3–5-merous; corolla tubular with spreading lobes. Fruit a succulent drupe.

Sometimes included in families Sambucaceae or Adoxaceae.

Sambucus australasica
Yellow Elderberry
Glabrous shrub to 3m tall. Leaves 10–25cm long; leaflets 5, terminal one longest, elliptic, to 10 x 3cm, soft; apex acuminate, base asymmetric; margins entire or toothed. Inflorescences 10–20cm across on long peduncles exceeding leaves. Flowers 3-merous, white to pale yellow, 3mm long. Drupe yellow, ± globose, 5mm diam. **HABITAT:** Rainforest margins, moist forest. **DIST:** Coast and ranges (e.g. Royal NP, Bulli Pass, Mt Tomah). Also NC, SC, NT, ST, WS, Qld, Vic. **FL:** Dec–Feb.

Sambucus gaudichaudiana. Shrub with 3–11 leaflets; lowest pair resembling leafy stipules. Flowers and fruit white. Occurs in Blue Mtns.

Sambucus australasica

LOGANIACEAE

Logania
Shrubs or herbs with opposite decussate simple leaves, the bases connected by a stipular ridge. Inflorescences usually axillary and cymose. Flowers tubular with 5 lobes, usually unisexual and white. Fruit a capsule, splitting into 2 carpels.

Logania albiflora
Erect shrub to 2m tall. Leaves narrow-ovate, 20–50 x 2–12mm, olive-green above, pale beneath; midvein prominent. Flowers few in short dense cymes, sweet-scented; corolla 2–3mm long. Capsule 4–7mm long. **HABITAT:** Forested hillsides and gullies. **DIST:** Coast and ranges. Also NC, SC, NT, ST, WS, SWP, Qld, Vic, LHI. **FL:** June–Oct.

Logania albiflora

Logania pusilla
Undershrub to 12cm tall with ridged stems. Leaves opposite, elliptic, 5–10 x 3–5mm, revolute, usually obtuse. Flowers solitary in axils; corolla 5–7mm long. **HABITAT:** Open forest, esp. on clay-shale soils. **DIST:** Uncommon. Coast, lower Blue Mtns (e.g. Castlereagh NR, Linden, Terrey Hills). Also NC, SC, NT, ST, Qld, Vic. **FL:** Sept–Oct.

Logania pusilla

Mitrasacme
Small herbs with opposite simple leaves. Flowers in lax terminal or axillary cymes, bisexual, white; calyx and corolla usually 4-lobed; corolla tube short with spreading lobes. Fruit a tiny capsule with 2 persistent styles.

Mitrasacme polymorpha
Mitre Weed
Herb to 25cm tall. Stems much-branched, with spreading hairs, sometimes scabrous. Leaves narrow-ovate, to 15 x 6mm, with recurved margins, slightly hairy. Flowers in terminal umbel-like clusters of 3–5. **HABITAT:** Heath, woodland, open forest. **DIST:** Common. Coast and ranges. Also NC, SC, ST, Qld, Vic, Tas. **FL:** Aug–Jan.

Mitrasacme pilosa is similar to *M. polymorpha* but the flowers occur singly on stalks to 4cm long.

Mitrasacme polymorpha

ASCLEPIADACEAE

Marsdenia
MILK VINE

Climbers or shrubs with milky or watery sap. Leaves opposite with minute glands at base of leaf-blade on midvein. Flowers in simple or compound umbels. Flowers 5-merous. Fruit a follicle; seeds numerous, with long silky hairs.

Marsdenia flavescens

Marsdenia flavescens
Hairy Milk Vine

Climber with pubescent stems to 4m long. Leaves oblong, to 10 x 4cm, glossy green above, densely pubescent and yellow-green beneath. Inflorescences of 2–6 umbels on a hairy peduncle shorter than the leaves; flowers 3mm long, pale yellow. Fruit a fusiform follicle, 4–6cm x 6–8mm. **HABITAT:** Dry rainforest, sheltered rocky sites. **DIST:** Coast, upper Blue Mtns. Also NC, SC, NT, Qld, Vic. **FL:** Dec–Jan.

Marsdenia rostrata
Common Milk Vine

Climber with stems to 10m long; latex white, copious. Leaves with broad-ovate lamina to 12 x 7cm, narrowing abruptly to extended obtuse apex, rounded at base, dark green above, paler below; stalk to 4cm long. Flowers pale yellow, in simple umbels to 3cm diam. Stigma long and beaked. Follicle ovoid, 5–6 x 2–4cm, inflated. **HABITAT:** Rainforest, sheltered gullies. **DIST:** Coast, less common CT. Also NC, SC, NT, ST, CWS, Qld, Vic, LHI. **FL:** Oct–Jan.

Marsdenia rostrata

Marsdenia suaveolens
Scented Marsdenia

Slender climber or shrub. Stems with milky latex and shortly hair. Leaves lanceolate, to 7 x 2.5cm, dark green above, paler below; stalk to 4mm long. Flowers white or creamy, in simple umbels to 2cm diam. Stigma short, covered by anther appendages. Follicle 5–10 x 1–1.4cm. **HABITAT:** Moist forest, sandstone gullies. **DIST:** Common coast and ranges. Also NC, SC. **FL:** Nov–Jan.

Marsdenia suaveolens

Tylophora

Small climbers. Leaves opposite, with minute glands at base of midvein. Flowers purplish, with 5 spreading lobes. Fruit a follicle.

Tylophora barbata
Bearded Tylophora
Stems glabrous with clear watery sap. Leaf blade ovate, to 5 x 3cm; stalks 7–20mm long. Flowers in 1–3 umbels, each with 3–5 flowers about 5mm across, with 5 purple-brown star-like petals with short hairs, a yellow stigma and a ring of 5 appendages forming a corona around staminal column. **HABITAT:** Rainforest, moist open forest. **DIST:** Chiefly coast (e.g. Royal NP, Mt Kembla, Douglas Park); Mt Tomah. Also NC, SC, ST, Vic. **FL:** Sept–Feb.

APOCYNACEAE

Melodinus
Climbers with milky latex. Leaves opposite or whorled, simple. Flowers in cymes, sometimes with only 1 or 2 flowers; 5-merous; calyx and corolla tubular; stamens in tube. Fruit a berry; seeds lacking hairs.

Melodinus australis

Melodinus australis
Southern Melodinus
Woody climber with glabrous twining stems. Leaves lanceolate-elliptic, to 10 x 3cm, glabrous, shortly stalked. Flowers 5–13 in axillary cymes 2.5cm long, yellow, 2–3mm long, with 5 spreading lobes. Berry red-orange, ± globose, 5–6mm long. **HABITAT:** Rainforest, on richer clay soils. **DIST:** Coast, Mt Kembla to Gosford (e.g. Bola Ck, Mt Keira). Also NC, Qld. **FL:** Dec–Apr.

Parsonsia
Woody climbers with watery ± yellowish sap. Leaves opposite, petiolate, simple. Flowers in cymes or panicles, usually 5-merous, tubular, with stamens exserted and united by anthers. Fruit a pod-like capsule splitting into 2 sections to release seeds with silky hairs.

Parsonsia brownii
Mountain Silkpod
Tall climber with pubescent young stems. Adult leaves lanceolate, to 13 x 4cm, glabrous, tapering at both ends, glossy above, paler beneath, with 9–15 pairs of lateral veins prominent. Flowers yellow, minutely hairy, with white hairs in throat. Capsule 5–8cm long. **HABITAT:** Rainforest, moist gullies. **DIST:** Upper Blue Mtns, Sthn H'lands. Also NC, SC, NT, ST, Vic, Tas. **FL:** Oct–Feb.

Tylophora barbata

Parsonsia straminea

Parsonsia straminea
Common Silkpod
Woody vine to 20m or more. Adult leaves elliptic, to 24 x 8cm, glossy above, paler and yellowish beneath; juvenile leaves 1–4cm long, cordate, purplish beneath. Flowers cream, hairy inside and outside in lower part. Capsule pendent, to 20 x 1cm. **HABITAT:** Rainforest, gullies, sheltered forest. **DIST:** Common. Coast and ranges. Also NC, SC, CWS, Qld. **FL:** Nov–Apr.

Parsonsia lanceolata. Twiner with lanceolate leaves and glabrous corolla. Rare in area. Recorded from The Crest at Bass Hill and Lansdowne Park.

Parsonsia brownii

MENYANTHACEAE

Nymphoides
Aquatic plants with round floating leaves and yellow bisexual flowers with fringed petals. Fruit a capsule, submerged at maturity.

Nymphoides geminata
Marshwort
Leaves floating or erect, entire, 3–8cm diam., with V-shaped indent at junction of stalk and lamina. Flowers paired or clustered on long stalks, yellow, 5-lobed, to 25mm diam. with fringed margins. **HABITAT:** Still water in lagoons, dams, etc. **DIST:** Cumberland Pl. (e.g. Castlereagh NR, Narellan) and Mountain Lagoon, Wingecarribee Swamp. Also NC, NT, ST, NWP, Qld, Vic. **FL:** Oct–Mar.

The white-flowered *Nymphoides indica* also occurs on the coast, but is rare locally.

Nymphoides geminata

Villarsia

Tufted swamp plant with emergent and/or floating leaves on long stalks. Flowers in loose panicles above leaves. Fruit an emergent capsule.

Villarsia exaltata
Yellow Marsh Flower
Erect glabrous herb. Leaf blades broadly ovate to 12 x 8cm. Flowers on leafless branched stems; petals 5, free, spreading, bearded inside at base. **HABITAT:** Swampy areas, margins of lagoons. **DIST:** Coast, Sthn H'lands (e.g. Kurnell, Maddens Plains, Penrose). Also NC, SC, ST, Qld, Vic, Tas. **FL:** Oct–Jan.

Villarsia exaltata

RUBIACEAE

Asperula
WOODRUFF
Dioecious herbs with 4-angled stems. Leaves appearing whorled due to leaf-like stipules. Flowers in terminal cymose inflorescences, white, 4-merous, with floral tube longer than lobes. Fruit fleshy, 2-lobed but not splitting into 2 mericarps.

Asperula conferta
Common Woodruff
Decumbent perennial to 20cm tall with numerous stems bearing short recurved hairs; internodes 4cm long. Leaves and stipules reflexed, in whorls of 5–6, linear, 3–9cm x 1mm. Flowers in terminal clusters, white or tinged pink. **HABITAT:** Grassland, woodland, on clay-shale soils. **DIST:** Cumberland Pl. (e.g. Mt Annan), higher t'lands (e.g. Moss Vale). Most of NSW. Also Qld, Vic, SA, NTerr. **FL:** Sept–Jan.

Asperula scoparia
Prickly Woodruff
Perennial to 15cm tall with short hairs on stems; internodes 5cm long. Leaves and stipules in whorls of 6, linear, to 8mm x 1mm, scabrous, acuminate. Flowers in terminal clusters of 3–5, white, 4mm long. **HABITAT:** Grassy understorey of open forest. **DIST:** Upper Blue Mtns, Sthn T'lands. Also NT, ST, Vic, Tas, SA. **FL:** Oct–Dec.

Asperula gunnii, from tablelands, has obovate leaves to 3mm wide.

Asperula scoparia

Asperula conferta

FLOWERING PLANTS

Coprosma quadrifida

Coprosma hirtella

Cyclophyllum longipetalum

Coprosma
CURRANT BUSH
Dioecious shrubs with opposite leaves. Male flowers with long filaments and 4–6 exserted anthers; female flowers with style divided to base into 2 branches. Fruit a currant-like drupe.

Coprosma hirtella
Rough Coprosma
Shrub to 2m tall. Leaves obovate, to 40 x 25mm, thick, dark green and scabrous above, light green below. Male flowers in dense clusters; female flowers in 3s. **HABITAT:** Woodland, esp. on rocky slopes. **DIST:** Upper Blue Mtns (e.g. Blackheath, Hassans Walls). Also SC, NT, ST, Vic, Tas. **FL:** Oct–Jan.

Coprosma quadrifida
Prickly Coprosma
Slender shrub to 3m tall with pubescent stems and many short side-branches ending in spines. Leaves narrow-ovate, to 15 x 5mm. Flowers solitary, terminal on axillary branches; corolla 3mm long. Fruit a red oblong drupe. **HABITAT:** Sheltered forests, rainforest margins. **DIST:** Illawarra (e.g. Bulli, Mt Keira), higher t'lands (e.g. Bundanoon, Mt Tomah). Also NC, SC, NT, ST, Vic, Tas. **FL:** Sept–Dec.

Cyclophyllum
Small trees or shrubs. Leaves opposite with interpetiolar stipules. Flowers in axillary cymes or clusters; calyx and corolla 4–6-lobed; stamens exserted. Fruit a drupe.

Cyclophyllum longipetalum
Coast Canthium
Tree to 10m tall. Leaves elliptic, to 10 x 4cm, apex blunt, dull and dark green above, paler below; domatia usually present. Flowers in clusters of 2–6, white; corolla tube 8mm long; lobes 5, spreading, fringed, 3–4mm long. Drupe compressed 10mm diam., red. **HABITAT:** Rainforest, in gullies and near ocean. **DIST:** Coast nth from Foxground (e.g. Bass Point, Yengo NP). Also NC, NT, Qld. **FL:** Jan–Feb.

Previously known as *Canthium coprosmoides*.

Galium
BEDSTRAW
Polygamous herbs with 4-angled stems. Leaves appearing whorled due to leaf-like stipules. Flowers in terminal and axillary cymes; corolla with 4 spreading lobes much longer than tube. Fruit dry, often with hooked bristles, splitting into 2 mericarps.

Galium gaudichaudii
Rough Bedstraw
Tufted perennial with rough trailing stems to 30cm long, brittle in lower part. Leaves and stipules in whorls of 4, sessile, linear, to 8 x 1.5mm, with rough hairs. Cymes exceeding leaves, with 1–7 cream flowers 1–2mm long. Fruit 1–2mm long. **HABITAT:** Grassland, woodland, open forest. **DIST:** Coast and ranges. Most of NSW. All other States. **FL:** Aug–Oct.

Galium propinquum
Maori Bedstraw
Decumbent perennial with weak stems to 30cm long, often prickly. Leaves and stipules in whorls of 4, elliptic ovate, to 10 x 4mm, upper surface slightly hairy. Cymes longer or shorter than leaves, with 1–5 (often 3) white to cream flowers 0.5mm long. **HABITAT:** Moist shaded forested sites. **DIST:** Coast and ranges. Also NC, SC, NT, ST, SWS, Qld, Vic, NZ. **FL:** Nov–Jan.

Galium binifolium is very similar to *G. propinquum* except that each whorl has 2 leaves larger than the 2 leaf-like stipules.

Morinda
Woody climbers. Leaves opposite, with interpetiolar stipules, often with domatia. Flowers in small heads in axillary or terminal inflorescences. Corolla tube cylindrical, 3–5-lobed. Fruit of drupes united into a head.

Morinda jasminoides
Morinda
Scrambling glabrous shrub. Leaves lanceolate, to 7 x 3cm, with domatia. Flowers yellow-white to purplish. Fruit orange-red, irregularly lobed, 10–15mm across. **HABITAT:** Rainforest, moist tall forest. **DIST:** Coast to lower Blue Mtns; Robertson. Also NC, SC, NT, Qld, Vic. **FL:** Nov–Dec.

Opercularia
STINKWEED
Herbs or subshrubs with ridged stems. Leaves opposite with stipules forming a sheath, with unpleasant smell when crushed. Flowers in globular heads on short recurved or long erect peduncles, tubular with 3–5 lobes. Fruit a compound capsule.

Opercularia aspera
Common Stinkweed
Stems ridged, to 1m tall, usually scabrous. Leaves lanceolate, to 4 x 2cm, scabrous; stalk 3–9mm long. Flowers bisexual; corolla 2–4mm long. Fruiting heads 5–10mm across. **HABITAT:** Forested gullies, creek banks. **DIST:** Common. Coast and ranges. Also NC, SC, NT, ST, WS, Qld, Vic. **FL:** Sept–Nov.

Opercularia hispida is covered with dense hairs. *Opercularia diphylla* is a glabrous weak herb with narrow leaves (to 8mm). *Opercularia varia* has many spreading stems from the main stem and leaves to 15 x 3mm.

Pomax
Subshrub with opposite leaves; stipules present. Flowers minute, with several fused into heads in lobed pedunculate cups arranged in terminal umbels. Monotypic genus.

Galium gaudichaudii

Galium propinquum

Morinda jasminoides

Pomax umbellata

Opercularia aspera

Pomax umbellata
Pomax
Much-branched plant with pubescent stems to 20cm tall. Leaves lanceolate, to 20 x 10mm. Flowers reddish, the empty cups often persisting. **HABITAT:** Open forest, woodland. **DIST:** Coast and ranges. Most of NSW. All mainland States. **FL:** Sept–Dec.

Psychotria
Shrubs or climbers. Leaves opposite, with domatia on lower surface; stipules united between stalk bases. Flowers bisexual, in terminal cymes or heads. Fruit a drupe.

Psychotria loniceroides
Hairy Psychotria
Shrub to 4m tall with white or rusty hairs. Leaves ovate to elliptic, to 10 x 4cm long, softly hairy, lateral veins prominent. Flowers white, 5mm long. Drupe yellow, ellipsoid, 6–8mm long, topped by persistent calyx. **HABITAT:** Rainforest, moist tall forest. **DIST:** Coast (e.g. Mt Kembla, Bola Ck, Ourimbah SF). Also NC, SC, NT, Qld. **FL:** Jan–Mar.

Psychotria loniceroides

CONVOLVULACEAE

Calystegia
Twining plants with alternate simple leaves. Flowers bisexual, in 1- or few-flowered cymes, 5-merous with large persistent bracteoles enclosing the calyx, funnel-shaped corolla and stigma with 2 lobes. Fruit a capsule.

Calystegia marginata
Glabrous twiner. Leaves with narrowly triangular lamina with hastate or sagittate base to 10cm long; stalk to 8cm long. Flowers solitary, white to pale mauve; corolla 2cm long; peduncle not longer than leaf stalk. Capsule 6mm long. **HABITAT:** Rainforest margins, moist tall forest. **DIST:** Coast and ranges. Also NC, SC, NT, Qld, Vic. **FL:** Oct–Jan.

Calystegia soldanella has broad kidney-shaped leaves and pink flowers. It grows in coastal dune vegetation. Rare (e.g. Point Potter, Bass Point).

Convolvulus erubescens

Convolvulus
BINDWEED
Twining plants with alternate simple leaves. Flowers axillary in 1- or few-flowered cymes, 5-merous with funnel-shaped corolla and bracteoles small and distant from the calyx. Fruit a capsule.

Convolvulus erubescens
Australian Bindweed, Pink Bindweed
Stems to 80cm long. Leaves softly hairy, very variable in shape, linear to triangular and deeply divided at the base into 2–6 narrow lobes; margins shallowly toothed or lobed. Flowers pink to white, to 15mm long; sepals ovate, pointed, softly hairy. **HABITAT:** Grassland, woodland on shale-clay and basalt soils. **DIST:** Scattered. Cumberland Pl., Illawarra. Most of NSW. All other States. **FL:** Sept–Mar.

The introduced weed *C. arvensis* has obtuse glabrous sepals.

Cuscuta
DODDER
Leafless annual parasitic plants, with yellowish or brownish stems draped over host shrubs, attaching themselves by special organs known as haustoria.

Cuscuta australis
Australian Dodder
Flowers in small cymose clusters, tubular with 5 obtuse lobes, white, 2mm long; stamens 5; styles 2. **HABITAT:** Parasitic on many herbs (e.g. *Persicaria*), in floodplain alluvium. **DIST:** Rare. Cumberland Pl. (e.g. Shanes Park, Longneck Lagoon), Kowmung R. Also NC, NWS, Qld, Vic. **FL:** Nov–Mar.

Calystegia marginata

Cuscuta australis

Dichondra repens

Polymeria calycina

Dichondra
Small creeping herbs with kidney-shaped leaves. Flowers axillary, bell-shaped, with 5 lobes.

Dichondra repens
Herb with creeping stems, rooting at nodes. Leaves entire, to 20mm long, on stalks to 6cm long, with soft weak hairs. Flowers greenish yellow, 2–4mm long, on stalks 1–5cm long. **HABITAT:** Dry rainforest, woodland, on clay-shale and basalt soils. **DIST:** Coast and ranges. Most of NSW. All other States (except NTerr). **FL:** Sept–Jan.

Polymeria
BINDWEED
Trailing or erect plants with funnel-shaped flowers; stigma divided into 4–8 lobes. Bracteoles minute and away from calyx.

Polymeria calycina
Polymeria
Stems trailing and twining, to 40cm long. Leaves elliptic, to 2–5cm x 5–20mm, entire, on stalks to 35mm long; base cordate. Flowers solitary, pink to violet with a lighter yellow throat, to 12mm long. Sepals unequal in size and overlapping. **HABITAT:** Open forest, woodland, in damp clay-shale soils. **DIST:** Coast (e.g. Burning Palms, Castlereagh NR, Pearl Bch). Also NC, SC, Qld, WA, NTerr. **FL:** Oct–Feb.

Wilsonia backhousei

Wilsonia
Subshrubs with simple fleshy leaves. Flowers tubular with sepals united.

Wilsonia backhousei
Glabrous perennial forming dense mats on landward side of estuarine salt marshes. Stems to 15cm tall. Leaves linear-lanceolate, to 15 x 2mm. Flowers solitary in upper axils; calyx tubular, 6mm long; corolla white, 10mm long. **HABITAT:** Salt marsh, sea-cliff tops. **DIST:** Rare. Georges R., Parramatta R. (e.g. Salt Pan Ck, Mill Ck), Clovelly cliffs. Also NC, SC, Vic, Tas, SA, WA. **FL:** Oct–Nov.

BORAGINACEAE

Austrocynoglossum
HOUND'S TONGUE
Perennial herbs with alternate petiolate leaves. Flowers mostly solitary on stems near leaf insertions; corolla narrowly funnel-shaped with 5 spreading lobes and scales in throat opposite lobes. Fruit dry, splitting into 4 mericarps.

Austrocynoglossum latifolium
Forest Hound's Tongue
Straggly plant with stems to 1m long; hairs lengthening into prickles. Leaves ovate, 3–5 x 2–5cm, dark green and densely hairy above; main veins hairy beneath. Corolla white, 4–5mm long. **HABITAT:** Moist forest, rainforest margins. **DIST:** Rare. Illawarra, Blue Mtns (e.g. Bulli Pass, Bilpin, Kowmung R.). Also NC, SC, NT, ST, Qld, Vic. **FL:** Dec–May.

Ehretia
Trees or shrubs with alternate leaves; flowers 5-merous, tubular. Fruit a drupe.
Sometimes placed in a segregate family, Ehretiaceae.

Austrocynoglossum latifolium

Ehretia acuminata
Koda
Tree to 20m tall, often deciduous. Leaves elliptic, to 15 x 5cm, glabrous, toothed. Flowers white, in large terminal panicles. Drupe with 2 cells, orange to yellow, 6mm diam. **HABITAT:** Rainforest, esp. along creeks. **DIST:** Coast (e.g. Ourimbah, Mt Keira, Razorback) and Kowmung. Also NC, SC, NT, CWS, Qld. **FL:** Nov–Feb.

New growth and flowers often covered in web of the caterpillar of the *Ethmia heliomela* moth.

Ehretia acuminata

Cyphanthera scabrella

Duboisia myoporoides

SOLANACEAE

Cyphanthera
RAY-FLOWER
Shrubs with alternate simple leaves. Flowers shortly tubular with 5 long spreading lobes and 4 stamens. Fruit a capsule.

Cyphanthera scabrella
Erect greyish shrub to 1m tall. Stems and leaves with glandular hairs. Leaves broadly ovate, 7 x 4mm. Flowers solitary or in cymes of 1–3; pedicels 5–10 mm long; corolla white or yellowish green, striped with indigo. **HABITAT:** Forest, in gullies. **DIST:** Endemic. Valleys of lower Blue Mtns (e.g. Bowen Ck, Erskine Ck). **FL:** Oct–Nov.

Duboisia
Glabrous shrubs or trees. Leaves alternate, simple, entire. Flowers bell-shaped or tubular, 5-lobed. Fruit a succulent berry.

Duboisia myoporoides
Duboisia, Corkwood
Tall shrub or small tree, with corky bark. Leaves oblanceolate, 5–10cm long, entire, light green above, slightly paler below. Flowers in loose terminal panicles; corolla white, tubular, to 7mm long, with 5 broad lobes; stamens 4. Berry globular, 8mm diam., purple-black. **HABITAT:** Forested gullies, rainforest margins. **DIST:** Coast (e.g. Macquarie Pass, Ourimbah, Audley). Also NC, SC, NT, Qld, NCal. **FL:** May–Nov.

Leaves of this species contain hyoscyamine and related alkaloids which were used in ophthalmology as a pupil dilator.

Nicotiana
NATIVE TOBACCO
Herbs or shrubs. Leaves basal and/or on stems. Inflorescences paniculate. Corolla tubular; stamens 5. Fruit a capsule.

Nicotiana forsteri
Herb to 1.5m tall. Leaves broad-elliptic, sparsely hairy, lamina to 25 x 14cm at base, much smaller on stems. Inflorescences much-branched, with dense glandular hairs. Flowers white, tubular, to 25mm long. **HABITAT:** Sheltered forest, on clay-shale and basalt soils. **DIST:** Coast, nth from Kiama (e.g. Jamberoo, Werrong). Also Grose V., NC, Qld, LHI. **FL:** Dec–Jan.

FLOWERING PLANTS

Solanum aviculare

Solanum
KANGAROO APPLE, NIGHTSHADE
Shrubs and herbs. Flowers rotate with 5 spreading lobes and orange stamens. Fruit a globular berry. About 125 species in Australia (95 native), 11 native species in Sydney district.

Solanum aviculare
Kangaroo Apple
Erect shrub to 4m tall, with green or purple stems, glabrous except new growth. Leaves variable, entire to deeply lobed, 15–30cm long. Flowers 3–4cm diam., blue-violet, with spreading acute lobes. Fruit ovoid to ellipsoid, to 25 x 15mm, yellow turning orange then red.
HABITAT: Sheltered forests, rainforest margins.
DIST: Blue Mtns, Sthn H'lands, Illawarra. Most of NSW. Also Qld, Vic, SA, WA, NZ, Pac. Iss.
FL: Nov–Dec.

Nicotiana forsteri

Solanum brownii

Solanum campanulatum

Solanum cinereum

Solanum brownii
Violet Nightshade
Shrub to 2m tall. Branches prickly, with grey stellate hairs. Leaves narrow-ovate to elliptic, entire, 8–12cm long, grey-green with prickles on midvein above, yellowish to rusty beneath. Flowers in short clusters of 5, purple, to 30mm diam. Fruit globular, to 28mm diam, pale yellow-green. **HABITAT:** Woodland, on hillsides. **DIST:** Scattered. Uncommon (e.g. Glen Davis, Mt Tomah, Wyong). Chiefly Hunter V.; also NC, NT, WS. **FL:** July–Oct.

Solanum celatum is similar to *S. brownii* but has dense hairs on the upper leaf surface and smaller fruits (16mm across). It occurs from Mt Kembla to Nowra and Bungonia NR.

Solanum campanulatum
Sprawling plant to 1m tall with stellate and simple hairs. Branches, leaves and calyces with spreading yellow prickles. Leaves ovate to elliptic, 12 x 9cm, with wavy irregular lobes, lower surface with dense long woolly hairs. Inflorescence 30mm long, 4–10-flowered, on peduncle 40mm long; flowers distinctly funnel- or bell-shaped, violet. Fruit globular, to 25mm diam., yellowish green, turning pale brown then black. **HABITAT:** Moist open forest. **DIST:** Coast and ranges (e.g. Bargo, Culoul Ra., Nepean R.). Also NC, NT, CWS, Qld. **FL:** Sept–Mar.

Solanum cinereum
Narrawa Burr
Undershrub to 70cm tall with yellow prickles. Branches with stellate and glandular hairs. Leaves broadly ovate, 6–10cm long, deeply lobed, green above, with close woolly white hairs below; petiole 5–15mm long. Flowers in cymes of 2–7 opposite leaves, purple, about 35mm across. Fruit globular, 15–20mm diam., mottled green to yellowish turning brown, with papery skin. **HABITAT:** Woodland, on hills and slopes. **DIST:** Scattered. Uncommon (e.g. Putty, Marsden Park). Also ST, NT, WS, NWP, Qld. **FL:** Chiefly Jan–Apr.

FLOWERING PLANTS

Solanum prinophyllum

Solanum prinophyllum
Forest Nightshade
Sprawling plant to 50cm tall with slender weak prickles. Stems purplish green with tiny purplish stellate hairs. Leaves broadly oblong in outline, 8 x 5cm, lobed, with irregular margins, thin, slightly discolorous. Flowers several on common peduncle, pedicellate, lilac-blue, 25–30mm across. Fruit globular, 15–20mm diam., mottled green or purple. **HABITAT:** Forest and rainforest margins. **DIST:** Coast and ranges. Also NC, SC, NT, ST, NWS, Qld, Vic. **FL:** Sept–Jun.

Solanum pungetium
Eastern Nightshade
Bushy plant to 1.5m tall with stout yellowish prickles and scattered pale stellate hairs. Leaves elliptic in outline, to 8 x 4cm, deeply lobed with irregular margins, slightly discolorous. Inflorescence extra-axillary, with 1–3 flowers on short stout axis; flowers pale purple, 20mm across. Fruit globular, to 30mm diam., marbled

Solanum pungetium

green turning yellow, finally dull green; pedicel 10–18mm long. **HABITAT:** Moist open forest, rainforest margins. **DIST:** Coast and ranges. Also NC, SC, ST, Vic. **FL:** Sept–May.

Solanum stelligerum

Solanum vescum

Solanum stelligerum
Devil's Needles
Shrub to 1.5m tall with scattered fine prickles on stems and stellate pubescence on most parts. Leaves lanceolate, 3–7 x 1–2.5cm, entire, glabrous and dark green with deep veins above, tomentose and paler or rusty beneath. Flowers in small cymes; corolla pale lilac, 20–25mm across, deeply incised, stellate. Fruit 5–10mm diam., bright red. **HABITAT:** Sheltered forest, rainforest margins, on richer soils. **DIST:** Coast. Also NC, SC, NT, ST, CWS, Qld. **FL:** June–Nov.

Solanum vescum
Gunyang
Spreading glabrous green shrub to 2m tall without prickles. Leaves variable, from narrow, entire and up to 15cm long to deeply lobed with lobes mostly 5–10cm x 8–12mm, lamina usually extending down petiole and decurrent on stems. Inflorescences to 12cm long, many-flowered; flowers on pedicels 20–25mm long, violet, 3–4cm across, with spreading lobes. Fruit, to 25mm diam. on pedicels 30–50mm long, smooth, greenish ivory. **HABITAT:** Open forest, hind-dune woodland. **DIST:** Uncommon. Coast and ranges. Most of NSW. Also Qld, Vic, Tas. **FL:** Aug–Oct.

SCROPHULARIACEAE

Derwentia
Shrubs or herbs with woody base. Leaves in opposite pairs distant on stems, often decussate. Inflorescences racemose. Calyx 4-lobed; corolla tubular with 1 broad upper lobe and 3 smaller spreading lower lobes; stamens 2, projecting beyond tube. Fruit a flattened 2-locular capsule.

Derwentia blakelyi
Glabrous shrub to 40cm tall. Leaves ovate, sessile, 25–50 x 10–20mm, concave above, margins with many shallow teeth. Racemes 10–40cm long. Flowers deep blue, about 10mm across. **HABITAT:** Open forest, woodland, swamp margins. **DIST:** Upper Blue Mtns (e.g. Clarence, Nullo Mtn), west to Peel R. (CT). **FL:** Nov–Dec.

Derwentia derwentiana subsp. *subglauca*
Derwent Speedwell
Herb to 1m tall, with several erect unbranched stems from rootstock. Leaves lanceolate with tiny marginal teeth, to 10–20 x 1–4.5cm, sessile. Flowers in racemes of 40–100, lavender to white with purple tinge. **HABITAT:** Forested slopes. **DIST:** Upper Blue Mtns (e.g. Jenolan Caves, Hassans Walls), west to Orange (CT). **FL:** Dec–Jan.

Derwentia perfoliata
Digger's Speedwell
Erect glaucous perennial to 1m tall. Leaves ovate, 2–5cm long, blue-green, with entire margins, joined by bases and encircling stem. Flowers in slender racemes of 25–70, deep blue. **HABITAT:** Forested slopes, esp. rocky areas. **DIST:** Higher tablelands (e.g. Rylstone, Clarence, Wingello). Also SC, ST, WS, Vic. **FL:** Nov–Jan.

Derwentia blakelyi

Derwentia perfoliata

Derwentia derwentiana subsp. *subglauca*

Euphrasia collina subsp. *speciosa*

Euphrasia

Herbs with opposite decussate leaves. Flowers in spike-like racemes; calyx 4-lobed; corolla 2-lipped, upper lip 2-lobed, lower lip spreading and 3-lobed; stamens 4, paired. Fruit a 2-valved capsule.

Euphrasia collina subsp. *speciosa*
Eye-bright
Perennial branching from base, with stems 20–40cm tall. Leaves oblong to wedge-shaped, sessile, to 13 x 7mm, with 2–5 pairs of teeth. Flowers sessile, outer surface of calyx glandular hairy; corolla bluish mauve. **HABITAT:** Heath, open forest, esp. in damp sandy soil. **DIST:** Upper Blue Mtns (e.g. Newnes SF, Wingello), coast (e.g. Wyong, Helensburgh). Also SC, ST. **FL:** Aug–Nov.

Subsp. *paludosa* has glabrous calyx and a yellow blotch often on the lower lip of the flower. It occurs in swamps of Sydney's eastern suburbs, Maddens Plains, upper Blue Mtns, t'lands and WS.

Gratiola

Creeping herbs from wet places. Leaves opposite, sessile, simple, 3-veined. Flowers 1 or 2 in upper axils; corolla ± 2-lipped, upturned; stamens 2. Fruit a 4-valved capsule.

Gratiola peruviana

Gratiola peruviana
Brooklime
Decumbent glabrous plant with creeping rootstock and stems ascending to 30cm. Leaves sessile, stem-clasping, ovate to elliptic, to 30 x 20mm; margins toothed. Flowers solitary in axils, pale pink; tube to 10mm long. **HABITAT:** Wet places. **DIST:** Rare on coast (e.g. Yarramundi). Mainly t'lands (e.g. Boyd Plat., Wingello). Also NC, SC, NT, ST, CWS, Qld, Vic, Tas, SA, WA, NZ, S. Amer. **FL:** Dec–Jan.

Mimulus
MONKEY-FLOWER
Creeping herbs, rooting at nodes. Leaves opposite. Flowers tubular, 5-lobed and 2-lipped, with 4 fertile stamens. Fruit a 2-valved capsule.

Mimulus repens
Creeping Monkey-flower
Glabrous prostrate plant with soft slightly succulent stems rooting at nodes. Leaves crowded, ovate, 5 x 3mm. Flowers single in upper axils, violet with yellow centre, like tiny snapdragons. **HABITAT:** Wet places, brackish coastal lagoons, shorelines. **DIST:** Coast (e.g. Deepwater Park, Kurnell). Also NC, SC, WP, Qld, Vic, Tas, SA, WA, NZ. **FL:** Nov–Mar.

Mimulus repens

Parahebe
Semi-woody perennials. Leaves opposite, toothed. Flowers with 4 spreading lobes; stamens 2, exserted from floral tube. Capsule 2-locular, flattened.

Parahebe lithophila
Trailing perennial with delicate glabrous stems to 40cm long. Leaves broad-ovate, to 25 x 15mm, with shallow acute teeth. Flowers in long axillary racemes of 3–18, pale lavender, streaked with violet. **HABITAT:** Shaded, damp cliff-faces. **DIST:** Endemic. Blue Mtns (e.g. McMahons L'out, Kanangra Walls, Pierces Pass). **FL:** Aug–Oct.

Parahebe lithophila

Veronica
Perennial herbs with opposite leaves. Flowers single or in racemes in axils; calyx usually 4-lobed; corolla shortly tubular with usually 4 spreading lobes; stamens 2, exserted. Fruit usually a flattened 2-locular capsule.

Veronica calycina
Hairy Speedwell
Small plant to 25cm tall with stolons rooting at nodes and erect flowering stems with ± spreading hairs 1–2mm long. Leaves broadly ovate with truncate base, to 3 x 2cm, margins with 5–25 obtuse teeth; stalk 2–20mm long. Flowers violet. **HABITAT:** Sheltered forest. **DIST:** Coast and ranges. Also SC, NT, ST, WS, Qld, Vic, Tas, SA, WA. **FL:** Dec–Jan.

Veronica calycina

Veronica plebeia

Veronica notabilis

Veronica notabilis
Forest Speedwell
Small plant with stolons and erect flowering stems to 40cm tall with soft ± spreading hairs. Leaves ovate, to 5 x 2cm, with 5–12 pairs of teeth. Flowers in racemes to 20cm long, pale lilac streaked with violet. **HABITAT:** Sheltered forest, in damp rock cracks. **DIST:** Illawarra (e.g. Mt Kembla) and Fitzroy Falls. Also SC, NT, ST, Vic, Tas. **FL:** Dec–Jan.

Veronica plebeia
Creeping Speedwell
Plant with stolons and flowering stems to 10cm tall. Stems with minute hairs. Leaves triangular, to 20 x 16mm; margins with 6–16 teeth. Flowers in loose racemes of 3–10, bluish mauve. **HABITAT:** Sheltered situations, esp. moist forests on heavier soils. **DIST:** Coast and ranges. Most of NSW. Also Qld, Vic, SA. **FL:** Sept–Feb.

BIGNONIACEAE

Pandorea
Woody climbers. Leaves opposite, pinnate with terminal leaflet. Flowers in showy cymose inflorescences, tubular, with 5 lobes and 4 stamens in 2 unequal pairs attached to base of tube. Fruit an elongated capsule containing flat, circular, winged seeds.

Pandorea pandorana
Wonga Vine
Climber with branches several metres long. Leaves with 5–9 ovate leaflets to 7 x 2cm (leaflets on juvenile plants 9–17 per leaf, much smaller, toothed). Floral tube to 25mm long, whitish with red spots in throat. **HABITAT:** Woodland to rainforest, esp. on rocky slopes. **DIST:** Common. Coast and ranges. Most of NSW. All mainland States, PNG, Malesia. **FL:** Aug–Sept.

Pandorea pandorana

GESNERIACEAE

Fieldia
Climbing semi-epiphytic plant. Leaves opposite, unequal in size. Corolla tubular. Fruit a succulent berry. Monotypic genus.

Fieldia australis
Fieldia

Branches and leaves with long hairs. Leaves elliptic-obovate, 30–70 x 10–25mm, or 10–30 x 4–10mm, paler beneath; margins irregularly toothed. Flowers solitary, pendent, bell-shaped, about 35mm long, pale cream. Fruit oblong to 20mm long, whitish with red flecks. **HABITAT:** On rocks and trees in rainforest. **DIST:** Coast and ranges (e.g. Carrington Falls, Mt Wilson). Also NC, SC, NT, ST, Qld, Vic. **FL:** Mar–May.

Fieldia australis

Utricularia dichotoma

Utricularia gibba

Utricularia lateriflora

LENTIBULARIACEAE

Utricularia
BLADDERWORT
Carnivorous herbs with modified stems functioning as traps for small water animals. Leaves in rosettes. Flowers in racemes on erect stems, bisexual; corolla tubular, 2-lipped, usually spurred at base.

Utricularia dichotoma
Fairies' Aprons
Stems to 20cm tall. Flowers solitary, in pairs or whorls of 3–4, or on stems to 30cm tall, with small upper lip and broad lower lip, 12–20mm long, violet with yellow centre. **HABITAT:** Wet sites and swampy heath. **DIST:** Common. Coast and ranges. Most of NSW. Also Qld, Vic, Tas, SA, WA. **FL:** Aug–Apr.

Utricularia gibba
Floating Bladderwort
Aquatic plant with leaves and stems submerged. Stems fine and interwoven. Leaves with hair-like segments interspersed with numerous small bladders about 1mm across. Flowers yellow on fine leafless stalks 2–15cm long above the water, each with upper lip ± equal to or longer than lower lip. **HABITAT:** Shallow water. **DIST:** Coast (e.g. Marley Lagoon, Thirlmere Lakes). Also NC, Qld, NTerr, trop. Amer. **FL:** Jan–Mar.

Utricularia lateriflora
Small Bladderwort
Stems to 20cm tall. Leaves to 30 x 0.5mm. Flowers 3–5, upper lip slightly exceeding the calyx, oblong; lower lip broad, pale lilac with yellow spot at base. **HABITAT:** Wet sand or peaty soils. **DIST:** Coast and Blue Mtns (e.g. Marley Lagoon, Bulli, Lawson). Also NC, SC, Qld, Vic, Tas, SA. **FL:** Chiefly Nov–Mar.

Utricularia uliginosa
Asian Bladderwort
Stems to 8cm tall. Leaves, when present, at base of stem, pale green, linear, 6–12mm long. Flowers scattered alternately along stem; small, lilac or white; upper lip obovate to 5mm long, shorter than the calyx; lower lip longer, broader (6–8mm wide) and convex. **HABITAT:** Mud, wet sand or in shallow water. **DIST:** Coast nth of Royal NP. Also NC, Qld, NTerr, WA, NCal, trop. Asia. **FL:** Chiefly Dec–Mar.

Utricularia australis. Floating plant with yellow flowers similar to *U. gibba* but with upper lip smaller than lower lip. Common in lagoons of Nepean R.

Utricularia uniflora. Occurs along wet creek banks from coast to ranges. Flowers mauve, one per stem.

Utricularia uliginosa

MYOPORACEAE

Eremophila
Shrubs and small trees. Leaves usually alternate. Flowers axillary, sessile or stalked, tubular with 5 lobes in 2 lips or rarely campanulate and ± regular; stamens 4, or 5 if flowers regular. Fruit dry or rarely succulent.

Eremophila debilis
Amulla, Winter Apple
Prostrate glabrous shrub. Stems to 1m long. Leaves lanceolate to elliptic, 3–8 x 1–2cm, dark green; margins entire or with a few scattered teeth. Flowers 1–2 in axils; corolla 8–11mm long, white to pink or mauve. Fruit ovoid, 6–9mm long, fleshy, white to reddish purple. **HABITAT:** Woodland, usually on clay soils. **DIST:** Cumberland Pl. (e.g. Doonside, Mt Annan). Also Glen Davis, NC, NT, WS, WP, Qld, NZ. **FL:** Oct–Apr.
Formerly known as *Myoporum debile*.

Eremophila debilis

Myoporum
Shrubs or small trees. Flowers axillary, shortly tubular with 5 lobes, white or violet; stamens 4, exserted. Fruit usually succulent and ± globose, rarely woody or dry or compressed.

Myoporum acuminatum
Northern Boobialla
Glabrous shrub or small tree with corky bark. Leaves elliptic, 5–15 x 1–3cm, discolorous, tapering to long-pointed apex; usually entire. Flowers in clusters of 3–8, white with purple dots, hairy inside. Fruit ovoid, purplish black, wrinkled, 4–7mm diam. **HABITAT:** Margins of coastal littoral rainforest and estuaries. **DIST:** Coast (e.g. Bass Point, Port Hacking, Patonga Creek). From Bega to NC, Qld. **FL:** July–Oct.

Myoporum boninense subsp. *australe*

Myoporum boninense subsp. *australe*
Boobialla
Glabrous shrub to 2m tall. Leaves ovate to oblanceolate, entire, 2–7cm long. Flowers 1–8 in axils, white, without spots; stalks 10–15mm long. Fruit globular, 5–9mm diam., pale purple, shiny. **HABITAT:** Heath on coastal cliffs, dune scrub. **DIST:** Coast (e.g. Pearl Bch, Jibbon Head, Minnamurra Spit). From Eden to Qld. **FL:** Mar–Aug.

Myoporum floribundum

Myoporum floribundum
Glabrous shrub to 3m tall with sparsely tuberculate branches. Leaves linear, 2–10cm x 1–2.5mm, pendent, entire, glabrous, with unpleasant smell. Flowers in clusters of 3–8; pedicels 1–3mm long; corolla 5–7mm diam., white, unspotted. Fruit ovoid, compressed, dry, 2–3mm wide. **HABITAT:** Open forest on rocky slopes. **DIST:** Rare. Scattered (e.g. Hill Top–Nattai). Also Snowy R. (ST), Vic. **FL:** Sept–Oct.

Myoporum acuminatum

Myoporum montanum
Western Boobialla, Water Bush
Very variable shrub to 3m tall. Leaves alternate, narrow-lanceolate, 3–10cm x 3–5mm long, entire, pointed, shortly stalked. Flowers in clusters of 1–5, white, 6–8mm across, unspotted or faintly spotted, hairy in upper tube; pedicels 5–15mm long. Fruit globular, purplish, 5–8mm diam. **HABITAT:** Woodland, esp. along rocky creek banks and hillsides. **DIST:** sw. Sydney (e.g. Razorback Mtn, Mt Annan), sthn Blue Mtns (e.g. Wombeyan Caves, Jenolan Caves). Most of NSW. Also Qld, Vic. **FL:** July–Nov.

Myoporum bateae. Rare glabrous tall shrub with narrow discolorous leaves, finely toothed. Bass Hill (The Crest), Illawarra district.

Myoporum montanum

ACANTHACEAE

Brunoniella
Perennial herbs with opposite leaves on grooved angular branches. Inflorescence various. Flowers tubular, broadening from base, with 5 large lobes; stamens 4, enclosed in tube. Fruit a cylindrical capsule.

Brunoniella australis
Blue Trumpet, Blue Yam
Herb with erect or spreading stems to 15cm tall from a fleshy white root. Leaf pairs often unequal; blades obovate to oblanceolate, 1–6cm x 5–25mm; stalk 2–15mm long. Flowers in sessile axillary clusters; calyx lobes linear, <1mm wide; corolla mauve-blue, 12–20mm across. **HABITAT:** Open forest, woodland, chiefly on shales and clays. **DIST:** Cumberland Pl. and lower Blue Mtns. Also NC, WS, NWP, Qld, NTerr. **FL:** Oct–Dec and Mar.

Brunoniella pumilo

Brunoniella pumilo
Dwarf Blue Trumpet
Herb to 10cm tall. Leaf pairs equal; blade elliptic, to 2 x 1cm long; stalk 1–5mm long. Flowers solitary or paired in upper axils; calyx lobes lanceolate, >1mm wide; corolla mauve or blue, 15–25mm across. **HABITAT:** Open forest, woodland, chiefly on shales and clays. **DIST:** Cumberland Pl., lower Blue Mtns. Also NC, WS, NWP, Qld, NTerr. **FL:** Oct–Dec and Mar.

Pseuderanthemum
Shrubs or herbs with opposite leaves. Inflorescence terminal or axillary, various. Flowers narrow-tubular, scarcely broadening; lobes 5, large, spreading; stamens 2, exserted; staminodes 2, enclosed. Fruit a capsule.

Pseuderanthemum variabile
Pastel Flower
Herb with stems 7–15cm tall. Leaves lanceolate, 3–5cm long. Flowers in terminal bracteate

Brunoniella australis

FLOWERING PLANTS

Pseuderanthemum variabile

racemes; corolla mauve, lower lobe spotted. Capsule to 14mm long. **HABITAT:** Sheltered forest, on clay-shale soils. **DIST:** Coast (e.g. Royal NP, Berry), lower Blue Mtns. Also NC, SC, Qld, NTerr. **FL:** Dec–Mar.

PLANTAGINACEAE

Plantago
PLANTAIN
Plants with basal rosette of leaves. Stems leafless. Flowers tiny, in dense terminal spikes; corolla scaly, tubular, 4-lobed, green, brown or purplish; stamens 4, prominent at flowering.

Plantago debilis
Slender Plantain
Leaves obovate, 3–15 x 1–4cm, 3–5 veined, ± entire or distantly toothed, softly hairy. Scape to 20cm long, spike to 10cm long. **HABITAT:** Moist forest, on clay-shale soils. **DIST:** Coast and ranges. Most of NSW. All other States. **FL:** Chiefly Dec–Mar.

Plantago gaudichaudii
Narrow-leaf Plantain
Perennial with thick fleshy taproot. Leaves linear to narrow-oblong, 7–20cm x 2–20mm, entire or with a few teeth, glabrous or hairy, 3-veined. Scape to 20cm long, spike to 12cm long. **HABITAT:** Grassland, woodland, on clay soils. **DIST:** Cumberland Pl. (e.g. Mt Annan, Doonside). Also NC, SC. Also t'lands and slopes, Qld, Vic, Tas, SA. **FL:** Chiefly Sept–Apr.

Plantago debilis

Plantago gaudichaudii

VERBENACEAE

Clerodendrum
Shrubs with opposite or whorled leaves. Flowers in loose clusters in terminal panicles. Calyx 5-lobed, tiny in flower, much enlarged in fruit. Corolla with slender tube and 5 spreading lobes; stamens 4, prominently exserted. Fruit a succulent or dry drupe in enlarged calyx.

Clerodendrum tomentosum
Hairy Clerodendrum
Tall shrub or small tree. Leaves broad-elliptic, to 12 x 4cm, with velvet indumentum on undersurface of young leaves; stalk 2–4cm long. Flowers numerous in cymose clusters; corolla white, 20–25mm long; lobes 4–7mm long, hairy. Drupe shiny black, 8mm diam., in enlarged red calyx. **HABITAT:** Sheltered forests, rainforest margins. **DIST:** Coast, lower Blue Mtns. Also NC, SC, NT, CWS, Qld, WA. **FL:** Oct–Nov.

Clerodendrum tomentosum (fruiting)

Gmelina leichhardtii

Gmelina
Trees or large shrubs with simple opposite leaves. Inflorescence terminal, paniculate. Corolla tubular, opening upwards into 5 lobes ± forming 2 lips. Stamens 4, with 2 longer than other 2, visible in open tube. Fruit a succulent drupe.

Gmelina leichhardtii
White Beech
Tree to 40m tall with grey bark and velvety-hairy new growth. Leaves ovate, to 20 x 10cm, green and glabrous above, pale and hairy below, with prominent raised veins; stalks 15–35mm long, velvety. Corolla 20mm long, white with purple and yellow markings. Drupe globular, 15–20mm diam., purplish blue, on enlarged calyx. **HABITAT:** Rainforest. **DIST:** Coast (e.g. Minnamurra Falls, Hacking R., upper Patonga Ck, Watagan Mtns). Also NC, Qld. **FL:** Oct–Jan.

Clerodendrum tomentosum

AVICENNIACEAE

Avicennia
Mangroves. Trees or shrubs with opposite leaves. Flowers clustered in upper axils, sepals 5; corolla shortly tubular with 4 or 5 lobes; stamens 4. Fruit a capsule with 2 valves.

Avicennia marina var. *australasica*
Grey Mangrove
Tree to 8m tall with erect aerating roots (pneumatophores). Leaves obovate, to 12 x 4cm, leathery, dark green and shiny above, paler below. Flowers orange; tube 2mm long; lobes 4mm long. Capsule 3cm long, germinating on tree before dropping into mud. **HABITAT:** Intertidal mud flats, estuaries. **DIST:** Coast. From tropics to sthn Australia. All mainland States, Pac. Iss, SE Asia. **FL:** Feb–Apr.

Two mangrove spp. occur in the Sydney area, with this sp. dominating the seaward zone. *Aegiceras corniculatum* occurs on the landward side and further up tidal rivers and creeks.

CHLOANTHACEAE

Chloanthes
Shrubs with dense woolly branched hairs. Leaves sessile, simple, opposite and decussate, or whorled. Flowers solitary in axils; corolla tubular, unequally lobed, with 2-lobed upper lip and 3-lobed lower lip; stamens 4. Fruit a dry drupe.

Chloanthes is sometimes included in Lamiaceae.

Chloanthes stoechadis
Common Chloanthes
Undershrub <1m tall. Leaves to 50 x 5mm, wrinkled above, white-woolly beneath; margins

Avicennia marina var. *australasica*

revolute. Flowers yellowish green, to 3cm long, stamens and style protruding. **HABITAT:** Heath, woodland, esp. near rocky outcrops. **DIST:** Coast and ranges. Also NC, SC, CWS, Qld, WA. **FL:** Chiefly July–Oct.

Spartothamnella
Shrubs. Leaves opposite, linear or scale-like. Flowers in short axillary cymes, tubular; stamens 4, exserted. Fruit a globose succulent drupe.

Spartothamnella is sometimes placed in Verbenaceae or Lamiaceae.

Spartothamnella juncea
Bead Bush
Glabrous shrub to 2m tall with 4-angled stems. Leaves reduced and scale-like, or narrow-elliptic and 3–18mm long on new shoots, falling early. Flowers whitish, 3–4mm long, divided almost to base into 5 narrow, spreading, acute lobes. Drupe orange, 3mm diam. **HABITAT:** Rocky sites, dry rainforests. **DIST:** Rare. Razorback Mtn (Camden), Mt Wilson. Also NC, WS, NWP, Qld. **FL:** Dec–Jan.

Chloanthes stoechadis

Spartothamnella juncea

LAMIACEAE

Ajuga
Herbs with opposite decussate simple leaves. Inflorescence with sessile cymes of 1–5 flowers in upper axils. Corolla tubular with large 3-lobed lower lip and tiny 2-lobed upper lip. Stamens 4, exceeding upper lip.

Ajuga australis
Austral Bugle
Softly hairy perennial, with leafy ascending stems to 30cm tall. Basal leaves obovate, to 12 x 4cm, tapering to stalk, broadly lobed; stem leaves smaller. Calyx 5-toothed; corolla purple, to 15mm long. Schizocarp splitting into 4 nutlets. **HABITAT:** Grassy woodland, esp. basalt and clay soils. **DIST:** Cumberland Pl., higher ranges. Most of NSW. Also Qld, Vic, Tas, SA. **FL:** Oct–Mar.

Ajuga australis

Hemigenia
Shrubs with opposite or whorled leaves. Flowers axillary, solitary or rarely clustered. Calyx 2-lipped or 5-toothed; corolla very shortly tubular, 2-lipped; stamens 4.

Hemigenia cuneifolia
Shrub mostly <1m tall. Leaves in whorls of 3, oblong, to 30 x 4mm, base cuneate, apex acute. Flowers single in upper axils; calyx lobes shorter than tube; corolla 8mm long, lilac blue. **HABITAT:** Open forest, often in rocky areas. **DIST:** Uncommon. CC (e.g. Buxton SRA, Burragorang Walls, Tallowa Dam, Colo Heights). Also NC, WS, Qld. **FL:** Sept–June.

Hemigenia cuneifolia

Hemigenia purpurea
Common Hemigenia
Shrub to 2m tall. Leaves in crowded whorls of 3, terete, channelled above, to 15mm long. Flowers solitary in axils; corolla 8–10mm long, lilac-blue. **HABITAT:** Heath, on sandy soils. **DIST:** Common in coastal heaths; also Blue Mtns (e.g. Woodford), SC. **FL:** July–Feb.

Hemigenia purpurea

Mentha diemenica

Mentha
MINT
Herbs with 4-angled branches and opposite leaves. Flowers clustered or solitary in upper axils; calyx tubular with 5 acute lobes; corolla shortly tubular, 4-lobed, the lower lip with 3 equal lobes, upper lip broader; stamens 4, exceeding tube.

Mentha diemenica
Slender Mint
Small herb with prostrate to ascending branches. Branches and calyx hairy. Leaves ovate, to 20 x 10mm. Flowers in clusters of 3–8; corolla 4–7mm long, pale purple. **HABITAT:** Open forest, woodland, on clay to sandy soils. **DIST:** Western Sydney (e.g. Mt Annan). Most of NSW. All States except WA. **FL:** Chiefly Nov–Apr.

Mentha satureioides
Creeping Mint
Small herb spreading by rhizomes, forming clumps with a few ascending branches. Branches and leaves glabrous or with short hairs. Leaves narrow-elliptic, to 30 x 6mm. Flowers in clusters of 3; corolla 3–5mm long,

Mentha satureioides

white or pink. **HABITAT:** Open forest, on clay soils. **DIST:** Chiefly western Sydney (e.g. Prospect Dam, Campbelltown, Razorback). Most of NSW. Also Qld, Vic, SA. **FL:** Nov–Mar.

Plectranthus

Herbs or shrubs with opposite leaves. Flowers in tiny clusters in narrow terminal inflorescences. Calyx 2-lipped, upper lip entire, lower lip with 4 acute teeth. Corolla tubular; upper lip 4-lobed, lower lip entire. Stamens 4.

Plectranthus parviflorus
Cockspur Flower
Shrub to 50cm tall with semi-succulent stems with short hairs. Leaves ovate, to 7 x 4cm, toothed, hairy. Corolla 6–12mm long, blue-mauve. **HABITAT:** Rocky areas in moist open forest, rainforest. **DIST:** Coast and ranges. Also NC, SC, NT, WS, Qld, Vic, Malesia, Pac. Iss. **FL:** Chiefly Aug–Mar.

Prostanthera
MINT-BUSH
Shrubs, often strongly scented, with square stems and opposite leaves. Flowers axillary or in terminal inflorescences. Calyx with 2 equal or unequal entire lips. Corolla with short tube (Sydney district) or long tube (inland and ranges) and ± erect 2-lobed upper lip and spreading 3-lobed lower lip. Stamens 4, in pairs; anthers 2-celled, often with 1–2 appendages.

Prostanthera askania
Tranquillity Mint-bush
Strongly aromatic shrub to 1m tall with soft downy hairs. Leaves ovate, to 25 x 16mm, coarsely toothed, dull green above, paler below, shortly stalked. Flowers in crowded terminal leafy clusters of 4–10; lower calyx lobe slightly

Plectranthus parviflorus

longer and narrower than upper lobe; corolla 12–14mm long, mauve to blue-mauve; anthers without appendage. **HABITAT:** Sheltered gullies, margins of rainforest. **DIST:** Endemic. Gosford–Ourimbah (e.g. Strickland SF, Askania Pk, Niagara Pk, Bouddi NP). **FL:** Sept–Nov.

Prostanthera caerulea
Shrub to 3m tall with 4-ridged sparsely hairy branchlets. Leaves narrow-ovate, to 6 x 2cm, toothed, glabrous, mid-green. Flowers in terminal racemes with leaves reduced to bracts or absent; corolla 10–12mm long, white to lilac or bluish; anther appendages shorter than anthers. **HABITAT:** Sheltered gullies, tall wet forests, mainly on sandstone below basaltic soils. **DIST:** Blue Mtns (e.g. Mt Irvine, Mt Banks, Mt Tomah). Also NC, SC, NT, Qld. **FL:** Oct–Nov.

Prostanthera askania

FLOWERING PLANTS

Prostanthera caerulea

Prostanthera cryptandroides

Prostanthera densa

Prostanthera cryptandroides
Sticky aromatic shrub to 1m tall, with hairy stems and glossy leaves. Leaves narrow-ovate, 6–9 x 2–3mm, lobed or toothed. Flowers solitary in upper axils, bracteoles not persisting; calyx with broad upper lobe 2–3mm long; corolla 13–15mm long, lilac to mauve with white markings and yellow spots in throat; anther appendages 1.5mm long, longer than anthers. **HABITAT:** Forested slopes and gullies, esp. at base of scree slopes. **DIST:** Rare. CT, CWS (e.g. Glen Davis). **FL:** Sept–Apr.

Prostanthera densa
Compact hairy shrub to 2m tall. Leaves crowded along branchlets, especially near ends, ovate, to 15 x 10mm, shortly stalked, somewhat succulent, dark green and rough above, paler below; margins recurved. Flowers axillary; calyx 6–8mm long, hispid; lobes ± equal; corolla 12–15mm long, mauve with white and orange markings in throat; anthers with 1 long appendage. **HABITAT:** Rocky slopes near sea. **DIST:** Cronulla, Marley (Royal NP). Also SC. **FL:** May–Dec.

Prostanthera denticulata

Prostanthera hirtula

Prostanthera granitica

Prostanthera hindii

Prostanthera denticulata
Rough Mint-bush
Scrambling aromatic shrub <1m tall with short hairs. Leaves irregular and ± distant on stems, narrow-ovate, 5–12 x 2–3mm, ± sessile, rough above; margins recurved. Flowers in short inflorescences in upper axils; buds with small deciduous bracts; calyx lobes reddish, hairy, broad, ± equal; corolla 7–10mm long, violet to lilac. **HABITAT:** Sheltered forest, esp. on Narrabeen shales. **DIST:** Endemic. Restricted to Mona Vale–Ku-ring-gai Chase–Cowan area. **FL:** Sept–Nov.

Prostanthera granitica
Shrub to 1m tall, with dense short hairs on branches. Leaves ovate, to 15 x 5mm, hirsute, light green, revolute, base ± cordate; hairs on upper surface often glandular. Flowers axillary; corolla 8–10mm long, purple; anthers without appendages. **HABITAT:** On and near pagoda rock formations. **DIST:** Wollemi NP (Glow Worm Tunnel), Capertee Valley. Also NT, WS. **FL:** Sept–Oct.

Prostanthera hindii
Erect shrub to 2m tall; branches densely hairy between ridges. Leaves ovate, 25 x 10mm, flat, paler beneath, often purplish, leaves subtending flowers smaller than others. Flowers in upper axils; calyx glabrous outside, densely hairy inside on lobes; corolla 10–14mm long, deep pinkish mauve to purple; anthers without appendages. **HABITAT:** Crevices of pagoda rock formations. **DIST:** Endemic. Blue Mtns (Clarence–Wolgan area) to e. of Rylstone. **FL:** Aug–Oct.

Prostanthera howelliae

Prostanthera hirtula
Spreading aromatic shrub to 2m tall. Branchlets and leaves with short rough hairs. Leaves narrow-elliptic, 20 x 5mm, entire, green and scabrous above, paler beneath; margins recurved. Flowers in loose terminal racemes with reduced leaves; calyx 5–7mm long, equally lobed; corolla 8–10mm long, mauve with purple throat; anther appendages short. **HABITAT:** Sheltered forest, often on rocks. **DIST:** Royal NP to ST (e.g. Maianbar, Carrington Falls, Belmore Falls). **FL:** Sept–Nov.

Prostanthera howelliae
Spreading aromatic shrub to 1m tall with short hairs and sessile glands. Leaf pairs nearly opposite, 15–20mm apart on stems; lamina narrow-ovate, to 10 x 2mm, rough, apex obtuse, margins recurved. Flowers axillary; bracteoles persistent; calyx 4–5mm long, hispid; lobes broad, upper lobe longer than lower; corolla 10mm long, deep mauve, with darker spots in throat; anthers without appendages. **HABITAT:** Woodland. **DIST:** Scattered (e.g. Sackville–Maroota, Bundanoon–Wingello). Also NC, WS, NWP, Qld, Vic. **FL:** Sept–Nov.

Prostanthera incana

Prostanthera incana
Velvet Mint-bush
Hoary shrub to 2m tall. Leaves ovate, 5–20 x 3–15mm, wrinkled, bluntly toothed; margins slightly recurved. Flowers in upper axils, often with reduced leaves; bracteoles persistent, 2–3mm long; calyx lobes broad, lower one slightly exceeding upper; corolla 6–9mm long, lilac-mauve; anther appendages short or absent. **HABITAT:** Sheltered forest, slopes near creeks. **DIST:** Blue Mtns, Sthn H'lands. Also Bomaderry Ck, NC, SC, Vic. **FL:** Aug–Nov.

Prostanthera incisa

Prostanthera lasianthos

Prostanthera linearis

Prostanthera incisa
Cut-leaf Mint-bush
Strong-smelling shrub to 2m tall. Branchlets 4-angled and ridged, densely covered with short curled hairs. Leaves broad-ovate, 10–25 x 4–10mm, toothed, green above, paler below, dotted with glands, narrowing to stalk 2–8mm long. Flowers solitary in upper axils. Calyx lobes broad, lower lobe longer than upper. Corolla 7–10mm long, lilac-mauve. Anther appendages absent or short. **HABITAT:** Moist, sheltered forest. **DIST:** Nth of Hawkesbury R., Blue Mtns. Also SC, NC. **FL:** Sept–Oct.

Prostanthera lasianthos
Victorian Christmas-bush
Shrub or small tree to 5m. Leaves lanceolate, 5–10 x 1–3cm, usually toothed, flat, glabrous, dark green above, paler below. Flowers in large, showy, ± leafless panicles or racemes; corolla 10–15mm long, pale mauve or white, spotted inside with purple; anthers with 1 long and 1 short appendage. **HABITAT:** Sheltered slopes, gullies, in richer soils. **DIST:** Coast and ranges. Also SC, NT, ST, NWS, Qld, Vic, Tas. **FL:** Dec–Jan.

Prostanthera linearis
Narrow-leaf Mint-bush
Slender glabrous shrub to 3m tall. Leaves oblong to linear, 15–35 x 2–3mm, obtuse, entire; stalk <2mm long. Flowers in terminal leafy or bracteate racemes; corolla 8–12mm long, whitish to mauve, often with yellow spots in throat; anthers with 1 long and 1 short appendage. **HABITAT:** Sheltered slopes and scrub, esp. along streams. **DIST:** Coast, higher ranges. Also SC, ST. **FL:** Aug–Mar.

Prostanthera ovalifolia
Oval-leaf Mint-bush
Shrub to 3m tall. Branchlets 4-angled, glandular, softly hairy, with distant leaf pairs. Leaves ovate, to 40 x 10mm, glabrous, discolorous, entire or sinuate, tapering to stalk 2–5mm long. Flowers in short racemes with leaves reduced or bract-like or deciduous at bud stage; bracteoles not persistent; corolla 6–10mm long, purple or mauve; anther appendages absent or very short. **HABITAT:** Forested hillsides and ridges, esp. among rocks. **DIST:** Scattered (e.g. Mooney Mooney, Wisemans Ferry, Mt Banks, Culoul Ra.). Also NC, SC, CWS, SWP, Qld, Vic. **FL:** Aug–Nov.

Prostanthera prunelloides

Prostanthera prunelloides
Prunella Mint-bush
Aromatic bushy shrub to 2m tall. Branchlets 4-ridged, glandular, hairy. Leaves broad-ovate, 20–55 x 6–30mm, entire or with a few rounded teeth; stalks to 15mm long. Flowers in compact bracteate terminal racemes, the bracts falling as flowers develop; bracteoles not persistent; corolla 12–15mm long, mauve to white with faint yellow dots in throat, softly hairy on outer surface; anthers with 1 long and 1 short appendage. **HABITAT:** Sheltered forested gullies, rocky slopes. **DIST:** Lower Blue Mtns (e.g. Grose L'out), Colo Plat. (e.g. Putty Rd). Also NC, SC, ST, CWS. **FL:** Dec–Jan.

Prostanthera rhombea
Open, weak shrub. Branches with stiff simple hairs and gland-tipped hairs. Leaves ± round to angular-ovate, 4–6 x 4–6mm, recurved, with spreading hairs along margins and dense glands below. Flowers axillary; bracteoles persistent, 1–2mm long; calyx hairy, dotted, ± equal in length; corolla 7–8mm long, mauve; anther appendages short or absent. **HABITAT:** Sheltered hillsides, gullies. **DIST:** Coast and ranges (e.g. Picton, Wondabyne, Katoomba, Bell). Also NC, WS, Qld, Vic. **FL:** Aug–Nov.

Prostanthera ovalifolia

Prostanthera rhombea

Prostanthera rotundifolia

Prostanthera rotundifolia
Round-leaf Mint-bush

Shrub to 3m tall. Branchlets terete, grooved longitudinally, with dense short curly hairs. Leaves rounded to broad-obovate, to 17 x 15mm, entire or obscurely lobed, glandular, dark green above, paler below; stalk 2–8mm long. Flowers axillary; bracteoles not persistent; calyx lobes broad, equal; corolla 10–15mm long, lilac to purple; anther appendages absent. **HABITAT:** Open woodland, often in rocky areas. **DIST:** Upper Blue Mtns, Sthn H'lands, Colo Heights. Also NC, NT, ST, WS, NWP, Qld, Vic, Tas. **FL:** Sept–Oct.

Prostanthera rugosa

Prostanthera rugosa

Shrub to 2m tall. Branchlets with short stiff white hairs. Leaves crowded, ovate, 3–8 x 2–5mm, thick, lobed, wrinkled; margins recurved. Flowers axillary; bracteoles <0.5mm long, persistent; calyx glandular, hairy, lower lobe longest; corolla 10–13mm long, mauve-purple; anther appendages unequal. **HABITAT:** Sheltered rocky sites. **DIST:** Endemic. Sthn H'lands (e.g. Berrima West, Little R. at Buxton). **FL:** Sept–Oct.

Prostanthera saxicola var. *saxicola*

Prostanthera saxicola var. *saxicola*
Slender Mint-bush
Shrub to 60cm tall with slightly hairy stems. Leaves flat, narrow-elliptic, to 8 x 3mm, entire, obtuse, thick. Flowers few, axillary; bracteoles persistent; corolla 8–10mm long, lilac or white; anthers with 1 appendage twice as long as anther cell. **HABITAT:** Shrubland, heath. **DIST:** Endemic. Coast (e.g. Lucas Heights, Heathcote Rd). **FL:** June–Oct.

Prostanthera saxicola var. *montana*
Undershrub to 30cm tall with prostrate to decumbent branchlets. Leaves more crowded than var. *saxicola*. Corolla 10–12mm long, white or mauve, with purple stripes in throat. **HABITAT:** Forest and heath. **DIST:** Blue Mtns (e.g. Narrow Neck, Kanangra Walls). Also SC, ST. **FL:** Dec–Jan.

Prostanthera scutellarioides
Shrub to 1m tall with ridged hairy branches. Leaves linear, to 20 x 2mm, glabrous, glandular, margins recurved. Flowers axillary, on short stalks; bracteoles persistent; upper calyx lobe slightly longer than lower lobe; corolla 7–9mm long, mauve to lilac or purple; anther appendages absent. **HABITAT:** Dry forest, woodland. **DIST:** Coast. Type form occurs in and around Castlereagh NR. Also NC, NT, ST, Qld. **FL:** Apr–Dec.

Prostanthera sieberi
Strongly aromatic shrub to 2m tall with hairy stems. Leaves glabrous, broad-ovate, 5–15mm long, small on new growth, coarsely toothed, narrowing to a long stalk. Flowers in upper

Prostanthera saxicola var. *montana*

Prostanthera scutellarioides

Prostanthera sieberi

Prostanthera violacea

Prostanthera violacea
Violet Mint-bush

Slender, aromatic, much-branched shrub to 2m tall. Branches with short stiff gland-tipped hairs. Leaves ovate to round, 2–6 x 2–5mm, crenate or few-lobed, wrinkled, discolorous, with stiff hairs; margins recurved. Flowers axillary; bracteoles persistent; corolla 6–8mm long, mauve to bluish; anther appendages absent. **HABITAT:** Sheltered forest, gullies, riverbanks. **DIST:** Chiefly lower Blue Mtns; also Cowan, Blackheath. Also WS, Qld, Vic. **FL:** Sept–Nov.

axils; calyx lobes obtuse, glabrous, ± equal; corolla purple to mauve; anther appendages short, unequal. **HABITAT:** Sheltered gullies. **DIST:** Coast, esp. Mt Kembla to Royal NP. Also lower NC. **FL:** Sept–Oct.

Not clearly separable from *P. incisa* and sometimes regarded as a variety of it.

Scutellaria
SKULLCAP

Herbs with 4-angular branches. Leaves opposite, toothed or lobed. Flowers solitary in upper axils; calyx tubular, ± 2-lobed, with prominent fold on upper side; corolla tubular, 2-lipped with unequal lobes; stamens 4.

Scutellaria humilis
Dwarf Skullcap

Stems ascending to about 15cm, hairy along ridges. Leaves on stalks 3–12mm long; lamina triangular-ovate, to 15 x 12mm, toothed. Flowers with pedicels as long as petiole of subtending leaf; corolla 6mm long, mauve, with lower lip slightly longer than upper lip. **HABITAT:** Woodland, on sheltered clay soils. **DIST:** Western Sydney (e.g. East Hills, Mt Annan, Razorback Mtn), upper Blue Mtns (e.g. Jenolan Caves). Also NC, SC, NT, ST, WS, Qld, Vic, Tas, SA. **FL:** Sept–Mar.

Scutellaria humilis

Teucrium

Herbs or shrubs with opposite leaves on 4-angled stems. Flowers in loose terminal compound inflorescences; calyx usually 5-lobed; corolla open-tubular with large lower lip and minute upper lip; stamens 4, longer than corolla tube.

Teucrium corymbosum
Forest Germander

Much-branched erect herb or subshrub to 1m tall. Leaves narrow-ovate, to 10 x 2cm, toothed, thin, glabrous above, with dense greyish hairs beneath. Inflorescence of axillary cymes with 3–9 flowers; corolla 8–12mm long, white, the lower lip with 1 lobe much longer than other 4. **HABITAT:** Sheltered forests, chiefly on richer soils. **DIST:** Localised, uncommon. Coast and ranges (e.g. Macquarie Pass, Mt Tomah, Colo R., Royal NP). Also WS, SWP, Vic, Tas, SA. **FL:** Dec–Apr.

Westringia

Shrubs with leaves in whorls of 3–5. Flowers with tubular calyx and corolla, the calyx 5-lobed, the corolla with erect 2-lobed upper lip and spreading 3-lobed lower lip. Fertile stamens 2; staminodes 2.

The 2 lower stamens reduced to staminodes help to separate *Westringia* from *Hemigenia* and *Prostanthera*.

Teucrium corymbosum

Westringia eremicola

Westringia eremicola
Slender Westringia
Shrub to 1.5m tall. Leaves in whorls of 3, rarely 4, linear, to 20 x 2mm, nearly sessile, hairy, recurved to revolute. Flowers axillary, shortly stalked; corolla 6–9mm long, lilac with purple-brown or orange spots in throat.
HABITAT: Rocky areas, often in dry gullies. **DIST:** Localised, uncommon (e.g. Kenthurst, Hill Top, Tallowa Dam). Also SC, NT, ST, WS, WP, Qld, Vic, SA. **FL:** Aug–Oct.

Westringia fruticosa
Coast Rosemary
Compact shrub to 1.5m tall. Leaves mostly in whorls of 4, narrow-lanceolate, 10–30 x 3–5mm, spreading, recurved, glabrous above, white-hairy below; stalk 1mm long. Flowers axillary, shortly stalked; calyx with white appressed hairs; corolla 12–14mm long, white, with pink or orange or brownish spots on lower lip. **HABITAT:** Heath and cliffs near the sea and along harbour foreshores. **DIST:** Common. Foreshore zone. Also NC, SC, LHI. **FL:** Some flowers all year.

Westringia fruticosa

Westringia longifolia
Long-leaved Westringia
Shrub 1–3m tall. Leaves in whorls of 3, linear, to 35 x 2mm, flat, shiny green with depressed midvein above, duller below. Flowers axillary, distinctly stalked; calyx 4–6mm long, sparsely hairy; corolla 8–10mm long, usually white, spotted inside with purple or pale brown.
HABITAT: Gullies, stream banks. **DIST:** Localised (e.g. Wheeny Ck, upper Georges R., Colo R., Bargo R.). Also NC, WS, Qld. **FL:** July–Dec.

Westringia longifolia

Wahlenbergia communis

Wahlenbergia gracilis

CAMPANULACEAE

Wahlenbergia
BLUEBELLS

Herbs with taproot. Leaves basal and/or on stems. Flowers solitary on long pedicels or in loose cymes; corolla campanulate or rotate. Fruit a capsule opening by apical slits, usually exceeded by persistent calyx lobes.

Wahlenbergia communis
Tufted Bluebell

Glabrous tufted perennial with numerous branched stems to 60cm tall. Leaves alternate, linear, to 80 x 6mm, ± entire. Flowers in cymes, blue, 10–15mm across. Capsule 4–9mm long. **HABITAT:** Woodland, grassland, esp. on clay soils. **DIST:** Coast to ranges. Most of NSW. Also other States (except Tas), PNG. **FL:** Peak Sept–Feb.

Wahlenbergia gracilis
Australian Bluebell

Multi-stemmed perennial to 50cm tall. Leaves tufted at base, obovate, to 6 x 1cm, linear and smaller on stems. Flowers usually blue, to 6mm across. **HABITAT:** Woodland, grassland, esp. on clay soils. **DIST:** Coast to ranges. Most of NSW. All other States (except WA), PNG, NZ, Pac. Iss. **FL:** Peak Oct–Apr.

Wahlenbergia luteola

Wahlenbergia luteola
Glabrous perennial with branched stems to 80cm tall. Leaves opposite, becoming alternate higher on stem, linear, to 50 x 4mm, ± entire. Flowers in cymes; corolla yellowish brown outside, blue inside; tube 3–5mm long; lobes 7–14mm long. Capsule glabrous, 5–12 x 2–4mm. **HABITAT:** Woodland, grassland, esp. on shale soils. **DIST:** Lithgow area (e.g. Little Hartley, Megalong V.). Also NT, ST, WS, WP, Vic, SA. **FL:** Peak Oct–Apr.

Wahlenbergia stricta subsp. *stricta*
Tall Bluebell
Multi-stemmed perennial with branched hairy stems to 90cm tall. Lower leaves opposite, obovate to linear, to 70 x 13mm, margins undulate, softly hairy. Flowers to 30mm across. Capsule glabrous or hairy, to 10mm long. **HABITAT:** Forest, woodland, grassland. **DIST:** Coast to ranges. Most of NSW. All other States (except NTerr), NZ. **FL:** Peak Oct–Jan.

Wahlenbergia stricta subsp. *stricta*

> **OTHER *WAHLENBERGIA* IN THE SYDNEY DISTRICT**
> All have alternate leaves and blue flowers.
> - *Wahlenbergia ceracea*. Flowers solitary. Upper Blue Mtns.
> - *Wahlenbergia graniticola*. Stems hirsute. Upper Blue Mtns.
> - *Wahlenbergia littoricola*. Leaves linear, stems glabrous. Coast, Sthn H'lands.
> - *Wahlenbergia multicaulis*. Leaves obovate below, linear above. Upper Blue Mtns; rare on coast. Corolla lobes 2–5mm long.
> - *Wahlenbergia planiflora*. Lower leaves obovate. Corolla lobes 6–14mm long. Coast, Sthn H'lands.

Isotoma axillaris

LOBELIACEAE

Isotoma
Herbs with tubular corolla and 5 spreading, ± equal lobes. Stamens 5, attached to corolla tube above middle; anthers fused around style. Fruit a capsule.

Isotoma axillaris
Rock Isotoma
Perennial with ascending stems to 40cm tall. Leaves to 10cm long, deeply divided into toothed linear lobes. Flowers axillary, solitary on stalks to 12cm long; corolla pale blue to mauve; tube 25mm long; lobes to 15–18mm long. Capsule 7–18mm long. **HABITAT:** Rock crevices, especially granite. **DIST:** Coast and ranges (e.g. Mt Gibraltar, Jenolan Caves), but rare in sandstone areas. Most of NSW. Also Qld, Vic. **FL:** Oct–June.

Isotoma fluviatilis subsp. *fluviatilis*

Isotoma fluviatilis subsp. *fluviatilis*
Swamp Isotoma
Scrambling or mat plant. Leaves alternate, elliptic, to 14 x 5mm, margins toothed. Flowers pale blue; yellow edged with deep mauve at base. **HABITAT:** Damp places, on clay soils. **DIST:** Coast (e.g. Agnes Banks, Narrabeen Lake), lower Blue Mtns (e.g. Springwood). Also NC, LHI. **FL:** Oct–Apr.

Lobelia alata

Lobelia dentata

Lobelia
Glabrous herbs with alternate leaves. Inflorescence terminal, racemose or flowers solitary. Corolla tubular, slit to base forming spreading 3-lobed lower lip and tiny 2-lobed upper lip. Stamens united into column. Fruit a capsule.

Lobelia alata
Angled Lobelia
Trailing or erect herb with angular ± winged stems. Leaves linear to obovate, 15–50 x 6–10mm, entire or with few teeth. Flowers solitary in axils of bracts; corolla 5–10mm long, white to pale blue, lobes of upper lip narrow, lobes of lower lip oblong, all similar. **HABITAT:** Damp places, esp. on wet rocks, drainage lines near sea. **DIST:** Common along coast; uncommon Blue Mtns. Also NC, SC, all States (except NTerr), NZ, LHI, S. Afr., S. Amer. **FL:** Nov–Mar.

A very compact form occurs at Kurnell (Potter Pt).

Lobelia dentata
Slender erect herb with stems to 40cm tall. Lower leaves ovate, to 40 x 10mm, deeply toothed or lobed. Racemes with 8–10 flowers on pedicels 20mm long; corolla bright blue, 15–25mm long; mid-lobe of lower lip narrow-oblong, distinctly longer than lateral lobes. **HABITAT:** Forest, woodland, on sandy soils. **DIST:** Coast and ranges. Also NC, SC, ST. **FL:** Chiefly July–Feb, peaking 6 months after fire.

Lobelia gibbosa

Lobelia gracilis

Lobelia gibbosa
Tall Lobelia
Slender erect herb to 60cm tall. Lower leaves mostly linear, to 70 x 4mm, ± entire. Flowers in long racemes, on pedicels, 12mm long; corolla pale blue, 15–25mm long, mid-lobe of lower lip similar to lateral lobes but slightly longer. Capsule gibbous on upper side. **HABITAT:** Forest, woodland, on sandy soils. **DIST:** Blue Mtns (e.g. Newnes Plat.), rare along coast. Also NC, SC, NT, ST, SWS, NWP, all States. **FL:** Chiefly Nov–Mar, peaking 6 months after fire.

Lobelia gracilis
Trailing Lobelia
Erect or decumbent plant, with stems to 30cm tall, often branched. Leaves linear to ovate, deeply cut and lobed or toothed. Racemes with up to 12 flowers; corolla deep blue, 10–15mm long, mid-lobe of lower lip broadly obovate, barely longer than lateral lobes. **HABITAT:** Open forest, on sandy soils. **DIST:** Coast nth from Royal NP. Also NC, NT, NWS, Qld. **FL:** Nov–May.

Lobelia trigonocaulis

Lobelia trigonocaulis
Forest Lobelia
Sparsely hairy creeping herb with angled stems to 30cm long. Leaves with heart-shaped toothed lamina to 35 x 35mm on stalk of similar length. Racemes with 3–6 flowers; corolla blue-mauve, 10–12mm long, lobes of lower lip oblong, all similar. **HABITAT:** Rainforest margins, sheltered creek banks. **DIST:** Wyong area, Watagan Mtns. Also NC, NT, Qld. **FL:** Nov–Feb.

Pratia
Herbs with alternate entire or toothed leaves, often 2-ranked. Flowers bisexual or unisexual, solitary on long stalks in upper axils; corolla split to base on upper side, formed into 2 unequal lips, upper lip 2-lobed, lower lip 3-lobed. Fruit globose or ovoid, fleshy.

Pratia concolor
Poison Pratia
Low plant with glabrous zigzag stems. Leaves ± sessile, distichous, oblong, to 30 x 15mm, with toothed margins. Flowers unisexual; corolla white to pink, tinged purple, 6–9mm long. **HABITAT:** Damp clay soils, usually along drainage lines. **DIST:** Rare. Western Sydney (e.g.

Pratia concolor

Doonside, Mt Annan). Also Wondabyne, NC, WS, WP, Vic, Qld. **FL:** Nov–Apr.

Pratia purpurascens
White Root
Glabrous ± prostrate herb with long white rhizomes. Leaves nearly sessile, elliptic, 10–25 x 5–10mm, toothed, usually purplish below. Flowers bisexual; corolla white to pale purple or bluish, 8–10mm long. **HABITAT:** Forest, woodland, in damp or sheltered sites. **DIST:** Coast to ranges. Also NC, SC, NT, NWS, Qld, Vic, LHI. **FL:** Nov–May.

Pratia purpurascens

Stylidium graminifolium

Stylidium laricifolium

Stylidium lineare

STYLIDIACEAE

Stylidium
TRIGGER PLANTS
Herbs or small shrubs. Leaves at base or alternate on stems. Flowers with sensitive column formed by fusion of 2 stamens with the style; corolla tubular with 5 lobes, the lower one much reduced and reflexed. Fruit a 2-valved capsule.

Stylidium graminifolium
Grass-leaf Trigger Plant
Perennial with one or more erect stems to 40cm tall. Leaves in single tuft at ground level, linear, to 20cm x 4mm. Flowers in racemes to 20cm long; corolla bright pink, 5–10mm long.
HABITAT: Heath, open forest, on sandy soils.
DIST: Coast and ranges. Also NC, SC, NT, ST, WS, Qld, Vic, Tas, SA. **FL:** Oct–Dec.

Stylidium productum

Stylidium laricifolium
Larch-leaf Trigger Plant
Undershrub 30–100cm tall. Leaves crowded on stems, linear, to 40 x 1mm, recurved. Flowers in panicles well above leaves; corolla pale pink, rarely white, 10–15mm long. **HABITAT:** Forests, often on hillsides, in sandy soils. **DIST:** Coast and ranges. Also NC, SC, NT, ST, WS, Qld, Vic. **FL:** Aug–Nov.

Stylidium lineare
Narrow-leaf Trigger Plant
Plant with slender red stem to 20cm tall. Leaves in small dense basal tuft, narrow-linear, to 55 x 1mm. Flowers on short stalks in short racemes, pale to deep pink, 5–8mm long. **HABITAT:** Heath, open forest, on sandy soils. **DIST:** Coast and ranges. Also SC, ST. **FL:** Sept–Apr.

Stylidium productum
Perennial with slender stem to 50cm tall. Leaves in tufts at intervals on lower part of stems, linear, to 20cm x 3mm, margins finely toothed. Flowers in racemes 5–15cm long; corolla deep pink, 4–8mm long. **HABITAT:** Open forest, esp. on sheltered slopes. **DIST:** Coast and ranges. Also NC. **FL:** Nov–Jan.

GOODENIACEAE

Coopernookia
Flowers in terminal leafy racemes. Corolla tube partially slit on upper side, with 3 spreading lobes between 2 erect lobes ± forming 2 lips; stamens free; indusium with bristles on lips. Fruit a capsule with 2–8 non-winged seeds.

Coopernookia barbata
Coopernookia
Erect sticky shrub to 60cm tall with glandular and stellate hairs. Leaves ± sessile, linear, to 25 x 2mm, revolute. Corolla pink-mauve to blue with conspicuous pale hairs in throat, to 15mm long; pedicels to 30mm long. **HABITAT:** Open forest. **DIST:** Coast (e.g. Putty, Thirlmere Lakes) and ranges (e.g. Yerranderie, Robertson, Glen Davis). Also SC, NT, ST, WS, Vic, Tas. **FL:** Chiefly Oct–Jan.

Coopernookia barbata

Dampiera

Flowers solitary or clustered or in terminal spikes; corolla tubular, partially split, with 3 spreading lobes between 2 erect lobes; anthers united around style; indusium glabrous around orifice. Fruit a nut.

Dampiera purpurea
Purple Dampiera
Soft multi-stemmed shrub to 1m tall with terete ribbed stems. Leaves elliptic, to 5 x 3cm, entire or toothed, with dense woolly brown hairs below. Buds and flowering branches with brownish hairs. Flowers in clusters of 3–5, purple with yellow centres, with dark hairs outside. **HABITAT:** Open forest, on sandy soils. **DIST:** Coast (sth of Broken Bay) and ranges. Also NC, SC, NT, ST, WS, Qld, Vic. **FL:** Oct–May.

Dampiera purpurea

Dampiera stricta
Blue Dampiera
Sparse herbaceous shrub to 50cm tall with angular glabrous stems. Leaves variable, narrow-elliptic, to 4cm long, often lobed or toothed. Flowers 1 or 2 on short stalks in axils, blue, rarely purple, with conspicuous rusty hairs outside. **HABITAT:** Open forest, heath. **DIST:** Common. Coast and ranges. Also NC, SC, NT, ST, Qld, Vic, Tas. **FL:** July–Dec.

Dampiera stricta

Goodenia bellidifolia subsp. *bellidifolia*

Goodenia
Herbs or shrubs. Inflorescences varied. Corolla tubular with partial slit, usually 2-lipped with 5 lobes, often with prominent anterior spur; stamens free; indusium usually with bristles on lips. Fruit a capsule with numerous flat winged seeds.

Goodenia bellidifolia subsp. bellidifolia
Herb with erect undivided stems to 50cm tall, often with long weak hairs. Leaves at or near base of stems, oblanceolate, 5–10cm, irregularly toothed, tapering to stalk. Flowers solitary or in small clusters, ± sessile, yellow, with appressed yellow hairs outside. **HABITAT:** Open forest, woodlands, swamps, in damp sandy soils. **DIST:** Coast and ranges. Also NC, SC, NT, ST, CWS, Vic. **FL:** Chiefly Nov–Feb.

Goodenia decurrens
Leafy undershrub to 60cm tall. Leaves sessile, decurrent, elliptic, 5–10 x 1–3cm, toothed. Flowers in leafless terminal branched inflorescences; corolla 15–18mm long, stellate-hairy outside. **HABITAT:** Sheltered forested slopes, often near cliffs. **DIST:** Blue Mtns (e.g. Mt Tomah, Blackheath, Upper Colo). Also CWS. **FL:** Oct–Feb.

Goodenia decurrens

Goodenia dimorpha var. *angustifolia*
Erect herb with branched ± leafless stems to 50cm tall. Leaves mostly basal, linear or obovate, 2–5cm x 2–10mm, entire or slightly toothed. Flowers in open panicles to 30–40cm long; corolla yellow, 12–15mm long, with stellate hairs on outside. **HABITAT:** Swampy heath. **DIST:** Endemic. Coast (Somersby to Waterfall). **FL:** Nov–Mar.

Var. *dimorpha* has spathulate leaves, 2–5cm long. It occurs mainly in Blue Mtns (also Cowan).

Goodenia dimorpha var. *angustifolia*

Goodenia glomerata
Stems to 50cm tall with long weak greyish hairs. Leaves mainly basal, smaller on lower stems, elliptic-oblanceolate, 8–16 x 1–2cm, tapering to indistinct stalk, toothed. Flowers sessile in spikes 12cm long; corolla 2–3cm long, with stellate hairs outside; ovary with long hairs. **HABITAT:** Damp soil pockets in flat rock outcrops. **DIST:** Bundanoon, Wingello SF. Mainly SC (e.g. Morton NP, Budawang NP). **FL:** Nov–Jan.

Goodenia hederacea subsp. *hederacea*
Ivy Goodenia
Prostrate herb with trailing hairy stems to 50cm or more long. Leaves mainly on stems, a few larger and basal, ± ovate, 1–10cm x 5–8mm, obscurely toothed or lobed, tapering to stalk. Flowers on slender stalks to 50mm long, in leafy racemes along most of stem; corolla yellow, 10–15mm long, with hairs inside and out. **HABITAT:** Open forest, woodland, grassland. **DIST:** Coast and ranges. Also ST, WS, Vic, Qld. **FL:** Sept–Apr.

Goodenia hederacea subsp. *hederacea*

Goodenia heterophylla subsp. *heterophylla*
Herb to 40cm tall with glandular or simple hairs. Basal leaves dying early. Stem leaves sessile, ovate, to 30 x 8mm, shallowly toothed or lobed, often with 2 larger lobes at base. Flowers in leafy thyrses or racemes on stalks to 25mm long; corolla yellow, 9–12mm long, ± glabrous inside. **HABITAT:** Open forest, scrubland. **DIST:** Coast, mainly nth of Sydney; Bulli Pass, Wentworth Falls. Also NC. **FL:** Chiefly Sept–Apr.

Goodenia heterophylla subsp. *heterophylla*

FLOWERING PLANTS

Goodenia glomerata

Goodenia ovata

Goodenia paniculata

Goodenia ovata
Hop Goodenia
Shrub to 1m tall. Leaves ovate, 2–8 x 1–4cm, toothed, light green, glabrous, sticky-shiny. Flowers in long leafy thyrses or racemes on stalks to 40mm long; corolla yellow, 10–20mm long, pubescent inside. **HABITAT:** Sheltered forest or rocky sites near sea, on clay-shale and volcanic soils. **DIST:** Common, sometimes forming large stands after fire. Coast and ranges. Most of NSW. Also Qld, Vic, Tas, SA. **FL:** Chiefly Oct–Dec.

Goodenia paniculata
Herb to 50cm tall with leaves in basal rosette, smaller on lower stems. Leaves obovate, to 2–10cm x 5–10mm, irregularly toothed, pubescent or glabrous. Flowers in loose racemes on leafless branches on stalks to 8cm long; corolla yellow, 10–15mm long, with glandular and simple hairs outside. **HABITAT:** Woodland, shrubland, in damp soils. **DIST:** Mainly coast (e.g. Wyong, Thirlmere Lakes, Albion Park); less common on ranges. Also NC, SC, NWS, Vic, Qld. **FL:** Chiefly Nov–Mar.

Goodenia rostrivalvis
Glossy glabrous shrub to 60cm tall. Leaves ± clustered on lower part of stem, obovate, 4–8 x 1–2cm, tapering to base, obscurely toothed. Flowers in loose thyrses or racemes on leafless scapes; corolla yellow, 14–18mm long, with yellow stellate and white cottony hairs outside. **HABITAT:** Damp benches along and below cliff-lines. **DIST:** Rare. Endemic. Central Blue Mtns (e.g. Lawson, Wentworth Falls). **FL:** Dec–Jan.

Goodenia stelligera
Glabrous herb with adventitious roots and erect stems to 40cm tall. Leaves in basal rosette, linear to oblanceolate, 5–25cm x 1–12mm, thick, glossy, entire or toothed. Flowers ± sessile, in spikes on unbranched stems; corolla yellow, 13–15mm long, with yellow stellate hairs and white cottony hairs outside. **HABITAT:** Heath, in deep moist sands. **DIST:** Coast (Broken Bay to Waterfall). Also NC, SC, ST, Qld. **FL:** July–Dec.

FLOWERING PLANTS | 439

Goodenia rostrivalvis

Goodenia stelligera

Scaevola aemula

Scaevola
FAN-FLOWER
Herbs or shrubs, often hairy. Corolla tube slit on upper side almost to base, opening out into 5 ± equal spreading lobes. Indusium visible above corolla, with bristles on lips.

Scaevola aemula
Fairy Fan-flower
Sprawling herb to 40cm tall, with yellowish hairs on stems. Leaves obovate, to 8 x 3cm, coarsely toothed, with scattered hairs. Flowers sessile and solitary in upper axils; corolla blue-mauve with yellow at base, 17–25mm long; indusium with long purple bristles on back and lips. **HABITAT:** Open forest. **DIST:** Sthn H'lands (e.g. Fitzroy Falls). Also NC, SC, ST, SA, Vic. **FL:** Jan–Mar.

Scaevola albida
Pale Fan-flower
Spreading herb with a few ascending stems to 40cm long. Leaves obovate, to 50 x 20mm, coarsely toothed. Flowers pale blue, sessile, in upper axils; indusium with short silvery hairs. **HABITAT:** Woodland, grassy headlands, esp. on clay and volcanic soils. **DIST:** Coast and ranges. Also NC, NT, ST, SA, Vic, Qld. **FL:** Nov–Feb.

Scaevola albida

Scaevola calendulacea
Dune Fan-flower
Mat-forming prostrate shrub. Leaves thick, semi-succulent, spathulate, to 50 x 25mm, entire or with a few teeth near apex. Flowers in upper axils; corolla light blue with yellow centre, 12–16mm long, with white hairs. Fruit a fleshy drupe to 10mm diam., creamy at first, turning purple. **HABITAT:** Coastal dunes and scrub. **DIST:** Coast (e.g. Avoca, Garie, Kurnell). Also NC, SC, Vic, Qld. **FL:** Sept–May.

Scaevola calendulacea

Scaevola hookeri
Creeping Fan-flower
Prostrate herb with hairy stems to 30cm rooting at nodes. Leaves sessile, ovate, to 50 x 15mm, entire. Flowers in leafy racemes, on stalks to 8mm long; corolla pink with yellow-brown centre, hairy outside. **HABITAT:** Wet rock ledges, swamp margins. **DIST:** Rare. Upper Blue Mtns (e.g. Govetts Leap, Newnes Plat.). Also NC, NT, ST, Vic, Tas. **FL:** Dec–Mar.

Scaevola hookeri

Scaevola ramosissima
Purple Fan-flower
Decumbent plant with wiry spreading branches to 40cm long with short stiff hairs. Leaves sessile, linear to lanceolate, to 2–6cm x 2–6mm, entire or toothed. Flowers in leafy thyrses or racemes, on stalks to 10cm long; corolla violet-purple, 15–25mm long, with short hairs outside and longer hairs inside. **HABITAT:** Heath, woodland, on sandy soils. **DIST:** Coast and ranges. Also NC, SC, NT, ST, Vic, Qld. **FL:** Chiefly Sept–Feb.

Scaevola ramosissima

Selliera radicans

Velleia lyrata

Selliera
Creeping herb of salt marsh edges, rooting at nodes. Flowers solitary or in condensed racemes in leaf axils at nodes; corolla with 5 equal lobes; stamens free; indusium with glabrous lips. Fruit an ovoid berry. Monotypic genus.

Selliera radicans
Perennial herb with prostrate glabrous stems to 50cm long. Leaves variable, narrow-obovate, 2–7cm x 2–15mm, shiny green, thick, entire. Flowers white with mauve and brown markings, 6–9mm long, on stalks to 20mm long. **HABITAT:** Sea-cliffs, rarely salt marshes. **DIST.** Coast sth of Lake Macquarie. Also SC, ST. **FL:** Oct–Mar.

Velleia
Herbs with basal leaves and stems forking 1 or more times, forks subtended (often encircled) by large bracteoles. Flowers in cymes in bracteole axils; corolla tubular, deeply slit, forming an erect 2-lobed upper lip and a spreading 3-lobed lower lip; stamens free; indusium with bristles on lips.

Velleia lyrata
Glabrous herb with ascending stems to 50cm tall. Leaves oblanceolate, 5–15 x 1–3cm, toothed. Bracteoles linear, 5–15mm long; sepals 3, ovate, 3–8mm long, leafy, green; corolla yellow, 10–15mm long. **HABITAT:** Heath, scrub, on moist sandy soils. **DIST:** Coast. Bulli Pass to Somersby (e.g. Royal NP, Wondabyne, Patonga). Also se. Qld. **FL:** Sept–Jan.

BRUNONIACEAE

Brunonia
Perennial herb with basal entire leaves and flowers in heads on upright stems. A monotypic genus, sometimes placed in Goodeniaceae.

The floral structure of *Brunonia* links the Goodeniaceae (style with an indusium) to the Asteraceae (tubular flowers in heads with involucral bracts).

Brunonia australis
Blue Pincushion
Stems hairy, usually leafless and unbranched, to 30cm tall. Leaves elliptic to spathulate, to 10cm x 15mm, with long silky hairs. Heads hemispherical, 13–25mm diam.; corolla bright blue, tubular, with 5 pointed lobes 3–4mm long; style protruding, with yellow indusium. **HABITAT:** Open forest, on sandy soils. **DIST:** Upper Blue Mtns (e.g. Bell, Newnes SF). Also ST, inland NSW. All States. **FL:** Nov–Jan.

Brunonia australis

ASTERACEAE

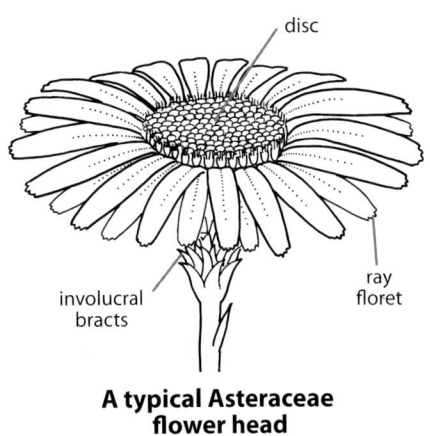

A typical Asteraceae flower head

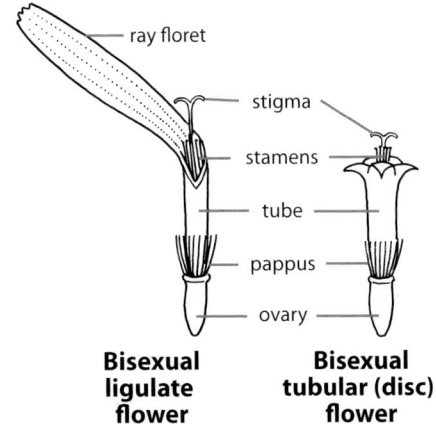

Bisexual ligulate flower

Bisexual tubular (disc) flower

Ammobium

Perennial herbs with woolly winged stems. Flowers tubular, surrounded by spreading papery bracts with green herbaceous claw. Fruit a 4-angled cypsela with a membranous crown.

Ammobium alatum

Herb 60–80cm high with branched silvery white 4-winged woolly stems. Leaves mostly basal, oblanceolate, 12–18cm long including stalk, silvery white beneath. Stem leaves fewer and smaller, decurrent. Heads in terminal corymbs, 2cm across; tubular florets yellow; bracts in several rows, white. **HABITAT:** Woodland, grassland. **DIST:** CT (e.g. Kanimbla V., Moss Vale). Also NT, ST, WS, Qld. **FL:** Dec–Feb.

Ammobium alatum

Brachycome graminea

Brachycome

Mostly small herbs with over 70 species Australia-wide. Leaves in basal rosette, sometimes also on stems. Flower heads solitary, often mauve or white. Ray flowers strap-like, female. Disc flowers tubular, bisexual. Cypselas usually with winged body, sometimes fringed with hairs; summit blunt; pappus of short hairs or absent.

Brachycome angustifolia

Brachycome angustifolia

Sprawling perennial to 35cm tall with stems spreading on ground and rooting at nodes. Leaves scattered on stems, variable, about 5cm long. Flower heads pink to mauve on leafless stalks; rays 6–10mm long. Cypselas winged, warty with minute pappus. Two varieties occur in the area. Var. *angustifolia* has narrow-lanceolate entire leaves. Var. *heterophylla* has broader deeply divided or toothed leaves about 9mm wide. **HABITAT:** Grassland, woodland, hind-dune forest. **DIST:** Var. *angustifolia*: coast and ranges to Vic, also Tas, SA. Var. *heterophylla*: coast to wstn slopes, also Vic. **FL:** Sept–Mar.

Brachycome graminea

Weak erect or sprawling plant 30–50cm tall; often spreading on ground and rooting at nodes. Leaves scattered on stems, entire, linear to oblanceolate, 6–12 x 7mm. Flower heads terminal with 20–30 mauve-pink or purple rays 9mm long surrounding the yellow disc. Cypselas smooth with minute pappus. **HABITAT:** Grassland in damp places. **DIST:** Blue Mtns and higher coastal areas (e.g. Thirlmere). Also SC, NT, ST, Qld, Vic, Tas, SA. **FL:** Oct–Feb.

Brachycome multifida
Cut-leaf Daisy
Sprawling or erect annual to 40cm tall with branched stems and finely divided leaves. Flower heads mauve, occasionally pink-white, with yellow centres; rays 7–10mm long. Cypselas rough, warty; wings absent; pappus minute, white. Two varieties occur in the area. Var. *multifida* has linear leaves 1mm wide, divided into segments 2cm long with pointed tips. Var. *dilatata* has leaves 2mm wide, more divided, with narrow-oblanceolate segments 2–2.5cm long on base and lower stems. **HABITAT:** Open forest and grassland. **DIST:** Var. *multifida*: coast to wstn plains, Vic, Qld. Var. *dilatata*: Coast, NT, Vic. **FL:** Sept–Feb, some flowers at other times.

Brachycome multifida

Brachycome ptychocarpa
Small annual daisy. Leaves in dense basal cluster, about 6cm long, finely divided with short narrow lobes. Flower heads 4–6mm across, solitary on leafless scape 10–12cm tall; rays pink to white. Cypselas brown, flattened with broad wings and short bristly pappus. **HABITAT:** Flat mossy rocks. **DIST:** Upper Blue Mtns (e.g. Boyd R. Crossing), t'lands, CWS, Vic. **FL:** Dec–Feb.

Brachycome ptychocarpa

Calomeria
Monotypic genus. Tall plants with large drooping panicles of tiny flower heads; involucres small; cypselas glabrous, without a pappus.

Calomeria amaranthoides
Plume Bush, Incense Plant
Uncommon but conspicuous strongly scented plant to >2m. Leaves alternate, stem-clasping, obovate, often crenate, glandular-hairy, 12–20 x 8–10cm, smaller on upper stems. Panicles with pendent tiny red-brown flower heads of 2–4 tubular florets enclosed in erect bracts to 4mm long. **HABITAT:** Sheltered forested slopes and rainforest margins, mainly on volcanic soils. **DIST:** Blue Mtns (Mt Hay, Mt Tomah), Sthn H'lands (Wingello). Also SC to Vic. **FL:** Dec–Apr.

Calomeria amaranthoides

Calotis
BURR-DAISY

Sprawling herbs or shrubs distinguished from *Brachycome* mainly by the cypselas, which are topped by rigid barbed awns, often also scales, and mature into spiny globular burrs.

Calotis cuneifolia
Bindi-eye, Blue Burr-daisy

Sprawling herb 25–60cm high with woody base and short stiff hairs. Leaves to 4cm long, variable in shape, base expanded around stem, narrowing above then expanding with 3–6 lobes near apex. Flower heads numerous, terminal on slender stems, 1–2cm across; ray flowers blue-mauve to white, numerous, disc yellow. Cypselas flattened, wedge- or V-shaped, topped by scales with smooth apex between 2 (rarely 4) spreading barbed spines. **HABITAT:** Grassland, open forest on heavier soils. **DIST:** Wstn Sydney, Blue Mtns (e.g. Nortons Basin, Lapstone, Hartley V.). Basically an inland species. Most of NSW. Also Qld, Vic, SA, NTerr. **FL:** Chiefly Sept–Dec.

Calotis dentex

Calotis dentex
White Burr-daisy

Shrub to 80cm tall with brown stems. Leaves stem-clasping, up to 8cm long, deeply and irregularly lobed; upper leaves often entire; basal leaves absent. Flower heads numerous on axillary peduncles up to 5cm long; ray flowers white, sometimes mauve, 10mm long; disc yellow 5–10mm diam. Cypselas reddish brown, flattened, wedge- or V-shaped, topped by scales with torn apex between 2 spreading spines 4–6mm long. **HABITAT:** Grassland, open forest, chiefly on clay soils. **DIST:** Coast and ranges. Also NC, NT, NWS, Qld. **FL:** Mainly Nov–Apr.

Calotis lappulacea
Yellow Burr-daisy

Much-branched perennial to 50cm tall. Basal leaves soon withering; stem leaves sessile, oblanceolate, up to 25 x 4mm, entire, toothed or deeply divided, becoming linear and bract-like on upper stems. Flower heads yellow, about 18mm across; ray flowers 3mm long, numerous. Cypselas flattened, V-shaped, warty, topped by 2 erect and several smaller spreading spines. **HABITAT:** Grassland, open forest, on heavier clay soils. **DIST:** Cumberland Pl., lower Blue Mtns (e.g. Narellan, Bringelly, Yerranderie). All States. **FL:** Most of year.

Calotis cuneifolia

FLOWERING PLANTS | 447

Calotis lappulacea

Cassinia, Ozothamnus

CASSINIA

Shrubs with alternate sessile entire leaves. Flower heads small, numerous, in large panicles or corymbs. Florets within heads with scales similar to involucral bracts surrounding each head (distinction from *Ozothamnus*). Florets all bisexual and tubular with 5 tiny lobes. Cypselas ovoid with pappus bristles fused into ring at their base.

Cassinia aureonitens

Cassinia aculeata
Dogwood, Dollybush

Shrub to 2m tall; branchlets with matted glandular hairs. Leaves 1–5cm x 0.5–2mm, dark green and sticky above, pale yellow-hairy beneath, margins slightly recurved. Inflorescence a dense corymb 10–12cm across. Flower heads 1–2mm diam. with 4–7 white to straw-coloured florets and several rows of white bracts alternately arranged. **HABITAT:** Open forest, woodland, heath on sandy or gravelly well-drained soils. **DIST:** Coast and ranges (e.g. Brisbane Water NP, Agnes Banks, Tallong, Clarence). Also NC, SC, ST, WS, SWP, Vic, Tas, SA. **FL:** Nov–Mar.

Cassinia aureonitens
Yellow Cassinia

Thin shrub 1–2m tall easily recognised by its bright yellow inflorescences. Stems and leaves minutely hairy. Leaves linear, 3–7cm x 2–4mm, flat with recurved margins, green above, paler and slightly yellowish beneath with midvein prominent. Inflorescence a dense flattish corymb up to 12cm across. Heads 1mm diam. with 5–6 yellow florets; involucral bracts yellow. **HABITAT:** Sheltered open forest and woodland. **DIST:** Coast and ranges. Also NC, SC. **FL:** Oct–Feb.

Cassinia cunninghamii

Erect shrub 1–2m tall with pale woolly-hairy stems. Leaves often crowded, linear, 20–35 x 1–1.5mm, margins rolled under to woolly undersurface. Inflorescence a dense, flat corymb about 12cm across. Heads 1mm diam. with 6 or 7 yellow florets; bracts in several rows ranked longitudinally (in lines), straw-coloured. **HABITAT:** Open forest. **DIST:** nw. Hornsby Plat. (e.g. Bucketty, Colo Heights, St Albans). Also NC, SC, CWS. **FL:** Aug–Dec.

Cassinia aculeata

Cassinia cunninghamii

Cassinia denticulata

Cassinia denticulata
Erect shrub to 2m tall. Leaves sessile, stem-clasping, broad-lanceolate, 10–25 x 3–6mm, entire or toothed, slightly sticky, smooth and shiny green above, pale or rusty tomentose beneath, often bent down near tip. Inflorescence a dense flattish corymb about 10cm across. Heads 2–4mm diam. with 10–14 pale yellow florets; bracts creamy white. **HABITAT:** Open forest on hillsides and gullies. **DIST:** Endemic. Coast and ranges. **FL:** Oct–Jan.

Cassinia longifolia
Shiny Cassinia, Cauliflower Bush
Spreading shrub to 3.5m with sticky aromatic foliage. Stems with short greyish glandular hairs. Leaves 3–8cm x 2–5mm, oblong-lanceolate with recurved margins, green and glabrous above, with matted whitish hairs beneath. Inflorescence 6–15cm across. Heads 1–1.5mm diam. with 5–8 creamy white florets; bracts white. **HABITAT:** Fertile soils in sheltered forest and rainforest margins, often in regrowth. **DIST:** Coast and ranges (e.g. Mooney Mooney Ck, Royal NP, Mt Tomah). Also SC, ST, Vic, Tas. **FL:** Nov–Apr.

Cassinia longifolia

Cassinia trinerva

Cassinia trinerva
Three-veined Cassinia
Shrub 2–3m tall or tree to 8m. Leaves narrow-lanceolate, 3–10cm x 3–8mm, flat, green and glabrous above, paler with minute glandular hairs and 3 prominent longitudinal veins beneath. Inflorescence flat-topped, 6–12cm across. Heads 1mm diam. with 3–4 creamy white florets; bracts white. **HABITAT:** Fertile soils in sheltered forests on slopes, often near rainforest. **DIST:** Coast, Sthn H'lands sth of Sydney (e.g. Macquarie Pass, Mt Kembla). Also NC, NT, SC, Vic, Tas. **FL:** Dec–Mar.

Resembles *C. longifolia* but easily distinguished from it by the 3-veined leaves.

Cassinia uncata
Sticky Cassinia, Bent Cassinia
Shrub to 3m tall; branchlets with short spreading glandular hairs. Leaves 15–30 x 0.5–1mm (to 50mm long in mountain forms) with recurved tip, appearing terete due to tightly revolute margins which obscure most of undersurface; upper surface smooth, rough or slightly sticky. Inflorescence a dense corymb 3–8cm across. Heads 1–2mm diam. with 5–6 creamy white florets; bracts in several rows, longitudinally ranked (in lines), white. **HABITAT:** Open forest, often in rocky areas. **DIST:** Coast and ranges (e.g. Agnes Banks, Clarence). Most of NSW, Vic, SA. **FL:** Dec–Mar.

Easily confused with *C. aculeata* but differs in leaves, inflorescence and arrangement of bracts.

Cassinia uncata

FLOWERING PLANTS

Ozothamnus adnatus

Ozothamnus argophyllus

Ozothamnus diosmifolius

OZOTHAMNUS
Shrubs with woody stems. Similar to *Cassinia* but florets within flower heads not accompanied by scales on receptacle. Involucral bracts around each head spreading, not erect or incurved as in *Cassinia*.

Ozothamnus adnatus
Shrub with white-woolly erect stems to 1m tall. Leaves linear, discolorous, 2–8 x 1mm, fused to stems in lower half, diverging in upper; margins revolute, obscuring outer surface. Inflorescence fairly small, open, flat. Heads 3–4mm diam. with 20–30 florets; bracts translucent, straw-coloured or opaque white. **HABITAT:** Open eucalypt forest on dry infertile slopes. **DIST:** Coast and Blue Mtns (e.g. Scotts Main Ra., Megalong V.). Also NT, ST, Vic. **FL:** Oct–Dec.

Ozothamnus argophyllus
Bushy aromatic shrub to 2m tall with white-hairy branches. Leaves lanceolate with intramarginal veins, 40–70 x 6–10mm, sticky, green above, grey-white or rusty beneath; stalk 3–4mm long. Inflorescence a dense domed corymb 3–4cm across. Heads 2–3mm diam. with 10–12 florets; outer bracts yellowish and ciliate, inner bracts white in upper part. **HABITAT:** Tall sheltered forest, rainforest margins. **DIST:** Chiefly coast (e.g. Illawarra escarpment, Robertson). Sth from Barrington Tops on coast and ranges to Vic, Tas, SA. **FL:** Oct–Nov.

Ozothamnus diosmifolius
Pill Flower, Ball Everlasting, White Dogwood
Common much-branched shrub to 2m or more. Leaves crowded, linear, 10–20 x 1–2mm, margins revolute obscuring undersurface, green and scabrous above, white-woolly beneath. Inflorescence a dense corymb to about 7cm across. Heads 3–6mm diam. with 20 or more florets; bracts broad, rounded, white or pink-tinged. **HABITAT:** Open forest. **DIST:** Coast to wstn plains. Also Qld. **FL:** Oct–Jan.

Ozothamnus ferrugineus

Celmisia sp. aff. *longifolia*

Ozothamnus ferrugineus
Tree Everlasting, Dogwood
Shrub 2–3m tall with grey-brown fibrous bark and open habit. Leaves decurrent on stems, linear-lanceolate, 30–50 x 2–3mm, margins slightly wavy and recurved, green above, white-hairy beneath. Inflorescence a dense domed corymb. Heads 2–4mm diam. with 4–7 florets; bracts translucent in lower part, white and rounded in upper part. **HABITAT:** Open forest, scrub on moist rich soils. **DIST:** Scattered, uncommon. Mainly upper Blue Mtns (e.g. Mt Tomah), coast and t'lands in NSW. Also Vic, Tas, SA. **FL:** Dec–Mar.

Celmisia
SNOW DAISY
Daisy, conspicuous in alpine areas, with silver-grey foliage and large flower heads in summer months. Ray florets white, strap-like, female; disc florets yellow, tubular, bisexual. Cypselas flattened-terete with pappus of long minutely barbed bristles.

Celmisia sp. aff. *longifolia*
Snow Daisy
Low perennial herb. Leaves basal, tufted with sheathing base, silvery white-tomentose beneath, to 25cm long. Flower heads on erect stalks longer than leaves; rays numerous, 1–2cm long in single row; involucre woolly, 2–3cm diam., with several rows of overlapping bracts. **HABITAT:** Grassland and open woodland, in wet soils or swampy margins. **DIST:** Blue Mtns (Wentworth Falls to Blackheath and Newnes Plat.) **FL:** Nov–Feb.

The local population(s) is (are) currently regarded as being part of a *C. longifolia* complex having a disjunct distribution from Newnes to the Kosciuszko region.

Chrysocephalum, *Helichrysum*, *Xerochrysum*
These 3 genera are everlastings or paper daisies formerly included in *Helichrysum*. In *Chrysocephalum* the involucral bracts are thin and translucent, occasionally opaque, often straw-coloured, and the margins are fringed or woolly. In *Helichrysum* they are papery and white or brightly coloured, and some or all have an elongated narrow flat base called the claw. In *Xerochrysum* they are leathery, opaque, usually yellow or golden, and have a short broad base. Receptacle scales absent in all 3.

CHRYSOCEPHALUM

Chrysocephalum apiculatum
Yellow Buttons, Common Everlasting
Variable perennial herb to 50cm high or forming a loose mat; stems and leaves usually with dense silvery grey indumentum. Leaves variable in shape, 20–50 x 10–25mm. Heads in terminal clusters, 5–10 x 7–15mm with bright yellow tubular flowers surrounded by scaly involucre of erect, radiating, yellow opaque bracts. **HABITAT:** Grassland, woodland, eucalypt forest, usually on Wianamatta shale clay soils. **DIST:** Throughout Aust. **FL:** Oct–Dec and at other times.

Some forms are similar to *C. semipapposum*.

FLOWERING PLANTS 453

Chrysocephalum apiculatum

Chrysocephalum semipapposum

Helichrysum calvertianum

Chrysocephalum semipapposum
Clustered Everlasting, Yellow Buttons
Variable erect perennial herb to 50cm high, woody near base. Stems and leaves of young plants with loose white hairs, glabrous on older plants, sometimes sticky. Leaves linear, 1–5cm x 1–2mm, shorter on upper stems; margins recurved. Heads in terminal clusters, 4–7 x 6–7mm; involucral bracts in 8–10 rows, outer bracts woolly. **HABITAT:** Grassland, woodland on clay soils and rocky hillsides, not on sandstone. **DIST:** Uncommon in Sydney district. Occurs throughout Aust. **FL:** Mainly Nov–Jan.

HELICHRYSUM

Helichrysum calvertianum
Much-branched subshrub to 20cm high with wiry stems. Leaves crowded, narrow, blunt, 6–9 x 1mm, glabrous, sticky; margins revolute. Heads solitary, 10–12mm across, with yellow centres and several rows of white or pink-tinged papery bracts; outer bracts scaly, brownish, slightly hairy near base. **HABITAT:** Sandstone escarpments in moist sandy ground. **DIST:** Endemic. Sthn H'lands (e.g. Fitzroy Falls, Carrington Falls, Berrima, near Mittagong). **FL:** Jan–Aug.

Helichrysum collinum
Herb to 70cm tall with grey-white woolly branches. Leaves oblong, 4–10cm x 2–10mm, green above, grey-woolly below. Flower heads terminal on ± leafless simple or branched peduncles to 10cm long, 20–25mm diam.; involucral bracts reflexed, scaly, acuminate, golden. Achenes with barbed bristles 5mm long. **HABITAT:** Woodland, often on heavier soils. **DIST:** Coast and ranges. Also WS, NWP, Qld. **FL:** Oct–Dec.

Helichrysum elatum
White Paper Daisy, Tall Everlasting
Erect shrub to 2m tall with white-woolly hairs on stems and branches. Leaves lanceolate, to 12 x 4cm, green above, greyish white tomentose or

Helichrysum collinum

Helichrysum elatum

floccose beneath. Heads terminal on stems exceeding foliage, 3cm across, with yellow centres surrounded by numerous white papery bracts, the innermost longest. **HABITAT:** Sheltered forests and slopes. **DIST:** Coast and ranges (e.g. Hacking R., Brisbane Water NP, Macquarie Pass, Thirlmere Lakes). Also NC, SC, NT, ST, Qld, Vic. **FL:** June–Nov.

Helichrysum rutidolepis
Pale Everlasting
Decumbent to erect perennial with hairy white stems, 20–40cm high. Leaves linear to lanceolate, 40–60 x 4–6mm long, stem-clasping, slightly recurved. Flower heads solitary on stems with reduced leaves, yellow, about 10mm across; involucral bracts scaly, spreading, brown to yellow, wrinkled, woolly. **HABITAT:** Grassland, sheltered forest, rainforest margins, esp. on rocky hillsides. **DIST:** Coast and t'lands, extending to SWP. Also Vic, SA. **FL:** Mainly Jan–May.

Helichrysum rutidolepis

Helichrysum scorpioides

Helichrysum scorpioides
Button Everlasting, Curled Everlasting
Sprawling perennial to 30cm high with long leafy woolly stems from leafy base. Leaves linear, stem-clasping, about 50 x 6mm, recurved, usually paler and woolly beneath, smaller and bract-like on upper stems. Flower heads solitary, terminal, 1.5–3cm across, bright yellow with an involucre of small yellow or brownish spreading bracts. **HABITAT:** Grassy forest, woodland, on clay soils, roadsides. **DIST:** Coast to w. slopes. Also Vic, Tas, SA. **FL:** Sept–Mar.

XEROCHRYSUM

Xerochrysum bracteatum
Yellow Paper Daisy, Golden Everlasting
Sparsely branched plant 20–70cm tall. Leaves lanceolate to oblanceolate, slightly stem-clasping, 2–8cm x 5–20mm, dark green, scabrous, nearly glabrous. Heads on long peduncles, 3cm across with bright yellow shiny stiff bracts surrounding yellow or orange tubular florets; outer bracts smaller and paler. **HABITAT:** Sheltered forests on fertile soils, often on roadsides. **DIST:** Throughout NSW. All mainland States. **FL:** Any time, peak in summer.

Xerochrysum viscosum
Sticky Everlasting
Multi-stemmed, sticky, erect or sprawling daisy 40–80cm tall with scabrous stems and varnished appearance. Leaves narrow-lanceolate, 3–9cm x 4–7mm. Heads 2–3cm across, bright yellow, with several rows of stiff shining yellow or slightly brownish bracts. **HABITAT:** Woodland on shallow stony soils on

Xerochrysum bracteatum

Xerochrysum viscosum

exposed sites. **DIST:** Chiefly wstn Blue Mtns (e.g. Hartley V.), also Buxton, Hill Top, Mt Gibraltar. Most of NSW. Also Qld, Vic. **FL:** Sept–Dec.

X. viscosum usually dies off in summer and re-shoots in autumn.

FLOWERING PLANTS

Cymbonotus lawsonianus

Cymbonotus
Herbs with leaves in a rosette. Flower heads yellow, solitary on short stems; ray florets female, in 1 row; disc florets bisexual. Achenes asymmetrical, obovoid; pappus absent.

Cymbonotus lawsonianus
Bears' Ears
Leaves ovate with long tapering base, 4–10 x 2–4cm, toothed to pinnatifid, white and woolly below. Heads 8–15mm diam., peduncles and bracts woolly. Achene 2.5mm long, pubescent, convex outside, concave on inner face. **HABITAT:** Dry forest, woodland. **DIST:** Most of NSW. Also Vic, Qld. **FL:** Nov–June.

Lagenifera
Small perennials with solitary flower heads, unbranched stems and a basal rosette of leaves. Ray flowers strap-like, female. Cypselas with neck or beak, not winged; pappus absent.

Lagenifera stipitata
Stems 30–40cm tall. Leaves obovate-spathulate, toothed, to 15 x 2cm. Ray florets 3–4mm long, mauve, pink or white; disc yellow; involucre <10mm across. **HABITAT:** Open forest, grassland. **DIST:** Coast and ranges from Vic to Qld. Also Tas, SA. **FL:** Sept–Apr.

Leptinella
Prostrate herbs. Leaves alternate on stems. Flower heads terminal, solitary on peduncles exceeding leaves. Disc florets sterile; style with terminal disc. Achenes without pappus.

Lagenifera stipitata

Leptinella longipes

Leptinella longipes
Leaves on stalk to 10cm long; lamina oblong-ovate, 3–8 x 1–2cm, deeply lobed. Heads 5mm diam. on stalks as long as leaves. Achenes sessile, ± compressed, to 2.5mm long, with thickened margins. **HABITAT:** Swamp margins. **DIST:** Coast. Also CT, Qld, Vic, Tas. **FL:** Mar–Apr.

Leucochrysum albicans subsp. *albicans*

Leucochrysum (formerly *Helipterum*)
SUNRAY
Herbs with alternate, flat, entire leaves, reduced and bract-like on upper stems. Flower heads solitary, terminal, with bisexual tubular florets and spreading papery or scaly involucral bracts having slender stipe at the base. Cypselas ellipsoid, glabrous, warty with pappus bristles feathery for entire length and breaking evenly above the base.

Leucochrysum albicans subsp. *albicans*
Hoary Sunray
Variable perennial 15–35cm tall with erect stems from a branched woody base. Young stems woolly; older stems often with persistent brown leaf bases. Leaves linear to narrow-oblong, 30–100 x 1–10mm. Heads 2–3cm across with yellow tubular florets and spreading bracts. Two varieties occur in the area. In var. *albicans* the inner (i.e. upper) bracts are yellow. In var. *tricolor* the inner bracts are white. **HABITAT:** Forest, woodland, grassy open country. **DIST:** Var. *albicans*: Upper Blue Mtns (e.g. Hartley, Lithgow, Jenolan Caves); Qld to Vic. Var. *tricolor*: Ranges south from Mudgee (e.g. Berrima, Bundanoon, Wingello). Also Vic, Tas. **FL:** Sept–Dec.

Leucochrysum graminifolium

Leucochrysum graminifolium
Small tufted perennial herb to 20cm high. Leaves grass-like to filiform, closely revolute, glabrous above, woolly beneath. Heads solitary on slender stalks 8cm long, 2–3cm across with yellow tubular florets and yellow spreading bracts. **HABITAT:** Exposed sites, often on or near pagoda rock formations. **DIST:** Lithgow to Newnes Plat. and Wolgan V. **FL:** Oct–Mar.

Olearia
DAISY-BUSH
Shrubs, sometimes tall, with large attractive flower heads. Ray florets in single row, female, usually white. Disc florets tubular, bisexual, yellow. Cypselas terete with pappus of long finely barbed bristles.

Olearia argophylla
Musk Daisy-bush, Native Musk
Shrub or tree to 9m tall. Leaves alternate, lanceolate, 8–14 x 3–4cm, green above, silvery below, toothed or entire, with musky smell. Heads small on long branched peduncles in

Olearia argophylla

conspicuous clusters, creamy white with 3–5 ray florets. **HABITAT:** Uncommon. Sheltered forested gullies, slopes. **DIST:** Blue Mtns, Sthn H'lands, Illawarra escarpment. Common in Vic, Tas. **FL:** Mainly Oct–Jan.

Olearia asterotricha

Olearia asterotricha
Rough Daisy Bush

Shrub to 1m. Leaves alternate, entire or lobed, 20 x 2–4mm, revolute, greyish, densely stellate-hairy. Heads terminal on leafy axillary stalks, mauve to white with yellow disc. **HABITAT:** Sheltered forest, especially rocky areas. **DIST:** Upper Blue Mtns, Sthn H'lands (e.g. Megalong V., Wallerawang, West Berrima). Also Vic. **FL:** Oct–July.

Olearia elliptica
Sticky Daisy-bush

Shrub to 2m tall. Leaves alternate, elliptic, 3–8 x 2–3cm on stalk 1cm long, glabrous, thick, shiny, usually viscid, pale green beneath. Flower heads numerous in terminal clusters exceeding leaves, 10–25mm across; ray florets white, 6–14 per head; involucre 6–8mm diam. **HABITAT:** Forest on protected slopes and in wet situations. **DIST:** Coast and ranges (e.g. Mt Keira, Fitzroy Falls, Wentworth Falls). Also NC, NT, WS, WP, Qld, LHI. **FL:** Oct–Feb.

Olearia erubescens
Silky Daisy-bush

Shrub to 1.5m. Leaves oblong to lanceolate with irregular pointed teeth or lobes, 30–80 x 7–15mm, green and glabrous above, pale with short simple hairs beneath, new growth often red. Heads 3–5 on axillary peduncles about 40mm long; rays few, white to pink-mauve; involucre 6–10mm diam. **HABITAT:** Forests. **DIST:** Coast and t'lands. Also Vic, Tas, SA. **FL:** Sept–Jan.

Olearia microphylla
Small-leaved Daisy-Bush, Snow Bush

Common shrub 1–2m tall. Leaves spathulate, 2–5mm long; margins revolute, grey-woolly beneath. Flower heads abundant but solitary, terminal on side-branches; rays 7–9, 3–5mm long, white; disc yellow, small, few-flowered. **HABITAT:** Forest, on shaley and sandy soils. **DIST:** Coast and ranges. Also Qld. **FL:** July–Nov.

Olearia myrsinoides
Blush Daisy-bush, Silky Daisy-bush

Spreading shrub to 1.5m with silky young branches. Leaves alternate, elliptic, about 15–20 x 4–8mm, entire or minutely toothed, green above, pale with simple hairs beneath. Heads 2–5 on short axillary peduncles; rays few, white; involucre 2–3mm diam. **HABITAT:** Open forest, woodland. **DIST:** Coast and t'lands, mainly on ranges from Qld border to Vic. Also Tas. **FL:** Nov–Mar.

Closely related to *O. erubescens*.

FLOWERING PLANTS

Olearia elliptica

Olearia erubescens

Olearia microphylla

Olearia myrsinoides

Olearia phlogopappa
Alpine Daisy-bush
Variable spreading shrub to 1.5m. Leaves mostly alternate, 30–50 x 5–9mm, entire or toothed, discolorous, lower surface grey stellate-tomentose. Heads about 20mm across, solitary on slender hairy or glabrous peduncles in axils or in lateral clusters; ray florets 14–30, white; disc florets 15–20, yellow. **HABITAT:** Eucalypt forest, woodland. **DIST:** Blue Mtns to Vic. Also Tas. **FL:** Oct–Mar.

Common in sub-alpine forests and Vic. but uncommon in Sydney district. Easily confused with *O. stellulata* which has flower heads in leafy panicles, 11–16 ray florets and 9–14 cream disc florets. *O. erubescens* is also similar but has simple hairs on undersurface of leaves and larger involucres 6–10mm diam.

Olearia quercifolia
Oak-leaved Daisy-bush
Shrub 0.5–1.5m tall. Leaves alternate, obovate with lobed margins, 35mm x 15–20mm, dark green and rough above, stellate-hairy beneath. Heads on long thick woolly peduncles in upper axils; ray florets 10–12, pale mauve to white. **HABITAT:** Drainage lines on hillsides, near swamps. **DIST:** Endemic. Blue Mtns (Wentworth Falls, Newnes Plat.) **FL:** Aug–Dec.

Considered at risk because of its limited distribution.

Olearia quercifolia

Olearia tomentosa
Shrub to 2m tall. Leaves ovate with sinuate lobes or teeth, 20–80 x 10–45mm, rough above, rusty-woolly beneath, deeply veined. Heads 2–3cm across on peduncles 4cm long; ray florets 6–8mm long, white, pink or bluish; involucre 6–10mm diam. **HABITAT:** Forests on sheltered slopes, also scrub or heath on estuarine headlands. **DIST:** Coast and lower Blue Mtns. Also NC, SC, ST. **FL:** Peak Sept–Dec.

Olearia viscidula
Shrub 1–2m tall, ± sticky-shiny. Leaves narrow-elliptic or ovate, 6–10cm x 4–7mm, pointed, with entire margins, green above, greyish white-felted beneath. Flower heads in axillary panicles, 8–15mm across; ray florets 8–10, white, 3mm long; involucre 5–6mm diam. **HABITAT:** Sheltered forest, rainforest margins. **DIST:** Scattered. Coast and Blue Mtns (e.g. Loftus, Thirlmere, Nattai, West Berrima). Coast to w. slopes; also Vic. **FL:** Aug–Nov.

Olearia phlogopappa

Olearia tomentosa

Olearia viscidula

Podolepis hieracioides

Podolepis jaceoides

Podolepis
Erect herbs with entire basal and stem leaves. Flower heads terminal on long wiry branched brownish stems. Florets of 2 types: inner tubular disc florets with 5 tiny lobes; and outer ligulate florets with 2–4 short or long narrow lobes. Involucral bracts in several rows, short, scaly, green or brownish.

Podolepis hieracioides
Erect perennial herb to 70cm tall. Basal leaves elliptic-oblanceolate to 16 x 2.5cm. Stem leaves linear, smaller, bract-like on upper stems. Heads in branched inflorescences of 2–6 or more; ligulate florets 15–20, 15–18mm long; involucre 15–20mm diam. with shiny green bracts. **HABITAT:** Forest and woodland with grassy understorey on clay soils. **DIST:** Upper Blue Mtns and Sthn H'lands (e.g. Hartley V., Bowral). Also NT, ST, Vic. **FL:** Nov–Dec.

Podolepis jaceoides
Showy Podolepis, Showy Copper-wire Daisy
Branched perennial with erect unbranched stems 30–80cm tall. Basal leaves lanceolate, green, 10–15 x 1–2.5cm. Stem leaves smaller, stem-clasping, recurved, rough above, hairy beneath, reduced to bracts on upper stems/peduncles. Flower heads usually solitary with 30–40 spreading yellow ray or ligulate florets to 25mm long; involucre 25–30mm diam. **HABITAT:** Grassland, woodland, usually on heavier clay soils. **DIST:** Coast and ranges. All of NSW. Also Qld, Vic, Tas, SA. **FL:** Sept–Dec.

Senecio
FIREWEED, GROUNDSEL

Cosmopolitan genus of about 2000 species, with more than 50 in Australia. Shrubs or herbs with alternate leaves and yellow flower heads. Some species with tubular disc florets and petal-like ray florets, others with tubular or tubular and filiform disc florets but no ray florets. Achenes terete with pappus becoming white, fluffy and conspicuous at fruiting.

Senecio lautus subsp. *dissectifolius*
Variable Groundsel

Soft shrub to 70cm tall. Leaves to 8cm long, deeply dissected into short linear or terete lobes with entire or toothed margins. Flower heads in loose terminal corymb, yellow, with 10–15 ray florets 5–10mm long and bell-shaped involucre 5–6mm diam. **HABITAT:** Open forest and woodland on rocky sites. **DIST:** Chiefly upper Blue Mtns and Sthn H'lands (e.g. Mt Gibraltar, Jenolan Caves road). Also NC, NT, WS, WP, all States. **FL:** May–Oct.

Senecio lautus subsp. *maritimus*

Senecio lautus subsp. *maritimus*
Coast Groundsel

Sprawling or creeping perennial. Leaves sessile, fleshy, narrow-obovate, 2–6cm long, toothed or entire. Flower heads yellow, with 12–20 ray florets 10–20mm long; involucre bell shaped, 5–8mm diam. **HABITAT:** Sand dunes and rocky sandy sites near the sea. **DIST:** Coast (e.g. Marley Bch, Kurnell), NC, SC, all States. **FL:** Some flowers most of year.

Senecio lautus subsp. *dissectifolius*

Senecio linearifolius

Senecio vagus subsp. *eglandulosus*

Senecio linearifolius
Fireweed Groundsel
Common shrub to 1.5m tall. Leaves variable, linear to lanceolate, 6–12cm x 8–20mm, with toothed, wavy or entire margins, stem-clasping or lobed at base. Flower heads numerous in compact terminal corymbs; ray florets 5–8mm long; involucre cylindrical, 3mm diam. **HABITAT:** Sheltered forest and rainforest margins, often in disturbed ground. **DIST:** Coast to WS, also Vic, Tas. **FL:** Some flowers all year.

Senecio vagus subsp. eglandulosus
Glabrous shrub 1m or more tall with distinctive stalked leaves 8–15 x 4–7cm, ovate in outline and deeply dissected by irregular pointed lobes or teeth. Flower heads in loose terminal clusters; ray florets 6–9, about 10mm long; involucre bell-shaped, about 7mm diam. **HABITAT:** Sheltered forests, gullies, rainforest margins. **DIST:** Coast and ranges. Also Qld. **FL:** Chiefly Oct–Jan.

Senecio velleioides
Erect herb or shrub to 1m tall. Leaves to 12 x 4cm, entire or irregularly toothed, stem-clasping with extended round or pointed

Senecio velleioides

ear-like bases. Flower heads in loose terminal clusters, 2–3 cm across; ray florets 8–10, 1cm long. **HABITAT:** Sheltered forested slopes, often near watercourses. **DIST:** Blue Mtns, Sthn H'lands and higher parts of coast (e.g. Macquarie Pass, Fitzroy Falls, Bundanoon). Also NC, SC, ST, CWS, Vic, Tas. **FL:** Oct–May.

Monocotyledons

See page 48 for main characteristics of monocotyledon plants.

ALISMATACEAE

Alisma
Aquatic perennial with erect stems and leaves. Inflorescence paniculate. Flowers bisexual, with 3 sepals, 3 petals, 6 stamens and superior ovary with numerous carpels. Fruit a dense head of nutlets.

Alisma plantago-aquatica

Ottelia ovalifolia

Alisma plantago-aquatica
Water Plantain
Leaf blade ovate, round at base, to 25 x 10cm, with 7 conspicuous longitudinal veins; stalk to 80cm long. Inflorescence whorled, to 60 x 40cm. Flowers 10mm diam., pale pink to white. **HABITAT:** Pools, swamps, streams. **DIST:** Cumberland Pl. (e.g. Longneck Lagoon), Illawarra (e.g. Gerringong), Blue Mtns (e.g. Glenbrook Ck). Also NC, NT, ST, SWS, SWP, Vic, Eur. **FL:** Dec–Feb.

Damasonium
Aquatic perennial with erect stems and leaves. Inflorescence paniculate. Flowers as in *Alisma*. Follicles forming a star-shaped head.

Damasonium minus
Star-fruit
Glabrous plant to 80cm tall. Leaf blades lanceolate to ovate, to 10 x 4cm, with 3–5 prominent veins; stalk to 30cm long. Inflorescence 30–50cm long. Flowers, white, 6mm diam. Aggregate fruit 10–12mm diam., with 6–9 triangular follicles joined at their bases. **HABITAT:** Shallow water, muddy margins of swamps. **DIST:** Cumberland Pl. (e.g. Castlereagh NR, Shanes Park) and Nattai V. Also NC, NT, WS, NWP, all other States. **FL:** Nov–Dec.

HYDROCHARITACEAE

Ottelia
Aquatic herbs with floating and submerged leaves. Some flowers above water, others submerged and self-pollinating. Fruit with numerous seeds.

Ottelia ovalifolia
Swamp Lily
Tufted perennial, rooted in mud. Leaf blades elliptic, to 15cm long; stalk to 1m long. Emergent flowers solitary, white with reddish base, to 4cm across; stamens 9–15, yellow. **HABITAT:** Lagoons, dams and shallow slow water. **DIST:** Coast and lower Blue Mtns, Berrima. Most of NSW. All mainland States. **FL:** Dec–Feb.

Damasonium minus

Vallisneria

Submerged freshwater dioecious herbs with ribbon-like leaves. Male flowers numerous in narrow membranous spathe; female flowers solitary in spathe.

Vallisneria gigantea
Eel-weed, Ribbon-weed
Leaves >2m x 2cm. Male flowers at base of plant, <1mm long, released from spathe to float to surface; female flowers 15–25mm long, floating, with slender coiled stalk which contracts after fertilisation dragging embryonic fruit to the mud. **HABITAT:** Larger streams, swamps and lakes. **DIST:** Uncommon. Coast (e.g. Hacking River, McCarrs Ck). Most of NSW. All mainland States. **FL:** Feb–Apr.

Vallisneria gigantea

JUNCAGINACEAE

Triglochin
Aquatic plants of freshwater streams or salt marsh. Leaves basal, linear, submerged, floating or semi-erect. Flowers in spikes; perianth segments 6, in 2 whorls.

Triglochin procerum
Water Ribbons
Aquatic plant with tubers and rhizomes. Leaves to 1m x 3cm, strap-like, the upper part usually floating (often erect in still water). Spike dense, to 30cm long. **HABITAT:** Freshwater swamps to 1m deep. **DIST:** Coast and ranges (e.g. Kurnell, Doonside, Megalong Valley). Also NC, SC, SWS, Qld, Vic, SA. **FL:** Sept–Dec.

Triglochin rheophilum (syn. *T. rheophila*)
Aquatic perennial with tuberous rhizomes. Leaves usually submerged, 40–400cm x 2–6mm, reddish, thin, often with undulate margins; leaf sheaths narrow, inrolled. Inflorescence with green rachis and pedicels. Fruits submerged, 10–16mm long, with usually 6 keeled carpels. **HABITAT:** Flowing fresh streams and rivers. **DIST:** Coast (e.g. Hacking R., O'Hares Ck). Also NC, SC, Qld, Vic, Tas. **FL:** Sept–Jan.

Triglochin microtuberosum has small (5–13mm) globose tubers clustered closely beneath the rhizome (e.g. Richmond, Dapto, Albion Park).

Triglochin striatum
Streaked Arrowgrass
Erect, slender grass-like plant. Leaves tufted, narrow-linear, to 30cm x 3mm. Flowers in narrow racemes 2–15cm long; perianth 2mm long, greenish. Compound fruit globose, with 3 fertile and 3 sterile carpels alternating, the fertile carpels eventually falling, the others remaining. **HABITAT:** Salt marsh, moist coastal cliffs. **DIST:** Coast. Also NC, SC, all other States, Pac. Iss, S. Afr., Amer. **FL:** Nov–Mar.

POTAMOGETONACEAE

Potamogeton
Freshwater plants with leaves of 2 types. Flowers bisexual in spikes above water, submerging after fertilisation.

Potamogeton tricarinatus
Floating Pondweed
Aquatic perennial with stems rising from rhizome. Lamina of floating leaves ovate, to 10 x 7cm, with 15–25 parallel veins, on long stalks. Submerged leaves narrow-elliptic, to 20 x 1cm, ± translucent, often undulate, ± sessile. Spikes dense, 2–4cm long on stout erect stems. **HABITAT:** Rivers, creeks, lagoons, dams. **DIST:** Coast and ranges. Most of NSW. All other States. **FL:** Oct–Mar.

Triglochin procerum

Triglochin striatum

Triglochin rheophilum

Potamogeton tricarinatus

Archontophoenix cunninghamiana

ARECACEAE

Archontophoenix
Monoecious trees with unbranched fibrous-woody trunks often ringed by old leaf scars. Leaves pinnate. Flowers in pendulous panicle or clustered spikes below crown; perianth 3-merous. Fruit globose, fleshy, with 1 seed.

Archontophoenix cunninghamiana
Bangalow Palm
Tree to 25m tall. Leaves 3–4m long with 80–100 leaflet pairs. Flowers pale lilac, in pendulous panicles. Fruit red, 12–15mm diam. **HABITAT:** Gully rainforest. **DIST:** Coast (e.g. Gosford–Watagan Mtns, Hacking R., Illawarra). Also SC, NC, Qld. **FL:** Dec–Apr.

Livistona
Leaves palmately divided (fan-shaped). Flowers bisexual in much-branched pendulous panicles among the leaves. Fruit globose, fleshy or dry, with 1 seed.

Livistona australis

Livistona australis
Cabbage Tree Palm
Tree to 30m tall. Leaves (fronds) 3–4m long; lamina 1–2m diam.; stalk 1.5–2m long with thorns or spines on margins. Flowers pale cream to yellowish, solitary or clustered. Fruit 15mm diam., black, hard. **HABITAT:** Rainforest margins, sheltered tall forests. **DIST:** Coast. Also SC, NC, Vic, Qld. **FL:** Sept–Jan.

ARACEAE

Alocasia
Herbs with leaves and 'flowers' resembling Arum Lily (*Zantedeschia*), the 'flower' consisting of a fleshy spike of tiny male and female flowers lacking petals and sepals (called a spadix), subtended by a large erect sheathing bract (called a spathe).

Alocasia brisbanensis
Cunjevoi, Spoon Lily
Tufted perennial with tuberous rhizome. Leaves 80–160cm long with thick spongy stalk to 90cm long; blade to 70 x 60cm, saggitate to ovate, light green with strong lateral veins. Spathe greenish yellow, 15–20cm long on peduncle about as long as leaf stalks. Spadix yellow, slightly shorter than spathe. Fruit an ovoid red berry, to 15mm long. **HABITAT:** Rainforest gullies. **DIST:** Coast (e.g. Macquarie Pass). Also NC, Qld, WA. **FL:** Dec–Mar.

Gymnostachys
Slender herb with creeping rhizome. Leaves distichous, linear. Flowers in flexible spadices clustered on upper stems and lacking subtending spathes, bisexual, 2-merous; perianth present. Fruit a 1-seeded berry. Monotypic genus.

Fibres were used by Aborigines to make fishing lines and by colonists to make twine.

Gymnostachys anceps
Settlers Flax
Tough perennial with leaves 1–1.5m x 10–15mm. Stems exceeding leaves, flattened, with rough margins; spadices to 12cm x 7mm, green turning blackish with flowers in various stages of development. Berry ovoid, blue-black, 6–12mm long. **HABITAT:** Rainforest, sheltered gullies. **DIST:** Coast and ranges. Also NC, SC, NT, Qld. **FL:** Oct–Jan.

Typhonium
Deciduous perennials, with underground stems. Inflorescence arum-like, purplish and foul-smelling.

Typhonium eliosurum
Leaves clustered, to 50cm tall; lamina 15–30cm across and long, hastate with 3 spreading narrow-triangular lobes; stalk to 30cm long. Inflorescence on peduncle shorter than leaves. Flower spike purple-black with slender sterile terminal appendage 5–15cm long. Sheathing bract greenish purple, acuminate, about 12cm long. **HABITAT:** Sheltered sites, creek banks. **DIST:** Endemic. Rare. Coast. Wyong to Nowra (e.g. Werrong Bch, Patonga). **FL:** Oct–Jan.

Typhonium brownii. Endemic. Similar to *T. eliosurum* but with conical appendage and truncate base. Occurs at Wheeny Ck, The Crest (Bass Hill), Culoul Ra.

Alocasia brisbanensis

Gymnostachys anceps

Typhonium eliosurum

COMMELINACEAE

Aneilema
Weak herbs with spirally arranged leaves sheathing the stems. Flowers bisexual in terminal or axillary reduced cymes partly enclosed by spathe-like bracts; sepals 3; petals 3; fertile stamens 2–3; staminodes 3. Fruit a capsule.

Aneilema acuminatum
Pointed Aneilema
Slender herb with stems to 30cm tall. Leaves lanceolate, 3–9cm long, apex acute. Flowers in narrow, open, mainly terminal panicles to 15cm long; petals white, 4–6mm long. Stamens and staminodes conspicuous. Capsule oblong, flattened, 4mm long. **HABITAT:** In or near rainforest. **DIST:** Uncommon. Coast (e.g. Mt Keira, Yarramalong). Also NC, NT, CWS, Qld. **FL:** Dec–Jan.

Commelina cyanea

Commelina
Trailing herbs with leaves sheathing the stems. Inflorescence axillary, enclosed in spathe-like bract. Flowers with 3 sepals, 3 petals; fertile stamens 3; staminodes 3.

Similar to the white-flowered weed *Tradescantia fluminensis* (Wandering Jew), which has 6 fertile stamens.

Commelina cyanea
Scurvy Weed
Glabrous prostrate herb with stems rooting at nodes. Leaves ovate, 3–8cm long. Flowers with 3 blue petals 12mm long; stamens yellow. **HABITAT:** Shaded forest; often in disturbed areas. **DIST:** Coast, lower Blue Mtns, Kowmung. Also NC, SC, NT, WS, NWP, Qld, LHI. **FL:** Dec–Apr.

Murdannia
Erect or prostrate herbs with glabrous sessile leaves. Flowers bisexual with 3 sepals and 3 petals, 2–3 fertile stamens and 3 staminodes. Fruit a 3-valved capsule.

Murdannia graminea
Murdannia
Erect plant to 60cm tall. Leaves mostly basal, linear, to 25 x 1cm, scabrous-pubescent. Flowers in open terminal panicles, sepals green, 7mm long; petals mauve to blue or white, 10mm long; stamens with winged and bearded filaments; staminodes with bearded filaments. Capsule ellipsoid, 6–10mm long. **HABITAT:** Woodland. **DIST:** Rare. Cumberland Pl. (e.g. Castlereagh NR, Shanes Park). Also NC, NT, WS, NWP, Qld, WA, NTerr. **FL:** Dec–Apr.

Aneilema acuminatum

FLOWERING PLANTS 473

Murdannia graminea

Pollia crispata

Xyris bracteata

Pollia
Erect or ascending herbs. Inflorescence terminal or in upper axils. Flowers with 3 petals and 3 sepals; stamens either 6 fertile or 3 fertile and 3 reduced or sterile; filaments glabrous. Fruit a nut.

Pollia crispata
Pollia
Glabrous perennial to 70cm tall with ascending stems rooting at nodes. Leaves stem-sheathing, narrow-ovate, to 15 x 3cm, undulate, acuminate. Inflorescence terminal, erect, dense, 3–5cm long; sepals green; petals blue or white, 4–6mm long. Nut ovoid, blue, shiny, 6mm long. **HABITAT:** In and near rainforest. **DIST:** Coast (e.g. Macquarie Pass, Wyong, Belmore Falls). Also NC, Qld. **FL:** Nov–Feb.

XYRIDACEAE

Xyris
Tufted rush-like plants of swampy sites. Leaves basal, linear to terete, glabrous. Inflorescence a compact globose-cylindrical spike on long scape. Flowers with many imbricate bracts; sepals 3, bract-like, unequal; petals 3, usually yellow. Fruit a capsule.

Xyris bracteata
Leaves linear, flat, 12–25cm x 2–4mm; margins pale, warty at base. Stems ± terete, to 60cm tall. Flower heads ovoid, 8–10mm x 5–8mm, with 8–12 flowers; bracts 7–10, upper bracts shorter

Xyris complanata

and darker than lower bracts. **HABITAT:** Wet heath, swampy depressions. **DIST:** Coast (e.g. West Head, Kurnell, Barren Gnds). Also SC, ST. **FL:** Nov–Jan.

Xyris complanata
Leaves linear, flat, 5–25cm x 1–4mm, often twisted; margins pale, warty. Stems flat, with 2–4 ribs, to 50cm long. Flower heads cylindrical to ovoid, to 22 x 8mm, with 2–6 keeled bracts; lateral sepals lacerate, golden brown; petals finely toothed. **HABITAT:** Swampy woodland, moist depressions. **DIST:** Coast nth of Bundeena, Cumberland Pl. Also NC, WS, Qld, WA, NTerr. **FL:** Dec–Feb.

Xyris juncea

Xyris gracilis
Slender Yellow-eye
Leaves linear, flat, 6–22cm x 1–2mm; margins pale, warty at base. Stems ± terete, to 55cm tall. Flower heads ellipsoid, to 8 x 4mm, with 3–4 flowers; upper bracts shorter and darker than lower bracts. **HABITAT:** Sedge swamps, wet heath. **DIST:** Coast and ranges. Also NC, SC, NT, Qld, Vic. **FL:** Nov–Feb.

Differs from *X. bracteata* in its narrower leaves, narrow flower heads and fewer flowers.

Xyris juncea
Dwarf Yellow-eye
Leaves flat to ± terete, 3–20cm x 1mm, ridged transversely; margins pale. Stems terete, 6–30cm tall. Flower heads rounded, to 7 x 7mm; bracts 4, lower pair longest, margins entire; fertile bracts (on flowers) fringed or torn. **HABITAT:** Swampy margins, wet heath. **DIST:** Chiefly coast (e.g. Wattamolla, Maddens Plains). Uncommon in Blue Mtns (e.g. Linden, Blackheath). Also NC, SC, Qld, Vic. **FL:** Oct–Mar.

Xyris gracilis

Xyris operculata

Eriocaulon scariosum

FLOWERING PLANTS | 477

Xyris ustulata

Xyris operculata
Tall Yellow-eye
Leaves subterete, 20–60cm x <1mm, shiny red-brown at base. Stems ± terete, 60–80cm tall. Flower heads obovoid, to 13 x 10mm; bracts numerous, in 5 vertical ranks, smaller towards base. **HABITAT:** Sedgeland, wet heath. **DIST:** Coast and Sthn H'lands. Also NC, SC, NT, ST, Qld, Vic, Tas, SA. **FL:** Sept–Jan.

Xyris ustulata
Leaves subterete, 30–50cm x 1mm, shiny red-black at base. Stems terete or angled, 30–100cm tall. Flower heads ovoid, to 18 x 12mm; bracts <20, loosely packed, ± reflexed, not ranked vertically. **HABITAT:** Sedgeland, wet heathy. **DIST:** Endemic. Illawarra, Sthn H'lands, Blue Mtns. **FL:** Oct–Dec.

ERIOCAULACEAE

Eriocaulon
Monoecious herbs. Leaves linear, tufted, sessile, sheathing at base. Flowers within scaly bracts in a head on scape usually exceeding leaves; tepals 4–6, in 2 whorls. Fruit a membranous capsule.

Eriocaulon scariosum
Common Pipewort
Glabrous herb. Leaves to 80 x 2mm. Flower heads hemispherical to spherical, 5–6mm diam.; outer flowers mostly female, inner rows male; subtended by hairy bracts; tepals

Flagellaria indica

membranous. **HABITAT:** In open swampy ground. **DIST:** Uncommon. Coast to ranges. Also NC, SC, NT, ST, CWS, Qld, Vic. **FL:** Nov–Feb.

FLAGELLARIACEAE

Flagellaria
Robust climber with bamboo-like stems to 15m long or more supported by coiled tendrils formed by twisted leaf tip. Monotypic genus.

Flagellaria indica
Whip Vine
Climber reaching high into canopy. Leaves alternate, narrow-lanceolate, to 30 x 2cm, with sheathing base and many parallel veins. Flowers in large terminal panicles, solitary and ± sessile in axil of subtending bract, fragrant; tepals 6, in 2 whorls, creamy white, 2mm long. Fruit a greenish red drupe, 5mm diam. **HABITAT:** Gully rainforest. **DIST:** Rare in Sydney district. Coast (e.g. Hacking R., Palm Jungle, Strickland SF). Also NC, Qld, WA, NTerr, Pac. Iss, Asia, E. Afr. **FL:** Nov–Feb.

Comparison of 4 families of grass-like plants

Character	Juncaceae (rushes)	Restionaceae (rushes)	Cyperaceae (rushes and sedges)	Poaceae (grasses)
leaves	at base of stems, terete to flat or reduced to a sheath	reduced to open sheaths on stems, overlapping on lower stem, distant on upper	3-ranked, basal or on lower stem, terete or with grass-like blade or reduced to sheathing scales	2-ranked, alternate, at nodes; consisting of sheath, ligule and blade
leaf sheaths	open or closed	open	closed, tubular	open, with overlapping margins
stems	terete, solid	solid or hollow, unbranched to many-branched	triangular or terete, solid, without nodes	terete, with hollow internodes and solid nodes
basic unit of inflorescence	cyme	usually a spikelet	spikelet	spikelet
spikelets	absent	1 to many-flowered, solitary or numerous, subtended by bracts; male and female spikelets often different	glumes in 2 ranks or imbricate and spirally arranged, most glumes bearing flowers, 1 to several lowest glumes empty	of 1 or more florets subtended by 2 lower empty glumes; florets consisting of 2 bracts (palea, lemma) and 1 reduced flower; glumes and bracts often awned
flowers	mostly bisexual	mostly unisexual (dioecious)	unisexual (monoecious), bisexual	mostly bisexual
perianth segments	6, in 2 whorls of 3, well-developed	6 or less, or absent, dry, glume-like (i.e. membranous, scaly or chaffy)	bristles, hairs, scales, or absent	usually 2, reduced (lodicules)
stamens	usually 6, sometimes 3	3 (male fls.), 0–3 staminodes (female fls.)	3, sometimes fewer (male and bisexual fls.)	3
styles/stigmas	styles 1 or 3, stigmas 3	styles 1–3, stigmas 1 or 3, (female fls.)	style 1, stigmas 2 or 3 (female fls.)	stigmas 2, occasionally 1 or 3
bracts per flower	1 or absent	1 (glume)	1 (glume)	2 (palea, lemma)
fruit	capsule with 1 to many seeds	3-sided capsule or nutlet	nutlet, flat or 3-angled	caryopsis (grain)
habitat	in water and waterlogged ground	infertile soils, periodically wet	mostly damp, wet or marshy areas	mostly on dry land in a variety of soils

RESTIONACEAE

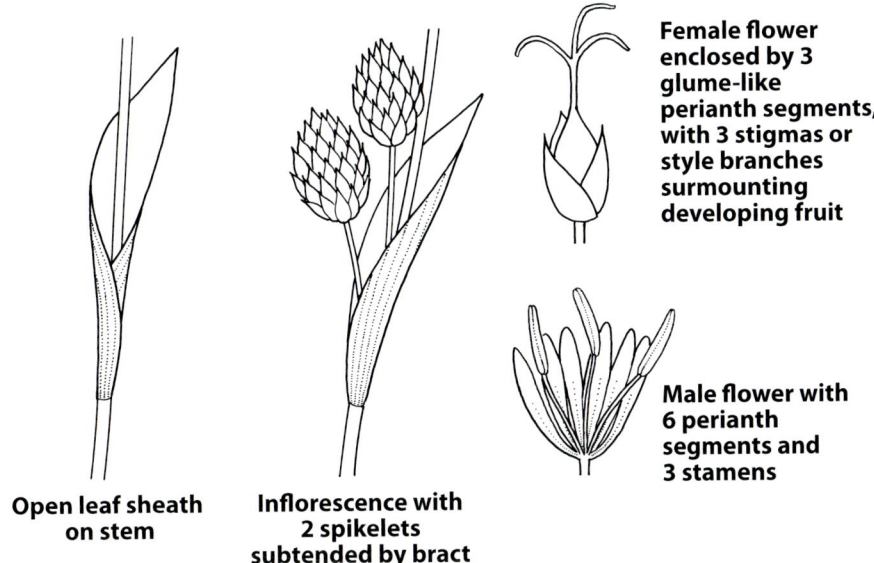

Open leaf sheath on stem

Inflorescence with 2 spikelets subtended by bract

Female flower enclosed by 3 glume-like perianth segments, with 3 stigmas or style branches surmounting developing fruit

Male flower with 6 perianth segments and 3 stamens

Empodisma
Stems thin, wiry, green and flexuose. Leaf sheaths distant on stems; upper sheaths green or pale with reflexed pointed tip. Spikelets axillary, females solitary and 1-flowered with 1 to 3 barren glumes, males solitary or several together with 1 or several flowers.

Empodisma minus
Spreading Rope-rush
Tufted herb with short erect glabrous stems in open sandy situations or >1m long, flexuose, with many branches forming a tangled mass in moist shady places. Leaf sheaths with narrowly pointed reflexed tip 1–6mm long and a few woolly hairs in axil. Spikelets sessile, few-flowered, 4–7mm long. Nut ovoid, 2mm long.
HABITAT: Boggy soil, creek banks, wet cliffs, heath and open forest. **DIST:** Common. Coast and ranges. Also Qld, Vic, SA, Tas, NZ.
FL: Sept–Nov and other times.

Empodisma minus

Hypolaena fastigiata

Hypolaena
Stems wiry, branched, striate, often flexuose; leaf sheaths closely appressed with erect tip. Male and female spikelets terminal, dissimilar. Male flowers with 3 stamens, females with 3-branched style. Nut with woody pericarp.

Hypolaena differs from *Empodisma* in stem and bract colour and its terminal inflorescences and from *Leptocarpus* in its striate flexuose wiry branched stems.

Hypolaena fastigiata
Tassel Rope-rush
Tufted herb with white-tomentose or greyish stems; leaf sheaths dark red-brown, closely appressed, striate, not reflexed at tip. Male flowers in narrow brown spikelets 5–7mm long on slender stalks in nodding or erect panicles; female flowers in narrow erect sessile spikelets 6–12mm long, solitary or 2–3 together. Nut ovoid, 4mm long. **HABITAT:** Dry or damp sandy heath. **DIST:** Coast. All States (except NTerr). **FL:** Aug–Dec.

Leptocarpus
Flowers dorsiventrally flattened with males and females in dissimilar terminal spikelets (NSW spp.). Male flowers usually with 3 stamens, females with 3-branched style. Fruit a nut.

Leptocarpus tenax
Slender Twine-rush
Herb with smooth, greyish green stems 60–100cm tall; leaves reduced to dark brown leaf sheaths with membranous tip weathering to aristate point. Male flowers in numerous narrow, stalked, purplish brown spikelets 4mm long in drooping panicles. Female spikelets reddish brown, >10mm long, nearly sessile, few, in erect inflorescences. Nut narrow-ellipsoid, 2–3mm long. **HABITAT:** Damp sandy soils in heath, sedgeland and low woodland. **DIST:** Mainly coast in NSW. All States. **FL:** Sept–Dec.

Lepyrodia, *Sporadanthus*
Tufted or creeping herbs; stems erect with sheathing scales or bracts overlapping at the base, becoming distant above. Flowers in narrow panicles, not in spikelets, each subtended by 1 or 2 tiny bracteoles. Male and female inflorescences similar. Fruit a capsule.

LEPYRODIA
Leaf sheaths striate; ovary 3-locular; styles 3.

Lepyrodia scariosa
Scale-rush
Dioecious herb, tufted or with short rhizome. Stems 30–90cm tall, smooth, glabrous,

Leptocarpus tenax

Lepyrodia scariosa

unbranched; leaf sheaths straw-coloured, 15–30mm long, loose, open. Male and female inflorescences similar; female flowers with conspicuous pale stigmas. **HABITAT:** Damp sandy or peaty ground in heath and open forest. **DIST:** Common. Coast and ranges. NSW and Qld. **FL:** Aug–Jan.

SPORADANTHUS
Leaf sheaths not striate; ovary 1- or 3-locular; styles 1 or 3.

Sporadanthus gracilis (formerly *Lepyrodia gracilis*)
Slender Scale-rush
Dioecious herb with short creeping rhizome. Stems mostly <60cm tall, occasionally much longer, erect, glabrous, with few to many branches; leaf sheaths brown, tight, closely appressed. Flowers in narrow panicles 3–10cm long; tepals 6, scaly, reddish brown, 2–3mm long, outer 3 shorter and narrower than inner 3. **HABITAT:** Wet sandy soils. **DIST:** Uncommon. Coast and ranges sth of Sydney. Also ST. **FL:** Chiefly Mar–Sept.

Sporadanthus gracilis

Baloskion fimbriatum

FLOWERING PLANTS

Baloskion gracile

Baloskion, *Chordifex*, *Eurychorda* (species formerly in *Restio*)

Dioecious perennial herbs, tufted or with creeping scaly and hairy rhizome. Stems green with sheathing scales (leaf sheaths), crowded at base, distant above. Male and female spikelets similar or dissimilar with 1 to many flowers. Bracteoles absent. Fruit a capsule.

BALOSKION
Pedicels fused to subtending glume. Male flowers usually with 6 tepals and 3 stamens; female flowers with 4 tepals and 2 staminodes. Capsule papery; seeds smooth.

Baloskion fimbriatum
Fringed Cord-rush
Herb with slender unbranched stems 20–80cm tall; leaf sheaths 10–15mm long, close-fitting, acutely pointed, upper ones with a tuft of fine hairs 1–4mm long at apex. Male and female spikelets similar in narrow inflorescences, purplish pink when fresh, 5–9mm long; glumes fringed with long hairs. **HABITAT:** Poorly drained, sandy peaty soils, often near swamps. **DIST:** Coast to ranges. Also SC, Qld. **FL:** Oct–Dec.

Baloskion australe is similar but has stouter stems 1.5–3mm diam. with rough wrinkled loose sheaths 15–30mm long. It occurs in upper Blue Mtns.

Baloskion gracile
Slender Cord-rush
Herb with slender stems 50–100cm x 0.5–1.5mm crowded on rhizome; leaf sheaths

Baloskion tetraphyllum subsp. *meiostachyum*

20–30mm long, acuminate, glabrous or with white hairs at apex. Spikelets in narrow inflorescences with subtending bract exceeding lowest spikelet. Male spikelets 5–10mm long, ellipsoid, brown, on slender stalks; female spikelets 6–16mm long, cylindrical, purplish, sessile. **HABITAT:** Wet sandy peaty soils, swamps and sedgeland. **DIST:** Endemic. Coast and Sthn H'lands. **FL:** Sept–Jan.

Baloskion pallens. Endemic. Occurs at Richmond, Agnes Banks, Castlereagh and Wyong. It has crowded sessile ellipsoid male and female spikelets of similar appearance subtended by short, abruptly pointed bracts with the apex displaced by the spikelets.

Baloskion tetraphyllum subsp. *meiostachyum*
Tassel-rush, Tassel Cord-rush
Tufted plant to 1.5m tall with many clusters of slender, bright green divided branchlets in axils of leaf sheaths. Spikelets 3–6mm long, numerous, in long loose narrow panicles above branchlets; male spikelets globose, female spikelets elliptic. **HABITAT:** Damp sandy ground in heath and open forest, near creeks. **DIST:** Coast from Milton to Qld. **FL:** Sept–Jan.

Chordifex dimorphus

Chordifex fastigiatus

CHORDIFEX
Male and female spikelets similar. Tepals 5 or 6; stamens 3; ovary usually 2-locular; styles 2. Seeds colliculate and ridged due to lines of convex cells and flat cells.

Chordifex dimorphus
Stems 30–100cm tall, divided near base into long weak flexuose branchlets; leaf sheaths 5–13mm long, broad, loose, glabrous, brown. Male spikelets numerous, ovoid, 4–6mm long with 3–8 flowers; female spikelets few, narrow, with 1 flower and several barren glumes. **HABITAT:** Sandy ground in heath and scrub. **DIST:** Endemic. Coast (Broken Bay to Cataract Dam). **FL:** Sept–Mar.

Chordifex fastigiatus
Stems 30–100cm tall, erect, branched, lacking clusters of barren branchlets; leaf sheaths short, closely appressed, dark brown. Spikelets narrow, 4–7mm long; male spikelets with several flowers; female spikelets with 1 fertile flower and several empty glumes. **HABITAT:** Damp sandy ground in heath and scrub. **DIST:** Coast and ranges (Budawangs to Blackheath and Gosford). Also Qld. **FL:** Sept–Dec.

EURYCHORDA
Stems flat; tepals 4; stamens or staminodes 2; ovary 2-locular; styles 2. Seeds finely patterned.

Eurychorda complanata
Flat Cord-rush
Stems unbranched, pale green, 30–90cm tall with a few tight pale leaf sheaths 10–30mm long. Inflorescence a narrow panicle of dense red-brown spikelets 6–12mm long; male spikelets with exserted anthers, ovoid, many-flowered; female spikelets ellipsoid, few-flowered. **HABITAT:** Wet sandy or peaty soils in heath and scrub. **DIST:** Mainly coastal; Blue Mtns (Carrington Falls). Also CT, Qld, Vic, Tas. **FL:** Nov–Dec.

CENTROLEPIDACEAE

Centrolepis
Small grass-like herbs inhabiting wet ground. Inflorescence reduced to a pseudanthium consisting of 1 or 2 male flowers with 1 stamen each and 2 or more female flowers with 1 carpel, fused together at base and subtended by 2 conspicuous sheathing bracts.

Centrolepis fascicularis
Perennial herb forming clumps to 8cm high and 12cm across. Leaves narrow-linear with blunt apex, 10–40 x <1mm, crowded. Scape terete, 2–6cm long; pseudanthium with 2–4 female flowers; bracts with long fine apex or awn, hairy, nearly equal. **HABITAT:** Damp peaty soils, seepage areas in heath and swamp margins. **DIST:** Coast and ranges. All States. **FL:** July–Jan.

The second Sydney sp., *Centrolepis strigosa*, is tufted rather than clump- or mat-forming. Its pseudanthium has 4–7 female flowers and hairy bracts with a short pointed tip or mucro.

Eurychorda complanata

Centrolepis fascicularis

JUNCACEAE

Juncus
RUSH

Wind-pollinated plants growing in or near water. Stems (culms) usually terete. Leaves glabrous, cylindrical or channelled or with internal septa, or flat or reduced to basal sheaths slit to the base. Flowers small, bisexual and usually clustered; tepals 6 in 2 whorls of 3, scaly, glume-like; stamens 6 or 3; style 3-branched. Fruit a capsule with numerous seeds.

Over 50 species occur in the Sydney district, many of them introduced, weed-like, difficult to identify or hybridising. Only a few representative native species are described.

Juncus continuus
Stems 50–120cm x 2–3.5mm, light yellow-green, striate, with continuous even pith. Leaves reduced to sheaths at base of stems; sheaths loose, shiny brown near base, dull and straw-coloured near top, with small erect acute or mucronate tip (equivalent to leaf blade). Inflorescence open, 4–10cm long; flowers clustered towards ends of branches; stamens 3. Capsule exceeding tepals, deep yellow-brown, smooth, shiny. **HABITAT:** Damp sandy soils. **DIST:** Coast and t'lands. Also WS, Qld, Vic, NTerr. **FL:** Peak Dec–Feb.

Juncus continuus

Juncus homocaulis
Stems terete or compressed, 5–30cm x 1mm. Leaves basal, channelled, to 15cm x 1mm. Inflorescences open, to 10cm long with subtending bract 2–15cm long. Flowers in several clusters of up to 30; outer tepals 4–6mm long, acuminate, straw- or red-brown. Capsule usually shorter than tepals, brown. **HABITAT:** Open forest, woodland, grassland. **DIST:** Coast and t'lands. Also WS, WP, Qld, Vic, SA. **FL:** Oct–Mar.

Juncus kraussii subsp. *australiensis*
Sea Rush
Stems in tussocks 1m or more tall, spreading over large areas of tidal marshland. Leaves dull-green, terete, stiff, pungent-pointed, 3mm diam., as tall as stems. Panicle loose, exceeded by erect rigid subtending bract to 30cm long. Flowers in small pale brown clusters; stamens 3 or 6. Capsule as long as tepals, pointed, shiny, dark red-brown to black. **HABITAT:** Saline and brackish marshes. **DIST:** Coast. All States, NZ. **FL:** Nov–Feb.

Juncus planifolius
Broad Rush
Tufted rush with terete or slightly compressed stems 10–60cm tall. Leaves shorter than stems; blade 5mm wide, thin, flat; sheath pale brown or pinkish. Inflorescence umbel-like, 3–12cm long, open; subtending bract shorter than inflorescence. Flowers in clusters of 8–30 at ends of inflorescence branches, with brown scaly tepals and 3 stamens. Capsule ellipsoid, shiny, brown, with remains of style base at top. **HABITAT:** Wet sandy and clayey ground, creek banks. **DIST:** Coast to WP (Pilliga). All States (except NTerr). **FL:** Nov–Mar.

Juncus prismatocarpus
Branching Rush
Tufted perennial with stems 10–40cm x 1–3mm, slightly flattened, firmer than leaves

Juncus homocaulis

Juncus kraussii subsp. *australiensis*

Juncus planifolius

Juncus prismatocarpus

and usually longer. Leaves on stems 1–2.5mm wide, flattened, hollow with cross-septa; septa sometimes incomplete. Inflorescence very open, subtended by septate bract to 14cm long, not exceeding inflorescence. Flowers in dense clusters at branches and ends of branches; tepals 3–5mm long, brown with scaly margins; stamens 3. Capsule tapering to top, much longer than tepals, brown. **HABITAT:** In or near water, wet ground. **DIST:** Coast and ranges. Also NWS, NWP, Qld, Vic, Tas, SA, NZ. **FL:** Nov–Mar.

Juncus usitatus
Common Rush, Tussock Rush
Tufted rush. Stems terete, to 1m x 1–2mm, mid-green, striate, easily compressed or bent; internal pith with many air spaces. Leaves reduced to sheaths, to 15cm long. Inflorescence 3–7cm long, subtended by erect needle-pointed bract to 25cm long. Flowers pale, solitary, along inflorescence branches; tepals straw-brown with broad scaly margins; stamens usually 3. Capsule longer than tepals, smooth, brown. **HABITAT:** Damp soil, banks of creeks and pools. **DIST:** Coast to WP. All mainland States, NZ, NCal. **FL:** Nov–Jan and other times.

This is not the only rush in the area having spaced, rather than clustered, flowers, but it is the commonest.

Luzula
Perennial herbs from moist cool habitats. Leaves mostly tufted at base, with closed sheath and flat grass-like linear blade. Flowers bisexual, in dense clusters subtended by 1–4 large bracts on erect culms (stems). Fruit a capsule with 1–3 seeds.

Luzula ovata
Perennial to 30cm tall with onion-like bulb. Leaves to 12cm x 4mm, soft, margins hairy, apex obtuse. Culms 10–30cm tall. Inflorescence a single ovate or globular head or several heads closely clustered on peduncles <10mm long; tepals acuminate, red-brown with pale margins. Capsule as long as tepals, deep red. **HABITAT:** Swampy sites. **DIST:** T'lands (e.g. Newnes SF). Also Vic. **FL:** Oct–Dec.

Juncus usitatus

Luzula ovata

CYPERACEAE

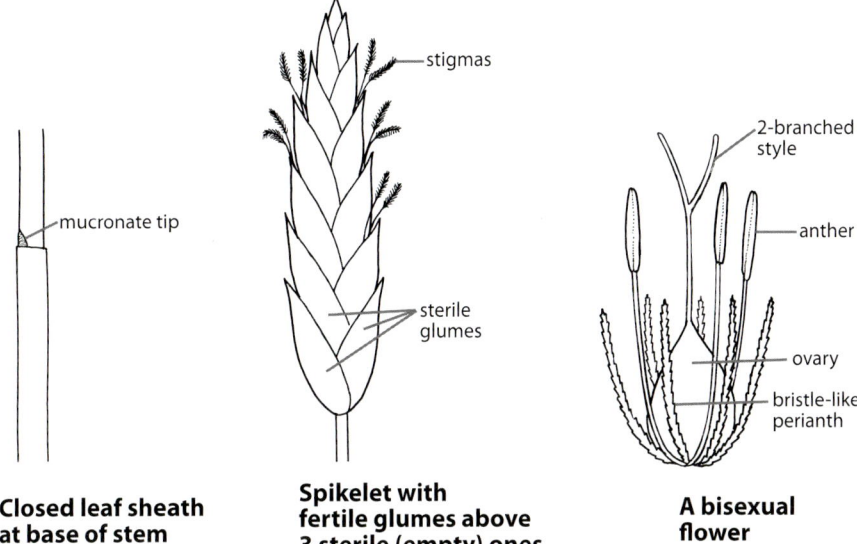

Closed leaf sheath at base of stem

Spikelet with fertile glumes above 3 sterile (empty) ones

A bisexual flower

Baumea
Perennials with terete or flattened pithy or septate stems. Leaves similar to stems or reduced to sheaths. Inflorescence paniculate, often contracted, occasionally large and open. Spikelets with 1–5 bisexual flowers; glumes 2-ranked. Flowers with 3 stamens and 3-branched style; tepals absent. Fruit a 3-ribbed nut with style base forming a conical apex.

Baumea articulata
Jointed Twig-rush
Stems terete, 1–2m x 10mm, hollow with white cross-partitions (septa). Leaves basal with pale sheaths, tapering to pointed apex, otherwise similar to stems. Inflorescence large, loose, drooping, 20–50cm long, brown; subtending bract shorter than inflorescence; spikelets 4cm long; glumes minutely scabrous on keel.
HABITAT: Coastal lagoons and swamps in water <1m deep. **DIST:** Coast (e.g. Marley Lagoon, La Perouse). Also NT, NWS, all States, Pac. Iss.
FL: Oct–Dec.

Baumea articulata

Baumea juncea

Baumea juncea
Bare Twig-rush
Slender herb spreading by long rhizomes with shiny, pale brown acuminate scales. Stems 40–70cm x 1–2mm, stiff, greyish. Leaves reduced to 2–3 mucronate sheaths at base of stems. Inflorescence spike-like, 3–10cm long; spikelets few, 4–5mm long, 1–2-flowered, lowest fertile. Nut ovoid, 2–4mm long with pubescent style base. **HABITAT:** Swamps, salt marshes, estuaries, lakes and damp heaths. **DIST:** Coast. All States except Vic, NZ, LHI, NCal. **FL:** Oct–Jan.

Baumea rubiginosa
Soft Twig-rush
Perennial with short scaly rhizome. Stems terete, angular or slightly flattened to 1m x 5mm. Leaves mostly basal, shorter and broader than stems, flattened, pungent-pointed, with loose pith; stem leaves 1 or 2, short. Panicle narrow; spikelets densely clustered on erect curved peduncles subtended by brown sheathing bracts, red brown with 5–6 fringed acuminate glumes. Nut shiny, orange-red, 3mm long. **HABITAT:** Wet or swampy ground. **DIST:** Coast to WS. All States, NZ, NCal, Malesia. **FL:** Oct–Dec and after fire.

Baumea teretifolia
Wrinkle-nut Twig-rush
Perennial with long rhizome. Stems 30–100cm x 1–3mm, terete, striate, without nodes. Leaves shorter, erect, acutely pointed, with long loose

Baumea rubiginosa

sheaths at base. Inflorescence compact, 5–15cm long, with dense clusters of brown spikelets on short erect peduncles in axils of brown sheathing bracts; spikelets 5cm long. Nut dull brown, 2–3mm long, longitudinally ribbed or wrinkled. **HABITAT:** Swamps and wet sandy peaty soils. **DIST:** Coast and Sthn H'lands. Also NT, Qld, PNG, NZ. **FL:** July–Dec.

Baumea tetragona
Square Twig-rush
Perennial with long rhizome. Stems terete to 4-sided, 30–100cm x 2mm. Leaves basal, shorter than stems, 4-sided, 3mm across, smooth or scabrous. Inflorescence a dense, erect, reddish brown panicle, 5–12cm long, with 1-flowered spikelets 4mm long. Nut indistinctly 3-ribbed, wrinkled. **HABITAT:** Swamps and streams in cool areas. **DIST:** Uncommon. Coast and ranges sth of Blue Mtns (e.g. Carrington Falls Swamp). Also SC, Qld, Vic, Tas, SA. **FL:** Sept–Nov.

Baumea teretifolia

Baumea tetragona

Bolboschoenus
Stout perennials with 3-angular stems, scabrous just below inflorescence. Leaves at base and at nodes on stems, long, linear. Inflorescence terminal, subtended by 2–3 unequal leafy bracts longer than inflorescence; spikelets large, ovoid, golden brown, on spreading smooth rays above a sessile cluster of spikelets. Fruit a nut; perianth bristles often persistent.

Bolboschoenus caldwellii
Stems tufted, 30–90cm x 1–3.5mm, with long keeled leaves, V-shaped in section, 6–8mm wide. Spikelets 1–3 on unequal rays and 3–6 in lower sessile cluster. Involucral bracts 5–20cm long. Flowers with 2-fid style. Nut pale or golden brown, 4mm long, flat or depressed both faces; bristles deciduous, half as long as nut. **HABITAT:** Boggy or periodically wet ground, beside creeks. **DIST:** Widespread but uncommon. NSW, all States. **FL:** Any time.

Bolboschoenus fluviatilis
Marsh Club-rush
Stems 1–2m x 5–10mm. Leaves 7–11mm wide, grooved centrally, recurved near margins. Inflorescence with 6 or more branches 2–10mm long, bearing clusters of 1–6 reddish brown ovoid spikelets 10–25mm long. Involucral

Bolboschoenus caldwellii

Bolboschoenus fluviatilis

bracts 15–25cm long. Flowers with 3-fid style. Nut pale to dark brown, 3-angled, 3–4mm long; bristles persistent, as long as nut. **HABITAT:** Shallow fresh or brackish water or wet ground. **DIST:** Coast to WP. Also Qld, Vic, NZ, Asia, N. Amer. **FL:** Oct–Mar.

More common and robust than *B. caldwellii* and usually growing in water.

Carex

Stems usually 3-angled with persistent leaf sheaths at base. Leaves basal and on stems, flat, often with scabrous margins. Inflorescence of 1-flowered unisexual spikelets with 1 glume; the 2 sexes in same or separate spikes. Spikes in panicles, umbels or solitary. Inflorescence bracts leaf-like, sheathing. Female flowers with flask-shaped utricle around ovary and (later) nut. Male flowers with 3 stamens, females with 2- or 3-branched style. Fruit a flat or 3-angular nut.

Carex longebrachiata

Carex appressa
Tall Sedge

Tufted sedge, sometimes covering large areas. Stems 1m tall, acutely triangular, finely scabrous on edges in upper part. Leaves flat, 4–5 (rarely to 12) mm wide with scabrous margins. Panicle 5–20cm long, narrow, with short brown spikes appressed to stem; lowest spike distant; upper spikes crowded. Nut biconvex. **HABITAT:** Swamp margins, creek banks and damp ground in open forest and rainforest. **DIST:** Coast to WP. All States, NG, Pac. Iss. **FL:** Oct–Dec.

Carex brunnea

Densely tufted plant with short rhizome. Stems 30–80cm tall, triangular in cross-section. Leaves basal, shorter than stem, 2–4mm wide. Inflorescence spreading or drooping, 10–20cm long. Spikes long-cylindrical on short stalks in clusters of 2–3 subtended by short subulate bracts. Utricle red-brown with prominent veins, pubescent with beak 1mm long and 2-pointed apex. **HABITAT:** Moist soil, usually in rainforest. **DIST:** Coast and ranges. Also SC, NC, Qld, sw. Pacific. **FL:** Nov–Mar.

Carex gaudichaudiana
Tufted Sedge

Sedge to >50cm tall, in spaced tufts on long rhizome. Leaves 4–5mm wide, erect, usually longer than stems, flat in lower part, V-shaped and minutely scabrous on keel and margins towards apex. Inflorescence 9–15cm long with 1–2 leafy bracts shorter than inflorescence. Spikes dense, brown to dark purplish brown; male spikes 1–3, narrow, above 1 or more broader female spikes. **HABITAT:** Swamps and stream banks. **DIST:** Coast to WS. Also Qld, Vic, Tas, NZ, PNG, Asia. **FL:** Sept–Apr.

Carex longebrachiata
Drooping Sedge

Spreading grassy sedge with stems to 80cm x 1.5mm, exceeded by flattened ridged and grooved leaves 2–4mm wide. Inflorescence 40–90cm long. Spikes 3–5cm long in bundles of 1–8 on long fine drooping stalks at nodes on stem; each node with erect leafy bract. Upper spikes all or mostly male, others all or mostly female. **HABITAT:** Alluvial soils of medium fertility in moist open forest and rainforest margins. **DIST:** Coast to WS. Also Vic. **FL:** Oct–Dec.

Carex pumila

Loosely tufted plant with long creeping rhizome. Stems short, barely exceeding leaf sheaths. Leaves to 50cm x 5mm, much longer than stems. Inflorescence with 3–6 spikes; lowest bract exceeding inflorescence. Terminal spike male, 2–4cm long, stalked, above 2–3 sessile mainly or entirely male spikes. Lower spikes 2–3cm long, mainly or entirely female. Female spikes 1.2cm diam., about twice as broad as males. Utricle greenish to golden brown, thick, corky. **HABITAT:** Moist depressions in coastal sand dunes. **DIST:** Coast. From SA to Qld; also Tas, LHI, NZ, Asia, S. Amer. **FL:** Sept–Dec.

FLOWERING PLANTS

Carex appressa

Carex brunnea

Carex gaudichaudiana

Carex pumila

Caustis

Perennials with terete striate stems. Leaves reduced to sheaths at base and on stems. Spikelets with 3–5 overlapping glumes all around axis, lower ones short and empty, upper 2 usually enclosing flowers. Perianth absent. Stamens 3–6. Style 3-branched. Fruit a nut with prominent style base.

Caustis flexuosa
Curly Sedge, Old Mans Whiskers
Plant 30–80cm tall with dark brown leaf sheaths. Upper sheaths subtending clusters of curved branchlets; branchlets divided several times, becoming fine and flexuose or coiled. Spikelets solitary on fine stalks resembling ultimate branchlets; upper flower bisexual, lower flower male, both with 3–4 pendulous stamens. **HABITAT:** Heath and open forest. **DIST:** Common. Coast to WS. Also Qld, Vic. **FL:** Sept–Dec.

Caustis flexuosa

Caustis pentandra
Plant to 2m tall from short stout rhizome; leaf sheaths striate, wrinkled, dark red-brown with subulate tip. Inflorescence branches with flat or angular inner faces, 2 or few fitting together to form terete cluster of branches, repeated at higher nodes. Spikelets in pairs subtended by short sheathing bract, 12–16mm long, 1 sessile, the other shortly stalked. **HABITAT:** Heath and dry open woodland. **DIST:** Coast and ranges. All States (except NTerr). **FL:** Apr–May, Sept–Dec.

Caustis recurvata
Tufted plant with short rhizome and stems to 70cm tall. Leaves reduced to dark red-brown sheaths. Inflorescence of functionally male and female spikelets on separate stems and of different appearance, subtended by broad striate sheathing bracts. 'Male' spikelets solitary or in short narrow clusters on erect branches. 'Female' spikelets solitary on short stalks with numerous crowded, curved, coiled or flexuose branchlets. **HABITAT:** Heath and dry open forest. **DIST:** Coast. Also CT, Qld. **FL:** Jul–Nov.

Var. *hirsuta* with stouter stems (to 5mm diam.) and hispid or scabrous inflorescences grows in coastal heath in Royal NP (e.g. Wattamolla).

Caustis pentandra

Caustis recurvata

Chorizandra cymbaria

Chorizandra sphaerocephala

Chorizandra
A genus of 6 species closely related to *Lepironia* but having spherical or ovoid inflorescences consisting of units (referred to here as spikelets) with a female flower above a few empty glumes and numerous glumes subtending a male flower with 1 stamen. Inflorescence appearing lateral due to erect inflorescence bract continuous with stem. Nut ridged, obovoid or globose.

Chorizandra cymbaria
Bristle-rush
Rush with creeping rhizome. Stems 60–100cm x 3–4mm, bright green, with transverse septa. Leaves few, terete, stem-like, usually shorter than stems. Inflorescence ovoid, about 13 x 10mm, enclosed on 1 side by sheathing base of floral bract; bract terete, tapering, 10cm long.

Spikelets very short with dark red-brown glumes. **HABITAT:** Damp marshy ground. **DIST:** Coast and ranges. Also Qld, Vic, WA, NCal. **FL:** Sept–Nov and after fire.

Chorizandra sphaerocephala
Round-headed Bristle-rush
Rush with creeping rhizome. Stems 50–100cm x 3–6mm, with transverse septa. Leaves few, terete, stem-like, usually shorter and narrower than stems. Inflorescence spherical, 7–12mm diam., not sheathed by subtending bract. Spikelets with dark red-brown glumes fringed with red hairs. Nut slightly flattened with about 8 longitudinal ridges. **HABITAT:** Damp marshy ground. **DIST:** Coast. Also CT, Qld, Vic. **FL:** Mainly Sept–Nov or after fire.

Cladium

Cladium is closely related to *Baumea* but has flat leaves in 3 ranks (not 2). Spikelets are in dense clusters (not panicles), with overlapping glumes (not distichous), and lack empty glumes above the 2 flowers (lowest and uppermost empty in *Baumea*). The flowers have 2 stamens (not 3) and a 2-fid style (not 3-fid).

Cladium procerum
Leafy Twig-rush, Tall Twig-rush

Plant with stout terete hollow stems 1–2m tall and narrow tapering leaves to 2.5m x 20mm with finely toothed margins and keel. Leaf blade continuous with smooth green stem sheath; new leaves occasionally shooting from sheaths. Inflorescence 20–40cm long, of dense corymbose clusters 4–5cm apart subtended by long leafy bracts. **HABITAT:** Swampy ground near coast. **DIST:** Widespread but uncommon. Coast (e.g. Yeramba Lagoon). Also Qld, Vic, SA, NTerr. **FL:** Nov–Apr.

Cladium procerum

Cyathochaeta

Rhizomatous perennials. Leaves mostly basal. Inflorescence a narrow panicle with distant nodes, longer than stem. Spikelets with spirally arranged glumes; bisexual flower with 4–5 bristles (tepals), 2 stamens and a 2-fid style.

Cyathochaeta diandra

Tufted herb with slender 1-noded stems to about 60cm tall. Leaves of similar length, tufted with many white fibres at base, stiff, channelled, pointed, twisted. Spikelets about 15mm long, brown with 4 narrowly pointed glumes, the 2 lower ones empty. Nut narrow, about 7mm long with style base of similar length at top and perianth bristles at base. **HABITAT:** Sandy soil, mostly in coastal heath, also open forest. **DIST:** Coast and ranges. Also Qld, Vic. **FL:** Irregular.

Sometimes common in early stages of regrowth after fire, but normally uncommon or overlooked.

Cyathochaeta diandra

Cyperus difformis

Cyperus
Annual or perennial herbs with 3-angular or terete stems without nodes. Leaves basal, rarely reduced to sheaths. Inflorescence subtended by leaf-like bracts. Spikelets in a cluster, spike, or on rays of simple or compound umbels with 2 to many flowers; glumes in 2 ranks, rarely spirally arranged. Flowers mostly bisexual with 2–3 stamens and 2–3-fid style; tepals absent. Nut 3-angular or biconvex.

About 40 native and introduced species in Sydney district. Only a few are shown.

Cyperus difformis
Variable Flat-sedge, Dirty Dora
Common annual weed of damp disturbed places. Stems to 50cm tall, soft-celled, sharply triangular with concave surfaces, often reddish at base. Leaves few, 2–4mm wide, usually shorter than stems. Inflorescence a simple umbel subtended by 1–3 bracts, the lowest longer than inflorescence and nearly erect; spikelets numerous in small dense globular clusters on long rays. Nut tiny, dark, 3-sided with deciduous 3-branched style. **HABITAT:** Shallow water or boggy ground. **DIST:** Throughout NSW. All States (except Tas), pantrop. **FL:** Dec–Apr.

Cyperus exaltatus
Tall Flat-sedge
Tussock-forming perennial 1–2m tall from a short rhizome. Stems smooth, 3-sided, 3–6mm across. Leaves as long as stems, with prominent lateral vein each side of midrib; margins scabrous. Inflorescence spreading, compound, subtended by 3–6 long finely serrated bracts. Spikelets 2–10 x 1–2mm in numerous brown spikes 2–6cm long clustered at ends of long rays up to 18cm long. Nut 3-angled, <1mm long, yellow-brown. **HABITAT:** In or near ponds, stream banks. **DIST:** Mainly Hawkesbury–Nepean region. CC, NC, WS, WP, all States except Tas, Afr., Asia. **FL:** Dec–Feb.

Cyperus polystachyos
Tufted perennial. Stems 30–60cm tall, firm, shiny green, 3-angled with convex to concave faces. Leaves flat, keeled, 2–3mm wide, much shorter than stems. Bracts 2–4, spreading, exceeding inflorescence. Spikelets densely crowded in clusters or umbels with short rays, green turning pale yellowish brown. Nut biconvex, shorter than glume, brown to blackish. **HABITAT:** Damp open ground. **DIST:** Coast. Also NWS, NWP. Widespread in Aust and tropics. **FL:** Mainly Jan–Apr.

Cyperus exaltatus

Cyperus polystachyos

Cyperus sanguinolentus

Annual or perennial forming a small tuft with 1 or few stems to 60cm tall. Leaves 1–3mm wide, shorter than or equalling stems. Umbels simple with 3–5 short rays or a cluster of spikelets, subtended by 1–3 long ribbony bracts. Spikelets 5–20 x 2–3mm, flat; glumes regularly arranged, broad, green with reddish purple or black markings near pale margins and apex. Nut biconvex, much shorter than glume, brown to blackish. **HABITAT:** Damp grassy areas, margins of swamps and lakes, stream banks, roadsides. **DIST:** Throughout NSW (except FWP). Widespread in Aust and tropics. **FL:** Dec–Apr.

Cyperus sanguinolentus

Eleocharis acuta

Eleocharis cylindrostachys

Eleocharis sphacelata

Eleocharis
Leafless annuals or perennials with septate terete or flattened stems, tufted or produced along rhizomes. Inflorescence a terminal spikelet without bracts; glumes spirally arranged, overlapping, often deciduous. Flowers bisexual, often with 6 perianth bristles, 1–3 stamens and 2- or 3-branched style. Fruit a nut topped by persistent style base.

Eleocharis acuta
Common Spike-rush
Rush with creeping rhizome and erect stems to about 60cm tall, tufted or spaced irregularly along it. Stems terete, 1–3mm diam., 3-sided immediately below inflorescence. Inflorescence dark brown or variegated, 15–30 x 3–7mm tapering to acute apex. Glumes acute with distinct midrib, persistent after flowering. Nut biconvex, yellow-brown, shorter than perianth bristles. **HABITAT:** In or near water, boggy ground, sometimes in dense stands. **DIST:** Throughout NSW. All States, NZ, PNG. **FL:** Oct–Apr.

Eleocharis cylindrostachys
Rush with short rhizome and densely tufted stems to 50cm tall. Spikelet pale brown, 10–25 x 1.5–3mm, cylindrical with obtuse apex. Glumes obtuse with distinct midrib, deciduous after flowering. Nut shiny brown, biconvex with ribbed margins; bristles barbed. **HABITAT:** In or near water, boggy ground. **DIST:** Wstn Sydney to NWP. Also Qld. **FL:** Nov–Apr.

Eleocharis sphacelata
Tall Spike-rush
Aquatic plant with stout erect stems from thick horizontal rhizome, emergent by about 1m, often forming extensive stands. Stems terete, hollow, septate, 5–15mm diam., terminated by tapering flower spike 3–6cm long. Glumes spirally arranged, overlapping, often with white stigmas or yellowish anthers emerging, finally falling away with fruit from the apex. **HABITAT:** Shallow fresh water. **DIST:** Widespread in NSW. All States, PNG. **FL:** Sept–Apr.

Fimbristylis dichotoma

Fimbristylis
Tufted annuals or perennials with narrow leaves at base of stems, sometimes short or reduced to sheaths. Spikelets many-flowered, in clusters or irregular umbels subtended by several unequal bracts. Characterised by a fringe of hairs on style.

Fimbristylis dichotoma
Common Fringe-rush
Slender perennial with flattened stems 30–70cm high and linear leaves 2–5mm wide, usually shorter than stems. Spikelets on long rays in umbels subtended by 2–5 leaf-like bracts often longer than inflorescence. Spikelets ovoid, 5mm long, with keeled brown glumes surrounding axis. Nut pale, shiny, biconvex, 1mm long with longitudinal and transverse ribs. **HABITAT:** Grassy understorey on clay soils in open forest. **DIST:** Uncommon. Coast to FWP but absent from CT. All States (except Vic, Tas). **FL:** Dec–May.

Fimbristylis ferruginea

Fimbristylis ferruginea
Perennial with short rhizome. Stems slightly flattened, striate, 30–60cm tall, 1.5–3mm diam. Leaves few, narrow, short or reduced to sheaths with fringed ligule. Spikelets in compact umbels with short rays subtended by 1 or 2 short linear bracts with 1 erect, light to dark brown, 8–15mm x 3–5mm with numerous keeled glumes. Nut biconvex, brown, smooth or finely striate. **HABITAT:** Saline and brackish swamps near the sea (e.g. Towra Point). **DIST:** Coast. Also NC, Qld, SA, WA, NTerr, pantrop. **FL:** Dec–Apr.

Gahnia aspera

Gahnia clarkei

Gahnia
Tufted perennials with leafy terete stems, the leaves ligulate, scabrous, linear, terete or inrolled, passing on upper stems into fine long bracts exceeding the inflorescence. Inflorescence large, paniculate with numerous spikelets. Spikelets usually with 2 flowers in upper glumes, the uppermost flower bisexual. Glumes overlapping (around axis), dark brown to black at maturity. Perianth absent. Fruit a nut.

Gahnia aspera
Rough Saw-sedge
Stems 40–60cm high, much shorter than leaves. Stems 2–4mm diam. Leaves flat or inrolled with pungent-pointed tip. Spikelets 8mm long, crowded in dense clusters in a narrow interrupted panicle. Nut 4–6mm x 3mm, bright red or brown-red, smooth, shiny, with dark acutely pointed apex. **HABITAT:** In or near rainforest. **DIST:** Coast to WP. Also Qld, Malesia, Polynesia. **FL:** Feb–May.

Gahnia clarkei
Tall Saw-sedge
Stems stout, leafy, to 2m or more tall. Leaves scabrous on back and margins, broad and flat near base, inrolled towards apex. Panicle compound, about 1m long, the primary branchlets 20–30cm long, drooping, each cluster subtended by a long leafy bract. Spikelets spaced along branchlets, 6mm long, dark brown, with 12–15 overlapping glumes. Nut ovoid, 3mm long, shiny orange-red, often hanging at maturity by short staminal filaments. **HABITAT:** Damp shady places, often near streams. **DIST:** Coast and CT. Also Qld, Vic, SA. **FL:** Irregular.

Gahnia grandis is similar to *G. clarkei* but the nuts are up to 5.5mm long, red-brown to dark brown and hang by staminal filaments up to 20mm long. It occurs in wet situations in upper Blue Mtns, often in the open.

Gahnia erythrocarpa
Stems to 120cm high with compound panicle 30–60cm long. Leaves flat with recurved margins, as long or longer than inflorescence. Spikelets 8mm long with 10–12 dark brown to black acuminate glumes, upper glumes smaller than lower. Nut flattened, ovoid, 5mm long, smooth, shiny, dark red-brown. **HABITAT:** Moist hillsides in open forest. **DIST:** Endemic. Coast (e.g. Brisbane Water NP, Waterfall). **FL:** Irregular.

This species resembles *G. sieberiana*, but is much less common and does not grow in swampy situations.

Gahnia erythrocarpa

Gahnia filifolia
Slender sedge. Stems to 50cm tall with inflorescence of similar length. Leaves 50–100cm x 1–2mm, channelled, scabrous; lower leaves with short dark open sheaths, upper leaves with long sheaths. Panicle compound, dark, narrow, with numerous branches, the clusters subtended by subulate bracts longer than the cluster. Spikelets crowded, dark grey or black, 6–9mm long; upper 2 glumes enclosing flowers shorter than others. Nut 3-angled, 3–3.5mm long. **HABITAT:** Wet to dry hillsides in scrub and open forest. **DIST:** Blue Mtns to Budawangs (e.g. Wentworth Falls L'out, Fairy Falls). **FL:** Aug–Dec.

Gahnia filifolia

Gahnia filum
Chaffy Saw-sedge
Sedge from saline habitats forming tussocks 1m high. Leaves not exceeding inflorescence, flat, 5mm wide, with 8–10 ridges on upper surface, smooth on back and margins, often inrolled and appearing terete; sheaths long, striate, brown. Panicle 30–50cm long, very narrow with clusters of erect branchlets subtended by long leaf-like bracts, higher bracts shorter. Spikelets 4–6mm long, numerous, straw-coloured, usually with 1 flower. Nut 3-sided, pale to dark brown. **HABITAT:** Margins of salt marshes. **DIST:** Nthn limit at Mill Ck, Salt Pan Ck, Sylvania Waters, Towra Point on Georges R. Also SC, Vic, Tas, SA. **FL:** Apr–Oct.

Gahnia filum

Gahnia melanocarpa

Gahnia microstachya

Gahnia melanocarpa
Black-fruit Saw-sedge
Stems about 1m tall with inflorescence about 40cm long. Stems stout, 4–5mm diam. Leaves flat with scabrous margins, longer than inflorescence. Inflorescence narrow with ascending branches. Spikelets 3–4mm long with 1 flower; upper 2 glumes shorter and broader than lower 2 or 3. Nut 3-sided, shiny, black, often hanging by staminal filaments. **HABITAT:** Usually in or near rainforest. **DIST:** Coast and t'lands. Also CWS, Qld, Vic. **FL:** June–July.

Gahnia microstachya
Slender Saw-sedge
Stems to 45 cm tall and inflorescence up to 60cm. Leaves as long as stems, about 3mm wide, striate, channelled or flat, with inrolled scabrous margins, filiform towards apex. Panicle narrow with spikelets on clustered branches, lowest cluster subtended by sheathing bract longer than cluster; higher clusters with short bracts. Spikelets evenly spaced, 2–4mm long with 5–7 nearly equal glumes. Nut 3-sided or slightly flattened, 2mm long, dark brown. **HABITAT:** Open woodland and forest. **DIST:** Chiefly ranges (e.g. Carrington Falls Swamp, Evans L'out). Also coast, SC, NT, CWS, Vic, Tas. **FL:** Mar–May.

A slender species, resembling *G. filifolia* but distinguishable by its shorter brown spikelets with all glumes nearly equal, and its smaller nut.

Gahnia sieberiana
Red-fruited Saw-sedge
Stems hollow, to 2m tall, occasionally much taller. Leaves arching, shorter than stems, 8–20mm wide with scabrous margins and back. Inflorescence 25–60cm long with many branches to 15cm long. Spikelets dense along branches, at first yellowish brown, dark brown to black at fruiting stage, with 5–8 glumes. Nut bluntly 3-angled, shiny, red, 3mm long, often hanging from spikelet by short staminal filaments. **HABITAT:** Sandy low-lying open ground and waterlogged areas. **DIST:** Coast and ranges. Also Qld, Vic, SA, PNG, NCal. **FL:** Sept–Dec and following fire.

Possibly the commonest saw-sedge in the Sydney district.

Gahnia sieberiana

Gahnia subaequiglumis

Gahnia subaequiglumis
Coarse perennial to about 1 x 1m. Stems 3–5mm diam., solid; bracts often exceeding inflorescence. Leaves as tall as inflorescence, scabrous, flat, inrolled towards apex, with fine long point; sheaths loose, dark brown. Spikelets dense in long narrow panicle, yellowish brown turning black at maturity, with 6–8 rough scabrous acuminate glumes. Nut obtusely 3-angled, 4–5mm long, shiny, red at maturity, often hanging from spikelet by short staminal filament. **HABITAT:** Periodically wet flat open ground. **DIST:** T'lands, uncommon on coast. Also Vic, Qld. **FL:** Irregular.

Gymnoschoenus
Tussocky perennials with rigid stems without nodes. Leaves basal. Inflorescence a dense globose head with numerous spikelets with 1 bisexual flower and 1 male flower above 4–6 empty glumes. Flowers with perianth of 2–3 bristles, 3 stamens and 3-fid style.

Gymnoschoenus sphaerocephalus
Button Grass, Button Bog-rush
Stems terete or slightly flattened, to 1m or more tall. Leaves linear, much shorter than stems, flat,

Gymnoschoenus sphaerocephalus

plano-convex or concave, 1.5–2.5mm across, smooth; sheaths to 20cm long, dark, striate, open. Inflorescence 15–20mm diam. with 3 broad orbicular involucral bracts at base. Spikelets subtended by small bracts with erose membranous margins. Glumes yellow-brown, papery. Stamens yellow. Style white. Nut obovoid, 3–4mm long, brown. **HABITAT:** Cool wet open places, often in heath and along streams. **DIST:** Scattered on coast and t'lands (e.g. Maddens Plains, Barren Grounds, Govetts Leap). Also NT, ST, SC, Vic, Tas, SA. **FL:** Oct–May.

A southern species, often forming extensive 'button grass plains' in high rainfall areas of Tasmania, reaching its northern limit in the Gibraltar Range east of Glen Innes.

Isolepis inundata

Isolepis

Mostly small tufted annuals with 3-angular or terete stems. Leaves few, narrow-linear or reduced to sheaths. Inflorescence terminal but often appearing lateral due to an erect bract. Spikelets in clusters or umbels. Perianth bristles absent. Fruit a smooth flattened or 3-angular nut.

Formerly included in *Scirpus*.

Isolepis inundata
Swamp Club-rush
Small tufted plant. Stems 10–30cm high terminating in cluster of 4–6 brown spikelets about 4mm long subtended by several bracts, frequently proliferous. Leaf blades short, slender or reduced to small point on sheath. **HABITAT:** Damp sandy or muddy ground. **DIST:** Coast and ranges. All States, NZ, S. Amer, Malesia. **FL:** Aug–Feb.

Isolepis nodosa
Knobby Club-rush
Common species with densely tufted stiff hard stems 40–90cm tall terminating in globular cluster of numerous sessile brown spikelets 5mm long; bract 2–4cm long. Leaves reduced to sheaths at base of stems. Nut shiny, brown-black. **HABITAT:** Coastal dunes, salt marsh, foreshores. **DIST:** Coast in all States, LHI. **FL:** Oct–Feb.

Lepidosperma

Tufted perennials with cylindrical, flat, concave or convex stems, often with sharp edges. Leaves basal, similar to stems or reduced to sheaths. Inflorescence a panicle or spike of sessile spikelets. Spikelets with 1 bisexual flower, 1 or more male flowers below it, and several outer empty glumes.

Lepidosperma concavum and *L. viscidum*
Sword-sedge, Hill Sword-sedge, Sticky Sword-sedge
These 2 spp. form an intergrading complex. Stems solid, flat, plano-convex or concave-convex, to 70cm x 2–10mm, margins acute, scabrous, often ciliate with resin particles on margins and sticky at the base in *L. viscidum*. Leaves similar to stems, but shorter, thinner and slightly narrower. Inflorescence 5–20 x 1.5–4cm long, tending to be short and broad with spikelets densely arranged (*L. concavum*) or long and narrow with spikelets loosely arranged (*L. viscidum*). Glumes minutely pubescent, mucronate or acute. **HABITAT:** Sandy soils in heath and open forest. **DIST:** Coast and ranges. Also Qld, Vic, Tas, SA. **FL:** Dec–Feb and after fire.

Isolepis nodosa

Lepidosperma concavum

Lepidosperma elatius
Tall Sword-sedge

Stems 1–2m x 6–12mm, biconvex with sharp scabrous margins. Leaves flat to concave-convex, 8–15mm wide, shorter and thinner than stems. Inflorescence loose, drooping, 30–60cm long with branches distant on axis; lowest involucral bract shorter than inflorescence. Spikelets 4–5mm long, brown, not clustered. Nut greyish or brown, 2.5–3mm long. **HABITAT:** Shady gullies, near rainforest. **DIST:** Coast and t'lands (e.g. Katandra Res.). Also Qld, Vic, Tas. **FL:** Dec–Feb.

Lepidosperma elatius

Lepidosperma forsythii
Stout Rapier-sedge

Stems arching, terete, 1m x 1.5–2mm, finely striate with loose pith in centre. Leaves reduced to long loose sheaths with short fine tip <1cm long. Inflorescence a very flexuose spike or reduced panicle to 6cm long with involucral bract 2–3cm long. Spikelets few, 8–12mm long; glumes 4–5, pale brown, the long mucro and subtending floral bract often reflexed at maturity. **HABITAT:** Wet or swampy ground. **DIST:** Uncommon. Coast (e.g. La Perouse, Maddens Plains). Also NC, Vic, Tas. **FL:** May–June.

Lepidosperma forsythii

Lepidosperma laterale

Lepidosperma limicola

Lepidosperma laterale
Variable Sword-sedge

Stems usually 50–100cm x 2–8mm, flat, plano-convex or concavo-convex, thin, shiny, with minutely scabrous cutting edges. Leaves thinner and shorter, with similar scabrous margins; sheaths not sticky. Panicle 6–25cm long, lowest bract usually about 6cm, occasionally much longer. Spikelets evenly spaced, not clustered; glumes minutely hairy, pointed. **HABITAT:** Moist sandy soils. **DIST:** Widespread, common. Coast to slopes and plains. All States. **FL:** Sept–Mar.

A number of forms, not always distinguishable, are included in this sp.

Lepidosperma limicola
Razor Sedge

Stems 100–120cm x 3–5mm, solid, striate, biconvex with sharp cutting edges. Leaves similar but shorter; sheaths not sticky. Panicle compact, 8–12cm long with lowest bract 5–10cm long. Spikelets crowded, narrow, sessile; glumes spirally arranged, glabrous, acutely pointed. **HABITAT:** Swamp or wet ground. **DIST:** Coast and ranges (e.g. Dancefloor Cave at Kanangra Walls, Braeside Walk at Blackheath). Also ST, NT, Vic, Qld. **FL:** Dec–Feb.

FLOWERING PLANTS

Lepidosperma longitudinale

Lepidosperma neesii

Lepidosperma urophorum

Lepidosperma longitudinale
Pithy Sword-sedge
Stems biconvex, 1–2m x 5–9mm, with soft loose pith, easily compressed, smooth with smooth margins; leaves similar but shorter with long acute apex, not sticky at base. Spikelets numerous, 6–8mm long, red-brown, crowded in narrow panicle 15–50cm long with short subtending bracts. Nut pale with dark brown markings. **HABITAT:** Swamps and lake margins. **DIST:** Coast (e.g. Agnes Banks, Thirlmere Lakes). All States. **FL:** Nov–Jan.

Lepidosperma neesii
Stiff Rapier-sedge
Stems 40–80cm x 1–2mm in sparse tufts, striate, terete, often with 1 (rarely 2) longitudinal groove(s), or weakly or strongly angled. Leaves 30–50cm tall, similar or slightly flattened or grooved. Panicle spike-like, 4–8cm long; lowest bract subulate, 3–4cm long. Spikelets numerous, 7mm long, clustered or in short dense spikes within short bracts; glumes dark, acute or acuminate, slightly pubescent. **HABITAT:** Damp heath. **DIST:** Coast and CT (e.g. Hill Top, Newnes Plat., Maddens Plains). Also NC, SC, Vic. **FL:** May–June.

Lepidosperma spp. are known as 'sword sedges' when the stems are broad and flattened, and 'rapier sedges' when they are slender and terete.

Lepidosperma urophorum
Stems terete, 80–150cm x 1–2mm, glabrous, smooth. Leaves shorter and narrower or reduced to dark sheathing bracts 10–15mm long with short lamina. Panicle 5–8cm long with short lowest bract, spike-like or with a few short erect branches. Spikelets 7–9mm long, few, appressed to branches, greyish or red-brown. **HABITAT:** Sandy soil, favouring shady gully slopes near rainforest. **DIST:** Coast and ranges. Also Vic. **FL:** July–Aug.

Lepidosperma filiforme is similar to *L. urophorum* but has pale or reddish leaf sheaths and a more open inflorescence with spikelets more spaced on slightly to strongly flexuous axes.

Lepironia articulata

Ptilothrix deusta

Schoenoplectus mucronatus

Lepironia
A monotypic genus related to *Chorizandra*, endemic and widespread in the Indian–Pacific region.

Lepironia articulata
Perennial with creeping rhizome. Stems rigid, septate, terete, grey-green, 1–2m x 4–9mm, terminating in tapering ellipsoid spike 2–3cm long, the spike appearing lateral due to subtending bract 2–6cm long. Spikelets numerous with 1 female flower above 15–25 mostly empty glumes. **HABITAT:** Mud or shallow fresh water in lakes and lagoons. **DIST:** Thirlmere Lakes to Moreton Bay in Qld. Also NT, Malesia, Madagascar. **FL:** Mar–May.

Ptilothrix
A monotypic genus endemic in eastern Australia.

Ptilothrix deusta
Small tufted plant. Stems usually 30–40cm x 1–2mm, terete, smooth, glabrous. Leaves basal, erect, linear, concave to channelled, to 30cm x 1mm. Inflorescence terminal, subtended by 2 dark sheathing bracts with pale membranous margins and unequal awn-like extensions to 5cm long. Spikelets numerous, 10–14mm long; glumes 6, scaly with dark midrib, exceeded by emergent yellowish perianth bristles. **HABITAT:** Open areas in heath, woodland and open forest. **DIST:** Coast and t'lands. Also Qld. **FL:** Sept–Nov; May–June after summer fires.

Schoenoplectus
Perennials with terete or 3-angular stems. Leaves basal, reduced to sheaths or with very short blades. Inflorescence terminal, sometimes appearing lateral. Spikelets in sessile cluster or irregular umbel. Fruit a 2- or 3-sided nut. Formerly included in *Scirpus*.

Schoenoplectus mucronatus
Stems 30–80cm tall, acutely 3-angled, leafless except for leaf sheaths at the base. Inflorescence a dense sessile cluster of pale yellow-brown spikelets 8–15mm long subtended by erect 3-angled bract 3–8cm long and continuous with stem; glumes acute, mucronate, striate, glabrous. Nut dark brown to black, shiny, exceeded by bristles. **HABITAT:** In shallow water and on creek and river banks. **DIST:** Coast, ne. NSW, nthn Aust. Also Malesia, Asia, Eur. **FL:** Nov–Apr.

Schoenoplectus validus

Schoenoplectus validus
River Club-rush
Stems terete, grey-green, soft, easily compressed, 1–2m x 4–9mm. Leaves reduced to sheaths, or uppermost with rudimentary blade. Inflorescence a large loose compound umbel longer than subtending bract. Spikelets brown, to 13 x 5mm; glumes fringed, with mucronate notched apex. Nut plano-convex, brown, not longer than barbed bristles. **HABITAT:** Margins of rivers, creeks and swamps, sometimes forming dense stands. **DIST:** Widespread (but absent from CT). All States, NZ, NCal, Americas. **FL:** Dec–Apr.

Schoenus
Annuals or perennials, mostly with leafless terete stems less than 1m high. Leaves basal, narrow-linear or reduced to short blade or sheath. Spikelets in terminal clusters, panicles, or solitary; glumes 2–8 in 2 opposite rows, dark, upper fertile glumes often on zig-zag axis. Flowers with perianth of bristles or scales or absent. Stamens usually 3. Style usually 3-branched. Fruit a 3-sided ribbed nut.

Schoenus apogon
Common Bog-rush, Fluke Bog-rush
Small tufted annual with slender striate terete stems 5–25cm tall. Leaves basal and on stems, linear or filiform, soft, to 15cm long. Inflorescence of several clusters of small spikelets subtended by leafy bracts. Spikelets brown to black, 4–6mm long with 2–4 empty acute glumes below 2–3 fertile obtuse glumes. Nut white. **HABITAT:** Damp to wet sandy soil. **DIST:** Widespread. Common in NSW (except WP). Also Qld, Vic, Tas, SA, NZ. **FL:** Nov–Dec.

Schoenus brevifolius
Zig-zag Bog-rush
Stems crowded on short rhizome, stout, stiff, erect, to 70cm tall. Leaves reduced to glabrous brown sheaths. Inflorescence a narrow panicle with 4–7 clusters of spikelets on short branches; bracts subtending branches short, brown, with erect point. Spikelets red-brown, 10mm long on short stalks with 3–5 fertile glumes above 2–4 shorter empty glumes. **HABITAT:** Damp sandy soils. **DIST:** Coast and t'lands. Also Qld, Vic, Tas, NZ, NCal. **FL:** Nov–Mar.

Schoenus ericetorum
Heath Bog-rush
Tufted perennial. Stems to 30cm tall, rigid, terete, grooved, with a terminal cluster of 1–14 spikelets. Leaves very short or reduced to sheaths, the sheaths brown with a short point and bearded opening. Spikelets 4–7mm long with 6–11 blackish glumes, some or all with woolly margins, upper 2 enclosing flowers, lower ones empty. **HABITAT:** Sandy and lateritic soils. **DIST:** Coast to WS, NWP. Also Qld, Vic, Tas, NZ, NCal. **FL:** Oct–Nov.

Schoenus apogon

Schoenus brevifolius

Schoenus ericetorum

Schoenus melanostachys
Black Bog-rush

Stems erect or weeping, 30–120cm long. Leaf sheaths dark brown to blackish, tubular, with reflexed point 5–8mm long, woolly at the top. Panicle 8–15cm long with spikelets on slender stalks, loosely clustered at nodes; glumes 7–11, dark brown to black with woolly or ciliate margins. **HABITAT:** Damp shady cool gullies. **DIST:** Coast and t'lands. Also Qld, Vic, Malesia. **FL:** Aug–Feb.

Depending upon conditions, this sp. varies from a small plant with a few stems 30–40cm tall bearing narrow spike-like inflorescences, to large spreading clumps on sandy creek banks and other sheltered damp places with large loose arching panicles of black spikelets on stems 1m or more long.

Schoenus melanostachys

Schoenus paludosus
Tufted annual. Stems slender, weak, 20–40cm tall with a loose panicle of reddish brown spikelets. Leaves narrow-linear, to 10cm long, chiefly basal, a few on stems passing into bracts on upper stem. Spikelets 2 or 3 on long stalks, pale brown, 4–5mm long; glumes 7–8, distichous on straight axis. **HABITAT:** Wet sandy soil, seepage areas in heath. **DIST:** Coast and CT. Also Qld. **FL:** Apr–June.

Schoenus turbinatus
Tufted perennial with slender grooved stems 30–40cm tall. Leaves 10–15cm long, filiform, inrolled, appearing terete. Inflorescence a dense head of 1-flowered pale red-brown spikelets 6mm long exceeded by 2–4 unequal scaly bracts with minutely scabrous margins, the longest about 6cm long. Nut 3-ribbed, dull brown or greyish, often exceeded by plumose bristles. **HABITAT:** Sandy soils in heath and open forest. **DIST:** Uncommon. Coast and CT. Also Vic, Tas. **FL:** Oct–Feb.

Schoenus villosus
Hairy Bog-rush
Stems rigid, terete, about 30cm x 1.5mm. Leaves with woolly dark striate sheaths at the base, 20cm x 1–2mm with 2–3 raised ribs and rough margins. Spikelets in 4–6 dense clusters at nodes, subtended by bracts longer or shorter than clusters, 8–14mm long with 2–6 flowers; glumes 7–14, pale brown to blackish, bearded with long hairs on margins. **HABITAT:** Moist sandy soils in heath and open forest. **DIST:** Coast and ranges. Also Qld. **FL:** May–June.

Schoenus turbinatus

Schoenus paludosus

Schoenus villosus

Scirpus

For long an unsatisfactory residue of heterogeneous elements, the Australian species of *Scirpus* were divided in 1981 into 5 segregate genera with only 1 species retained in *Scirpus*. It has leaves at nodes, leafy involucral bracts, scabrous inflorescence branches and short spikelets with numerous flowers in which the long perianth bristles become contorted at the fruiting stage.

Scirpus polystachyus
Large-headed Club-rush

Stems terete or slightly angular, 50–150cm x 4–6mm, stout, pithy. Leaves basal and at nodes, to 60cm x 7mm, striate on back, scabrous on keel and margins, with finely pointed apex. Inflorescence compound with many branches; bracts leafy, smaller than stem leaves. Spikelets 4–8mm long with brown glumes 4mm long. Nut pale brown, 1–1.5mm long, exceeded by bristles 2–3 times as long. **HABITAT:** Creek margins and boggy or wet sandy ground. **DIST:** Uncommon. T'lands. Also NC, SC, Qld, Vic. **FL:** Oct–Nov.

Scirpus polystachyus

POACEAE

There are about 150 species of native grasses belonging to 55 genera in the Sydney district. Many require examination with a microscope or lens to determine their identity. The species below are a small sample of common or noticeable species, selected to illustrate the main features of grasses.

Anisopogon
Erect tufted grasses. Inflorescence an open panicle with few spikelets. Lemma with 3 awns, central awn robust and twisted, 2 lateral awns shorter.

Anisopogon avenaceus
Oat Spear Grass
Stems 1–2, arching, to 1m tall. Leaf blade inrolled, ribbed on upper surface, finely pointed; ligule flat, <1mm long. Spikelets pendulous on fine stalks with 2 narrow outer glumes, 4–5cm long. Lemma densely hairy; longest awn 9cm long, twisted below bend.
HABITAT: Open forest, scrubby heath, on sandy soils. **DIST:** Common. Coast and ranges. Also SC, ST, Qld, Vic. **FL:** Oct–Nov.

Anisopogon avenaceus

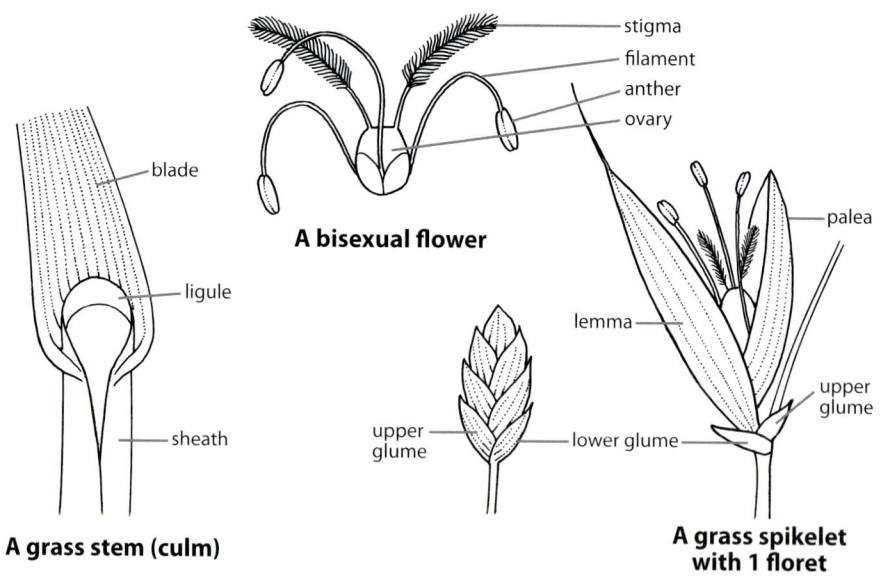

Aristida

Inflorescence a narrow or open panicle. Spikelets with 1 flower. Lemma terminating in a 3-branched awn.

Aristida ramosa
Purple Wire Grass
Tussocky perennial to 1m tall, often branching along stems. Leaves 1–3mm wide, flat or inrolled, ± scabrous. Panicle 10–20cm long, narrow, with loosely appressed branchlets. Glumes <half length of lemma. Lemma pale or purplish, 7–10mm long; awns all similar, 9–13mm long. **HABITAT:** Woodland, scrub, on clay-shale soils. **DIST:** Common. Coast and ranges. Also NC, SC, NT, WS, NWP, Qld, Vic, SA, WA. **FL:** Nov–Jan.

Aristida warburgii

Austrodanthonia tenuior

Aristida warburgii
Three-awned Spear Grass
Tufted perennial to 80cm tall. Leaves 1–2mm wide, folded or rolled inwards, bristly, minutely scabrous. Panicle with few branches, open, interrupted, to 20cm long. Glumes unequal. Lemma with purple markings, twisted on column below awns; awns 20–40mm long, dissimilar, middle awn bent. **HABITAT:** Open forest, heathy scrub. **DIST:** Coast and ranges. Also NC, SC, NT, NWS, Qld. **FL:** Oct–Nov.

Austrodanthonia, Notodanthonia

Fine-leaved tussock grasses. Spikelets with numerous florets (flowers). Lemmas 2-lobed, with short lateral awns on lobes and a longer twisted and/or bent awn between lobes; hairs on lobes in rows and/or tufts. Formerly included in *Danthonia*.

Austrodanthonia tenuior
Wallaby Grass
Erect perennial to 90cm tall with glabrous or hairy leaves 1–3mm wide. Panicles to 20cm long with numerous spikelets; spikelets 4–6-flowered. Glumes with purplish margins, 12–15mm long, longer than lemma. Lemma with dense hair tufts in 2 rows; central awn 7–10mm long. **HABITAT:** Open forest, on sandy or clay soils. **DIST:** Common. Coast and ranges. Most of NSW. Also Qld, Vic, SA, NZ. **FL:** Oct–Apr.

Aristida ramosa

Notodanthonia longifolia

Notodanthonia longifolia
Long-leaved Wallaby Grass
Tufted perennial with noded culms to 80cm high and slender, inrolled, slightly scabrous leaves to 35cm long. Panicles 5–15cm long, with ± erect branches, and many spikelets; spikelets with 5–6 loosely packed flowers. Glumes 8–12mm long. Lemma body 2mm long, with obscurely tufted hairs to 3mm long below sinus above shorter hairs over most of body; central awn 7–9mm long, slightly exceeding lateral awns. **HABITAT:** Woodland, forest, on rocky and sandy soils. **DIST:** Common. Coast and ranges. Most of NSW. Also Qld, Vic, Tas. **FL:** Oct–Jan.

Austrostipa
Perennials with single-flowered spikelets. Lemma with long fine awn which is spirally twisted and bent when mature. Spikelets separating below lemma to release slender sharp-pointed 'seed'.

Austrostipa ramosissima

Austrostipa pubescens
Tall Spear Grass
Open plant about 1m tall with pubescent stem nodes and loose panicles to 30cm long. Leaves 20–30cm long, narrow, inrolled, with short fringed ligule. Glumes unequal, 15–25mm long, exceeding lemma body. Lemma hairy, 10–15mm long, pimply and sparsely hairy; awn to 10cm long, bent twice. **HABITAT:** Open forest, woodland, heath, on sandy soils. **DIST:** Coast and ranges. Also NC, SC, NT, ST, Qld. **FL:** Oct–Dec.

Austrostipa ramosissima
Stout Bamboo Grass
Large cane-like grass to 2m tall with branches clustered or whorled at nodes. Leaf blade 1–10mm wide, scabrous. Panicle lax, narrow, 20–50cm long with many whorled branches. Spikelets on slender stalks. Glumes 2–3mm long. Lemma 2–3mm long, glabrous, pimply, dark brown; awn 3cm long, scabrous, fairly straight. **HABITAT:** Moist forested gullies (occasionally dry scrub). **DIST:** Coast and ranges. Most of NSW. Also Qld, LHI. **FL:** Oct–Mar.

Austrostipa pubescens

Chloris ventricosa

Chloris

Tufted perennials. Inflorescence with 2–10 or more spreading branches with spikelets on 1 side of the axis or rachis. Spikelets solitary, ± sessile, with 2–7 florets.

Chloris ventricosa
Tall Chloris

Glabrous plant to 1m tall with 3–4 branches 4–10cm long. Leaf sheath and ligule scabrous or with long hairs; blade flat, 2–3mm wide. Glumes unequal, 1–4mm long. Spikelets with 2 florets, lower spikelet fertile with lemma to 5mm long and awn to 17mm, upper spikelet sterile with lemma to 2.5mm long and awn to 9mm. **HABITAT:** Woodland, on clay-shale and alluvium. **DIST:** Common. Coast and ranges (e.g. Cumberland Pl., Albion Park, Glen Davis). Also NC, SC, NT, WS, NWP, Qld, Vic, SA, Eur., N. Amer. **FL:** Nov–Mar.

Cymbopogon

Tufted perennials with rigid stems. Leaves basal and on stems. Inflorescence a narrow panicle with spikelets in paired racemes subtended by a spathe-like bract, the racemes becoming reflexed at maturity and giving a barbed wire appearance.

Cymbopogon refractus
Barbed Wire Grass

Perennial to 1m tall with culms branching at nodes; nodes glabrous, purplish. Inflorescence ± glabrous, blue-green turning red-brown at maturity. Racemes with 2–6 spikelet pairs, lower spikelet sessile and fertile with 2 florets, upper spikelet stalked and sterile with 2 empty glumes. **HABITAT:** Open forest, woodland. **DIST:** Coast and ranges (e.g. Cumberland Pl., Hill Top). Most of NSW. Also Qld, Vic, NTerr. **FL:** Jan–May.

Dichelachne

Inflorescence often dense and spike-like. Spikelets with 1 floret. Lemma shorter or ± equal to glumes, with long bent awn attached below apex and often twisted in lower part.

Dichelachne crinita
Long-hair Plume Grass

Slender tufted perennial to 1m tall with unbranched erect stems. Leaves basal and on stems; blade 2–7mm wide; ligule to 5mm long. Panicle dense, 10–20cm long, slender with erect spikelets, later spreading slightly. Awns fine, silky, 25–55mm long, tinged purple. **HABITAT:** Woodland, open forest, grassland. **DIST:** Coast and ranges. Also NC, SC, NT, ST, WS, Qld, Vic, SA, WA, Pac. Iss. **FL:** Oct–Dec.

Dichelachne micrantha is similar to *D. crinita* but with narrower panicle and awns 15–20mm long.

Dichelachne rara
Rare Plume Grass

Tufted perennial to 1m tall with glabrous stems. Inflorescence open; spikelets relatively few, not hidden by awns; awns to 17mm long, bent and twisted. **HABITAT:** Woodland, grassland, on clay-shale soils. **DIST:** Scattered. Coast and ranges. Also SC, NT, ST, CWS, Vic, Tas. **FL:** Nov–Jan.

Cymbopogon refractus

Dichelachne crinita

Dichelachne rara

Echinopogon
Inflorescence a dense ovoid to oblong-cylindrical panicle. Spikelets with 1 floret, nearly sessile, separating above the glumes at maturity; lemma ± as long as glumes, 5–11-nerved, with short lateral projections and central awn, or entire with single long awn.

Echinopogon caespitosus
Tufted Hedgehog Grass
Tufted perennial with slender scabrous stems about 60cm tall, not bent at basal nodes. Leaves few, short, flat, 2–5mm wide, mostly basal. Panicle 1–10 x 1–5cm (incl. awns), continuous or interrupted in lower part. Lemma 5–7-nerved; awn 2–15mm long, scabrous.
HABITAT: Open forest, woodland. **DIST:** Common. Coast and ranges. Also NC, SC, NT, ST, NWS, Qld, PNG. **FL:** Oct–Jan.

Echinopogon ovatus has shorter ovoid inflorescences.

Echinopogon caespitosus

Entolasia

Inflorescence a very narrow panicle with erect branches. Spikelets with 2 florets, the lower one sterile. Glumes and lemma of similar size and appearance; lemma with fine silky white hairs, without awn.

Entolasia marginata

Entolasia marginata
Bordered Panic
Straggling perennial to 70cm tall, rooting at lower nodes and often branching at upper ones. Leaf blade narrow-ovate, flat, acute to acuminate, 2–12cm x 2–5mm; marginal veins thickened, pale; midvein raised beneath. Inflorescence short with spikelets often pressed close to axis. Upper glume and lower lemma similar with incurved pointed tips and enclosing shorter hairy upper lemma. **HABITAT:** Forests, preferring damp shaded sandy soils. **DIST:** Coast and ranges. Also Qld, Vic. **FL:** Nov–Mar.

Entolasia stricta
Wiry Panic
Straggling perennial with wiry stems, unbranched or branching at nodes among clusters of short leaves. Leaves grey-green, hairy and 2–5mm wide or green, glabrous and 1–2mm wide; blades at right angles to stems. Panicles 2–10cm long with spikelets in short erect racemes; spikelets 2–3.5mm long. **HABITAT:** Heath, scrub, woodland, open forest, on sandy soils. **DIST:** Common. Coast and ranges. Also NC, SC, NT, ST, NWS, Qld. **FL:** Oct–Mar.

Imperata
Erect coarse grass, spreading by long rhizomes, regenerating rapidly after fire. Inflorescence a dense cylindrical panicle with long fine silky white hairs surrounding spikelets and on glumes. Spikelets paired, with 2 florets.

Entolasia stricta

Imperata cylindrica

Joycea pallida

Imperata cylindrica
Blady Grass
Common perennial occurring in patches. Leaves sheathing on lower stems, tough, erect, to 100 x 2cm with minutely scabrous margins. Inflorescence cylindrical, to 25cm long; glumes 3–7-nerved, 3–4mm long; hairs 10–15mm long; lemmas 1–4mm long, awnless. **HABITAT:** Forests in open areas on sandy soils. **DIST:** Coast and lower Blue Mtns. Also NC, SC, NT, WS, all other States, SE Asia. **FL:** Nov–Feb.

Joycea
Inflorescence paniculate. Spikelets flat, with 3–6 bisexual florets; glumes ± equal, awnless. Lemma deeply 2-lobed with 1 awn in sinus between lobes; hairs on lobes not tufted or clearly in rows, mainly in lower part. Formerly included in *Danthonia*.

Joycea pallida
Silvertop Wallaby Grass, Red-anther Wallaby Grass
Tufted perennial with inrolled bristly leaves; sheath loose; ligule fringed, 1–6mm long. Panicle spreading, to 35cm long, often purplish. Glumes 10–16mm long. Lemma (incl. lobes) 6–9mm long; awn 7–17mm long, twisted near base. **HABITAT:** Open forest, woodland, heath. **DIST:** Common. Coast and ranges. Most of NSW. Also Vic. **FL:** Sept–Mar.

Microlaena stipoides

Microlaena
Slender perennials. Inflorescence racemose or a spike-like panicle. Spikelets with 3 florets, the outer 2 reduced to sterile lemmas with long awns, the fertile lemma shorter and awnless but narrowly pointed, separating as a unit above the glumes.

Microlaena stipoides
Weeping Grass
Tufted plant to 60cm tall with short flat linear leaves along stems. Panicle narrow, 10–17cm long, ± 1-sided. Glumes <1mm long with ring of hairs inside. Sterile lemmas rough on keel, 3–7-nerved; upper one 8–9mm long with awn 15–20mm long. **HABITAT:** Damp shaded areas of forest, woodland, on clay-shale soils. **DIST:** Common. Coast and ranges. Most of NSW. All other States, Malesia, NZ, Pac. Iss. **FL:** Any time of year.

Oplismenus

Creeping shade-loving perennials. Leaf blades broad, contracting to ligule and sheath. Inflorescence of racemes ± alternate on stems; racemes with several spikelets on silky axis; spikelets with 2 florets; outer glume with long awn.

Oplismenus aemulus

Creeping perennial with trailing leafy stems to 20cm tall. Leaves lanceolate, to 45 x 12mm, with undulate margins. Lower racemes 4–5cm long; spikelets crowded, 3mm long. Lower glume shorter than spikelet, 5-nerved; awn 7mm long; upper (inner) glume pointed. **HABITAT:** Moist, shaded forest, rainforest margins. **DIST:** Common. Coast and ranges. Also NC, SC, WS, Qld, Vic, PNG. **FL:** Dec–May.

Oplismenus imbecillis is also common. Its leaves are <7mm wide and not undulate, racemes are <2mm long and the inner glume is obtuse.

Oplismenus aemulus

Phragmites

Tall reed-like perennials. Inflorescence a large dense panicle.

Phragmites australis
Common Reed

Stems stout, noded, 2–4m tall. Leaves flat, linear, 20–60 x 1–3cm, with bases tightly sheathing and overlapping on stems. Panicles 15–30cm long, initially green, becoming purplish brown as bracts colour, finally fluffy and whitish at maturity due to long silky hairs on spikelet axes. **HABITAT:** Shallow fresh and brackish water. **DIST:** Chiefly coast. Most of NSW. All other States, cosmop. **FL:** Jan–Apr.

Poa

Tufted plants, often forming large tussocks. Spikelets with a number of florets (flowers). Lemmas keeled, often hairy towards the base and without awns.

Poa labillardieri
Tussock Grass

Dense perennial tussocks with stems to 1m tall with flat or inrolled leaves to 3mm wide, scabrous on the back. Inflorescence erect to 25cm tall, purplish grey, narrow at first, but opening at maturity into a large spreading panicle with spikelets occurring along the upper parts of branches. **HABITAT:** Open forest, woodland, gullies, creek banks, on clay-shale soils. **DIST:** Coast and ranges. Most of NSW. Also Qld, Vic, Tas, SA. **FL:** Chiefly Mar–Apr.

Phragmites australis

FLOWERING PLANTS

Poa labillardieri

Spinifex sericeus

Spinifex
Beach grasses with vigorous stolons. Plants of 2 types: one producing heads of female or bisexual spikelets, the other producing spikelets of male flowers.

Spinifex sericeus
Hairy Spinifex
Perennial plant with hairy grey inrolled leaves. Male inflorescences of 2-flowered spikelets in racemes or spikes in groups of 4–6 with spathe-like bracts. Bisexual or female inflorescences of spikelets in sessile racemes at centre of large spiny spherical heads, becoming detached at maturity. **HABITAT:** Coastal sand dunes, beaches. **DIST:** Coast. Also NC, SC, Qld, Vic, Tas, NZ, Pac. Iss. **FL:** Oct–Nov.

Themeda
Tufted grasses. Spikelets in condensed racemes subtended by a spathe-like bract with 1 to several of these subtended by a larger spathe, these at intervals along scape and forming a panicle. Spikelets sessile or stalked, male, sterile or bisexual in a complex arrangement.

Themeda triandra (syn. *T. australis*)
Kangaroo Grass
Reddish brown tufted perennial to 1m high. Leaves basal and on stems, 20–40cm x 3–5mm, flat or folded along midrib. Inflorescence to 25cm long with subtending sheaths ± glabrous. Fertile spikelets sessile, about 8mm long with a lower sterile floret and an upper bisexual floret,

Themeda triandra

the latter with an erect awn on the lemma 4–6cm long. **HABITAT:** Grassy woodland, grassy heath on coastal headlands, esp. on heavier clay soils. **DIST:** Common. Coast and ranges. Most of NSW. All other States, PNG. **FL:** Dec–Mar.

Sparganium subglobosum

Typha orientalis

SPARGANIACEAE

Sparganium
Aquatic perennials. Emergent leaves linear, with sheathing base. Flowers in unisexual, stalked or sessile clusters along the stem, the female clusters lower on stems.

Sparganium subglobosum
Bur-reed
Rush-like plant with creeping rhizome. Leaves mostly basal, also on stems, progressively smaller above, 60–80cm x 3–6mm, often 3-angled, apex acute. Inflorescence unbranched or with 1–2 short branches; clusters 2–20, males to 10mm diam., females to 20mm, burr-like at fruiting. Individual fruit 5–6mm long, pointed with persistent style. **HABITAT:** Freshwater ponds, slow streams, in shallow water or mud. **DIST:** Localised. Coast (e.g. Longneck Lagoon), Sthn H'lands (e.g. Tallong). Also NC, SC, NT, ST, SWS, Qld, Vic, NZ. **FL:** Nov–Feb.

TYPHACEAE

Typha
Aquatic perennials with extensive rhizomes. Leaves stem-sheathing in 2 rows, erect, linear. Flowers in cylindrical spikes. Two native species in Australia and one introduced species.

Typha orientalis
Broad-leaf Cumbungi, Bulrush
Perennial with erect stems to 3m. Leaves in 2 rows, to 2m x 30mm wide. Spikes chestnut brown; male spikes above female spikes, separated by 0–5cm of bare axis. Female spikes 8–30 x 1–4cm, producing masses of fluffy seeds distributed by wind. **HABITAT:** Lagoons, backwaters and other fresh water sites. **DIST:** Coast and ranges. Most of NSW. All other States, Malesia, NZ. **FL:** Nov–Apr.

Typha domingensis is much less common in Sydney region than *T. orientalis*. It has leaves and female spikes <20mm wide.

Alpinia caerulea

Libertia paniculata

ZINGIBERACEAE

Alpinia
Perennial herbs. Leaves ± distichous, with long sheathing base and stalked or sessile lamina. Inflorescence a thyrse or spike. Flowers bisexual, tubular, 3-merous. Fruit a fleshy capsule.

Alpinia caerulea
Native Ginger
Erect perennial with leafy stems to 1.5m tall. Leaves glabrous, to 40 x 10cm. Flowers white with yellow, in short clusters to 30cm long. Fruit globular, 10cm diam., with blue outer covering when mature. **HABITAT:** Rainforest, on moist but well-drained soil, in shaded sites with some filtered sunlight. **DIST:** Rare. Gosford district (e.g. Maitland Bay). Also NC, Qld. **FL:** Nov–Dec.

Alpinia arundelliana has smaller leaves than *A. caerulea*, rose-red flowers and dark blue-black fruit. Recorded from Gap Ck (Olney SF) and Matcham.

IRIDACEAE

Libertia
GRASS-FLAG
Tufted herbs with rhizomes. Leaves basal, distichous and folded along the midvein. Flowers with very short floral tube and 3 petals slightly to distinctly larger than sepals; stamens 3; style deeply divided, with 3 linear lobes. Fruit a 3-valved ovoid capsule.

Libertia paniculata
Branching Grass-flag
Leaves linear, 25–60cm x 4–12mm, flat, overlapping, bases in same plane on opposite sides of scape. Scapes with numerous loose clusters of 3–6 white flowers on side-branches. Petals 8–13mm long, longer and broader than sepals. Stamens with conspicuous yellow anthers. **HABITAT:** Damp shaded slopes and rainforest margins. **DIST:** Coast and ranges. From Vic to Qld, mainly coast. **FL:** Sept–Oct.

Libertia pulchella

Libertia pulchella
Pretty Grass-flag
Small grass-like plant usually <25cm tall. Leaves <20cm x 5mm, flat with bases overlapping. Flowers white in irregular loose clusters on leafless scapes longer than leaves. Petals 3–5mm long, similar to or slightly larger than sepals. Stamens with conspicuous yellow anthers. **HABITAT:** Mossy and wet rock ledges, near waterfalls and along shaded rocky creeks. **DIST:** Blue Mtns (e.g. Dantes Glen, Lawson). T'lands, from Barrington Tops to Vic. Also Tas, NZ, PNG. **FL:** Nov–Dec.

Patersonia
NATIVE IRIS, PURPLE-FLAG
Small tufted herbs with rhizomes. Leaves linear, usually overlapping in a plane on lower scape. Floral tube narrow with 3 broad petal-like sepals and 3 tiny petals at the top, and 3 yellow stamens and 3-branched style emergent from the tube. Plants produce a succession of fragile short-lived flowers on sunny days from individual bract-pairs held within 2 larger brownish spathe bracts.

Patersonia fragilis
Short Purple-flag
Leaves basal, 15–50cm x 1–6mm, flat to almost cylindrical, green to blue-green, sometimes twisted. Scapes with basal leaf-like bracts, usually much shorter than leaves. Spathe bracts 25–45mm long, green to pale brown, similar in texture to leaves. Flowers mauve-blue; sepals 10–25mm long; tube glabrous. **HABITAT:** Damp sandy soils in heath and woodland. **DIST:** Coast and ranges. Also NC, SC, ST, Vic, Tas, SA, Qld. **FL:** Sept–Dec.

Patersonia glabrata
Leafy Purple-flag
Leaves on stems, to 50cm x 2–5mm, flat, finely veined, glabrous with silky hairs on lower margins. Scapes slender, 10–40cm long, glabrous or with short silky hairs at the base. Spathe bracts 4–6cm long, brown, smooth, silky-hoary, not woolly. Flowers pale violet. **HABITAT:** Heath and woodland on sandy, shaley soils, also laterite. **DIST:** Coast and t'lands. From Eden to se. Qld. **FL:** Sept–Nov.

Patersonia fragilis

Patersonia glabrata

Patersonia longifolia

Patersonia longifolia
Dwarf Purple-flag
Leaves basal, to 40cm x 1–2mm, bluish green, biconvex, not grooved, often twisted, rather lax, margins with inwards-pointing hairs (i.e. across leaf surface). Scapes to 15cm long, leafless. Spathe bracts 25–35mm long, dark-coloured, silky-woolly when young. Flowers purple-mauve. **HABITAT:** Heath, woodland, open forest. **DIST:** Coast and ranges. From Hunter R. to SC, ST and Genoa R. (Vic). **FL:** Oct–Dec.

This plant is similar to *P. sericea* and is sometimes regarded as a subspecies of it but generally has fewer and narrower leaves and a shorter scape with shorter and narrower spathe bracts. The inwards-pointing hairs on the leaf margins are characteristic.

Patersonia sericea
Silky Purple-flag
Densely tufted plant with basal leaves to 40cm x 2–6mm. Leaves flat to almost terete, grooved, with woolly lower margins. Scapes to 40cm long, leafless and silky-woolly at the top when young. Spathe bracts 30mm long, dark-coloured, silky-woolly when young. Flowers deep violet to mauve. **HABITAT:** Heath, woodland, open forest. **DIST:** Coast, t'lands and slopes. From Qld to Vic. **FL:** July–Dec.

Patersonia sericea

LILIACEAE

The 15 genera included here in Liliaceae are placed in 6 segregate families in *Flora of New South Wales*, which follows a different system of classification.

Alania
Clumped herb with branched stems rooting at nodes. Leaves filiform, crowded along stems bearing umbel-like racemes of white flowers on filiform peduncles. Flowers with 6 equal tepals and 6 stamens with long anthers. Fruit a capsule with black seeds.

Alania endlicheri

Alania endlicheri
Alania
Perennial herb with stems to 30cm long. Leaves to 12cm x 0.5–1mm, minutely toothed, acute at tip, expanded and brown at base. Umbels 20–30-flowered on scapes to 9cm long. Flowers subtended by scaly brown bracts on scape; tepals white, 3mm long, 3- or 5-veined, persisting in fruit. **HABITAT:** Wet rock faces in sheltered gullies. **DIST:** Restricted. Blue Mtns (e.g. Wentworth Falls, Kings T'land, Bowen Ck Rd, Mt Coricudgy) with unusual occurrence along Piles Ck (Brisbane Water NP). **FL:** Nov–Dec.

Arthropodium
VANILLA LILY
Tuberous perennial herbs with basal linear leaves, often withering early. Flowers on erect, simple or branched scapes with 1–9 per node; tepals not twisted after flowering, outer 3 narrower than inner 3. Stamens 6, with bearded filaments. Capsule with black seeds.

Arthropodium millefrorum
Pale Vanilla Lily
Branching slender herb with flat, often glaucous leaves 20–50cm x 4–10mm. Flowers in clusters of 2 or more, white to pale lilac on slender pedicels 8–15mm long; tepals 5–9mm long. Staminal filaments with yellow beard above middle. **HABITAT:** Damp

Arthropodium millefrorum

grassy slopes, open woodland and forest. **DIST:** Widespread on moderately fertile soils throughout NSW (except FWP). Also Qld, Vic, SA, Tas. **FL:** Nov–Dec.

Arthropodium minus (Small Vanilla Lily) also occurs in the area. It is smaller in all its parts with white to rich purple flowers borne singly on pedicels <8mm long. Its staminal filaments are bearded almost to the base.

Blandfordia
CHRISTMAS BELL
Tufted perennial herbs. Leaves linear, pointed at tip, slightly sheathing at base. Inflorescence a terminal raceme with 1–20 pendent flowers. Flowers tubular with 6 short rounded lobes and 6 stamens as long as the tube. Fruit a 3-angled capsule with numerous small brown seeds.

Blandfordia cunninghamii
Mountain Christmas Bell
Stout plant with basal tuft of leaves to 1m x 7–12mm with smooth translucent margins. Flowers in dense clusters of 7–20 or more on scape to 80cm long; floral tube tapering to short pedicel, to 50 x 20–30mm; stamens attached to tube below middle. **HABITAT:** South-facing wooded slopes. **DIST:** Endemic. Blue Mtns (e.g. Wentworth Falls, Blackheath, Mt Tomah), Mt Kembla. **FL:** Nov–Dec.

Blandfordia cunninghamii

Blandfordia grandiflora

Blandfordia nobilis

Blandfordia grandiflora
Northern Christmas Bell
Erect plant similar to *B. nobilis* but larger in most respects. Leaves to 70cm x 3–5mm, margins minutely rough, slightly thickened. Flowers in clusters of 3–10 or more on scape to 80cm long; floral tube 35–50 x 25–35mm, intermediate in shape between other 2 spp.; stamens attached to lower half of tube.
HABITAT: Damp sandy/peaty soils in heath, swamp margins. **DIST:** Coast and ranges. From Hawkesbury R. to Gympie (Qld). **FL:** Dec–Jan.

Blandfordia nobilis
Christmas Bell
Erect plant with basal tuft of linear leaves about 60cm x 3–5mm; leaf margins minutely rough and slightly thickened. Flowers in loose clusters of 3–10 or more on scape to 70cm long; floral tube 20–30 x 7–10mm, cylindrical in upper two-thirds with slight constriction below lobes, tapering abruptly to slender pedicel 30–35mm long; stamens attached to tube at middle.
HABITAT: Open forest, damp heath with sedges, swamp margins. **DIST:** Coast. Hawkesbury R. to SC and ST. **FL:** Dec–Feb (sometimes earlier).

Intermediate forms, possibly hybrids with *B. grandiflora*, occur between Hawkesbury R. and Gosford.

Bulbine
LEEK LILY
Perennial or annual herbs with linear succulent leaves sheathing at the base. Flowers in racemes on stout scapes, the lower flowers opening first; tepals 6, in 2 whorls of 3, all similar, each with a central stripe on outer surface; filaments of 3 or all 6 stamens bearded. Fruit a rounded capsule.

Bulbine bulbosa
Native Leek, Golden Lily
Perennial herb, usually with bulb. Leaves to 40cm long, slightly scabrous, narrow, channelled, hollow. Scapes 1 or 2, to 50cm tall. Flowers numerous; tepals 10–20mm long; all 6 stamens bearded. Capsule 6mm across, globose, 3-celled. **HABITAT:** Grassland and woodland on shale or clay soils and rocky outcrops. **DIST:** Cumberland Pl. and higher ranges (e.g. Wallerawang, Jenolan Caves, Wombeyan Caves Rd). All States except WA. **FL:** Oct–Nov.

This sp. dies back in summer and produces new leaves following autumn and winter rains.

Bulbine semibarbata, an annual with tepals 4–6mm long and only 3 stamens bearded, occurs near Kiama and Wallerawang.

Bulbine glauca, a non-bulbous perennial with 2 or more glaucous scapes, occurs in Megalong Valley.

FLOWERING PLANTS | 533

Bulbine bulbosa

Burchardia

Perennial herbs with tuberous roots and basal leaves. Flowers solitary or in umbels of up to 20 on stout scapes with 1–4 bracts. Tepals 6, in single whorl, stamens 6; ovary superior, 3-angled; style 3-branched. Fruit an elongated angular capsule splitting into 3 carpels.

Burchardia umbellata
Milkmaids

Herb with erect unbranched scape 15–45cm tall, 1 or 2 linear leaves to 40cm x 4mm at the base and 1 or 2 smaller bracts higher up. Tepals spreading, obovate, 5–8mm long, white; anthers purple; ovary red; style short. **HABITAT:** Wet heath, grassland, woodland. **DIST:** Coast and ranges. Also NC, SC, ST, CWS, SWS, all States. **FL:** Sept–Nov.

Burchardia umbellata

Caesia
GRASS LILY
Tufted perennial herbs with linear leaves crowded at base of stems. Flowers in panicles on a simple or branched axis. Tepals 6, 3-nerved, equal, often reflexed when open, spirally twisted after flowering; stamens 6, glabrous, alternately long and short. Capsule 3-lobed, with black seeds.

Caesia parviflora var. *parviflora*
Blue Grass Lily
Herb 20–50cm tall with leaves to 5mm wide. Tepals to 9mm long, white, occasionally pinkish or bluish, with green or purple veins. Filaments greenish white; anthers yellow. **HABITAT:** Grassland and woodland. **DIST:** Coast and Blue Mtns. Also NC, SC, SWS, Qld, Vic, Tas. **FL:** Nov–Jan.

Caesia parviflora var. *vittata*
Blue Grass Lily
Herb 20–50cm tall, larger in all its parts than var. *parviflora*. Leaves to 40cm x 8mm. Tepals to 9mm long, pale to dark blue with darker veins. Filaments blue with white bands at base; anthers yellow. **HABITAT:** Grassland and woodland. **DIST:** Coast, ranges and slopes in NSW. Also Qld, Vic. **FL:** Nov–Feb.

Dianella
FLAX LILY
Glabrous perennial herbs. Leaves linear, forming flattened sheaths in their lower part, fused (occluded) or folded lengthwise immediately above, then open to the tip, the open portion grooved above, keeled along midrib, flat or recurved to revolute. Inflorescence cymose, on stems emerging from leaf sheaths. Tepals 6, all similar, often reflexed. Stamens with glabrous filaments, thickened at the top; anthers large. Fruit a shiny blue-purple berry.

This genus is under revision. It is likely that a number of forms and newly discovered populations from Colo–Putty, Megalong Valley, Dharawal NR and Wattamolla will be described as new species.

Dianella caerulea
Paroo Lily
Tufted plant with different forms, some reaching 1–2m in height or forming mats. Leaves 10–70cm x 5–25mm, closely and strongly veined, flat for most of length with finely serrated margins and keel. Flowers on

Caesia parviflora var. *parviflora*

spreading branched stems exceeding leaves; pale to dark blue (greenish blue or greenish white in some forms), 6–11mm long, sepals 5–7-veined, petals 3–5-veined; stamens with yellow or yellow-brown anthers 2–3 times longer than thickened part of orange filament. **HABITAT:** Heath, forest. **DIST:** Coast to WS. Also Qld, Vic, Tas. **FL:** Oct–Feb.

D. caerulea is a complex with some forms distinct enough to be regarded as varieties. Six occur in the Sydney district. *D. congesta* is similar but has leaves with smooth margins and midrib and a short congested inflorescence with few flowers among the foliage. It forms mats on sand near the sea in a few places (e.g. Wattamolla, Kurnell).

Caesia parviflora var. *vittata*

Dianella caerulea

Dianella longifolia var. *longifolia*

Dianella longifolia var. *longifolia*
Smooth Flax Lily
Tufted perennial herb to 1.5m high. Leaves 20–80cm x 5–25mm, sheathing for less than half length, glaucous or grey-green, purplish or brownish at base; veins obscure, margins and keel smooth or slightly scabrous, recurved or revolute. Inflorescence exceeding foliage. Flowers usually blue, 5–10mm long, sepals 5-veined, petals 3-veined. Stamens with orange filaments and pale yellow anthers 3–5mm long above filament swelling 1–2.5mm long.
HABITAT: Open forest, woodland and grassland on clayey soils, near creeks. **DIST:** Coast and ranges to WP. All States. **FL:** Oct–Jan.

Dianella prunina

FLOWERING PLANTS | 537

Dianella revoluta var. *revoluta*

Dichopogon fimbriatus

Dianella prunina
Tufted perennial with narrow base, not forming mats by spreading underground. Leaves to 90cm long, purplish red near base, folded and fused in lower half, opening to blade 10–35mm wide in upper part. Inflorescence much exceeding foliage. Tepals blue to violet, 7–10mm long, 5-veined. Stamens with brown-yellow or blue anthers longer than yellow filaments. **HABITAT:** Tall eucalypt forests. **DIST:** Endemic. Wollombi–Putty Rd to lower Blue Mtns and Sthn H'lands. **FL:** Oct–Dec.

Dianella tasmanica grows in similar places. It is a clumped plant <1m tall with similar leaves having finely serrated margins and keel. The inflorescence is shorter or slightly longer than the foliage and the flowers are lavender-violet with yellow anthers. The berry is narrow and 12–25mm long.

Dianella revoluta var. *revoluta*
Spreading Flax Lily, Black-anther Flax Lily
Tufted perennial herb to 80cm high, often forming clumps or mats. Leaves 15–85cm x 4–12mm, grey-green or glaucous, sometimes reddish at base, margins and midrib usually smooth. Inflorescence exceeding or level with foliage. Flowers mid to dark blue or violet, 5–10mm long, sepals 5–7 veined, petals 5-veined. Stamens with pale yellow-brown or black anthers >3 times as long as thickened part of yellow filament. **HABITAT:** Eucalypt forest, woodland. **DIST:** Coast and ranges. Most of NSW (except FWP). All States. **FL:** Oct–Jan.

Dichopogon
CHOCOLATE LILY
Tuberous perennial herbs with linear leaves at base often surrounded by previous year's growth. Flowers lilac or mauve, chocolate- or vanilla-scented, with 3 broad wavy-edged petals alternating with 3 narrower sepals. Stamens with short glabrous filaments and long anthers with hairy basal appendages. Capsule with black seeds.

Dichopogon fimbriatus
Nodding Chocolate Lily
Erect herb usually 50–60cm tall. Leaves flat, grass-like, often withering before flowering commences. Flowers erect or nodding on curved articulate stalks in clusters of 2–6 per node, lilac, 12–25mm across; anthers dark purple with hairy yellow appendages. **HABITAT:** Grassland, open forest. **DIST:** Higher elevations (e.g. Wingello/Tallong), also coast (e.g. Doonside, Burragorang V., Agnes Banks), t'lands to plains in NSW. Also Qld, Vic, SA. **FL:** Oct–Jan.

Dichopogon strictus is rare in the Sydney region. Its flowers occur singly on the stems.

Laxmannia gracilis

Laxmannia
WIRE LILY

Wiry herbs with flowers in compact terminal head-like umbels surrounded by an involucre of dry membranous bracts. Flowers with 6 similar tepals and 6 stamens, 3 erect and 3 fused to inner tepals (petals). Fruit a triangular capsule.

Laxmannia gracilis
Slender Wire Lily

Slender wiry plant with erect stems to 15cm tall, often with slender brown stilt-like roots. Leaves linear, 20–50 x 1–2mm, tufted at base and at intervals along stems. Flowers sessile on long leafless axes; white to pink, sepals shorter than petals. **HABITAT:** Open forest, woodland. **DIST:** Coast and ranges. Also NWS, NWP, Qld, Vic. **FL:** July–Nov.

Laxmannia compacta is compact and prostrate with the sepals equal to, or longer than, petals.

Schelhammera
LILAC LILY

Soft herbs. Leaves alternate, 2-ranked, sessile, with longitudinal veins. Flowers solitary or in few-flowered cymes. Tepals 6, inner 3 slightly broader than others; stamens 6; style 3-branched. Fruit a capsule.

Schelhammera undulata
Lilac Lily

Prostrate or erect herb with stems about 20cm long. Leaves 3–5cm x 8–15mm, ovate-lanceolate with cordate base and undulate margins. Flowers on slender terminal or axillary stalks, 10–15mm across; tepals lilac; stamens with purple anthers. Capsule ovoid, wrinkled, 5–8mm across. **HABITAT:** Shaded forest. **DIST:** Coast. From Vic nth to Lismore. **FL:** Chiefly Oct–Nov.

Sowerbaea
RUSH LILY

Herbs with onion-like leaves and stout scapes exceeding leaves. Flowers in simple or compound umbels subtended by bracts. Tepals 6, elliptic, concave, papery, 1-nerved, not twisted after flowering. Capsule obscurely 3-lobed, with brown or black seeds.

Sowerbaea juncea
Rush Lily, **Vanilla Plant**

Tender herb with leaves to 40cm x 2mm, terete, flattened near tip. Scape to 45cm long, usually unbranched, smooth. Flowers pink-lilac, rarely white, with 3 yellow stamens alternating with 3 staminodes lacking anthers. **HABITAT:** Wet or intermittently waterlogged shallow sandy sites in heath. **DIST:** Coast and ranges. Also ST, coast from Qld to Vic. **FL:** Oct–Dec.

Schelhammera undulata

Sowerbaea juncea

Stypandra glauca

Stypandra
BLUE LILY

Monotypic genus. Perennial with erect or ascending leafy stems. Flowers nodding; tepals 6, all similar, petaloid, spreading or reflexed. Stamens 6, shorter than tepals; filaments densely hairy above a kinked midpoint; anthers becoming recurved or coiled after pollen release. Capsule 3-valved; seeds tiny, black, with ridged margins.

Stypandra glauca
Nodding Blue Lily

Herbaceous or shrubby perennial to 1.5m tall. Leaves linear with sheathing bases, arranged alternately on stem in 1 plane, 5–20cm x 5–15mm, smooth, often slightly glaucous. Flowers in loose terminal cymose clusters, pale to deep blue, about 20–30mm across, tepals narrowly elliptic, 5-nerved; stamens yellow-woolly on upper part of filament. **HABITAT:** Forest, in rocky ground. **DIST:** Coast and ranges. Throughout NSW. Also Qld, Vic, SA, WA. **FL:** Chiefly July–Oct.

Thelionema

A genus of 3 species formerly included in *Stypandra* but having erect flowers and differing in the stamens and seeds.

Thelionema caespitosum
Tufted Blue Lily

Tufted herb 30–90cm tall. Leaves 10–45cm x 5–10mm, sheathing in 1 plane from base; lower surface papillose with prominent parallel veins and keeled midrib. Inflorescence exceeding leaves, branched, with leaf-like bracts becoming smaller higher on stems. Tepals 8–14mm long, blue, rarely yellowish white. Staminal filaments yellow-woolly almost to base. Capsule obovoid, 4–10mm long. **HABITAT:** Open forest and rocky outcrops. **DIST:** Coast and ranges. Also NC, SC, ST, Qld, Vic, SA, Tas. **FL:** Oct–Dec.

Thelionema umbellatum

Tufted herb 30–40cm tall. Leaves 15–35cm x 2–4mm with a few parallel veins on lower surface. Inflorescence bracteate with 3–10 creamy white flowers, shorter than to barely exceeding leaves. Tepals 5–8mm long. Stamens with yellow-hairy filaments. Capsule obovoid, 5–7mm long. **HABITAT:** Heath on shallow sandy and peaty soils and on rock ledges. **DIST:** Coast and ranges (e.g. Kurnell, Narrow Neck). Also SC, ST, Vic, Tas. **FL:** Sept–Dec.

Thysanotus
FRINGE LILY

Erect or (rarely) twining herbs. Leaves mainly basal, linear, often absent or withered at flowering. Flowers with 3 narrow entire sepals and 3 broad fringed petals. Capsule 3-valved, globular, with several black seeds.

Thysanotus juncifolius

Erect perennial to 40cm tall. Leaves annual, few, to 10cm long, often absent. Flowering stems grooved, hairy near base, glabrous in upper part with V-shaped branching; umbels 1–5-flowered. Flowers mauve; sepals 5-veined, 1.5–2mm wide; petals 10–14 x 4–5mm with fringe 3–4.5mm long. **HABITAT:** Dry sandy soils in heath and woodland. **DIST:** Chiefly coast; also Blue Mtns; NC, SC, Vic, SA. **FL:** Oct–Jan.

Thelionema caespitosum

Thelionema umbellatum

Thysanotus juncifolius

Thysanotus patersonii

Thysanotus tuberosus subsp. *tuberosus*

Thysanotus patersonii
Twining Fringe Lily

Delicate leafless twiner with tuberous roots; stems produced annually, much-branched, terete, hairy at base. Flowers solitary at end of short branchlets; petals 8–10 x 4–5mm, fringe 1mm long. **HABITAT:** Open forest on sandy soils. **DIST:** Rare and restricted in Sydney district (Tallong and higher ranges). Coast to plains in NSW. All States. **FL:** Oct–Nov.

Thysanotus tuberosus subsp. *tuberosus*
Common Fringe Lily

Erect perennial to 40cm tall. Leaves shooting annually from tuberous roots, present but often withering at flowering, 20–40cm long, linear becoming terete near apex. Stem leaves (if present) small, dried, few. Flowers in open paniculate inflorescence in umbels of 1–8 on much-branched stems; tepals purple, 7–17mm long; sepals 2–2.5mm wide; petals 5–7mm wide with fringe 5mm long. **HABITAT:** Heath, woodland. **DIST:** Coast and ranges. Throughout NSW (except FWP). Also Qld, Vic, SA. **FL:** Oct–Dec.

Although the flowers are short-lived, opening on sunny days and lasting only 1 day, the plant produces flowers over several months.

Thysanotus virgatus

Tricoryne elatior

Tricoryne simplex

Thysanotus virgatus
Erect perennial to 30cm with fibrous roots, usually without basal leaves. Flowering stems branched, with scabrous ridges and mostly 2-flowered umbels. Tepals 10–12mm long; sepals 6–7-veined, 3–4mm wide; petals 5–6mm wide with fringe 4–5mm long. **HABITAT:** Periodically dry lateritic and gravel areas in heath and low open woodland. **DIST:** Endemic. Royal NP to Mt Keira. **FL:** Oct–Dec.

Tricoryne
RUSH LILY
Small herbs with fibrous roots and narrow erect leaves, mainly from the base, reduced on stems. Flowers yellow; tepals spirally twisted after flowering; stamens with tuft of yellow hairs below anthers. Fruit splitting into 1–3 smaller 1-seeded mericarps or cocci.

Tricoryne elatior
Yellow Rush Lily
Perennial with wiry much-branched stems 10–50cm tall. Leaves linear, to 10cm x 3.5mm, scale-like on upper branches. Flowers in umbels of 2–10; tepals 3-veined, 10–14mm long; outer 3 oblong; inner 3 elliptic, shorter and broader. **HABITAT:** Grassland, woodland, open forest. **DIST:** Coast and ranges. Throughout NSW (except FWP). Also Qld, Vic, SA, NT, WA. **FL:** Dec–Feb.

Tricoryne simplex
Herb with unbranched or lightly branched stems <30cm tall. Leaves linear, to 25cm x 5mm. Flowers in umbels of 6–20 or more; tepals 3- or 4-veined, 5–9mm long; inner tepals slightly broader than outer. **HABITAT:** Grassland, woodland, open forest. **DIST:** Coast. Also NC, SC. **FL:** Sept–Dec.

XANTHORRHOEACEAE

Lomandra
MAT-RUSH
Perennial herbs with distichous linear or terete leaves ascending from sheathing bases on (usually) short stems. Inflorescence a spike, raceme or panicle, usually shorter than leaves. Flowers unisexual with males and females on separate plants, or bisexual; tepals 6 in 2 whorls, scaly; stamens 6; ovary superior, 3-locular. Fruit a capsule opening by splitting of outer walls. *Lomandra* is sometimes assigned to the segregate families Lomandraceae or Dasypogonaceae.

Lomandra brevis
Tufted Mat-rush
Perennial with dense tussock about 15cm tall. Leaves 1mm wide with inrolled margins; bases purplish above whitish sheath, tip smoothly pointed. Flowers yellow, drying black, borne singly on erect narrowly and sparsely branched panicles to 10cm tall. Male flowers globose, 2–3 x 3–4mm on stalk 3mm long; female flowers elongated, 3mm long on stalk 1mm long. **HABITAT:** Sheltered damp sandy situations in open forest. **DIST:** Endemic. Coast to lower Blue Mtns. **FL:** Sept–Dec, with some flowers Mar–May.

Lomandra brevis

Lomandra confertifolia subsp. *rubiginosa*
Perennial. Leaves channelled-concave, to 40cm x 2mm, margins scabrous, apex with 2–3 teeth; sheaths reddish brown with torn margins. Male inflorescence a narrow panicle or spike of clusters shorter than the flattened scape; cluster bracts spiny. Female inflorescences very short. Flowers yellow, often purplish, sessile, elongated to tubular, about 4mm long, with 2 brown floral bracts. **HABITAT:** Open forest on dry sandy soils. **DIST:** Coast and ranges, from Gloucester (NC) to Vic. Also CWS. **FL:** July–Nov.

Subsp. *pallida* occurs in rocky sandstone soils on or near the Gt Div. Ra. (e.g. Wolgan Gap, Glen Davis). It has shorter, flat leaves, often curved and lax, with whitish sheath margins and flowers with whitish bracts.

Lomandra cylindrica
Needle Mat-rush
Tufted perennial often with 2 leaves per flowering stem. Leaves terete or semi-terete, 15–60cm x 1–2mm, twisted, pointed at tip. Flowers spaced on stems, not clustered, yellow, globular bell-shaped; males 2mm long on stalks 3mm long; females 3mm long on stalks 1mm long; sepals shorter than petals. **HABITAT:** Open forest, on sandy and gravelly soils. **DIST:** Coast and ranges. Also ST, SC, Vic. **FL:** Sept–Dec.

Lomandra filiformis subsp. *filiformis*
Wattle Mat-rush, Iron Grass
Tufted plant with narrow incurved or inrolled leaves to 30cm x 3mm, with numerous whitish or purplish brown shreds at base; tip ragged, blunt. Inflorescence branched or unbranched on smooth axis; bracts tiny. Flowers borne singly, yellow, not blackening when dried; tepals unequal. Male flowers globose, 1.5–2mm long on stalk 3mm long; female flowers elongated, 3mm long, sessile, in shorter inflorescences. **HABITAT:** Open forest on sandy soils. **DIST:** Coast to WS, NWP. Also Qld, Vic. **FL:** Aug–Dec.

This subsp. intergrades with the 2 other subsp. in the area. It is similar to *L. brevis* but taller and less densely tufted and has smaller male flowers. It grows on drier, more open sites.

Lomandra filiformis subsp. *coriacea*. Leaves flat or almost so, firm and leathery. Male inflorescences with scabrous axis. Widespread from coast to ranges.

Lomandra confertifolia subsp. *rubiginosa*

Lomandra cylindrica

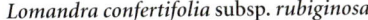
Lomandra filiformis subsp. *filiformis*

Lomandra fluviatilis
Perennial herb forming large tussocks to about 50cm high. Leaves to 60cm x 1–3mm, channelled-concave with 2–3 teeth at apex and dark red-brown, shiny, shredded sheath margins at base. Male inflorescences narrow with opposite or whorled branches on flattened scapes 2–3 times as long; female inflorescences shorter, mostly unbranched. Flowers sessile in clusters with conspicuous spiny cluster bracts and short brown floral bracts; sepals scaly and purplish or yellow; petals cream. **HABITAT:** Sandy creek beds in flood zone, creek banks, between large rocks near creeks. **DIST:** Endemic. Coast and ranges. **FL:** Jan–Mar, but irregular.

Lomandra glauca
Pale Mat-rush
Spreading herb with tufted leafy branches from decumbent stems, sometimes forming mats. Leaves 8–20cm x 1–2mm, blue-grey, flat to concave, often twisted, with white or pale brown shreds and tissue lattice at the base. Flowers yellow, sessile, 2–3mm long; males clustered in panicles to 12cm long on scape obscured by leaf bases; females in a solitary cluster on short stalk. **HABITAT:** Open forest on dry sandy and rocky soils, heath. **DIST:** Common. Coast and ranges. Also NC, SC, ST, CWS. **FL:** Sept–Nov.

Lomandra gracilis
Perennial with numerous tufted leaves. Leaves channelled or semi-terete, to 40cm x 1–2mm, strongly veined, with purplish brown shreds at base. Flowers, globose, borne singly on branched stems; tepals unequal, not spreading, yellow, about 3mm long. Male flowers on stalks 3–5mm long, female flowers on stalks 1–2mm long. **HABITAT:** Open forest on sandy soils in rocky situations. **DIST:** Coast to Blue Mtns. Also Qld. **FL:** Sept–Nov.

Lomandra longifolia
Spiny-headed Mat-rush
Variable species with tough strap leaves and large inflorescences with long spiny bracts at base of flower clusters. Leaves mostly 40–80cm x 8–12mm, much narrower in some forms, flat or slightly concave, apex with 2–3 pungent teeth. Male and female inflorescences with similar distichous branching; culms 5–6mm

Lomandra fluviatilis

wide, biconvex. Flowers in whorled clusters on side-branches, sessile, 3–4.5mm long; tepals yellow with purplish centres; sepals shiny; petals dull. **HABITAT:** Common in many situations including sand dunes, grassy headlands, exposed lateritic ridges, heath, open forest, creek banks and open areas in rainforest. **DIST:** Coast to WS. Also Qld, Vic, Tas. **FL:** Aug–Dec.

This species is not always distinct and narrow-leaf forms are sometimes difficult to distinguish from several others in the area, particularly *L. confertifolia* subsp. *pallida*.

Lomandra micrantha subsp. *tuberculata*
Small-flowered Mat-rush
Tufted perennial herb. Leaves stiff, erect, linear, convex or flat and inrolled, 30–40 x 1–3mm, rounded at tip, not shredding at the base. Flowers in branched inflorescences single or 2–3 together; male flowers to 3mm long, distinctly stalked; female flowers to 6mm long, sessile; tepals spreading, yellow or reddish, all similar. **HABITAT:** Open forest on sandy soils. **DIST:** Coast sth of Gosford to Blue Mtns. Also ST, Vic, SA, WA. **FL:** Mar–Sept.

FLOWERING PLANTS

Lomandra glauca

Lomandra gracilis

Lomandra longifolia

Lomandra micrantha subsp. *tuberculata*

Lomandra montana

Lomandra montana
Tufted perennial herb. Leaves about 50cm x 2–4mm, glossy green, thin, flat or slightly concave; base with a few dull shreds; apex with 3 pointed teeth, central tooth longest. Inflorescence unbranched, 5–8cm long on flattened scape about 25cm long. Flowers tubular, 3–5mm long, sessile in whorled clusters with conspicuous cluster bracts and short brown floral bracts; sepals scaly, green, yellow or purplish; petals creamy white. **HABITAT:** Hillsides and cliffs in moist forests, beside rocky creeks near waterfalls. **DIST:** Blue Mtns (e.g. Lawson, Grose R., Mt Wilson, Bowens Ck), w. to Rylstone. **FL:** Nov–Dec.

Lomandra multiflora
Many-flowered Mat-rush
Tufted perennial herb. Leaves 30–70 x 2–4mm, grey-green, firm, thick, flat or slightly concave, with few shreds at base, rounded at apex. Scapes terete. Male inflorescence longer than scape, branched or unbranched with inconspicuous cluster bracts and floral bracts. Female inflorescences similar, unbranched. Male flowers in whorls, bell-shaped, on slender stalks to 12mm long; sepals purplish, petals yellow. Female flowers similar, sessile. **HABITAT:** Woodland and open forest, mostly on clay soils. **DIST:** Coast to slopes and plains in NSW. From Cape York to w. Vic, outlier on Gove Pen. (NTerr). **FL:** Sept–Jan.

Lomandra multiflora

Lomandra obliqua

Lomandra obliqua
Twisted Mat-rush
Decumbent plant with numerous branched stems and leaves evenly spaced along some or most of stem. Leaves 20–40 x 1–2mm, spreading, flat, often twisted, acutely pointed, green or greyish. Flowers tubular, yellow with purple tinges; male flowers in sessile clusters on short erect axes; female flowers in solitary cluster on stem or short scape. **HABITAT:** Open forest on shallow sandy and rocky soils. **DIST:** Coast and ranges. From Jervis Bay to Qld. **FL:** Sept–Nov.

FLOWERING PLANTS

Xanthorrhoea
GRASS TREES

Stout perennials with underground or above-ground caudex covered in old leaf bases. Leaves linear, tapering to apex, flat or triangular or quadrangular in section, with ridged or concave faces. Inflorescence a large spike above a smooth erect scape. Flowers embedded in bracts of 2 kinds, creamy white with 6 tepals, 6 stamens and 3-locular ovary. Capsules shiny brown, hard, pointed, protruding from bracts after flowering. Hybridisation occurs extensively. Flowering stimulated by fire, the different species often having different latent periods before flowering.

Xanthorrhoea arborea

Trunk to 2m, often bent or branched, black, crusty; old leaf bases retained on upper part. Leaves flat with small ridges, to 1.5m x 8mm, dull greenish grey to glaucous (mid-green in Colo–Nepean district). Flower spike slightly hairy but not velvety, spike to 1.5m long; scape to 1.6m (rarely to 2m). **HABITAT:** Rocky sites with eucalypt shade on slopes and escarpments. **DIST:** Sandstone areas of coast and t'lands from Stanwell Tops to Wollemi NP. **FL:** Nov–Jan.

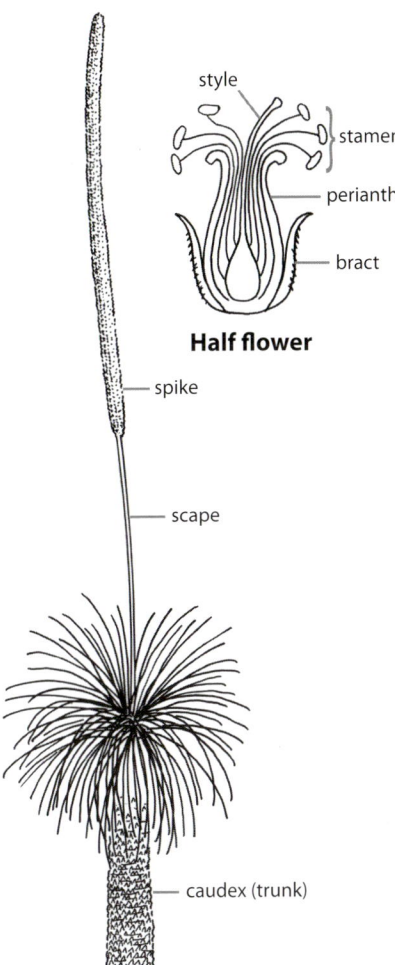

Xanthorrhoea arborea

Xanthorrhoea australis

Xanthorrhoea concava

Xanthorrhoea australis
Austral Grass Tree
Trunk 1–3m tall, stout, often branched, black, with skirt of old leaves. Leaves to 1m x 1–3mm, quadrangular or flattened quadrangular in section, greyish or silvery. Flower spike 90–200 x 3–6cm on stout scape 35–50cm long. Floral bracts light brown, sharply pointed. **HABITAT:** Rocky hillsides, often on limestone and volcanic rock. **DIST:** Sthn H'lands (e.g. Bundanoon), Sthn Blue Mtns (e.g. Wombeyan Caves). Coast from Nowra to Vic, SA, Tas. **FL:** Nov–Dec.

X. australis is uncommon and restricted in the Sydney district. It hybridises with *X. media* near Putty and with *X. latifolia* near Wyong. It could easily be confused with *X. johnsonii*, which occurs in the Howes Valley region.

Xanthorrhoea concava
Plant with 1 to several crowns from underground trunk. Leaves 3–6mm wide, flat or concave above, blue-green or glaucous. Flower spike creamy light brown and velvety at flowering due to long pale hairs on bracts and sepals; spike to 1m long; scape to 2.5m x 10mm. **HABITAT:** Heath, swamps, open forest on sandy and clay soils. **DIST:** Sth of Sydney and Sthn H'lands (e.g. Menai, Nepean Dam, Hill Top, Bundanoon, Wingello), ST, SC to Eden. **FL:** Chiefly Oct–Dec.

Xanthorrhoea fulva
Plant with 1 to several crowns from underground trunk. Leaves 2–3mm wide, depressed-triangular in section, usually blue-green or glaucous. Flower spike similar in appearance to *X. concava*, slender, <30cm long on young plants, stout and up to 60cm long on old plants; scape to 1.6m x 5–20mm. **HABITAT:** Heaths, low dunes and wet places on sandy soils. **DIST:** Nth and nw. of Sydney (e.g. Wyong to Warnervale, Putty) and t'lands (Barren Grounds). Also NC, Qld. **FL:** Chiefly Sept–Dec.

Xanthorrhoea fulva

Xanthorrhoea latifolia

Xanthorrhoea macronema

Xanthorrhoea latifolia
Trunk simple or branched, underground or 1m or more high. Leaves 2.5–5.5mm wide, quadrangular to nearly flat, bright glossy green, not glaucous. Flower spike nearly glabrous with cluster bracts around flowers obscure or absent over most of spike; spike 0.5–1.2m long; scape 1.5–2m x 10–15mm. **HABITAT:** Open forest on sandy and gravelly clay soils, coastal heath on sand. **DIST:** Nth of Gosford (e.g. Doyalson, Dora Creek, Munmorah SCA). Also NC, Qld. **FL:** July–Oct.

Xanthorrhoea macronema
Trunk underground. Leaves loosely tufted, about 1m x 2–4mm, triangular in section, ribbed and grooved with rough margins, glossy green, not glaucous. Spike glabrous, 3–12cm long on slender scape up to 1.6m x 5mm. Flowers large, yellow; petals larger than floral bracts; stamens prominent giving 'bottlebrush' appearance to flowering spike. **HABITAT:** Shaded eucalypt forest on coastal sand. **DIST:** Coast (e.g. Helensburgh, Garawarra, Wyong, Tuggerah, Bateau Bay, Munmorah). Also NC, Qld. **FL:** July–Aug.

Xanthorrhoea media
Trunk unbranched, underground or to 60cm above ground. Leaves <3mm wide, triangular to quadrangular in section, glossy green. Flower spike with short acute floral bracts, nearly glabrous, one-third to half length of scape; spike 30–80cm long; scape to 2m x 8–12mm. **HABITAT:** Dry rocky ridges on Hawkesbury sandstone; occasionally in drier coastal heath. **DIST:** Endemic. Widespread. Coast and ranges. **FL:** Chiefly Aug–May.

X. media resembles *X. resinifera* (see comment that species). Trunked hybrids occur with *X. australis* in the Putty–Howes Valley region. Possible hybrids with *X. minor* occur in wstn Sydney.

Xanthorrhoea minor subsp. *minor*
Small Grass Tree
Trunk branched underground, producing several scapes on one plant. Leaves triangular in section, usually concave, 2–3.5mm wide, green,

Xanthorrhoea media

Xanthorrhoea minor subsp. *minor*

not glaucous. Flower spike glabrous or slightly hairy, not velvety, less than half as long as scape; spike 5–12cm long; scape 25–50cm x 2–5mm. **HABITAT:** Open forest on seasonally waterlogged clay soils. **DIST:** Endemic. Wstn parts of coast, and lower Blue Mtns (e.g. Holsworthy, Picnic Point, Agnes Banks, Glenbrook). **FL:** Oct–Nov.

Xanthorrhoea resinifera
Trunk underground or barely above ground. Leaves 3–3.5mm wide, quadrangular or convex-triangular in section, often blue-green or glaucous. Flower spike velvety dark brown due to matted hairs on floral bracts; spike to 1.6m long; scape to 2m x 10–30mm. **HABITAT:** Heath and woodland in wet places on sandy soils. **DIST:** Coast and Blue Mtns mainly sth of Sydney. Also SC, Vic. **FL:** June–Jan.

X. resinifera is easily confused with *X. media*, but the latter grows in drier situations and its flower spike is shorter relative to the scape and not velvety.

Xanthorrhoea resinifera

SMILACACEAE

Eustrephus
Vines with tuberous roots, twining stems without tendrils and unisexual flowers on separate plants. Leaves alternate with longitudinal veins. Flowers solitary or in small cymes in upper leaf axils; sepals 3, entire, petals 3, fringed. Fruit a 3-lobed orange capsule with numerous black seeds.

Eustrephus latifolius
Wombat Berry
Glabrous branching climber reaching several metres. Leaves distant on stems, sessile, twisted at the base, 5–10cm long, dull green, without a distinct midvein but with several longitudinal veins raised on undersurface. Flowers in groups of 4–6 on slender stalks 8–15mm long, usually white, rarely pinkish or mauve. **HABITAT:** Shaded moist forests. **DIST:** Coast to WS. Also Qld, Vic, PNG, NCal, Malesia. **FL:** Oct–Nov.

Geitonoplesium
Wiry climber. Leaves shortly stalked, glossy above, dull beneath; midvein distinct. Tepals 6, entire. Fruit a black berry with numerous black seeds.

Geitonoplesium cymosum

Geitonoplesium cymosum
Scrambling Lily
Glabrous branching climber to several metres. Leaves alternate, often in a plane on zigzag stems, 5–10cm x 5–15mm, tapering to apex, rounded, asymmetric and twisted at the base. Flowers white, 10–20mm across. **HABITAT:** Shaded moist forests. **DIST:** Coast and ranges. Also Qld, Vic, NCal, Malesia. **FL:** Oct–Dec.

Ripogonum
SUPPLEJACK
Climbers without tendrils. Leaves glabrous or hairy, prickly or smooth. Leaf veins longitudinal and reticulate, prominent. Flowers with 6 tepals in 2 whorls, 6 stamens and superior ovary with sessile stigma. Fruit a black berry with few seeds.

Ripogonum album
White Supplejack
Large climber with glabrous prickly branches. Leaves in whorls of 3 or opposite or alternate, ovate, 8–15 x 3–5cm; stalks twisted, curved, 7–13mm long. Flowers greenish white in axillary racemes or spikes. **HABITAT:** Rainforest. **DIST:** Coast (e.g. Bola Ck, Burning Palms in Royal NP). Also Vic, Qld, PNG. **FL:** Sept–Oct.

Ripogonum fawcettianum
Small Supplejack
Shrubby climber, without prickles or sparsely prickly, softly hairy on branches, leaf stalks and

Eustrephus latifolius

Ripogonum fawcettianum

Ripogonum album

young shoots. Leaves opposite, 5–10cm x 3–3.5cm, rounded or cordate at base; stalks <7mm long. Flowers white in axillary racemes or spikes. **HABITAT:** Coastal rainforest. **DIST:** Coast nth of Sydney (e.g. Ourimbah, Bouddi NP, Strickland SF). Also NC. **FL:** Oct–Nov.

Smilax
SARSAPARILLA
Climbers with alternate leathery leaves, often with 2 coiled tendrils at base of leaf stalks. Flowers in axillary umbels, male and female flowers on separate plants. Fruit a shiny black berry with 1–2 seeds.

Smilax australis

Smilax australis
Lawyer Vine, Austral Sarsaparilla
Climber with tough prickly stems. Leaves oblong to broadly elliptic with 5 prominent longitudinal veins and finer reticulate veins, 5–12 x 4–8cm, green and glabrous on both surfaces. Flowers reddish in bud; tepals greenish white, 4–5mm long, spreading. Fruit 8mm diam. **HABITAT:** In and near rainforest and shaded gullies. **DIST:** Coast and ranges. Also CWS, Qld, Vic, LHI. **FL:** Chiefly Sept–Jan.

Smilax glyciphylla

Smilax glyciphylla (syn. *S. glycophylla*)
Sweet Sarsaparilla
Common climber with non-prickly stems. Leaves ovate to lanceolate, 4–8 x 2–3cm, with 3 prominent longitudinal veins, green above, glaucous-grey below. Flowers with greenish white to cream tepals 3–4mm long, scarcely opening. Fruit 7mm diam. Young leaves and berries with sweet taste. **HABITAT:** Sheltered forests in gullies or tops, river flats in valleys, rarely on open sites. **DIST:** Coast. Also SC, NC, NT, Qld. **FL:** Oct–Dec.

Dioscorea transversa

Cordyline stricta

DIOSCOREACEAE

Dioscorea
YAM
Tuberous herbs with twining stems without stipules or tendrils. Leaves usually alternate and simple. Flowers unisexual, on separate plants; female flowers with inferior ovary developing into a capsule. (Smilacaceae has superior ovary developing into a berry.)

Dioscorea transversa
Native Yam
Slender glabrous climber. Leaves alternate on stalks 1–9cm long, broadly hastate to cordate at base, finely pointed at apex, 6–10 x 2–6cm, with 5–7 prominent longitudinal veins. Flowers very small in narrow inflorescences 3–10cm long. Fruit brown, 3-winged, 3cm across. **HABITAT:** Rainforest and moist eucalypt forest. **DIST:** Coast nth from Stanwell Park. Also NC, NT, Qld, NTerr. **FL:** Sept–Oct.

AGAVACEAE

Cordyline
Glabrous shrubs. Leaves covering stems on young plants, tufted at apex on older plants leaving lower stems ringed by leaf bases. Panicles often large on long scapes. Flowers bisexual with 6 tepals in 2 whorls of 3; stamens 6; ovary superior. Berry variously coloured, with black seeds.

Cordyline is sometimes classified in the Asteliaceae family.

Cordyline stricta
Narrow-leaf Palm Lily
Palm-like shrub to 2–3m tall (in Sydney district); stems 20–25mm diam. Leaves linear, scarcely narrowing to indistinct stalk or base, 30–50 x 1–2cm, margins smooth. Flowers purplish blue in panicles to 40cm long. Fruit a black berry 10–15mm diam. **HABITAT:** Rainforest in moist sheltered gullies. **DIST:** Lower Blue Mtns (e.g. Cabbage Tree Ck, Wheeny Ck), Colo R., Hawkesbury R. Also NC, Qld. **FL:** Nov–Jan.

FLOWERING PLANTS

Doryanthes excelsa

Doryanthes

Large slow-growing plants with very short stems bearing large leaves in crowded basal rosette. Inflorescence a large globose compound raceme (Sydney district) with large leafy bracts on stout scape bearing smaller sheathing leaves. Tepals 6, in 2 whorls united at base; stamens 6, large, partly fused to tepals; ovary inferior, 3-locular. Fruit a capsule with numerous brown seeds.

A genus with 2 species, sometimes classified as sole genus in family Doryanthaceae.

Doryanthes excelsa
Gymea Lily, Flame Lily, Spear Lily

Leaves fibrous, 1–2m x 7–10cm. Inflorescence to 70cm diam. with several to numerous flowers subtended by large dark red bracts; scape 3–5m tall; tepals red or pinkish red, fleshy, 10cm long; stamens shorter with red filaments and anthers opening in 2 slits exposing green pollen; style 10cm long, red. Capsule ovoid, brown, 7–10cm long. **HABITAT:** Open forest, on deep sandy soils. **DIST:** Coast from Bulli to Georges R. and Broken Bay to Karuah (absent between); isolated populations near Grafton (NC). **FL:** Aug–Nov.

PHILYDRACEAE

Philydrum

Monotypic genus. Aquatic perennial herb with woolly flower spikes. Flowers subtended by 2 large bracteoles; tepals 4, in 2 whorls, the outer pair ovate, hairy outside and vertically arranged, the inner pair smaller, lateral.

Philydrum lanuginosum
Woolly Frogmouth

Erect plant 1–2m tall. Leaves 20–60cm long, crowded and distichous at base of stems, a few alternate on lower stems, bract-like on upper stems, often reddish-tinged. Flowers yellow, sessile in ovate-acuminate sheathing bracteoles, the 2 outer tepals resembling the open beak of frogmouths (*Podargus* spp.). Capsule 3-locular with reddish brown seeds. **HABITAT:** Shallow freshwater and boggy margins. **DIST:** Chiefly coast. From Vic to Qld; also NTerr, Malesia. **FL:** Nov–May.

Philydrum lanuginosum

Crinum pedunculatum

AMARYLLIDACEAE

Crinum
Bulbous herbs. Leaves narrow- to broad-linear, persistent or deciduous, erect and spreading or decumbent. Inflorescence a large umbel with 2 spathe-like bracts and numerous small flower bracts. Flowers bisexual, with long slender tube opening to 6 long narrow spreading or reflexed lobes; stamens 6, with slender filaments and large versatile anthers; ovary inferior, 3-locular, with filiform style. Fruit globose, with several large seeds.

This genus is sometimes retained in the Liliaceae family.

Crinum pedunculatum
Swamp Lily
Robust plant with broad leaves from base or on above-ground extension of bulb (pseudostem) to 40cm tall. Leaves 50–100 x 5–12cm, margins entire, smooth. Scape leafless, 1–1.5m tall with 10–25 white flowers with pink to red or purple stamens. **HABITAT:** Estuaries on low-lying damp sandy ground. **DIST:** Uncommon in Sydney region (e.g. Werrong, Minnamurra Spit). SC to Qld, Timor, PNG, Pac. Iss. **FL:** Nov–Dec.

HAEMODORACEAE

Haemodorum
BLOODROOT, BLOOD-LILY
Glabrous plants with orange-red bulbous rootstock. Leaves sheathing, mostly basal, reducing to bracts on stems. Inflorescence variously branched. Flowers dark reddish brown with 6 free tepals in 2 whorls; stamens 3, attached to petals; ovary becoming rounded above. Capsule 3-lobed, with winged seeds.

Haemodorum corymbosum
Rush-leaf Bloodroot
Erect perennial with 3–4 terete or compressed leaves about 50cm x 1–2mm. Flowers in several dense terminal corymbose clusters on scape 50–60cm tall, black, 12–15mm long. **HABITAT:** Heath and woodland. **DIST:** Coastal Sydney, Agnes Banks. Also SC, ST. **FL:** Nov–Jan after fire.

Haemodorum planifolium
Strap-leaf Bloodroot
Erect perennial with about 6 flat strap leaves, mostly 30–50cm x 2–6mm, at base of brown-black scape >60cm tall. Flowers clustered in open or dense inflorescences, black, 11–16mm long; stamens with orange-yellow anthers. **HABITAT:** Heath and woodland. **DIST:** Coast and ranges from Batemans Bay to Qld. **FL:** Nov–Jan after fire.

HYPOXIDACEAE

Hypoxis
Perennial herbs, formerly included in Amaryllidaceae but developing from a corm instead of a bulb, and flowers solitary or in racemes (rather than umbels). They have 6 stamens (3 in Iridaceae) and an inferior ovary (superior in Liliaceae).

Hypoxis hygrometrica
Golden Star, Golden Weather-glass
Herb seldom >20cm tall. Leaves grass-like, grooved above, 10–18cm x 1mm, sprinkled with long silky hairs. Tepals spreading, 7–11mm long, yellow; anthers yellow, arrowhead-shaped, attached centrally. Fruit an ellipsoidal capsule 2–5mm long. **HABITAT:** Moist to periodically wet grassy or open woodland. **DIST:** Coast to WS. Also Qld, Vic, Tas. **FL:** Nov–Mar.

Haemodorum corymbosum

Hypoxis hygrometrica

Haemodorum planifolium

Var. *hygrometrica* has 1 flower per stalk and glabrous sepals. Var. *villosisepala* has 2–3 flowers per stalk and hairy sepals. *Tricoryne simplex* is similar but has flowers in umbels and the tepals are spirally twisted after flowering.

Acianthus caudatus

ORCHIDACEAE

Acianthus, Nemacianthus
Tiny terrestrial herbs with solitary heart-shaped leaf at base. Flowers in a terminal raceme. Segments narrowly pointed with extended tip. Dorsal sepal erect or bent over column; lateral sepals narrow, forward-pointing; labellum variously shaped.

Acianthus caudatus
Mayfly Orchid
Stem 10–15cm tall with up to 9 purple flowers. Dorsal sepal to 40mm long, erect. Lateral sepals to 25mm long, fine, spreading forward. Column bent forward above labellum. Labellum ovate, 7mm long, finely pointed. **HABITAT:** Open forest, woodland, on sandy soils. **DIST:** Coast and ranges. Also NC, SC, Qld, Vic, Tas, SA. **FL:** July–Sept.

This sp. was reclassified as *Nemacianthus caudatus* in 2002; the change is not universally accepted.

Acianthus fornicatus
Pixie Caps
Stem to 20cm tall with up to 10 flowers. Flowers translucent, green, with some pink or purple shading. Dorsal sepal broad-ovate, 10mm long, finely pointed, hooded over column. Lateral sepals narrow, diverging. Labellum 5 x 3mm, ovate with apical point, pimply near apex. **HABITAT:** Open forest, woodland, on sandy soils. **DIST:** Common. Coast and ranges. Also NC, SC, Qld. **FL:** May–Aug.

Acianthus fornicatus

Acianthus pusillus
Gnat Orchid
Stem 5–20cm tall with 1–18 flowers 8–12mm long. Dorsal sepal narrow-obovate with finely pointed tip, about 7 x 2.5mm not hooded closely over column. Petals reflexed. Labellum narrow-oblong, 4–4.5 x 2–3mm. Column bent forward. **HABITAT:** Sheltered moist sites. **DIST:** Coast and ranges. Also NC, SC, T'lands, Qld, Vic, Tas, SA. **FL:** Mar–July.

Acianthus exsertus is similar but has flowers 12–16mm long; petals reflexed; labellum elliptic, 5–6 x 3.5–4mm long; column only slightly curved forward. It flowers in Mar–May.

Acianthus pusillus

Adenochilus
Terrestrial herbs with fleshy rhizome. Leaves 1 per shoot, sessile. Flowers solitary. Dorsal sepal broad and hooded.

Adenochilus nortonii
Stem 10–15cm tall; leaf solitary at base or on lower part of stem, glabrous, ovate to cordate, to 30 x 10mm. Flower 20mm across, white. Lateral sepals and petals lanceolate, spreading, finely pointed. Labellum with red markings and yellow and white calli down the centre. **HABITAT:** Mossy clay banks and wet rock crevices. **DIST:** Blue Mtns (e.g. above Wentworth Falls, Blackheath). Also Kellys Falls (Stanwell Tops); NT. **FL:** Nov–Dec.

Bulbophyllum
Epiphyte or lithophyte with creeping rhizomes bearing small pseudobulbs. Leaflet often arising from pseudobulb. Flowers solitary or several in raceme, very small.

Bulbophyllum exiguum
Autumn Bulbophyllum
Mat-forming plant with rounded furrowed pseudobulbs, 5–8mm diam. Leaves solitary on pseudobulbs, oblong-linear, flat, pointed, 10–40mm long. Flower 10mm across, creamy yellow, in racemes of up to 5. **HABITAT:** Rainforest on trunks of trees and on rocks. **DIST:** Chiefly coast. Also lower Blue Mtns, Sthn H'lands, NC, SC, NT, Qld. **FL:** Mar–May.

Bulbophyllum shepherdii
Wheat-leaved Orchid
Mat-forming plant, with creeping rhizomes. Pseudobulbs ovoid, 2mm diam. Leaves succulent, narrow, channelled, 2–4cm long. Inflorescence 1-flowered; flower 8mm across, cream with yellow tips. Labellum deep orange. **HABITAT:** Rainforest on trunks of trees and on rocks. **DIST:** Coast. Also NC, SC, NT, Qld. **FL:** Sept–Nov.

Adenochilus nortonii

Bulbophyllum exiguum

Bulbophyllum shepherdii

Calanthe

Terrestrial orchid. Roots fleshy with pseudobulbs. Leaves scattered or in a tuft. Flowers in raceme; sepals and petals free; labellum 4-lobed. A large tropical and subtropical genus with only 1 species as far south as Sydney.

Calanthe triplicata

Calanthe triplicata
Christmas Orchid

Sturdy plant to 1.5m tall, often growing in colonies. Leaves erect, to 90 x 18cm, obovate to lanceolate, pleated, with up to 10 leaves on lower stems. Inflorescence with 18–40 white flowers. Flowers to 30mm across; sepals and petals spreading; labellum with spreading lateral lobes and deeply divided midlobe. **HABITAT:** Shaded damp sites, usually in rainforests. **DIST:** Coast, nth from Illawarra (e.g. Somersby, Ourimbah Ck). Also NC, NT, Qld, Malesia, Asia. **FL:** Oct–Jan.

Caleana

Terrestrial orchids with labellum hinged to the column by an elongated claw, resembling the head and neck of a duck. Contact by an insect with labellum irritates the claw, causing it to snap downwards, forcing the insect to rub against the column.

Caleana major
Flying Duck Orchid

Slender stem to 40cm high with 1–4 flowers. Leaf solitary, narrow-lanceolate, to 12cm x 8mm, sessile. Flower 10mm across. Lateral sepals linear, greenish, pointing backwards;

Caleana major

dorsal sepal pointing downwards. Labellum lamina 8 x 6mm, smooth, shiny purple. **HABITAT:** Swampy heath, woodland, open forest. **DIST:** Coast and ranges. Also NC, SC, NT, ST, Qld, Vic, Tas, SA. **FL:** Sept–Jan.

Calochilus campestris

Calochilus paludosus

Calochilus

Terrestrial glabrous herbs with erect stems and a single long narrow channelled basal leaf. Flowers few to many in terminal raceme. Dorsal sepal hooded. Labellum conspicuously bearded with metallic-lustrous hairs, frequently ending in a ribbony filament.

Calochilus campestris
Copper Beard Orchid

Stem to 50cm tall. Flowers pale green with purplish lines; labellum with long coppery-red-purple hairs and a short naked strap-like apex, upper basal part of labellum shiny purple without hairs. Column with a black gland on each side, not connected by a coloured ridge. **HABITAT:** Open forest, woodland, in sand or clay soils. **DIST:** Coast. Most of NSW, Qld, Vic, Tas. **FL:** Sept–Nov.

Calochilus paludosus
Red Beard Orchid

Stem to 35cm tall. Flowers green with purplish lines; labellum with long, spreading, lustrous dark red hairs and long elongated naked tip; upper basal part of labellum with purple calli. Column lacking black gland on each side, but connected by coloured ridge. **HABITAT:** Open forest, woodland, on sandy soils. **DIST:** Coast and ranges. Also NC, SC, ST, WS, Qld, Vic, Tas, NZ. **FL:** Sept–Nov.

Calochilus robertsonii
Purplish Beard Orchid

Stem to 45cm tall. Flowers green with purplish lines; labellum with long lustrous purple hairs which are short and gland-like at base, tip short. Column wings with a shiny black gland on each side, connected by coloured ridge. **HABITAT:** Woodland, heath, on sandy soils. **DIST:** Coast and ranges. Most of NSW. All other States, NZ. **FL:** Sept–Oct.

Chiloglottis

Small terrestrial herbs with 2 leaves at base. Flowers solitary on short stem. Dorsal sepal narrow, erect; petals and lateral sepals spreading or deflexed. Labellum broad with characteristic arrangements of calli on upper surface. Column slender, hooded by dorsal sepal.

Calochilus robertsonii

Chiloglottis chlorantha
Stem to 5cm tall. Leaves ovate, 4 x 2cm. Flower solitary, green-yellow or brownish, about 3cm across, with spreading segments. Labellum heart-shaped, 12mm long, with 10–16 stalked calli in pairs on either side of midline and 1 larger sessile callus on midline. **HABITAT:** Moist sheltered areas. **DIST:** Uncommon. Coast (e.g. Jamberoo Mt), upper Blue Mtns (e.g. Boyd Plat.). Also SC. **FL:** Sept–Nov.

Chiloglottis chlorantha

Chiloglottis formicifera
Ant Orchid
Stem to 10cm tall. Leaves ovate, 6 x 2.5cm. Flower solitary, greenish brown, about 2cm across. Lateral sepals spreading, linear, with a terminal 1mm gland. Petals broader, bent down and back. Labellum with narrow base and broad obtusely rounded tip. Labellum with numerous shiny black calli (resembling ants) from base to apex; a conspicuous double-headed callus is located at the rear near claw. **HABITAT:** Open forest, in sheltered moist sites. **DIST:** Coast and ranges. Also NC, SC, NT, ST, WS. **FL:** Aug–Oct.

Chiloglottis formicifera

Chiloglottis seminuda
Autumn Bird Orchid
Stem to 5cm tall. Leaves ovate, 85 x 18mm. Flower solitary, narrow, green to reddish, about 15mm across. Dorsal sepal narrow spoon-shaped, ending in a long point. Lateral sepals linear, bent downwards and ending in a small gland. Petals oblong, swept backwards. Labellum diamond-shaped with red glandular hairs at base; calli black, large and sessile or small and stalked. **HABITAT:** Open forest, in sheltered moist sites. **DIST:** Scattered. Sthn H'lands (e.g. Fitzroy Falls, Bundanoon). Also Wedderburn, Woodford and ST. **FL:** Feb–Apr.

Corunastylis (Genoplesium)
Terrestrial orchids with solitary cylindrical leaf. All but 1 species of *Genoplesium* were reclassified as *Corunastylis* in 2002, but this change is not universally accepted.

Corunastylis archeri
Variable Midge Orchid
Stem 10–20cm tall with 10–20 flowers crowded on the upper part. Flowers to 5mm across with green or reddish sepals and petals. Dorsal sepal ovate. Lateral sepal lanceolate, divergent but horizontal, 5mm long. Petals small, 3mm long, ovate. Labellum elliptic, reddish purple, with small tip, margins hairy. **HABITAT:** Moist sites in open forest, woodland. **DIST:** Uncommon. Coast and ranges. Also NC, SC, NT, ST, Vic, Tas. **FL:** Dec–Mar.

Corunastylis fimbriata
Fringed Midge Orchid
Stem slender, 15–30cm tall with 5–30 flowers in an open spike. Flowers with green lateral sepals 5mm long. Dorsal sepal and petals acuminate, ciliate, pale with crimson lines; petals shorter than sepal. Labellum linear-oblong, to 4mm long with numerous crimson hairs. **HABITAT:** Woodland, heath, on sandy soils. **DIST:** Coast and ranges. Also NC, SC, NT, Qld. **FL:** Jan–Apr.

Corunastylis woollsii
Stem 10–25cm tall with 5–35 nodding flowers in a dense spike. Flowers 4mm across, dark purple. Petals and dorsal sepal ovate, fringed. Labellum oblong with pointed tip and fringed margins. **HABITAT:** Damp soils and moss, in woodland and heath. **DIST:** Scattered. Coast and ranges (e.g. Heathcote, Carrington Falls). Also SC. **FL:** Mar–Apr.

Chiloglottis seminuda

Corunastylis fimbriata

Corunastylis woollsii

Corunastylis archeri

Corybas aconitiflorus

Corybas pruinosus

Corybas
Terrestrial orchids with solitary ovate, cordate or orbicular horizontal leaf. Flower solitary, reddish purple, hooded, sessile or nearly so.

Corybas aconitiflorus
Spurred Helmet Orchid
Stem short. Leaf almost round, dark green above, purplish below. Flower purplish, 15–22mm long, strongly hooded, with pointed tip. Labellum pink or purple, concealed under dorsal sepal. Base of labellum with 2 short hollow spurs visible below dorsal sepal. **HABITAT:** In ground litter, forested gullies and heath. **DIST:** Coast and ranges. Also NC, SC, Qld, Vic, Tas. **FL:** Apr–June.

Corybas pruinosus
Toothed Helmet Orchid
Flower erect on short stem, 20mm long, grey-green, spotted with maroon. Dorsal sepal broad, hooded. Labellum with rounded lamina, margins maroon, with coarse linear teeth. **HABITAT:** In ground litter, moist open forest and rainforest margins. **DIST:** Coast, lower Blue Mtns. Also NC, SC. **FL:** Apr–June.

Corybas fimbriatus is similar to *C. pruinosus*, but has a reddish boss at the back of the fringed labellum.

Cryptostylis
Terrestrial glabrous orchids. Labellum large, red-brown or purple, the other segments small and green. Pollinated by wasps of the family Ichneumonoidea which are attracted by scent and the resemblance of the labellum to the female insect (see picture of *C. leptochila*).

Cryptostylis erecta
Bonnet Orchid
Stem 15–40cm tall. Flowers 2–10, each 30mm long, marked with purple or red-brown. Labellum broad, hooded above column, ± enclosed in basal part. Sepals and petals linear, spreading. **HABITAT:** Open forest on sandy slopes. **DIST:** Chiefly coast; recorded Katoomba. Also NC, SC, Vic. **FL:** Nov–Feb.

Cryptostylis hunteriana
Leafless Tongue Orchid
Saprophytic leafless plant. Stem 15–40cm tall. Flowers 5–10, each 25mm long. Labellum deep maroon, densely hairy, oblong with recurved margins, pointing upwards. Sepals and petals linear and spreading. **HABITAT:** Low heath, often in damp sandy soils. **DIST:** Rare. Sthn area of Lake Macquarie, Loftus Hts (Royal NP). Also NC, SC, NT, Vic. **FL:** Dec–Feb.

Cryptostylis erecta

Cryptostylis hunteriana

Cryptostylis leptochila

Cryptostylis leptochila
Small Tongue Orchid
Stem 15–40cm tall. Flowers 4–15, each 20mm long. Labellum oblong with recurved margins, purple, finely hairy, strongly curved upwards and backwards. Sepals and petals linear and spreading. **HABITAT:** Scrub, open forest, on sandy soils. **DIST:** Blue Mtns and Sthn H'lands. Also SC, NT, ST, Qld, Vic, Tas. **FL:** Dec–Mar.

Cryptostylis subulata
Large Tongue Orchid
Stem 15–80cm tall with 2–8 flowers each 30mm long. Labellum oblong with strongly incurved or inrolled margins ± forming a tube, red-

Cryptostylis subulata

brown. Sepals and petals linear, spreading. **HABITAT:** Woodland, on sandy soils. **DIST:** Coast and ranges. Also NC, SC, NT, ST, Qld, Vic, Tas, SA, NZ. **FL:** Oct–Mar.

Cyanicula

A genus of about 10 species previously placed in *Caladenia*, mostly with blue flowers. Differs from *Caladenia* in nature of its tubers. One species in Sydney district.

Cyanicula caerulea
Blue Fairy

Stem 10–15cm tall with a single blue flower to 25mm across. Dorsal sepal erect. Lateral sepals and petals to 20mm long, in a spreading fan-shape. Labellum 3-lobed, with dark blue bands on a yellow tip. Calli in 2 rows, with yellow heads, shorter on midlobe. **HABITAT:** Open forest, woodland, with grassy understorey. **DIST:** Coast and ranges. Most of NSW. Also Qld, Vic, SA, WA. **FL:** July–Sept.

Cymbidium

Epiphytic orchids with several long leaves and many-flowered inflorescence. Dorsal sepal similar to lateral sepals and petals. Labellum 3-lobed. One species in Sydney district.

Cymbidium suave

Large clumped orchid with extensive fleshy root system. Leaves strap-like, 20–40 x 1–2cm. Flowers in racemes 10–35cm long, golden or greenish yellow, blotched with red. Sepals and

Cymbidium suave

petals broad and spreading. Labellum reddish brown near base, yellow at tip. **HABITAT:** Hollow limbs and tree trunks, in open forest, woodland. **DIST:** Coast and ranges. Also NC, SC, NT, WS, Qld. **FL:** Oct–Dec.

Dipodium

Terrestrial saprophytic orchids, parasitic on roots of surrounding plants. Leaves reduced to scales. Flowers with spreading lateral petals and sepals, usually spotted. Three species in Sydney district.

Dipodium punctatum
Hyacinth Orchid

Stem 40–80cm tall, purple-red, with 14–60 flowers. Flowers 18mm across; pedicel crimson, not spotted; petals and sepals flat, pale pink with darker spots; labellum crimson, with central band of pink hairs. **HABITAT:** Open forest, woodland. **DIST:** Coast and ranges. Also NC, SC, NT, ST, WS, Qld, Vic. **FL:** Dec–Apr.

Dipodium variegatum

Stem 25–80cm tall, greenish, with 2–50 flowers. Flowers 20cm across; pedicel and ovary light green with maroon spots; petals and sepals flat, cream with maroon blotches; labellum mauve, with central band of white hairs. **HABITAT:** Open forest, heath. **DIST:** Chiefly coast; uncommon in Blue Mtns. Also NC, SC, NT, ST, WS, Qld, Vic. **FL:** Aug–Jan.

Dipodium roseum has dark reddish brown stems, unspotted pedicels and light pink petals and sepals with darker spots and recurved tips. Flowers Dec–Jan.

Cyanicula caerulea

Dipodium punctatum

Dipodium variegatum

Diuris
Terrestrial glabrous herbs with erect stems and basal leaves. Dorsal sepal erect above column; lateral sepals narrow-linear, downwards-pointing. Lateral petals ear-like.

Diuris alba
White Doubletails
Stem 15–40cm tall with 2–7 flowers. Flowers white, blotched with lilac. Petals 15 x 8mm with dark claw. Dorsal sepal ovate, bent forward. Lateral sepals 30mm long, green, often crossed. **HABITAT:** Moist sandy coastal heath and grassy areas in woodland. **DIST:** Coast nth of Wamberal (e.g. Munmorah SCA). Also NC, NT, NWS, Qld. **FL:** Aug–Nov.

Diuris alba

Diuris aurea
Golden Doubletails

Stem 30–60cm tall with 2–5 golden yellow flowers. Dorsal sepal ovate with pale brown patches; lateral sepals 10–16mm long, parallel or weakly crossed. Labellum 3-lobed, with folded midlobe; lateral lobes <half as long as midlobe. **HABITAT:** Woodland, moist grassland. **DIST:** Scattered. CC, CT (e.g. Castlereagh NR, Dharug NP, Thirlmere). Also NC, ST. **FL:** Sept–Oct.

Diuris aurea

Diuris sulphurea

Diuris maculata
Leopard Orchid

Stem to 40cm tall with 1–10 flowers. Flowers yellow, blotched with dark brown, chiefly on underside of petals. Dorsal sepal short, ovate, with obtuse apex; lateral sepals 14–16mm long, greenish brown, strongly crossed, backwards-pointing. **HABITAT:** Woodland, in grassy areas. **DIST:** Scattered in wstn Sydney (e.g. Holsworthy, Casula); Bulli–Appin Rd. Also NC, SC, CT. **FL:** July–Oct.

Diuris sulphurea
Tiger Orchid

Stem 30-60cm tall with 1–7 flowers. Flowers sulphur-yellow with brown markings on labellum and dorsal sepals. Dorsal sepal ovate, bent forward; lateral sepals about 20mm long, brown with yellow tip. Labellum diamond-shaped with central ridge from base to middle, continuing as fold to apex. **HABITAT:** Open forest, woodland, in grassy areas. **DIST:** Coast and ranges. Most of NSW. Also Qld, Vic, Tas, SA. **FL:** Sept–Dec.

Dockrillia

Many *Dendrobium* species from Australia and PNG were reclassified as *Dockrillia* in 1981. All are epiphytic or lithophytic plants. There are 6 species in the Sydney district.

Dockrillia linguiformis
Tongue Orchid

Lithophyte. Stems 3–30cm long, thick, often hidden, clinging in patches to rock surface. Leaves ovate, 2–3cm long, with 3–4 longitudinal grooves on upper surface, thick, leathery. Flowers white with narrow sepals and petals. Labellum deeply 3-lobed with purple markings; midlobe narrow, recurved and with undulate margins. **HABITAT:** Rocks, in eucalypt forest. **DIST:** Common. Coast. Also NC, SC, NT, ST, NWS, Qld. **FL:** Sept–Oct.

Diuris maculata

Dockrillia linguiformis

Dockrillia pugioniformis
Dagger Orchid

Epiphyte. Stems long, wiry, forming tangled mass on trunks and branches. Leaves ovate, pungent-pointed, 5 x 2cm. Flowering stem usually with 1 flower. Flower 25mm across, pale green to pale brown. Lateral sepals spreading, broad. Labellum recurved, white with purple markings. **HABITAT:** On trees (rarely rocks) in cool rainforest. **DIST:** Coast and ranges (e.g. Cambewarra, Robertson, Mt Wilson). Also NC, SC, NT, ST, Qld. **FL:** Sept–Nov.

Dockrillia pugioniformis

Dockrillia striolata
Streaked Rock Orchid

Lithophyte with mats of spreading stems. Leaves terete, tapering to apex and base, to 4–7cm long, shallowly grooved. Flowers 1–2 on stalks from leaf bases, green to yellowish. Sepals and petals with red lines on outer surface. Labellum 3-lobed, midlobe recurved with white undulate-crisped margins. **HABITAT:** Rock surfaces in rainforest, sheltered open forest. **DIST:** Coast and ranges (e.g. Glenbrook, Mt Banks, Bundanoon). Also SC, ST, CWS, Vic, Tas. **FL:** Sept–Nov.

Dockrillia striolata

Dockrillia teretifolia

Dockrillia teretifolia
Pencil Orchid, Rat's-tail Orchid
Epiphyte. Stems long, wiry, extensively branched. Leaves pendulous, terete, 10–30cm long. Flowers 4–15, 35mm across, cream with purple centre. Sepals and petals to 30 x 3mm. Labellum 3-lobed, with red marks, curved; midlobe ruffled. **HABITAT:** On trees (less commonly on rocks); on *Casuarina glauca* near estuaries, in cool rainforest elsewhere. **DIST:** Chiefly coast (e.g. Woy Woy, Royal NP). Also NC, SC, CWS, Qld. **FL:** July–Oct.

Dockrillia fairfaxii (from Blue Mtns) is similar to *D. teretifolia* but has only 1–4 flowers per cluster.

Eriochilus
Terrestrial orchid with single basal leaf. Flowers few on slender stem. Lateral sepals larger than lateral petals, spreading. Labellum clawed at base, convex, short, hairy. Two species in Sydney district.

Eriochilus autumnalis
Parson's Bands
Stem to 12cm tall. Leaf ovate, to 15mm long, upper surface hairy, prominently veined, lower surface purplish. Flowers 1–3, 12mm across,

Eriochilus autumnalis

Erythrorchis cassythoides

FLOWERING PLANTS

Gastrodia sesamoides

Gastrodia
Leafless terrestrial saprophytes with erect unbranched stems and tubular flowers. One species in Sydney district.

Gastrodia sesamoides
Potato Orchid, Cinnamon Bells
Stem brown, to 70cm tall, with a few scale-like bracts. Flowers up to 20 in terminal raceme, pendulous, pale cinnamon brown on outside, white inside; perianth tube 9–12mm long with 5 short lobes exposing the labellum. **HABITAT:** Cool moist forests, esp. in well-drained humus-rich soils. **DIST:** Coast and ranges. Also NC, SC, ST, WS, Qld, Vic, Tas, SA, NZ. **FL:** Sept–Jan.

Glossodia
Terrestrial orchids with 1 basal hairy leaf. Flowers 1–2, purple to blue, on an erect stem. Sepals and petals free, similar, spreading. Labellum with 1–2 basal calli.

Glossodia major
Waxlip Orchid
Stem 20–30cm tall with 1 mauve flower 40mm across. Lateral sepals and petals 15–30mm long. Labellum ovate, 8–11mm long, with white base and mauve tip. **HABITAT:** Open forest, woodland, heath. **DIST:** Coast and ranges. Most of NSW. Also Qld, Vic, Tas, SA. **FL:** Aug–Sept.

white with pink markings. Dorsal sepal hooded over column. Lateral sepals to 10mm long. Labellum white with tufts of red hairs. **HABITAT:** Heath, woodland, in moist sandy soils. **DIST:** Coast and ranges (e.g. Jannali Res., Lane Cove, Glenbrook). Nowra to NC, Qld. **FL:** Mar–May.

Eriochilus cucullatus has a glabrous leaf, green below and white to pink flowers with lateral sepals 10–17mm long. It flowers Jan–May.

Erythrorchis
Leafless terrestrial saprophyte with 1 or more branched stems rising from a thick fleshy rhizome, climbing by attaching itself by sucker-like aerial rootlets.

Erythrorchis cassythoides
Climbing Orchid
Stems brown, to 6m long, usually <2m. Flowers in panicles at nodes along stems, yellow and white with greenish brown tints. Sepals and petals spreading, 12–15mm long. Labellum to 12mm long, white with crisped margins. **HABITAT:** Open forest, in sandy soils. **DIST:** Scattered along coast. Also NC, NT, Qld. **FL:** Oct–Dec.

Glossodia major

Glossodia minor
Small Waxlip Orchid
Stem to 15cm tall with 1 deep violet flower 20mm across. Lateral sepals and petals 8–15mm long. Labellum ovate, 3–5mm long, not white at base. **HABITAT:** Heath, scrub. **DIST:** Chiefly coast but extending to CSW (e.g. Mittagong). Also NC, SC, Qld, Vic. **FL:** Aug–Sept.

Glossodia minor

Liparis
Mostly epiphytes with crowded pseudobulbs, spreading narrow-lanceolate leaves and small yellowish green flowers in racemes, often with unpleasant smell.

Liparis reflexa
Yellow Rock Orchid
Pseudobulbs short, ovoid. Leaves 10–30 x 1–3cm, channelled, emerging on opposite sides of the flowering stem. Flowers 6–25 in loose racemes 10–30cm long. Sepals and petals narrow, 10–15mm long, at first spreading, later reflexed; labellum sharply bent back about the middle; callus orange. **HABITAT:** Rock faces in shaded forest and rainforest. **DIST:** Coast and ranges. Also NC, SC, NT, CWS. **FL:** Mar–May.

Liparis reflexa

Microtis unifolia

Lyperanthus
Terrestrial orchids, usually with a solitary leaf. Sepals and petals narrow, spreading; dorsal sepal hooded over column; labellum short, curved, 3-lobed, concave.

Lyperanthus suaveolens
Brown Beaks
Stem 15–35cm tall, sheathed at base by erect linear-lanceolate leaf. Flowers 2–8, 25mm across, yellowish brown, fragrant, each within a large subtending bract. Lateral sepals downcurved; petals erect. Labellum about 10mm long, yellow towards tip. **HABITAT:** Open forest, woodland. **DIST:** Coast. Also NC, SC, CT. **FL:** Aug–Sept.

Microtis
Terrestrial orchids with 1 long basal cylindrical leaf enclosing stem below inflorescence. Flowers in an erect spike, green.

Microtis unifolia
Common Onion Orchid
Stem to 60cm tall with few to many flowers. Dorsal sepal hooded, enclosing petals; lateral sepals ovate, rolled back. Labellum oblong to 2.5mm long with notched apex; calli deep green, 2 near base, 1 towards apex. **HABITAT:** Woodland, heath, swamp margins. **DIST:** Coast and ranges. Most of NSW. All States (except NTerr), NZ, Pac. Iss, E. Asia. **FL:** Sept–Dec.

Lyperanthus suaveolens

Orthoceras strictum

Orthoceras

Terrestrial orchid with 2–5 leaves on lower stem. Flowers in a raceme, each with a large subtending bract and lateral sepals much longer than dorsal sepal and lateral petals. One species in Australia.

Orthoceras strictum
Horned Orchid

Stem to 60cm tall with 1–9 flowers. Flowers almost sessile, green-brown. Dorsal sepal acutely pointed and hooded over column. Lateral sepals linear, spreading, 3cm long. Petals tiny, hidden under dorsal sepal. Labellum 3-lobed, brown with yellow centre. **HABITAT:** Moist sandy soils in heath and scrub. **DIST:** Coast and ranges. Also NC, SC, NT, ST, WS, Qld, Vic, Tas, SA. **FL:** Dec–Jan.

Petalochilus, Stegostyla

These 2 genera were formed by the break-up of *Caladenia*. *Petalochilus* has flowers which are not glandular-hairy outside. *Stegostyla* has hairy flowers. The new names are not universally recognised. See also *Cyanicula*.

PETALOCHILUS

Petalochilus carneus
Pink Fingers

Stem to 20cm tall with 1–2 flowers. Flowers 20–24mm across, pink. Lateral sepals and petals elliptic, spreading. Labellum 3-lobed, with transverse red bands and yellow recurved tip; calli in 2 rows, with yellow heads. **HABITAT:** Woodland, on sandy soils. **DIST:** Coast and ranges. Also NC, SC, NT, ST. **FL:** Aug–Oct.

Petalochilus catenatus
White Fingers

Stem to 30cm tall with a single white or pinkish flower 30mm across. Dorsal sepal greenish white; lateral sepals and petals lanceolate, 20mm long. Labellum 3-lobed, white, with golden tip. Calli in 2 rows with yellow heads. **HABITAT:** Open forest, woodland, on sandy to clay soils. **DIST:** Chiefly coastal. Also NC, SC, Qld, Vic, NZ. **FL:** Aug–Oct.

Petalochilus pictus is similar to *P. catenatus*, but has a column marked with a red blotch. It flowers in May–June.

Petalochilus carneus

Petalochilus catenatus

STEGOSTYLA
Stegostyla dimorpha
Stem to 30cm tall with 1–3 flowers. Flowers 35mm across, white on inside, brownish with glandular hairs on back. Dorsal sepal hooded. Lateral sepals and petals lanceolate, 20mm long. Labellum weakly 3-lobed, white, with pink tip; margins with small teeth. Calli in 4 rows, clubbed, white to yellow, becoming smaller towards tip. **HABITAT:** Open forest, woodland, in swampy, grassy sites. **DIST:** Upper Blue Mtns (e.g. Bell, Clarence). Also ST, WS, Vic. **FL:** Sept–Oct.

Stegostyla dimorpha

Stegostyla gracilis
Musky Caladenia
Stem to 40cm tall with 2–4 flowers. Flowers 30mm across, with musky odour, white on inside, brown-mauve or red-green on back. Dorsal sepal hooded. Lateral sepals and petals lanceolate, 12–20mm long. Labellum white with purple tip, weakly 3-lobed; lateral lobes entire, midlobe fringed. Calli in 4 rows with white or yellow heads. **HABITAT:** Open forest, woodland, on sandy soils. **DIST:** Upper Blue Mtns (e.g. Hassans Walls, Clarence). Also NT, ST, Vic, Tas, SA. **FL:** Sept–Nov.

Stegostyla gracilis

Stegostyla testacea

Stegostyla testacea
Honey Caladenia
Stem to 18cm tall with 1–3 flowers. Flowers 17mm across, glandular, yellow becoming brown at tips; outside surface darker brown. Dorsal sepal hooded. Lateral sepals and petals to 12mm long, usually drooping. Labellum ovate, recurved, with maroon tip; margins toothed and with stalked calli. Central calli in 4 rows with white or yellow heads near the base, pink near midlobe. Column with pink stripes. **HABITAT:** Woodland, scrub, in stony and sandy sites. **DIST:** Scattered on coast and t'lands (e.g. Glenorie, Hill End, Penrose). Also NC, SC, Vic, Tas. **FL:** Sept–Feb.

Plectorrhiza
Epiphytic orchids. Flowers in lateral racemes; labellum with backwards-pointing spur.

Plectorrhiza tridentata
Tangle Orchid
Small plant with tangled aerial roots. Leaves narrow elliptic, to 10cm long. Racemes with 3–8 flowers. Flowers brown to green, fragrant. Sepals and lateral petals ovate. Labellum white with green and brown markings, 3-lobed, with prominent lateral lobes. **HABITAT:** Branches of rainforest trees, esp. along creeks. **DIST:** Coast and ranges (e.g. Hacking R., Mt Kembla, Mt Tomah). Also NC, SC, NT, Qld, Vic. **FL:** Sept–Dec.

Plectorrhiza tridentata

Papillilabium beckleri is similar to *Plectorrhiza tridentata* but with smaller cupped flowers (2.5mm) and a labellum with tiny bumps on surface. It flowers Sept–Nov.

Prasophyllum
Terrestrial orchids with erect stem partly sheathed by solitary terete leaf. Flowers numerous in a spike. Dorsal sepal lanceolate, concave. Petals short, narrow, forward-pointing. Flowers with dorsal sepal lowermost.

Prasophyllum brevilabre
Short-lip Leek Orchid
Stem 10–25cm tall with 6–30 flowers. Flowers greenish brown, about 10mm across. Dorsal sepal ovate. Petals linear to lanceolate. Labellum conspicuous, white, margins undulate, bent back almost 180°; callus plate white with deep central channel. **HABITAT:** Woodland, moist heath. **DIST:** Coast, t'lands, WS. Also Qld, Vic, Tas. **FL:** Aug–Dec.

Prasophyllum elatum
Tall Leek Orchid
Robust plant often growing in colonies. Stem to 1.2m tall with up to 60 flowers. Flowers greenish mauve with white labellum. Dorsal sepal ovate. Petals lanceolate. Labellum broadly ovate, bent weakly backwards, with undulate margins and long green callus plate. **HABITAT:** Heath, scrub, on sandy soils. **DIST:** Coast and ranges. Also NC, SC, CWS, NWP. All States (except NTerr). **FL:** Aug–Nov.

Prasophyllum flavum
Yellow Leek Orchid
Stem 60–90cm tall with up to 50 flowers. Flowers yellowish green, about 10mm across. Dorsal sepal ovate and bent forward. Petals lanceolate. Labellum ovate, with undulate margins, gently bent backwards; callus plate green, triangular, channelled. **HABITAT:** Open forest, on both moist and drier sandy soils. **DIST:** Coast and ranges. Also NC, SC, NT, ST, NWS, Qld, Vic, Tas. **FL:** Nov–Jan.

Prasophyllum brevilabre

Prasophyllum elatum

FLOWERING PLANTS

Prasophyllum flavum

Prasophyllum patens

Prasophyllum patens
Broad-lip Leek Orchid
Stem 10–30cm tall with up to 40 flowers. Flowers greenish, 10–15mm across. Lateral sepals linear to lanceolate and widely diverging. Dorsal sepal ovate to lanceolate. Petals lanceolate, often with a dark stripe. Labellum ovate, with undulate margins, strongly recurved; callus plate to 7mm long, greenish yellow. **HABITAT:** Heath, on sandy soils. **DIST:** Coast and ranges. Most of NSW. Also Qld, Vic, Tas, SA, NZ. **FL:** Sept–Dec.

Prasophyllum striatum
Streaked Leek Orchid
Stem 10–15cm tall with up to 10 flowers. Flowers with unpleasant odour, to 7mm across, with green sepals and white petals, all with red-brown stripes. Dorsal sepal triangular, acute. Petals lanceolate. Labellum ovate, strongly recurved, white, with wavy margins; callus plate white, extending nearly to apex. **HABITAT:** Damp heath, swamp margins. **DIST:** Coast and ranges. Also NC, SC, ST. **FL:** Apr–June.

Prasophyllum striatum

Pterostylis

Terrestrial orchids with leaves in basal rosette or on stem. Flowers mostly green, often with shades of brown or red. Galea formed by fusion of dorsal sepal and lateral petals hooded over column and labellum. Lateral sepals erect or deflexed, joined in their lower part forming a V-shaped cleft (sinus), their upper parts free and often tapering to fine points.

Pterostylis acuminata

Pterostylis curta

Pterostylis acuminata
Sharp Greenhood
Stem 15–24cm tall. Flower solitary, 25mm long, translucent green and white with some light brown. Apex of galea acuminate. Lateral sepals bulging at the sinus then produced into long erect filiform points. Labellum linear with acuminate tip and strongly curved downwards. **HABITAT:** Open forest, on sandy soils. **DIST:** Coast, lower Blue Mtns. Also NC, SC, Qld. **FL:** Apr–May.

Pterostylis baptistii
King Greenhood
Stem 30–40cm tall. Flower solitary, 60mm long, white and green with brown at apex. Galea erect, ± horizontal or decurved in upper part, tip acuminate. Lateral sepals broad and bulging at sinus then swept back. Labellum oblong, ridged with narrow tip. **HABITAT:** Woodland, scrub, on sheltered slopes. **DIST:** Scattered. Coast. Also NC, SC, Qld, Vic. **FL:** Aug–Oct.

Pterostylis concinna
Trim Greenhood
Stem to 25cm tall. Flower solitary, 20mm long, green with white stripes and brown tints at sinus and galea. Galea erect, decurved in upper part, tip acute. Lateral sepals erect, very fine, to 20mm long. Labellum oblong, with broad notch visible above wide sinus. **HABITAT:** Open forest, shrubland, esp. in moist sites. **DIST:** Coast. Also NC, SC, CWS, Vic, Tas, SA. **FL:** June–Sept.

Pterostylis curta
Blunt Greenhood
Stem 15–25cm tall. Flower solitary, 35mm long, white and green, often with brown at tip. Galea erect, upper part curving forward, tip acute. Lateral sepals very wide at sinus, then bent back slightly, with short acuminate tips rarely rising above galea. Labellum oblong, brown, twisted. **HABITAT:** Open forest, on sandy hillsides. **DIST:** Coast and ranges. Most of NSW. Also Qld, Vic, Tas, SA, Pac. Iss. **FL:** June–Sept.

Pterostylis baptistii

Pterostylis concinna

Pterostylis daintreana (syn. Pharochilum daintreanum)

Stem to 30cm tall with 3–5 bracts and 3–7 flowers. Flowers to 10mm long, translucent, green and white, nodding. Galea with fine tip. Lateral sepals narrow, pointing forward and down. Labellum short, oblong, dark brown, 3-lobed. **HABITAT:** On moist mossy sandstone outcrops. **DIST:** Coast and ranges. Also NC, SC, NT, Qld. **FL:** Mar–July.

Pterostylis grandiflora
Cobra Greenhood

Stem to 30cm tall with spreading stem leaves and basal rosette. Flower about 35mm long, green and white, with bronze-brown wings (petals). Lateral sepals erect, 35mm long; bulging at sinus. Labellum brown, narrow, visible above sinus. **HABITAT:** Moist shady gullies in open forest. **DIST:** Coast and ranges. Also NC, SC, Qld, Vic, Tas. **FL:** Apr–July.

Pterostylis hildae
Hilda's Greenhood

Stem to 15cm tall. Flower solitary, 20mm long, white and green with brown tints at tip. Galea leaning forward with short tip. Lateral sepals short, swept up and back. Labellum elliptic, brown, protruding through sinus. **HABITAT:** Rainforest margins, moist open forest. **DIST:** Uncommon. Illawarra (e.g. Mt Kembla), Sthn H'lands (e.g. Bundanoon). Also NC, SC, NT, Qld. **FL:** Aug–Sept.

Similar to *P. curta* but with smaller flowers and a shorter, non-twisted labellum.

Pterostylis laxa

Stem to 40cm tall. Flower solitary, slightly nodding, 25mm long, green with white stripes and brown tints on galea. Galea with filiform point to 18mm long overhanging sinus. Lateral sepals erect, very fine, curving back, to 40mm long. Labellum lanceolate, reddish brown, protruding through sinus. **HABITAT:** Grassy slopes in open forest. **DIST:** Upper Blue Mtns, Sthn H'lands. Also NC, NT, ST, CWS, Vic. **FL:** Feb–May.

Pterostylis daintreana

Pterostylis longifolia
Tall Greenhood

Stem 15–40cm tall with 5–7 stem leaves 3–10cm long. Flowers 1–7, pale translucent green and white, about 15mm long on short pedicel. Labellum oblong, pale green with black central line, with bristles around lobes, notched at tip. **HABITAT:** Open forest, in moist sites. **DIST:** Coast and ranges. Also NC, SC, NT, ST, CWS, Qld, Vic, Tas, SA. **FL:** Apr–Aug.

Pterostylis nutans
Nodding Greenhood

Stem 15–30cm tall. Flower solitary, very nodding, 25mm long, translucent with green and some brown. Lateral sepals crossing galea near tip. Labellum ovate, curved, pubescent on upper surface. **HABITAT:** Open forest, scrub, in sheltered damp sites. **DIST:** Common. Coast and ranges. Most of NSW. Also Qld, Vic, Tas, SA, NZ. **FL:** May–Oct.

Pterostylis grandiflora

Pterostylis hildae

Pterostylis laxa

Pterostylis longifolia

Pterostylis nutans

Pterostylis parviflora
Tiny Greenhood

Stem slender, 10–25cm tall, with basal rosette and up to 4 stem leaves. Flowers 1–5 facing stem, 7mm long, shiny green with white stripes. Galea brown, narrow, ± erect. Lateral sepals rarely exceeding galea and tight against it. Labellum narrow, not visible through sinus. **HABITAT:** Open forest, woodland, heath. **DIST:** Coast and ranges. Most of NSW. Also Qld, Vic, Tas, SA. **FL:** Mar–May.

Pterostylis pedunculata
Maroonhood

Stem slender to 25cm tall. Flower solitary, 15mm long, green and white striped. Lateral sepals and galea tip brownish. Galea erect, with upper part bent forward and acute tip horizontal. Lateral sepals filiform, erect or divergent. Labellum ovate; tip visible in sinus. **HABITAT:** Sheltered sites of tall forest, rainforest margins. **DIST:** Coast and ranges. Also NC, SC, NT, Vic, Tas. **FL:** July–Oct.

Pterostylis parviflora

Pterostylis pedunculata

Pterostylis erecta is similar to *P. pedunculata*, except flower is 20mm long and tip of galea points slightly up (not horizontal). It flowers Aug–Sept.

Pterostylis saxicola

Stem to 25cm long, with 2–4 sheathing stem leaves and 2–6 flowers. Flowers 13mm long, reddish brown on petals, translucent green on galea, semi-erect. Galea with upturned tip, 3mm long. Lateral sepals ovate with narrow filaments, curved forward and divergent. Labellum broad-ovate, dark red-brown, with fine bristle-like hairs. **HABITAT:** Woodland, esp. among rocks. **DIST:** Endemic. Rare. Wstn Sydney (e.g. Picnic Point, Kentlyn). **FL:** Sept–Nov.

Rimacola

Monotypic genus related to *Lyperanthus*. Leaves few to several, basal and cauline. Inflorescence a crowded terminal raceme with 6–18 flowers. Flowers with free petals and sepals.

Rimacola elliptica
Green Beaks

Stem to 25cm, arching. Leaves elliptic, to 10 x 3cm. Flowers green with reddish markings, to 2cm across. Dorsal sepal hooded over column and labellum, often recurved towards apex. Labellum dark brownish red, shorter than other segments. **HABITAT:** Dripping ledges of sandstone cliffs and caves. **DIST:** Endemic. Blue Mtns (e.g. Linden, Wentworth Falls), coast (e.g. Dural, Middle Hbr). **FL:** Nov–Dec.

Sarcochilus

Epiphytic and lithophytic orchids. Leaves scattered along stems in 2 rows. Sepals and petals similar. Labellum with hollow pouch in front.

Sarcochilus australis
Butterfly Orchid

Leaves narrow-elliptic, falcate, 2–10cm x 3–12mm. Flowers 2–14 in racemes 3–15cm long, greenish brown, 15mm across. Sepals and petals narrow-obovate, spreading, green-brown. Labellum spurred, with oblong lateral lobes, white with purple stripes. **HABITAT:** On tree branches in gullies and rainforest. **DIST:** Coast (e.g. Ku-ring-gai Chase NP). From Hunter R. to SC, ST, Vic, Tas. **FL:** Oct–Dec.

Pterostylis saxicola

Sarcochilus australis

Rimacola elliptica

Sarcochilus falcatus

Spiranthes sinensis

Sarcochilus hillii

Sarcochilus falcatus
Orange-blossom Orchid
Leaves thick, narrow-oblong, to 15 x 2cm long, curved. Flowers 1–12 in raceme to 10cm long, white, 30mm across. Sepals and petals elliptic. Labellum with orange and purple markings. **HABITAT:** On trunks and branches of rainforest trees. **DIST:** Coast and ranges (e.g. Mt Kembla, Mt Wilson). Also NC, SC, NT, ST, Qld, Vic. **FL:** July–Nov.

Sarcochilus hillii
Little Gem Sarcochilus
Epiphyte with extensive root system. Flowering stem to 10cm, pendulous, bearing 2–6 flowers. Leaves linear, to 8cm long, channelled, spotted. Flowers white to pink, often cupped. Labellum 2–4mm long with densely hairy white midlobe. **HABITAT:** Trunks and branches, esp. on *Backhousia myrtifolia*, in rainforest along creeks. **DIST:** Coast (e.g. Pittwater, Kentlyn, Hill Top). Also CT, NC, SC, NT, Qld. **FL:** Oct–Jan.

Sarcochilus olivaceus. Epiphyte with green flowers. From rainforests (e.g. Bulli Pass, Wentworth Falls). It flowers Nov–Dec.

Thelychiton speciosus

Tetrabaculum tetragonum

Tetrabaculum, Thelychiton, Tropilis

New genera created by revision of *Dendrobium*. Not universally accepted.

Tetrabaculum tetragonum (syn. *Dendrobium tetragonum*)
Tree Spider Orchid
Pseudobulbs succulent, swollen, 4-angled, to 40cm long. Leaves 2–5 at apex of pseudobulb, elliptic, to 8cm x 25mm. Racemes short, 1–5-flowered. Flowers greenish to dull yellow, 25mm across; sepals narrow-triangular, with red margins; labellum midlobe with incurved margins. **HABITAT:** On rainforest trees, esp. *Trochocarpa laurina*, often near creeks. **DIST:** Coast (e.g. Hacking R., Mt Kembla, Somersby). Also NC, Qld. **FL:** Sept–Nov.

Thelychiton speciosus (syn. *Dendrobium speciosum*)
Rock Orchid
Large clumped orchid. Mature plants with erect thick pseudobulbs, to 40cm long, topped by 2–5 thick dark ovate leaves to 23 x 8cm. Inflorescences to 60cm long with many yellow fragrant flowers, 35mm across. Sepals and petals narrow. Labellum 3-lobed with purple spots and stripes. **HABITAT:** Rocky outcrops, in sheltered but often sunny, rather dry places. **DIST:** Chiefly coast; Blue Mtns (Glenbrook). Also NC, SC, CWS, Vic. **FL:** Sept–Oct.

Spiranthes
Terrestrial orchids with basal leaves, stem bracts and a distinctive spiral raceme of flowers.

Spiranthes sinensis
Ladies' Tresses
Stem rising 15–40cm from cluster of linear leaves, with 2–3 stem bracts and a long spiral plait of closely packed, tiny, bright pink and white flowers. Petals and dorsal sepal 4–5mm long, hooded over column. Lateral sepals spreading. Labellum with white fringe. **HABITAT:** Woodland, sedgelands, swamp margins, creek banks. **DIST:** Coast and ranges. Also NC, SC, NT, ST, WS, Qld, Vic, Tas, SA. **FL:** Dec–Mar.

Tropilis aemula

Tropilis aemula (syn. *Dendrobium aemulum*)
Ironbark Orchid
Epiphyte with succulent pseudobulbs, to 30 x 1cm. Leaves ovate, leathery, to 60 x 25mm. Racemes to 10cm long with up to 12 flowers, white to yellow, ageing to pink. Labellum white or yellow with red markings. **HABITAT:** On trees, esp. *Eucalyptus paniculata* and *Backhousia myrtifolia*, in and near rainforest. **DIST:** Chiefly coast (e.g. Upper Colo, Mill Ck, Woodford–Glenbrook). Also NC, SC, Qld. **FL:** July–Oct.

Thelymitra carnea

Thelymitra
Terrestrial herbs with erect stems and 1 basal stem-clasping leaf. Sepals and petals free, similar. Column short with prominent wings and 2 lateral lobes often terminating in a tuft of hairs. Flowers open mostly around midday in warm, sunny, still weather.

Thelymitra carnea
Pink Sun Orchid
Stem 10–20cm tall with erect channelled terete leaf at base and 2 small stem bracts. Flowers 1–4, salmon pink, about 10mm across. Column pinkish cream with yellow 3-lobed apex; lateral lobes lacking hair tufts. **HABITAT:** Woodland, heath, on moist sandy soils. **DIST:** Coast and ranges. Also NC, NT, Qld, Vic, Tas, SA, NZ. **FL:** Sept–Nov.

Thelymitra ixioides
Dotted Sun Orchid
Stem to 50cm tall. Leaf linear, to 20 x 1cm. Flowers numerous, 30–50mm across, blue with dark blue spots; column 3-lobed, lateral lobes with white terminal hair tufts, central lobe short with crest consisting of 2–3 rows of calli. **HABITAT:** Open forest, heath, on sandy soils. **DIST:** Coast and ranges. Also NC, SC, ST, Qld, Vic, Tas, SA, NZ. **FL:** Aug–Nov.

Thelymitra media var. *media*
Tall Sun Orchid
Stem to 90cm tall. Leaf lanceolate, 30cm x 18mm, ribbed, fleshy. Flowers numerous, to 30mm across, blue, unspotted, with light pink or blue streaks. Apex of column yellow subtended by black band, lateral lobes with terminal white hair tufts. **HABITAT:** Swamp margins. **DIST:** Coast and ranges. Hawkesbury R. to Vic. **FL:** Sept–Jan.

Thelymitra pauciflora
Slender Sun Orchid
Stem to 50cm tall. Leaf linear, keeled, to 20cm long. Flowers 3–4 (rarely to 15), 10–20mm across, pale blue to pink, often not opening fully. Column hooded, dark towards the top except for yellow cleft at front; lateral lobes bent upwards ending in white hair tufts. **HABITAT:** Open forest, woodland, heath. **DIST:** Coast and ranges. Most of NSW. Also Qld, Vic, Tas, SA, WA, NZ. **FL:** Aug–Jan.

FLOWERING PLANTS | 591

Thelymitra ixioides

Thelymitra media var. *media*

Thelymitra pauciflora

Thelymitra venosa
Veined Sun Orchid
Stem to 70cm tall. Flowers 1–10 in loose raceme, 25mm across, blue or mauve, with dark blue veins. Column yellow, winged, not hooded over anther; midlobe short with small crowded calli, the 2 lateral lobes coiled without hair tufts. **HABITAT:** Damp areas in heath, mountain bogs and below cliff faces. **DIST:** Endemic. Blue Mtns (e.g. Mt Banks, Blackheath, Porters Pass). **FL:** Oct–Dec.

Thelymitra venosa

Glossary

acuminate: tapering gradually to a prolonged point.
acute: pointed, sharp.
anther: the part of a flower bearing the pollen.
apical: at the apex or top.
apiculum: a small abrupt discrete point (adj. apiculate).
appressed: pressed closely against; lying flat against.
aril: the swollen part of the stalk attaching seed to pod.
awn: a bristle-like projection.
axil: the angle between stem and leaf.
axillary: in the axil.

basal (also radical): attached or grouped at the base.
basifixed: of anthers, attached at or by the base.
bipinnate: twice pinnately divided, i.e. with the primary divisions (pinnae) again divided into leaflets (pinnules).
bipinnatifid: twice pinnatifid, i.e. with the primary lobes again lobed.
bract: modified leaf associated with a flower or inflorescence.
bracteate: with bracts, bearing bracts.
bracteole: small bract-like structure on the stalk (pedicel) or calyx of a flower.

callus: thickened or hardened tissue (pl. calli).
calyx: collective term for the sepals of a flower.
campanulate: bell-shaped.
capsule: a dry fruit formed from 2 or more carpels and splitting at maturity to release the seeds.
carpel: a unit of the female part of a flower (gynoecium) in which the ovules are fertilised and develop into seeds; usually divisible into ovary, stigma and style.
caudex: a thick erect trunk.
cauline: on the stem or old wood (hence 'cauliflorous').
ciliate: having a margin fringed with fine hairs; like an eyelash.
concolorous: of leaves, having the same colour on both surfaces.
connate: referring to organs of the same kind which are fused together.
cordate: heart-shaped.
coriaceous: leathery.
corolla: the collective term for the petals of a flower.
corymbose: a terminal inflorescence in which the flower stalks become shorter towards the top so that all flowers are borne at about the same level.
cotyledon: the first 1 or 2 leaves of the embryo; a seed-leaf.
crenate: scalloped, with small rounded teeth.
crenulate: minutely crenate.
crownshaft: a smooth section of tubular leaf bases below the crown of leaves of some pinnate-leaved palms.
culm: the flowering stem of plants with basal leaves such as grasses, sedges, rushes, terrestrial orchids.
cyme: an inflorescence in which a flower is at the end of the main axis and further flowers occur on lateral axes below it (adj. cymose).
cypsela: a dry, one-seeded fruit of Asteraceae.

decumbent: with branches spreading along the ground and tips growing upwards.
decurrent: of leaves, the base or stalk extending downwards forming a wing or ridge on the stem.
decurved: gradually curved downwards.
decussate: of opposite leaves, with each pair at right angles to the preceding and succeeding pair.
dentate: toothed.
denticulate: finely toothed.
dichotomous: dividing equally into two, forking.
dicotyledon: a flowering plant whose embryo has 2 seed-leaves.
digitate: radiating from a single point like the fingers of a hand.
dimorphic: having two forms, e.g. in ferns where sterile and fertile fronds differ.
dioecious: with male and female flowers on different plants.
discolorous: of leaves, having a different colour on each surface.

distichous: in two opposite rows in the same plane.
domatia: small depressed structures in the axils (junction) of the mid vein and main lateral veins on the undersurface of some leaves.
dorsifixed: of anthers, attached at or by the back.
drupe: a succulent 1-seeded fruit with the seed enclosed in a stony layer.

emarginate: with a shallow notch at the apex.
endemic: with the natural distribution confined to a particular geographic region.
epicalyx: a whorl of bracts on the pedicel below the flower and resembling a second calyx.
epiphyte: a plant growing on another plant, but not parasitic.
exsert: protruding from, as stamens projecting beyond the floral tube.
extra-axillary: not in the axil.

falcate: of leaves, substantially curved, more or less sickle-shaped.
fertile: of fern fronds, bearing sporangia.
filament: the stalk of a stamen.
filiform: thread-like.
flexuose: bent from side to side in a zigzag pattern.
floret: (1) a small flower; (2) a grass flower including the lemma and palea which enclose it.
1-foliolate: a compound leaf with only the terminal leaflet developed.
follicle: a dry fruit formed from 1 carpel which splits along one of its edges.
frond: the 'leaf' of a fern, consisting of stipe (stalk) and lamina (blade).
funicle: the stalk of an ovule.
fusiform: narrower at both ends than in middle.

galea: perianth segment or segments of an orchid, shaped like a helmet.
gibbous: with a hump or swelling.
glabrous: without hairs.
glaucous: blue-green with a waxy whitish bloom which can be partly rubbed off.
glume: a small chaffy bract at the base of most grass spikelets.
globular, globose: nearly spherical (3-dimensional).

hastate: shaped like an arrow-head, i.e. narrow-triangular with spreading basal lobes.
hemiparasitic: partly parasitic on another organism and partly self-supporting. For plants, usually containing chlorophyll and able to undergo photosynthesis.
hispid: with stiff bristly hairs.
hoary: with dense covering of whitish or greyish hairs.

imbricate: with edges overlapping.
incurved: bent inwards along the margins, concave above.
indusium: a pollen cup on the end of the style of Goodeniaceae; also the tissue covering the sorus of a fern.
inflorescence: the arrangement in which flowers are borne on a plant.
interjugary: on the rachis of bipinnate leaves between successive pairs of pinnae or pinnules (*Acacia*); cf. jugary.
intramarginal: of a leaf, inside and close along the margin.
involucre: a ring of bracts surrounding an inflorescence or several flowers.
involute: rolled upwards and inwards.

jugary: on the rachis of bipinnate leaves between the bases of oppositely paired pinnae or pinnules (*Acacia*); cf. interjugary.

labellum: (1) a lip; (2) the middle petal of an orchid.
lamina: the blade of a leaf.
lanceolate: shaped like a lance-head; long and narrow, tapering at both ends, broadest below the middle, pointed at the tip.
lemma: the outer (lower) of 2 bracts enclosing a grass flower.
lenticels: small raised spots on a surface, often corky, through which gases are exchanged.
lignotuber: a woody swelling of the trunk, partly or wholly below ground.
ligulate: strap-shaped.
ligule: a membranous or hairy appendage at the junction of a leaf-sheath and blade of grass; the corolla limb of a ray floret of Asteraceae.
linear: long and narrow with nearly parallel sides.
lithophyte: growing on rocks.
loculus: a compartment of an ovary.

Malesia: a botanic-geographic region consisting of Malaysia, Indonesia, the Philippines and New Guinea.
marginal: on or along the margin or edge.
membranous: thin and translucent.
mericarp: a fruitlet formed by the splitting of a schizocarp into separate carpels (also coccus, pl. cocci).
-merous: a suffix indicating the numeral which divides evenly into the number of parts present in each whorl of a flower, e.g. a 5-merous flower generally has 5 parts in each whorl, or multiples thereof in some.
monocotyledon: a flowering plant whose embryo has only 1 seed-leaf.
monoecious: having unisexual flowers with males and females on the same plant.
monotypic: of a genus, having only 1 species.

nerves: the longitudinal veins of a phyllode.

obcordate: heart-shaped leaf, attached at the pointed end and notched at the tip.
oblanceolate: lance-shaped but broader above the middle.
oblate: almost circular with breadth slightly longer than length (2-dimensional), a leaf shape.
oblong: with the sides parallel, relatively broader than linear.
obovate: similar to ovate, but attached by the narrow end.
obtuse: rounded or broadly pointed at the apex.
operculum: the cap covering the flower bud in *Eucalyptus* (also calyptra).
orbicular: circular.
ovary: the base of a carpel enclosing the ovules.
ovate: of leaves, a 2-dimensional egg-shape, broadest below the middle.
ovoid: a 3-dimensional egg-shaped solid.
ovule: an egg which develops into a seed after fertilisation.

palea: the inner (upper) of 2 bracts enclosing a grass flower.
palmate: with several leaflets or veins spreading from the same point (also digitate).
panicle: a branched inflorescence (adj. paniculate).
pappus: tuft or ring of hairs or scales at the top of the cypsela of the Asteraceae.

pedicel: the stalk of a flower.
peduncle: the stalk of an inflorescence from the base to the first branch or flower.
peltate: of leaves, with stalk attached on lower surface away from margin.
penniveined: of leaves, more or less parallel veins which diverge from the midrib to the margin.
perianth: the calyx and corolla of a flower.
petiole: the stalk of a leaf (adj. petiolate).
phyllode: a flattened enlarged leaf stalk which looks and functions like a leaf (as in *Acacia*).
pilose: with long soft simple hairs.
pinna: the first division of a compound leaf (pl. pinnae).
pinnate: a compound leaf with leaflets diverging from a common axis.
pinnatifid: of leaves and leaflets, cut into lobes, but not as deeply as pinnatisect.
pinnatisect: cut into lobes nearly to the midrib.
pinnule: a leaflet of a bipinnate leaf.
plano-convex: flat on one side, convex on the other.
polygamous: having bisexual and unisexual flowers on the same plant.
procumbent: lying or spreading along the ground.
proliferous: capable of developing into a separate plant, or having adventitious buds with this capability.
pruinose: with a waxy powdery whitish coating which can be easily rubbed off.
pubescent: covered with short downy hairs.
pungent: sharp and stiffly pointed.

raceme: a simple inflorescence with stalked flowers along the axis.
rachis: the axis of an inflorescence or of a pinnate leaf (also rhachis).
radical: see basal.
recurved: curved downwards or with margins curved downwards.
reflexed: bent sharply downwards or backwards.
reticulate: with a network of veins.
revolute: of leaves, with margins rolled downwards and touching the undersurface.
rhizome: an underground stem, usually horizontal.
rotate: of a corolla, shortly tubular at the base, widely flared and shallowly lobed above.

GLOSSARY

saprophyte: a plant deriving its nutrition from dead organic matter; usually lacking chlorophyll.
scabrous: rough to the touch (also scabrid).
scale: (of ferns) a flat papery outgrowth on the surface of rhizomes, stipes and rachises.
schizocarp: a dry fruit which splits into individual carpels (see mericarp).
sepal: one of the outer segments surrounding a flower; collectively known as the calyx.
septate: having cross-partitions.
septum: a partition.
serrate: toothed like a saw.
serrulate: finely toothed.
sessile: without a stalk.
sinuate: with wavy (in and out) margins; cf. undulate: wavy (up and down).
sinus: a gap or notch between two segments.
sorus: a cluster of sporangia in a fern.
spadix: a spicate inflorescence on a thick axis, often surrounded by a spathe (pl. spadices).
spathe: a large bract at the base of a spadix and sheathing it.
spathulate: spoon-shaped.
spike: an unbranched inflorescence with sessile flowers.
spinescent: ending in a spine; spine-like.
sporangium: a spore-producing organ (pl. sporangia).
sporophylls: a leaf-like structure which bears one or more sporangia.
stamen: the male organ of a flower, consisting of a filament and anther.
staminode: a sterile stamen, i.e. lacking an anther.
sterile: barren; (in ferns) lacking sporangia.
stigma: the part of the carpel which receives the pollen, usually at the tip of the style.
stipe: the stalk of a fern frond, or a stalk-like elongation below an ovary or a pod.
stipule: one of 2 scaly or leafy appendages occurring at the base of leaves or leaf stalks in some dicotyledons (adj. stipulate, stipular).
stolon: a prostrate stem, rooting at the nodes.
striate: marked with lines or ridges.
strobilus: a cluster of sporophylls.
subulate: gradually tapering to a fine point.
subshrub: an undershrub, usually small, soft-wooded, weak or sparse.
subtend: to stand below or close to.
subterete: nearly terete, i.e. oval in cross-section, slightly flattened.

tendril: part of a plant modified into a climbing organ.
tepal: a perianth segment of a flower which is not clearly differentiated into sepals and petals. The plural often used (as here) to refer to petals and sepals collectively.
terete: cylindrical, round in cross-section.
terminal: at the apex.
terrestrial: growing in the ground.
thyrse: a compound inflorescence with a mixed type of branching.
tomentum: a covering of dense matted hairs, like felt.
triad: a 3-flowered inflorescence, or unit of an inflorescence, consisting of a terminal flower and 2 opposite lateral flowers.
tripinnate: of fern fronds, thrice divided pinnately; the first two divisions referred to as pinnae (primary and secondary), the third as pinnules or ultimate segments.
truncate: ending abruptly as if cut off.
tuber: a swollen underground storage organ formed from the stem.
tuberculate: with small wart-like swellings.

ultimate segments: the finest divisions (excluding lobes).
umbel: an inflorescence in which the flower stalks diverge from one point at the top of a peduncle.
undulate: wavy (up and down); cf. sinuate, with wavy (in and out) margins.
urceolate: urn-shaped, i.e. narrowing below the rim then widening before tapering to the base.

valvate: meeting edge to edge but not overlapping.
vein: a distinct line of tissue in a leaf or of colour in a petal or sepal.
venation: the arrangement of veins in a leaf or frond.
versatile: of anthers, with a pivotal attachment near the midpoint.
villous: with dense, long, weak hairs.
viscid: sticky, and often shiny.

Leaf and flower shapes

Simple leaves

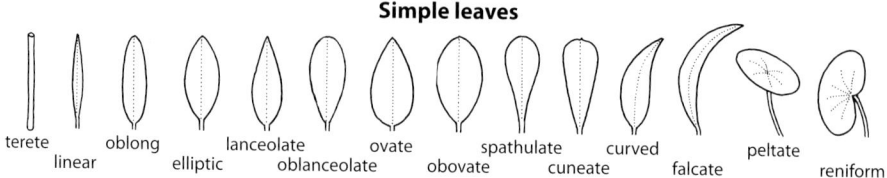

terete, linear, oblong, elliptic, lanceolate, oblanceolate, ovate, obovate, spathulate, cuneate, curved, falcate, peltate, reniform

Compound leaves

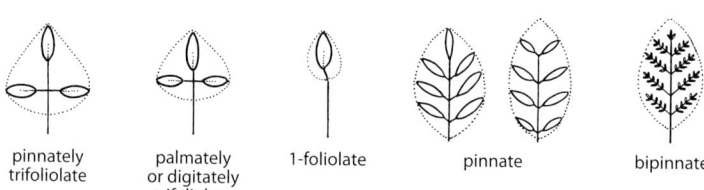

pinnately trifoliolate, palmately or digitately trifoliolate, 1-foliolate, pinnate, bipinnate

LEAF TIPS

truncate, mucronate, hooked, emarginate (notched), rounded, obtuse, acute, acuminate

LEAF BASES

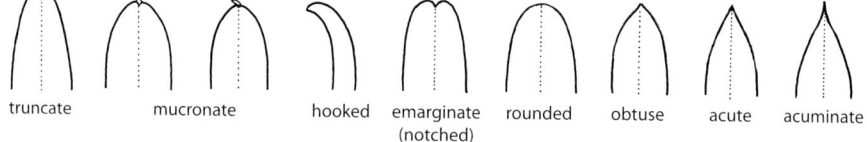

truncate, cordate, hastate, cuneate, obtuse, sagittate

INFLORESCENCES

FLOWER STRUCTURE

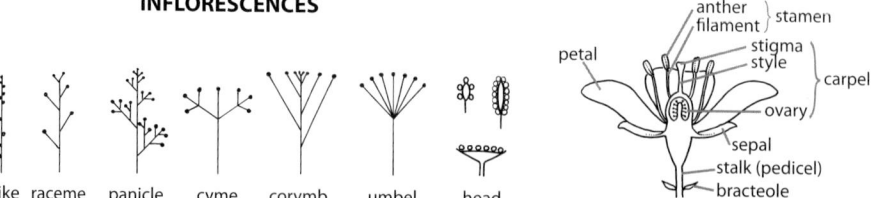

spike, raceme, panicle, cyme, corymb, umbel, head

anther, filament, stamen, petal, stigma, style, carpel, ovary, sepal, stalk (pedicel), bracteole

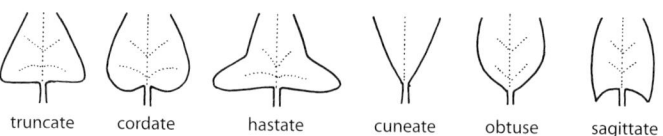

Common name index

Acronychia
  Common 317
  Yellow 337
Alania 530
Amulla 406
Aneilema, Pointed 472
Aotus, Common 166
Apple
  Argyle 268
  Black 113
  Broad-leaved 238
  Narrow-leaved 236
  Rough-barked 237
  Smooth-barked 237
  Winter 406
Apple Berry
  Climbing 132
  Hairy 132
Arrowgrass, Streaked 468
Ash
  Black 249
  Blue Mountains 246
  Blue Mountains Mallee 251
  Blueberry 81
  Budawang 250
  Cliff Mallee 252
  Faulconbridge Mallee 251
  Giant Mallee 250
  Narrow-leaved Mallee 252
  Pigeonberry 80
  Red 370
  Silvertop 249
  Whipstick Mallee 249
  Yellow 372
  Yellow-top 246
Austral Bugle 412

Baeckea
  Swamp 276
  Tall 274
Ballart
  Cherry 360
  Pale 360
Bangalay 258

Banksia
  Coast 206
  Hairpin 208
  Heath 205
  Hill 208
  Large-leaf 207
  Old Man 208
  Saw 208
  Silver 206
  Swamp 206, 207
  Wallum 204
Bead Bush 411
Beard-heath
  Blunt 104
  Coast 104
  Lance-leaf 103
  Prickly 102
  Small-leaved 104
Bear's Ears 457
Bedstraw
  Maori 388
  Rough 388
Beech
  Brown 52, 367
  Silky 367
  White 410
Beefwood, Scrub 233
Bidgee-widgee 136
Bindi-eye 446
Bindweed
  Australian 390
  Pink 390
Bitter-pea
  Broad-leaf 172
  Gorse 173
  Slender 172
Black-eyed Susan 345
Blackbutt 244
Blackthorn, Native 132
Blackwood 162
Bladderwort
  Asian 405
  Floating 405
  Small 405
Bleeding Heart 120
Bloodroot
  Rush-leaf 558
  Strap-leaf 558

Bloodwood
  Brush 116
  Red 238
  Yellow 239
Blue Gum
  Mountain 257
  Sydney 257
Blue Fairy 570
Blue Pincushion 442
Blue Trumpet 408
  Dwarf 408
Blue Yam 408
Bluebell
  Australian 426
  Tall 427
  Tufted 426
Bog-rush
  Black 513
  Button 505
  Common 512
  Fluke 512
  Hairy 514
  Heath 512
  Zig-zag 512
Bolly Gum 53
  White 53
Bolwarra 49
Bonewood 372
Boobialla 406
  Northern 406
  Western 408
Boronia
  Barker's 320
  Dean's 320
  Dwarf 323
  Fraser's 320
  Ledum 323
  Milkwort 324
  Narrow-leaved 319
  Pale Pink 320
  Pinnate 324
  Rose 325
  Small-leaved 323
  Soft 323
  Sticky 318
  Swamp 324
  Sydney 323
  Waxy 323
Bossiaea
  Matted 167
  Spiny 168

Bottlebrush
  Alpine 280
  Crimson 278
  Green 278
  Lemon 278
  Megalong 286
  Narrow-leaved 278
  Pine-leaved 278
  Red 278
  Stiff 280
  White 280
  Willow 280
Box
  Apple 263
  Blue 270
  Coast White 265
  Grey 268
  Paddy's River 265
  White-topped 265
  Yellow 272
Bracken, Common 28
Brake
  Jungle 29
  Tender 28
Bramble
  Broad-leaf 136
  Forest 137
Bristle-rush 496
  Round-headed 496
Brittlebush 117
Brooklime 400
Brookweed, Creeping 115
Broom, Native 204
Broom-heath
  Prickly 107
  Tree 106
Brown Barrel 246
Brown Beaks 576
Brown Jack 52
Bulbophyllum, Autumn 562
Bulrush 526
Bur-reed 526
Burr
  Narrawa 396
  Sheep's 136
Burr-daisy
  Blue 446
  White 446
  Yellow 446

Burrawang 42
Bush Lawyer 136
Bush-pea
  Fine-leaf 201
  Large-leaf 194
  Rough 201
  Spreading 199
Buttercup
  Common 56
  Hairy 57
  River 56
Button Grass 505

Caladenia
  Honey 579
  Musky 579
Candlebark 267
Canthium, Coast 387
Carraway, Australian 355
Cascarilla, Native 118
Cassia, Southern 164
Cassinia
  Bent 450
  Shiny 449
  Sticky 450
  Three-veined 450
  Yellow 448
Cauliflower Bush 449
Cedar
  Pencil 351
  Red 317
  White 316
Celerywood 350
Celtis, Native 58
Cheese Tree 118
Cherry
  Blue 307
  Brush 306
  Native 360
Chloanthes, Common 411
Chloris, Tall 520
Christmas Bell 532
  Mountain 530
  Northern 532
Christmas Bush 128
  Victorian 419
Churnwood 367
Cinnamon Bells 575
Clerodendron, Hairy 410
Climbing Panax 350
Cloak Fern, Bristly 30

Clubmoss
  Bushy 19
  Scrambling 18
  Slender 18
Club-rush
  Knobby 506
  Large-headed 515
  Marsh 491
  River 512
  Swamp 506
Coachwood 128
Cockspur Flower 414
Cockspur Thorn 60
Coffee Bush 117
Comb Fern
  Branched 22
  Forked 22
Coneseeds 209
  Long-leaf 209
  Sprawling 209
Conesticks
  Common 232
  Prickly 232
  Stalked 231
Coopernookia 433
Copper-wire Daisy, Showy 463
Coprosma
  Prickly 387
  Rough 387
Coral Fern
  Pouched 23
  Scrambling 23
Coral-pea, Dusky 187
Cord-rush
  Flat 484
  Fringed 483
  Slender 483
  Tassel 483
Corkwood 52, 394
Correa
  Coastal 326
  Common 327
  Mountain 326
  White 326
Crabapple 129
Cranberry, Native 106
Cranesbill, Australian 343
Creamy Candles 369
Crinkle Bush 226
Crowea
  Lance-leaf 328
  Small 328
Crowfoot, Blue 341

Cryptandra
  Silky 372
  Spiny 372
Cumbungi, Broad-leaf 526
Cunjevoi 471
Currant-bush 360
Currant, Native 360
Cypress, Mueller's 45

Daisy
  Cut-leaf 445
  Showy Copper-wire 463
  Snow 452
  see also Burr-daisy; Paper Daisy
Daisy-bush
  Alpine 462
  Blush 460
  Musk 459
  Oak-leaved 462
  Rough 460
  Silky 460
  Small-leaved 460
  Sticky 460
Dampiera
  Blue 434
  Purple 434
Daphne, Native 334
Darling Pea, Smooth 203
Devil's Needles 398
Devil's Twine 54
Dirty Dora 498
Dodder, Australian 390
Dogwood 186, 448, 452
  White 451
Dollybush 448
Doubletails
  Golden 572
  White 571
Doughwood, Hairy-leaved 337
Drumsticks
  Broad-leaf 224
  Dawson's 224
  Fletcher's 225
  Narrow-leaf 224
  Prostrate 225
Duboisia 394

Eel-weed 468
Eggs and Bacon 178
Elderberry Panax 351

Elderberry, Yellow 380
Elkhorn 36
Euodia, White 337
Everlasting
  Ball 451
  Button 456
  Clustered 454
  Common 452
  Curled 456
  Golden 456
  Pale 455
  Sticky 456
  Tall 454
  Tree 452
Eye-bright 400

Fairies' Aprons 405
Fan Fern, Spreading 24
Fan-flower
  Creeping 441
  Dune 440
  Fairy 440
  Pale 440
  Purple 441
Featherwood 131
Felt Fern, Rock 36
Fern
  Bat's-wing 26
  Bird's Nest 33
  Crepe 20
  Finger 36
  Fragrant 36
  Gristle 38
  Hare's Foot 31
  Jointed 31
  Kangaroo 36
  King 20
  Necklace 33
  Screw 28
  Sickle 30
  Strap 38
  Umbrella 24
  see also Cloak Fern, Bristly; Comb Fern; Coral Fern; Fan Fern, Spreading; Felt Fern, Rock; Filmy Fern, Common; Fork Fern, Skeleton; Ground Fern; Lady Fern, Austral; Maidenhair Fern; Rasp Fern; Rock Fern, Poison; Shield

COMMON NAME INDEX 599

Fern; Tree Fern;
  Water Fern; Wedge
  Fern, Lacy
Fieldia 403
Fig
  Deciduous 60
  Moreton Bay 59
  Port Jackson 60
  Rusty 60
  Sandpaper 59
  Small-leaved 59
Filmy Fern, Common 25
Fireweed 464
Five-corners
  Green 112
  Long-leaf 110
  Red 112
Flame Pea, Eastern 170
Flame Tree, Illawarra 82
Flannel Flower 352
  Lesser 353
Flat-pea, Handsome 191
Flat-sedge
  Tall 498
  Variable 498
Flax, Settlers 471
Flintwood 88
Fork Fern, Skeleton 17
Frangipani, Native 133
Fringe Myrtle, Common 290
Fringe-rush, Common 501
Frogmouth, Woolly 557
Fuchsia, Native 327

Geebung
  Broad-leaf 228
  Hairy 227
  Lance-leaf 228
  Laurel 228
  Myrtle 230
  Narrow-leaf 228
  Nodding 230
  Pine-leaf 231
  Prostrate 227
  Soft 228
Geranium, Wild 343
Germander, Forest 424
Ginger, Native 527
Glycine, Twining 179
Golden Star 558
Golden-tip 183

Goodenia
  Hop 438
  Ivy 436
Grape, Slender 379
Grass
  Barbed Wire 520
  Blady 522
  Button 505
  Iron 544
  Kangaroo 525
  Long-hair Plume 520
  Long-leaved Wallaby 518
  Oat Spear 516
  Purple Wire 517
  Rare Plume 520
  Red-anther Wallaby 523
  Silver-top Wallaby 523
  Stout Bamboo 518
  Tall Spear 518
  Three-awned Spear 517
  Tufted Hedgehog 521
  Tussock 524
  Wallaby 517
  Weeping 523
Grass-flag
  Branching 527
  Pretty 528
Grass Tree
  Austral 550
  Small 552
Green Beaks 586
Greenhood
  Blunt 582
  Cobra 584
  Hilda's 584
  King 582
  Nodding 584
  Sharp 582
  Tall 584
  Tiny 586
  Trim 582
Grevillea, Carrington Falls 217
Ground-berry 93
Ground Fern
  Common 26
  Harsh 27
  Lacy 26
Groundsel
  Coast 464
  Fireweed 465
  Variable 464

Guinea Flower
  Bundled 74
  Climbing 78
  Erect 77
  Hairy 79
  Prickly 72
  Rough 72
  Shining 74
  Silky 77
  Stalked 77
  Trailing 74
  Twining 73
Guioa 315
Gum
  Black 262
  Bolly 53
  Brittle 263
  Cabbage 261
  Camden White 266
  Drooping Red 261
  Forest Red 262
  Grey 260
  Gully 266
  Hard-leaved Scribbly 256
  Maiden's 265
  Mallee Snow 250
  Manna 266
  Mountain 266
  Mountain Blue 257
  Mountain Grey 265
  Mountain Spotted 263
  Mountain Water 308
  Nepean River 266
  Ribbon 266
  Round-leaved 257
  Rusty 237
  Scribbly 254, 256
  Snappy 256
  Snow 249
  Spotted 239
  Swamp 262
  Sydney Blue 257
  Sydney Red 237
  Water 307, 308
  White 254
  White Bolly 53
  White Sally 249
  Wolgan Snow 250
Gunyang 398

Hakea
  Broad-leaved 220
  Dagger 222
  Small-fruited 222
  Willow-leaved 222

Heath
  Blunt-leaf 97
  Coral 97
  Cranberry 93
  Daphne 94
  Fuchsia 96
  Mueller's 97
  Peach 106
  Pine 94
  Pink Swamp 109
  Swamp 98
  Tree 112
  Urn 106
  see also Beard-heath;
  Broom-heath
Heath-myrtle
  Flax-leaf 276
  Fringed 305
  Rosy 276
  Short-leaved 274
  Twiggy 274
Hemigenia, Common 412
Heronsbill, Blue 341
Hickory 161
  Broad-leaved 161
  Mountain 162
  Two-veined 160
Hillock Bush 284
Holly, Native 115, 192
Honey Flower 225
Honey-myrtle
  Blush 284
  Bracelet 282
  Mealy 288
  Red 284
  Swamp 288
  Thyme 288
Hop Bush
  Camfield's 312
  Common 314
  Fern-leaf 312
  Narrow-leaf 315
  Pinnate 313
  Sticky 315
  Thread-leaf 313
Hound's Tongue, Forest 393
Hovea
  Long-leaf 184
  Narrow-leaf 184
  Velvet 184
Hydrangea, Native 131

Illawarra Flame Tree 82
Incense Plant 445

Indigo, Native 185
Ironbark
  Beyer's 270
  Broad-leaved 268
  Grey 270
  Mugga 272
  Narrow-leaved 268
Iron Grass 544
Isotoma
  Rock 428
  Swamp 428

Jackwood 52
Jam Tarts 106
Joywood, Lesser 68

Kangaroo Apple 395
Kangaroo Thorn 150
Kanooka 308
Kerrawang 86
Koda 393
Knotweed
  Hairy 71
  Pale 70
  Slender 70
  Spotted 71
Kunzea
  Crimson 296
  Mt Cookem 296
  Pink 296
  Rock 296
  Small-leaved 296
Kurrajong, Brush 83

Ladies' Tresses 589
Lady Fern, Austral 34
Lantern-bush 87
Leek, Native 532
Lillypilly 235
Lily
  Black-anther Flax 537
  Blue Grass 534
  Common Fringe 542
  Flame 557
  Golden 532
  Gymea 557
  Leek 532
  Lilac 538
  Narrow-leaf Palm 556
  Nodding Blue 540
  Nodding Chocolate 537

Lily *cont.*
  Pale Vanilla 530
  Paroo 534
  Rush 538
  Scrambling 554
  Slender Wire 538
  Smooth Flax 535
  Spear 557
  Spoon 471
  Spreading Flax 537
  Swamp 466, 558
  Tufted Blue 540
  Twining Fringe 542
  Yellow Rush 543
Lobelia
  Angled 429
  Forest 431
  Tall 430
  Trailing 430
Lomatia
  Holly-leaved 226
  River 226
Love Creeper 344

Mahogany
  Bastard 240
  Blue Mountains 259
  Broad-leaved White 240
  Large-fruited Red 258
  Red 259
  Southern 258
  Swamp 258
  White 240
Maidenhair Fern
  Common 29
  Giant 30
  Rough 30
Maiden's Blush 82
Mallee Ash
  Blue Mountains 251
  Cliff 252
  Faulconbridge 251
  Giant 250
  Narrow-leaved 252
  Whipstick 249
Mallee, Port Jackson 250
Mangrove
  Grey 411
  River 114
Maroonhood 586
Marsdenia, Scented 383
Marsh Flower, Yellow 386

Marshwort 385
Mat-rush
  Many-flowered 548
  Needle 544
  Pale 546
  Small-flowered 546
  Spiny-headed 546
  Tufted 544
  Twisted 548
  Wattle 544
Matchheads, Pink 344
Melodinus, Southern 384
Messmate 244
Milk Vine
  Common 383
  Hairy 383
Milkmaids 533
Milkwort, Heath 344
Mint
  Creeping 413
  Slender 413
Mint-bush
  Cut-leaf 419
  Narrow-leaf 419
  Oval-leaf 420
  Prunella 420
  Round-leaf 421
  Rough 416
  Slender 422
  Tranquillity 414
  Velvet 417
  Violet 423
Mistletoe
  Drooping 363
  Golden 366
  Jointed 366
  Paper-bark 362
  She-oak 362
Mitre Weed 382
Mock Olive
  Large 380
  Smooth 380
Monkey-flower, Creeping 401
Morinda 388
Mountain Devil 225
Mulberry, Native 51
Murdannia 472
Murrogun 52
Musk, Native 459
Muttonwood 114
  Brush 114
Myall, Coast 144
Myrtle
  Brown 290
  Grey 278

Myrtle *cont.*
  Ironwood 284
  Narrow-leaf 273
  Ridge 284
  see also Fringe Myrtle, Common; Heath-myrtle; Honey-myrtle

Nardoo 41
Native Rose 325
Needle Bush, Silky 222
Nettle, Scrub 62
New Zealand Spinach 65
Nightshade
  Eastern 397
  Forest 397
  Violet 396

Oak
  Forest 64
  River 64
  Swamp 65
Olax 358
Old Man's Beard 55
Old Man's Whiskers 494
Olive Plum, Red 368
Oliveberry, Black 80
Orangebark, Narrow-leaved 369
Orange Thorn 133
Orchid
  Ant 565
  Autumn Bird 566
  Bonnet 568
  Broad-lip Leek 581
  Butterfly 586
  Christmas 563
  Climbing 575
  Common Onion 576
  Copper Beard 564
  Dagger 573
  Dotted Sun 590
  Flying Duck 563
  Fringed Midge 566
  Gnat 561
  Horned 578
  Hyacinth 570
  Ironbark 590
  Large Tongue 569
  Leafless Tongue 568
  Leopard 572
  Mayfly 561
  Orange-blossom 588
  Pencil 574
  Pink Sun 590

COMMON NAME INDEX 601

Orchid *cont.*
  Potato 575
  Purplish Beard 564
  Rat's-tail 574
  Red Beard 564
  Rock 589
  Short-lip Leek 580
  Slender Sun 590
  Small Tongue 569
  Small Waxlip 576
  Spurred Helmet 568
  Streaked Leek 581
  Streaked Rock 573
  Tall Leek 580
  Tall Sun 590
  Tangle 580
  Tiger 572
  Tongue 572
  Toothed Helmet 568
  Tree Spider 589
  Variable Midge 566
  Veined Sun 591
  Waxlip 575
  Wheat-leaved 562
  Yellow Leek 580
  Yellow Rock 576

Palm
  Bangalow 470
  Cabbage Tree 470
  Palm Lily, Narrow-leaf 556
Panax
  Climbing 350
  Elderberry 351
Panic
  Bordered 522
  Wiry 522
Paperbark
  Broad-leaved 287
  Deane's 282
  Flax-leaf 286
  Prickly-leaved 288
  Rosy 284
  Scented 288
  Swamp 284
Paper Daisy
  White 454
  Yellow 456
Parson's Bands 574
Parrot-pea
  Showy 179
  Smooth 176
Passionflower, Red 92
Passionfruit, Native 92
Pastel Flower 408

Pea *see* Bitter-pea; Bush-pea; Coral-pea, Dusky; Darling Pea, Smooth; Flame Pea, Eastern; Flat-pea, Handsome; Parrot-pea; Shaggy-pea; Twining-pea, Purple; Wedge-pea
Peach, Native 58
Pennywort
  Forest 354
  Swamp 354
Pepper, Water 70
Pepperbush
  Brush 49
  Mountain 49
Peppermint
  Narrow-leaved 253
  River 254
  Sydney 254
  Urn-fruited 254
Phebalium
  Alpine 333
  Everlasting 333
  Toothed 330
  Scaly 332, 333
  Shining 330
Philotheca 336
Phylotta
  Common 191
  Dense 191
  Dwarf 190
Pigface 65
Pill Flower 451
Pincushion, Blue 442
Pine
  Black Cypress 44
  Dwarf Mountain 45
  Plum 47
  Port Jackson 45
  Wollemi 47
Pink Buttons 296
Pink Fingers 578
Pink Tip 280
Pinkwood 116, 129
Pipewort, Common 477
Pittosporum
  Sweet 134
  Yellow 134
Pixie Caps 561
Plantain
  Narrow-leaf 409
  Slender 409
  Water 466

Platysace
  Heath 356
  Lance-leaf 356
  Narrow-leaf 356
Plum, Black 113
Plume Bush 445
Plumwood 129
Podolepis, Showy 463
Pollia 474
Polymeria 392
Pomaderris
  Hazel 376
  Rusty 374
  Velvet 374
  Woolly 374
Pomax 389
Pondweed, Floating 468
Poranthera
  Heath-leaved 122
  Small 122
Possumwood 131
Powder Bark 263
Pratia, Poison 431
Pricklefoot 354
Prickly Moses 150
Primrose, Water 308
Princes Feathers 71
Psychotria, Hairy 389
Purple-flag
  Dwarf 529
  Leafy 528
  Short 528
  Silky 529
Purslane, Pink 66

Quandong, Hard 81
Quince, Native 311

Rapier-sedge
  Stiff 509
  Stout 507
Raspberry, Native 137
Rasp Fern 40
  Small 40
Raspwort
  Common 309
  Creeping 309
  Rough 310
Reed, Common 524
  see also Bur-reed
Ribbon-weed 468

Rice-flower
  Curved 124
  Slender 124
  Smooth 124
  Tall 124
Rock Fern, Poison 30
Rope-rush
  Spreading 479
  Tassel 480
Rose
  Dog 130
  Native 325
  River 130
Rosella, Native 87
Rosemary, Coast 425
Rosewood
  Bastard 316
  Scentless 316
Running Postman 186
Rush
  Branching 486
  Broad 486
  Common 488
  Sea 486
  Tussock 488
  see also Bog-rush; Bristle-rush; Bulrush; Club-rush; Cord-rush; Fringe-rush, Common; Mat-rush; Rope-rush; Scale-rush; Spike-rush; Tassel-rush; Twig-rush; Twine-rush, Slender
Rusty-petals 84
  Joyce's 84
  Red 85
  Shrubby 85
  Small 85
Rusty Pod 184

Sally
  Black 252
  Little 252
  Narrow-leaved 252
  White 249
Saloop 68
Saltbush
  Coastal 69
  Ruby 69
Samphire 69
Sandlewood 361
Sarcochilus, Little Gem 588

Sarsaparilla
  Austral 555
  Sweet 555
Sassafras 50
  Oliver's 52
  Southern 50
Satinheart, Green 329
Satinwood 332
Saw-sedge
  Black-fruit 504
  Chaffy 503
  Red-fruited 504
  Rough 502
  Slender 504
  Tall 502
Scale-rush 480
  Slender 481
Scaly-bark 260
Scurvy Weed 472
Seablite 69
Sea Celery 353
Sedge
  Curly 494
  Drooping 492
  Razor 508
  Tall 492
  Tufted 492
  see also Flat-sedge;
    Rapier-sedge; Saw-
    sedge; Sword-sedge
Selaginella, Swamp 19
Senna
  Australian 164
  Smooth 164
Settlers Flax 471
Shaggy-pea
  Common 189
  Heart-leaved 189
  Netted 192
  Tall 188
She-oak
  Black 63
  Drooping 64
  Scrub 63
Shield Fern
  Creeping 35
  Mother 35
  Shiny 34
  Trim 34
Silkpod
  Common 385
  Mountain 384
Skullcap, Dwarf 424
Snow Bush 460
Snow Gum 249
  Mallee 250
  Wolgan 250

Snow Wood 164
Snow in Summer 286
Sourbush
  Dwarf 359
  Leafless 361
  White 359
Speedwell
  Creeping 402
  Derwent 398
  Digger's 398
  Forest 402
  Hairy 401
Spider Flower
  Green 214
  Grey 211
  Pink 218
  Prickly 212
  Red 214, 219
Spike-rush
  Common 500
  Tall 501
Spinifex, Hairy 525
Spleenwort
  Mother 33
  Willow 33
Sprengelia, Rock 109
Spurge
  Blunt 120
  Broom 115
  Thyme 120
St John's Wort, Small 80
Stackhousia
  Coast 370
  Leafless 369
  Slender 370
Star-bush 318
Star-fruit 466
Star-hair
  Broad-leaf 349
  Long-leaf 350
  Woolly 349
Starwort
  Forest 67
  Prickly 67
Stinging Tree, Giant 61
Stinkweed, Common 388
Stinkwood 338
Stonecrop, Australian 135
Storksbill, Native 343
Stringybark
  Blaxland's 241
  Blue-leaved 242
  Brown 241
  Camfield's 241
  Heart-leaved 241

Stringybark *cont.*
  Narrow-leaved 243
  Privet-leaved 244
  Sandstone 243
  Scrub 306
  Thin-leaved 242
  White 243
  Yellow 240
Sundew
  Common 127
  Forked 126
  Pale 126
  Pygmy 126
Sunray, Hoary 459
Supplejack
  Small 554
  White 554
Swamp Heath, Pink 109
Sword-sedge 506
  Hill 506
  Pithy 509
  Sticky 506
  Tall 507
  Variable 508
Symphionema
  Mountain 233
  Swamp 233

Tallowwood 272
Tamarind, Native 312
Tassel-rush 483
Tea-tree
  Coast 300
  Creek 300
  Flaky-barked 304
  Grey 300
  Peach-blossom 304
  Pink 304
  Prickly 298
  Round-fruited 304
  Round-leaf 303
  Slender 304
  Small-leaf 302
  Spider 298
  Swamp 300
  Woolly 298
  Yellow 302
Tick Bush 295
Tick-trefoil
  Rusty 174
  Slender 174
Traveller's Joy 55
Tree Fern
  Prickly 25
  Rough 25
  Soft 26

Trefoil, Australian 187
Trigger Plant
  Grass-leaf 432
  Larch-leaf 433
  Narrow-leaf 433
Tuckeroo 311
Trumpet, Blue 408
  Dwarf 408
Turpentine 306
  Brush 290, 306
Twig-rush
  Bare 490
  Jointed 489
  Leafy 497
  Soft 490
  Square 490
  Tall 497
  Wrinkle-nut 490
Twine-rush, Slender 480
Twining-pea, Purple 183
Tylophora, Bearded 384

Umbrella Tree 351

Vanilla Plant 538
Velvet-flower 83
Vine
  Anchor 51
  Gum 127
  Headache 56
  Large-leaf Staff 368
  Lawyer 555
  Pearl 57
  Pepper 55
  Round-leaf 57
  Snake 58
  Staff 368
  Whip 477
  Wonga 402
  see also Milk Vine;
    Water Vine
Violet
  Bank's 90
  Ivy-leaved 91
  Sandstone 91
  Showy 90
  Slender 89
  Tree 89

COMMON NAME INDEX

Waratah 234
  Monga 234
Water Bush 408
Water Fern
  Fishbone 38
  Hard 40
  Swamp 38
Water Gum 307, 308
  Mountain 308
Water-milfoil, Common 310
Water Ribbons 468
Water Vine 379
  Five-leaf 379
Water Primrose 308
Wattle
  Black 128, 141
  Blunt-leaf 146
  Boomerang 157
  Box-leaved 157
  Bynoe's 148
  Cedar 140
  Chalker's 157
  Coast 147
  Dorothy's 158
  Downy 143
  Early Green 139
  Fan 147
  Fern-leaved 140
  Flax 158
  Fringed 158
  Golden Prickly 148

Wattle *cont.*
  Gosford 160
  Green 141
  Hamilton's 158
  Hedgehog 148
  Hop 155
  Kowmung 158
  Long-leaf 146
  Lunate-leaved 158
  Maiden's 146
  Myrtle 160
  Ploughshare 149
  Red-stemmed 162
  Rough Hairy 153
  Rush-leaved 155
  Sally 144
  Sickle 160
  Silver 139
  Silver-stemmed 143
  Spike 146
  Spreading 149
  Sunshine 144
  Swamp 153
  Sweet-scented 163
  Sydney Golden 144
  Sydney Green 143
  Three-veined 150
  Tiny 150
  Varnish 157
Wax-flower
  Box-leaf 334

Wax-flower *cont.*
  Fairy 335
  Hairy 334
  Long-leaf 334
  Pink 329
  Rock 336
  Silky 335
  Spreading 334
Weather-glass, Golden 558
Wedding Bush 122
Wedge Fern, Lacy 28
Wedge-pea
  Broad-leaf 182
  Dainty 180
  Large 180
  Pale 180
  Pinnate 182
  Red 182
Westringia
  Long-leaved 425
  Slender 425
Whalebone Tree 61
White Fingers 578
White Root 431
Wilga
  Brush 329
  Scrub 329
Wilkiea, Veined 51
Willow-herb, Smooth 308
Wombat Berry 554

Woodruff
  Common 386
  Prickly 386
Woody Pear 235
Woollybutt 260
  Camden 265

Xanthosia, Rock 358

Yam
  Blue 408
  Native 556
Yellow Buttons 452
Yellow-eye
  Dwarf 475
  Slender 475
  Tall 477
Yellowwood 317, 337
Yertchuk 247

Zieria
  Downy 339
  Hairy 341
  Murphy's 341
  Sandfly 341
  Tree 338

Scientific name index

Abrophyllum ornans 131
Abutilon oxycarpum 87
Acacia
  amblygona 147
  amoena 157
  asparagoides 148
  baueri subsp. *aspera* 150
  baueri subsp. *baueri* 150
  binervata 160
  binervia 144
  brownii 148
  buxifolia 157
  bynoeana 148
  chalkeri 157
  clunies-rossiae 158
  dealbata 139
  decurrens 139
  dorothea 158
  echinula 148
  elata 140
  elongata 153
  falcata 160
  falciformis 161
  filicifolia 140
  fimbriata 158
  floribunda 144
  fulva 141
  genistifolia 149
  gordonii 153
  gunnii 149
  hamiltoniana 158
  hispidula 153
  implexa 161
  irrorata 141
  ixiophylla 153
  jonesii 141
  juncifolia 155
  leucolobia 157
  linifolia 158
  longifolia 144
  longissima 146
  lunata 158
  maidenii 146
  mearnsii 141
  meiantha 160
  melanoxylon 162
  myrtifolia 160
  obliquinervia 162
  obtusifolia 146

Acacia cont.
  oxycedrus 146
  paradoxa 150
  parramattensis 143
  parvipinnula 143
  penninervis 162
  prominens 160
  ptychoclada 155
  pubescens 143
  quadrilateralis 150
  rubida 162
  saliciformis 163
  schinoides 143
  sophorae 146
  stricta 155
  suaveolens 163
  terminalis 144
  trinervata 150
  ulicifolia 150
  undulifolia 155
  verniciflua 157
Acaena
  echinata 136
  novae-zelandiae 136
  ovina 136
Acanthaceae 408
Acianthus
  caudatus 561
  exsertus 561
  fornicatus 561
  pusillus 561
Acmena smithii 235
Acronychia oblongifolia 317
Acrophyllum australe 127
Acrotriche
  divaricata 93
  aggregata 93
Actinotus
  forsythii 352
  gibbonsii 352
  helianthi 352
  minor 353
Adenochilus nortonii 562
Adiantaceae 29
Adiantum
  aethiopicum 29
  atroviride 29
  formosum 30
  hispidulum 30

Aegiceras corniculatum 114
Agavaceae 556
Aizoaceae 65
Ajuga australis 412
Alania endlicheri 530
Alchornea ilicifolia 115
Alectryon subcinereus 311
Alisma plantago-aquatica 466
Alismataceae 466
Allocasuarina
  diminuta 62
  distyla 63
  glareicola 64
  littoralis 63
  nana 63
  paludosa 64
  portuensis 64
  torulosa 64
  verticillata 64
Almaleea incurvata 166
Alocasia brisbanensis 471
Alphitonia excelsa 370
Alpinia
  arundelliana 527
  caerulea 527
Alternanthera denticulata 68
Amaranthaceae 68
Amaryllidaceae 558
Ammobium alatum 443
Amperea xiphoclada 115
Amyema
  cambagei 362
  congener 362
  gaudichaudii 362
  miquelii 363
  pendulum 363
Amylotheca dictyophleba 363
Aneilema acuminatum 472
Angophora
  bakeri 236
  costata 237
  crassifolia 236
  euryphylla 237
  floribunda 237

Angophora cont.
  hispida 238
  inopina 236
  subvelutina 238
Anisopogon avenaceus 516
Aotus ericoides 166
Aphanopetalum resinosum 127
Apiaceae 352
Apium prostratum 353
Apocynaceae 384
Araceae 471
Araliaceae 349
Araucariaceae 47
Archontophoenix cunninghamiana 470
Arecaceae 470
Aristida
  ramosa 517
  warburgii 517
Arthropodium
  milleflorum 530
  minus 530
Arthropteris
  beckleri 31
  tenella 31
Asclepiadaceae 383
Asperula
  conferta 386
  scoparia 386
  gunnii 386
Aspleniaceae 33
Asplenium
  australasicum 33
  bulbiferum 33
  difforme 33
  flabellifolium 33
  polyodon 33
Asteraceae 443
Asterolasia
  buckinghamii 318
  correifolia 318
Astroloma
  humifusum 93
  pinifolium 94
Astrotricha
  floccosa 349
  latifolia 349
  ledifolia 350
  longifolia 350

SCIENTIFIC NAME INDEX 605

Atherosperma
  moschatum 50
Athyriaceae 34
Atkinsonia ligustrina 365
Austrocynoglossum
  latifolium 393
Austrodanthonia tenuior 517
Austromyrtus
  acmenoides 273
  tenuifolia 273
Austrostipa
  pubescens 518
  ramosissima 518
Avicennia marina 411
Avicenniaceae 411

Babingtonia
  densifolia 274
  virgata 274
Backhousia myrtifolia 278
Baeckea
  brevifolia 274
  diosmifolia 274
  imbricata 274
  linifolia 276
Baloghia inophylla 116
Baloskion
  australe 483
  fimbriatum 483
  gracile 483
  pallens 483
  tetraphyllum 483
Banksia
  aemula 204
  cunninghamii 205
  ericifolia 205
  integrifolia subsp.
    integrifolia 206
  integrifolia subsp.
    monticola 206
  marginata 206
  oblongifolia 206
  paludosa 206
  penicillata 206
  robur 207
  serrata 208
  spinulosa var. *collina* 208
  spinulosa var.
    spinulosa 208
Bauera
  microphylla 130
  rubioides 130

Baueraceae 130
Baumea
  articulata 489
  juncea 490
  rubiginosa 490
  teretifolia 490
  tetragona 490
Bertya pomaderroides 116
Beyeria viscosa 116
Bignoniaceae 402
Billardiera
  mutabilis 132
  scandens 132
Blandfordia
  cunninghamii 530
  grandiflora 532
  nobilis 532
Blechnaceae 38
Blechnum
  ambiguum 38
  camfieldii 38
  cartilagineum 38
  indicum 38
  minus 38
  patersonii 38
  wattsii 40
Bolboschoenus
  caldwellii 491
  fluviatilis 491
Boraginaceae 393
Boronia
  algida 318
  anemonifolia 318
  anethifolia 319
  barkeriana 320
  deanei 320
  floribunda 320
  fraseri 320
  ledifolia 323
  microphylla 323
  mollis 323
  nana 323
  parviflora 324
  pinnata 324
  polygalifolia 324
  rubiginosa 325
  serrulata 325
  thujona 326
Bossiaea
  buxifolia 167
  ensata 167
  heterophylla 167
  kiamensis 168
  lenticularis 168
  neo-anglica 168
  obcordata 168

Bossiaea cont.
  prostrata 168
  rhombifolia 168
  scolopendria 170
  stephensonii 170
Brachychiton acerifolius 82
Brachycome
  angustifolia 444
  graminea 444
  multifida 445
  ptychocarpa 445
Brachyloma daphnoides 94
Breynia oblongifolia 117
Brunonia australis 442
Brunoniaceae 442
Brunoniella
  australis 408
  pumilo 408
Bulbine
  bulbosa 532
  glauca 532
  semibarbata 532
Bulbophyllum
  exiguum 562
  shepherdii 562
Burchardia umbellata 533
Bursaria
  longisepala 132
  pinosa 132

Caesalpiniaceae 164
Caesia
  parviflora var.
    parviflora 534
  parviflora var. *vittata* 534
Calandrinia
  calyptrata 66
  eremaea 66
  pickeringii 66
Calanthe triplicata 563
Caleana major 563
Callicoma serratifolia 128
Callistemon
  citrinus 278
  linearifolius 278
  linearis 278
  pinifolius 278
  pityoides 280
  rigidus 280
  salignus 280
  shiressii 280
  subulatus 280

Callitris
  endlicheri 44
  muelleri 45
  rhomboidea 45
Calochilus
  campestris 564
  paludosus 564
  robertsonii 564
Calochlaena
  dubia 26
Calomeria
  amaranthoides 445
Calotis
  cuneifolia 446
  dentex 446
  lappulacea 446
Calystegia
  marginata 390
  soldanella 390
Calytrix tetragona 290
Campanulaceae 426
Caprifoliaceae 380
Carex
  appressa 492
  brunnea 492
  gaudichaudiana 492
  longebrachiata 492
  pumila 492
Carpobrotus glaucescens 65
Caryophyllaceae 67
Cassine australis 368
Cassinia
  aculeata 448
  aureonitens 448
  cunninghamii 448
  denticulata 449
  longifolia 449
  trinerva 450
  uncata 450
Cassytha
  glabella 54
  paniculata 54
  pubescens 54
Cassythaceae 54
Casuarina
  cunninghamiana 64
  glauca 65
Casuarinaceae 62
Caustis
  flexuosa 494
  pentandra 494
  recurvata 494
Cayratia clematidea 379
Celastraceae 368

Celastris
  australis 368
  subspicata 368
Celmisia sp. aff. *longifolia* 452
Celtis paniculata 58
Centella asiatica 354
Centrolepidaceae 485
Centrolepis
  fascicularis 485
  strigosa 485
Cephalaralia
  cephalobotrys 350
Ceratopetalum
  apetalum 128
  gummiferum 128
Cheilanthes
  austrotenuifolia 30
  distans 30
  sieberi 30
Chenopodiaceae 68
Chiloglottis
  chlorantha 565
  formicifera 565
  seminuda 566
Chloanthaceae 411
Chloanthes stoechadis 411
Chloris ventricosa 520
Chordifex
  dimorphus 484
  fastigiatus 484
Choretrum
  candollei 359
  pauciflorum 359
Choricarpia leptopetala 290
Chorizandra
  cymbaria 496
  sphaerocephala 496
Chorizima parviflorum 170
Christella dentata 32
Chrysocephalum
  apiculatum 452
  semipapposum 454
Cinnamomum oliveri 52
Cissus
  antarctica 379
  hypoglauca 379
  opaca 379
  sterculiifolia 379
Citriobatus pauciflorus 133
Citrolella moorei 367
Cladium procerum 497
Claoxylon australe 117

Clematis
  aristata 55
  glycinoides 56
Clerodendrum
  tomentosum 410
Clusiaceae 80
Comesperma
  defoliatum 344
  ericinum 344
  sphaerocarpum 344
  volubile 344
Commelina cyanea 472
Commelinaceae 472
Commersonia fraseri 83
Conospermum
  ellipticum 209
  ericifolium 209
  longifolium 209
  taxifolium 209
  tenuifolium 209
Convolvulaceae 390
Convolvulus erubescens 390
Coopernookia barbata 433
Coprosma
  hirtella 387
  quadrifida 387
Cordyline stricta 556
Correa
  alba 326
  lawrenciana 326
  reflexa 327
Corunastylis
  archeri 566
  fimbriata 566
  woollsii 566
Corybas
  aconitiflorus 568
  fimbriatus 568
  pruinosus 568
Corymbia
  eximia 239
  gummifera 238
  maculata 239
Crassula sieberiana 135
Crassulaceae 135
Crinum pedunculatum 558
Croton verreauxii 118
Crowea
  exalata 328
  saligna 328
Cryptandra
  amara 370
  ericoides 371

Cryptandra cont.
  propinqua 372
  spinescens 372
Cryptocarya
  glaucescens 52
  microneura 52
  obovata 52
  rigida 52
Cryptostylis
  erecta 568
  hunteriana 568
  leptochila 569
  subulata 569
Cunoniaceae 127
Cupaniopsis
  anacardioides 311
Cupressaceae 44
Cuscuta australis 390
Cyanicula caerulea 570
Cyanthochaeta diandra 497
Cyathea
  australis 25
  cooperi 25
  leichhardtiana 25
Cyatheaceae 25
Cyclophyllum
  longipetalum 387
Cyclosorus interruptus 32
Cymbidium suave 570
Cymbonotus lawsonianus 457
Cymbopogon refractus 520
Cyperaceae 489
Cyperus
  difformis 498
  exaltatus 498
  polystachyos 498
  sanguinolentus 499
Cyphanthera scabrella 394

Damasonium minus 466
Dampiera
  purpurea 434
  stricta 434
Darwinia
  biflora 291
  camptostylis 291
  diminuta 292
  fascicularis subsp.
    fascicularis 292
  fascicularis subsp.
    oligantha 292

Darwinia cont.
  glaucophylla 292
  grandiflora 293
  leptantha 293
  peduncularis 295
  procera 295
  taxifolia subsp.
    macrolaena 295
  taxifolia subsp.
    taxifolia 295
Davallia solida var.
  pyxidata 31
Davalliaceae 31
Daviesia
  acicularis 171
  alata 171
  corymbosa 172
  latifolia 172
  leptophylla 172
  mimosoides 172
  squarrosa 173
  ulicifolia 173
Dendrobium
  aemulum 590
  speciosum 589
  tetragonum 589
Dendrocnide excelsa 61
Dendrophthoe vitellina 365
Dennstaedtia davallioides 26
Dennstaedtiaceae 26
Derwentia
  blakelyi 398
  derwentiana 398
  perfoliata 398
Desmodium
  brachypodium 174
  rhytidophyllum 174
  varians 174
Dianella
  caerulea 534
  congesta 534
  longifolia 535
  prunina 537
  revoluta 537
  tasmanica 537
Dichelachne
  crinita 520
  rara 520
Dichondra repens 392
Dichopogon
  fimbriatus 537
  strictus 537
Dicksonia antarctica 26
Dicksoniaceae 26
Dilleniaceae 72

SCIENTIFIC NAME INDEX 607

Dillwynia
  acicularis 174
  brunioides 175
  elegans 175
  floribunda 176
  glaberrima 176
  parvifolia 176
  phylicoides 176
  ramosissima 177
  retorta 178
  rudis 178
  sericea 179
  sieberi 179
  tenuifolia 179
Dioscorea transversa 556
Dioscoreaceae 556
Diospyros
  australis 113
  pentamera 113
Diplazium australe 34
Diploglottis australis 312
Dipodium
  punctatum 570
  roseum 570
  variegatum 570
Diuris
  alba 571
  aurea 572
  maculata 572
  sulphurea 572
Dockrillia
  fairfaxii 574
  linguiformis 572
  pungioniformis 573
  striolata 573
  teretifolia 574
Dodonaea
  boroniifolia 312
  camfieldii 312
  falcata 313
  multijuga 313
  pinnata 313
  triquetra 314
  truncatiales 314
  viscosa subsp.
    angustifolia 315
  viscosa subsp.
    angustissima 315
Doodia
  aspera 40
  australis 41
  caudata 40
  linearis 41
Doryanthes excelsa 557
Doryphora sassafras 50
Dracophyllum secundum 94

Drosera
  auriculata 126
  binata 126
  burmannii 127
  glanduligera 127
  peltata 126
  pygmaea 126
  spathulata 127
Droseraceae 126
Dryopteridaceae 34
Duboisia myoporoides 394

Ebenaceae 113
Echinopogon
  caespitosus 521
  ovatus 521
Ehretia acuminata 393
Einadia
  hastata 68
  nutans 68
  polygonoides 68
  trigonos 68
Elaeocarpaceae 80
Elaeocarpus
  holopetalus 80
  kirtonii 80
  obovatus 81
  reticulatus 81
Elatostema reticulatum 61
Eleocharis
  acuta 500
  cylindrostachys 501
  sphacelata 501
Emmenosperma
  alphitonioides 372
Empodisma minus 479
Enchylaena tomentosa 69
Endiandra
  discolor 52
  sieberi 52
Entolasia
  marginata 522
  stricta 522
Epacris
  calvertiana var.
    calvertiana 95
  calvertiana var.
    versicolor 95
  coriacea 95
  crassifolia 96
  longiflora 96
  microphylla 97
  muelleri 97

Epacris cont.
  obtusifolia 97
  paludosa 98
  pulchella 98
  purpurascens var.
    onosmiflora 98
  purpurascens var.
    purpurascens 98
  reclinata 100
  rigida 100
  sparsa 100
Epilobium
  billardierianum 308
Eremophila debilis 406
Ericaceae 93
Eriocaulaceae 477
Eriocaulon scariosum 477
Eriochilus
  autumnalis 574
  cucullatus 575
Eriostemon australasius 329
Erodium crinatum 341
Eryngium vesiculosum 354
Erythrorchis cassythoides 575
Eucalyptus
  acmenoides 240
  agglomerata 242
  aggregata 262
  amplifolia 261
  apiculata 252
  baueriana 270
  benthamii 266
  beyeriana 270
  bicostata 265
  blaxlandii 241
  botryoides 258
  bridgesiana 263
  burgessiana 251
  camfieldii 241
  capitellata 241
  cinerea 268
  consideniana 247
  copulans 252
  crebra 268
  cunninghamii 252
  cypellocarpa 265
  dalrympleana 266
  deanei 257
  dendromorpha 250
  elata 254
  eugenioides 242
  fastigata 246
  fibrosa 268

Eucalyptus cont.
  globoidea 243
  gregsoniana 250
  haemastoma 254
  imitans 243
  laophila 252
  ligustrina 244
  longifolia 260
  luehmanniana 246
  macarthurii 265
  maidenii 265
  mannifera 263
  melliodora 272
  microcorys 272
  moluccana 268
  moorei 252
  muelleriana 240
  multicaulis 249
  notablis 259
  obliqua 244
  oblonga 243
  obstans 250
  oreades 246
  ovata 262
  paniculata 270
  parramattensis 261
  pauciflora 249
  pilularis 244
  piperita subsp.
    piperita 254
  piperita subsp.
    urceolaris 254
  punctata 260
  quadrangulata 265
  racemosa 256
  radiata 253
  resinifera 259
  robusta 258
  rossii 254
  rubida 267
  saligna 257
  scias 258
  sclerophylla 256
  sideroxylon 272
  siderphloia 268
  sieberi 249
  smithii 266
  sparsifolia 243
  squamosa 260
  stellulata 251
  stricta 251
  tereticornis 262
  umbra 240
  viminalis 266
Eucryphia moorei 129
Eucryphiaceae 129
Euphorbiaceae 115

Euphrasia
  collina subsp.
    paludosa 400
  collina subsp. *speciosa* 400
Eupomatia laurina 49
Eupomatiaceae 49
Eurychorda complanata 484
Euryomyrtus ramosissima 276
Eustrephus latifolius 554
Exocarpos
  cupressiformis 360
  strictus 360

Fabaceae 165
Ficus
  coronata 59
  macrophylla 59
  obliqua 59
  rubiginosa 60
  superba var. *henneana* 60
Fieldia australis 403
Fimbristylis
  dichotoma 501
  ferruginea 501
Flacourtiaceae 88
Flagellaria indica 477
Flagellariaceae 477

Gahnia
  aspera 502
  clarkei 502
  erythrocarpa 502
  filifolia 503
  filum 503
  grandis 502
  melanocarpa 504
  microstachya 504
  sieberiana 504
  subaequiglumis 505
Galium
  binifolium 388
  gaudichaudii 388
  propinquum 388
Gastrodia sesamoides 575
Geijera
  latifolia 329
  salicifolia 329
Geitonoplesium cymosum 554

Genoplesium see *Corunastylis*
Geraniaceae 341
Geranium
  graniticola 343
  homeanum 342
  neglectum 342
  potentilloides 343
  solanderi 343
Gesneriaceae 403
Gleichenia
  dicarpa 23
  microphylla 23
  rupestris 23
Gleicheniaceae 23
Glochidion
  ferdinandi var. *ferdinandi* 118
  ferdinandi var. *pubens* 118
Glossodia
  major 575
  minor 576
Glycine
  clandestina 179
  microphylla 180
  tabacina 180
Gmelina leichhardtii 410
Gompholobium
  glabratum 180
  grandiflorum 180
  huegelii 180
  latifolium 182
  minus 182
  pinnatum 182
  uncinatum 182
Gonocarpus
  micranthus subsp. *micranthus* 309
  micranthus subsp. *ramosissimus* 309
  teucrioides 309
Goodenia
  bellidifolia 435
  decurrens 435
  dimorpha var. *angustifolia* 436
  dimorpha var. *dimorpha* 436
  glomerata 436
  hederacea 436
  heterophylla 436
  ovata 438
  paniculata 438
  rostrivalvis 438
  stelligera 438
Goodeniaceae 433

Goodia lotifolia 183
Grammitidaceae 36
Grammitis
  billardierei 36
  stenophylla 36
Gratiola peruviana 400
Grevillea
  acanthifolia 210
  arenaria 210
  aspleniifolia 210
  baueri 211
  buxifolia subsp. *buxifolia* 211
  buxifolia subsp. *ecorniculata* 211
  caleyi 211
  capitellata 212
  diffusa subsp. *diffusa* 212
  diffusa subsp. *constablei* 212
  diffusa subsp. *filipendula* 212
  juniperina 212
  kedumbensis 212
  laurifolia 212
  linearifolia 212
  longifolia 214
  mucronulata 214
  oldei 214
  oleoides 214
  parviflora 216
  patulifolia 216
  phylicoides 217
  raybrownii 217
  rivularis 217
  sericea subsp. *riparia* 218
  sericea subsp. *sericea* 218
  shiressii 218
  speciosa 219
  sphacelata 219
Grossulariaceae 131
Guioa semiglauca 315
Gymnoschoenus sphaerocephalus 505
Gymnostachys anceps 471

Haemodoraceae 558
Haemodorum
  corymbosum 558
  planifolium 558

Hakea
  bakeriana 220
  constablei 220
  dactyloides 220
  dohertyi 221
  gibbosa 221
  laevipes 221
  microcarpa 222
  pachyphylla 222
  propinqua 222
  salicifolia subsp. *angustifolia* 222
  salicifolia subsp. *salicifolia* 222
  sericea 222
  teretifolia 222
Haloragaceae 309
Haloragis heterophylla 310
Haloragodendron
  gibsonii 310
  lucasii 310
Hardenbergia violacea 183
Hedycarya angustifolia 51
Helichrysum
  calvertianum 454
  collinum 454
  elatum 454
  rutidolepis 455
  scorpioides 456
Hemigenia
  cuneifolia 412
  purpurea 412
Hibbertia
  acicularis 72
  aspera subsp. *aspera* 72
  bracteata 72
  cistiflora 73
  dentata 73
  diffusa 74
  (East Heathcote) 79
  empetrifolia 74
  fasciculata 74
  (Howes Swamp) 79
  linearis 74
  (Maddens Plains) 79
  (Megalong Valley) 79
  (Menai) 79
  (Minnehaha Falls) 79
  monogyna 74
  nitida 74
  obtusifolia 74
  pedunculata 77
  praemorsa 77

SCIENTIFIC NAME INDEX

Hibbertia cont.
  riparia 77
  rufa 78
  saligna 78
  scandens 78
  serpyllifolia 78
  vestita 79
Hibiscus
  diversifolius 87
  heterophyllus 87
  splendens 87
  sturtii 87
Histiopteris incisa 26
Homalanthus see
  Omalanthus
Hovea
  heterophylla 184
  linearis 184
  longifolia 184
  pannosa 184
  purpurea 184
  speciosa 185
Howittia trilocularis 88
Hybanthus
  monopetalus 89
  vernonii 89
Hydrocharitaceae 466
Hydrocotyle
  geraniifolia 354
  tripartita 354
Hymenanthera dentata 89
Hymenophyllaceae 25
Hymenophyllum
  cupressiforme 25
Hymenosporum flavum 133
Hypericum
  gramineum 80
  japonicum 80
Hypolaena fastigata 480
Hypolepis muelleri 27
Hypoxidaceae 558
Hypoxis hygrometrica 558

Icacinaceae 367
Imperata cylindrica 523
Indigofera australis 185
Iridaceae 527
Isolepis
  inundata 506
  nodosa 506
Isopogon
  anemonifolius 224
  anethifolius 224

Isopogon cont.
  dawsonii 224
  fletcheri 225
  prostratus 225
Isotoma
  axillaris 428
  fluviatilis 428

Jacksonia scoparia 186
Joycea pallida 523
Juncaceae 486
Juncaginaceae 468
Juncus
  continuus 486
  homocaulis 486
  kraussii 486
  planifolius 486
  prismatocarpus 486
  usitatus 488

Kennedia
  prostrata 186
  rubicunda 187
Keraudrenia corollata 83
Korthalsella rubra 366
Kunzea
  ambigua 295
  cambagei 296
  capitata 296
  parvifolia 296
  rupestris 296
  sp. E 296

Lagenifera stipitata 457
Lambertia formosa 225
Lamiaceae 412
Lasiopetalum
  ferrugineum 84
  joyceae 84
  macrophyllum 85
  parvifolium 85
  rufum 85
Lastreopsis
  acuminata 34
  decomposita 34
  hispida 34
  microsora 35
Lauraceae 52
Laxmannia
  compacta 538
  gracilis 538
Legnephora moorei 57

Leionema
  dentatum 330
  diosmeum 330
  lamprophyllum 330
Lentibulariaceae 405
Lepidosperma
  concavum 506
  elatius 507
  filiforme 509
  forsythii 507
  laterale 508
  limicola 508
  longitudinale 509
  neesii 509
  urophorum 509
  viscidum 506
Lepironia articulata 511
Leptinella longipes 457
Leptocarpus tenax 480
Leptomeria acida 360
Leptopteris fraseri 20
Leptospermum
  arachnoides 298
  continentale 298
  emarginatum 298
  grandifolium 298
  juniperinum 298
  laevigatum 300
  macrocarpum 300
  morrisonii 300
  myrtifolium 300
  obovatum 300
  parvifolium 302
  polyanthum 302
  polygalifolium 302
  rotundifolium 303
  spectabile 303
  sphaerocarpum 304
  squarrosum 304
  trinervium 304
Lepyrodia scariosa 480
Leucochrysum
  albicans 459
  graminifolium 459
Leucopogon
  amplexicaulis 101
  appressus 101
  ericoides 102
  esquamatus 102
  exolasius 102
  fletcheri 102
  fraseri 102
  juniperinus 102
  lanceolatus 103
  microphyllus 104
  muticus 104
  parviflorus 104

Leucopogon cont.
  setiger 104
  virgatus 104
Libertia
  paniculata 527
  pulchella 528
Liliaceae 530
Lindsaea
  linearis 28
  microphylla 28
Lindsacaceae 28
Liparis reflexa 576
Lissanthe
  sapida 106
  strigosa subsp. *strigosa* 106
  strigosa subsp. *subulata* 106
Litsea reticulata 53
Livistona australis 470
Lobelia
  alata 429
  dentata 429
  gibbosa 430
  gracilis 430
  trigonocaulis 431
Lobeliaceae 428
Logania
  albiflora 382
  pusilla 282
Loganiaceae 382
Lomandra
  brevis 544
  confertifolia subsp. *pallida* 544
  confertifolia subsp. *rubiginosa* 544
  cylindrica 544
  filiformis subsp. *coriacea* 544
  filiformis subsp. *filiformis* 544
  fluviatilis 546
  glauca 546
  gracilis 546
  longifolia 546
  micrantha 546
  montana 548
  multiflora 548
  obliqua 548
Lomatia
  ilicifolia 226
  myricoides 226
  silaifolia 226
Loranthaceae 362
Lotus australis 187
Ludwigia peploides 308

Luzula ovata 488
Lycopodiaceae 18
Lycopodiella
  cernua 18
  lateralis 18
Lycopodium
  deuterodensum 19
Lyperanthus suaveolens 576

Maclura cochinchinensis 60
Macrozamia
  communis 42
  elegans 42
  spiralis 42
Malvaceae 87
Marsdenia
  flavescens 383
  rostrata 383
  suaveolens 383
Marsilea
  hirsuta 41
  mutica 41
Marsileaceae 41
Maytenus silvestris 369
Melaleuca
  armillaris 282
  biconvexa 282
  capitata 282
  deanei 282
  decora 284
  ericifolia 284
  erubescens 284
  hypericifolia 284
  linariifolia 286
  megalongensis 286
  nodosa 286
  parvistaminea 286
  quinquenervia 287
  sieberi 287
  squamea 288
  squarrosa 288
  styphelioides 288
  thymifolia 288
Melia azedarach 316
Meliaceae 316
Melichrus
  procumbens 106
  urceolatus 106
Melicope micrococca 337
Melodinus australis 384
Menispermaceae 57
Mentha
  diemenica 413
  satureioides 413

Menyanthaceae 385
Micrantheum
  ericoides 118
  hexandrum 118
Microlaena stipoides 523
Micromyrtus
  blakelyi 305
  ciliata 305
  minutiflora 305
Microsorum
  pustulatum 36
  scandens 36
Microstrobos fitzgeraldii 45
Microtis unifolia 576
Mimosaceae 138
Mimulus repens 401
Mirbelia
  baueri 187
  platyloboides 187
  rubiifolia 188
  speciosa 188
Mitrasacme
  pilosa 382
  polymorpha 382
Monimiaceae 50
Monotaxis linifolia 120
Monotoca
  elliptica 106
  ledifolia 107
  scoparia 107
Moraceae 59
Morinda jasminoides 388
Muellerina
  celastroides 365
  eucalyptoides 366
Murdannia graminea 472
Myoporaceae 406
Myoporum
  acuminatum 406
  bateae 408
  boninense 406
  floribundum 406
  montanum 408
Myriophyllum
  variifolium 310
Myrsinaceae 114
Myrsine
  howittiana 114
  variabilis 114
Myrtaceae 235

Nemacianthus caudatus 561
Nematolepis squamea 332
Neolitsea dealbata 53
Nicotiana forsteri 394
Notelaea
  longifolia 380
  neglecta 380
  ovata 380
  venosa 380
Notodanthonia longifolia 518
Notothixos subaureus 366
Nymphoides
  geminata 385
  indica 385
Nyssanthes erecta 68

Ochrosperma
  oligomerum 276
Olacaceae 358
Olax stricta 358
Oleaceae 380
Olearia
  argophylla 459
  asterotricha 460
  elliptica 460
  erubescens 460
  microphylla 460
  myrsinoides 460
  phlogopappa 462
  quercifolia 462
  stellulata 462
  tomentosa 462
  viscidula 462
Omalanthus
  populifolius 120
  stillingifolius 120
Omphacomeria acerba 361
Onagraceae 308
Opercularia
  aspera 388
  diphylla 388
  hispida 388
  varia 388
Oplismenus
  aemulus 524
  imbecillis 524
Orchidaceae 561
Oreomyrrhis eriopoda 355
Orthoceras strictum 578

Osmundaceae 20
Ottelia ovalifolia 466
Oxylobium
  arborescens 188
  cordifolium 189
  ellipticum 189
Ozothamnus
  adnatus 451
  argophyllus 451
  diosmifolius 451
  ferrugineus 452

Palmeria scandens 51
Pandorea pandorana 402
Papillilabium beckleri 580
Parahebe lithophila 401
Pararchidendron pruinosum 164
Parsonsia
  brownii 384
  lanceolata 385
  straminea 385
Passiflora
  cinnabarina 92
  herbertiana 92
Passifloraceae 92
Petalochilus
  carneus 578
  catenatus 578
  pictus 578
Patersonia
  fragilis 528
  glabrata 528
  longifolia 529
  sericea 529
Pelargonium australe 343
Pellaea
  falcata 30
  nana 30
  paradoxa 30
Pennantia cunninghamii 367
Peperomia
  blanda var. *floribunda* 55
  tetraphylla 55
Peperomiaceae 55
Persicaria
  decipiens 70
  elatior 70
  hydropiper 70
  lapathifolia 70

SCIENTIFIC NAME INDEX

Persicaria cont.
orientalis 71
praetermissa 71
prostrata 71
strigosa 71
subsessilis 71
Persoonia
acerosa 227
chamaepitys 227
hirsuta subsp. evoluta 227
hirsuta subsp. hirsuta 227
isophylla 228
lanceolata 228
laurina subsp. intermedia 228
laurina subsp. laurina 228
laurina subsp. leiogyna 228
levis 228
linearis 228
mollis 228
myrtilloides 230
nutans 230
oblongata 230
oxycoccoides 230
pinifolia 231
Petalochilus
carneus 578
catenatus 578
pictus 578
Petrophile
canescens 231
pedunculata 231
pulchella 232
sessilis 232
Phebalium
squamulosum subsp. argenteum 332
squamulosum subsp. lineare 332
squamulosum subsp. ozothamnoides 333
squamulosum subsp. squamulosum 333
Philotheca
buxifolia subsp. buxifolia 334
buxifolia subsp. obovata 334
hispidula 334
myoporoides 334
obovalis 335
reichenbachii 336

Philotheca cont.
salsolifolia 336
scabra subsp. latifolia 336
scabra subsp. scabra 336
trachyphylla 336
Philydraceae 557
Philydrum lanuginosum 557
Phragmites australis 524
Phyllanthus
gasstroemii 120
hirtellus 120
Phyllota
grandiflora 190
humifusa 190
phylicoides 191
squarrosa 191
Pimelea
curviflora 124
glauca 124
latifolia 124
ligustrina subsp. hypericina 124
ligustrina subsp. ligustrina 124
linifolia subsp. collina 125
linifolia subsp. linifolia 124
linifolia subsp. linoides 125
spicata 125
Piper novae-hollandiae 55
Piperaceae 55
Pittosporaceae 132
Pittosporum
revolutum 134
undulatum 134
Planchonella australis 113
Plantaginaceae 409
Plantago
debilis 409
gaudichaudii 409
Platycerium bifurcatum 36
Platylobium
formosum subsp. formosum 191
formosum subsp. parviflorum 191

Platysace
clelandii 355
ericoides 356
lanceolata 356
linearifolia 356
stephensonii 356
Plectorrhiza tridentata 580
Plectranthus parviflorus 414
Poa labillardieri 524
Poaceae 516
Podocarpaceae 45
Podocarpus
elatus 47
spinulosus 47
Podolepis
hieracioides 463
jaceoides 463
Podolobium
ilicifolium 192
scandens 192
Pollia crispata 474
Polygalaceae 344
Polygonaceae 70
Polymeria calycina 392
Polyosma cunninghamii 131
Polypodiaceae 36
Polyscias
elegans 350
murrayi 351
sambucifolia 351
Polystichum
australiense 35
proliferum 35
Pomaderris
adnata 373
andromedifolia 373
aspera 376
brunnea 376
discolor 376
elliptica 373
eriocephala 376
ferruginea 374
intermedia 374
lanigera 374
ledifolia 374
ligustrina 376
mediora 378
phylicifolia subsp. ericoides 378
phylicifolia subsp. phylicifolia 378
prunifolia 378
velutina 374

Pomax umbellata 389
Poranthera
corymbosa 121
ericifolia 122
microphylla 122
Portulacaceae 66
Potamogeton tricarinatus 468
Potamogetonaceae 468
Prasophyllum
brevilabre 580
elatum 580
flavum 580
patens 581
striatum 581
Pratia
concolor 431
purpurascens 431
Primulaceae 115
Prostanthera
askania 414
caerulea 414
cryptandroides 415
densa 415
denticulata 416
granitica 416
hindii 416
hirtula 417
howelliae 417
incana 417
incisa 419
lasianthos 419
linearis 419
ovalifolia 420
prunellioides 420
rhombea 420
rotundifolia 421
rugosa 421
saxicola var. montana 422
saxicola var. saxicola 422
scutellarioides 422
sieberi 422
violacea 423
Proteaceae 204
Pseudanthus pimeleoides 122
Pseuderanthemum variabile 408
Psilotaceae 16
Psilotum nudum 16

Psychotria loniceroides 389
Pteridaceae 28
Pteridium esculentum 28
Pteris
  tremula 28
  umbrosa 29
Pterostylis
  acuminata 582
  baptistii 582
  concinna 582
  curta 582
  daintreana 584
  erecta 586
  grandiflora 584
  hildae 584
  laxa 584
  longifolia 584
  nutans 584
  parviflora 586
  pedunculata 586
  saxicola 586
Ptilothrix deusta 511
Pultenaea
  aristata 192
  blakelyi 192
  canescens 192
  capitellata 194
  daphnoides 194
  divaricata 194
  echinula 195
  elliptica 202
  ferruginea var. *deanei* 195
  ferruginea var. *ferruginea* 196
  flexilis 196
  glabra 196
  hispidula 196
  linophylla 199
  microphylla 199
  paleacea 199
  parviflora 199
  polifolia 200
  retusa 200
  rosmarinifolia 200
  scabra 201
  stipularis 201
  subspicata 201
  tuberculata 202
  villosa 202
  viscosa 202
Pyrrosia
  confluens 36
  rupestris 36

Quintinia sieberi 131

Ranunculaceae 55
Ranunculus
  inundatus 56
  lappaceus 56
  plebeius 57
Restionaceae 479
Rhagodia condolleana 69
Rhamnaceae 370
Rhodamnia rubescens 306
Rhytidosporum procumbens 135
Ricinocarpos pinifolius 122
Rimacola elliptica 586
Ripogonum
  album 554
  fawcettianum 554
Rosaceae 136
Rubiaceae 386
Rubus
  moluccanus 136
  nebulosus 136
  parvifolius 137
  rosifolius 137
Rulingia
  dasyphylla 86
  hermanniifolia 86
  prostrata 86
Rupicola
  apiculata 108
  ciliata 108
  sprengelioides 108
Rutaceae 317

Sambucus
  australasica 380
  gaudichaudiana 380
Samolus repens 115
Santalaceae 359
Santalum obtusifolium 361
Sapindaceae 311
Sapotaceae 113
Sarcochilus
  australis 586
  falcatus 588
  hillii 588
  olivaceus 588
Sarcocornia quinqueflora 69

Sarcomelicope simplicifolia 337
Sarcopetalum harveyanum 57
Scaevola
  aemula 440
  albida 440
  calendulacea 440
  hookeri 441
  ramosissima 441
Schelhammera undulata 538
Schizaea
  bifida 22
  dichotoma 22
  rupestris 22
Schizaeaceae 22
Schizomeria ovata 129
Schoenoplectus
  mucronatus 511
  validus 512
Schoenus
  apogon 512
  brevifolius 512
  ericetorum 512
  melanostachys 513
  paludosus 514
  turbinatus 514
  villosus 514
Scirpus polystachyus 515
Scolopia braunii 88
Scrophulariaceae 398
Scutellaria humilis 424
Selaginella uliginosa 19
Selaginellaceae 19
Selliera radicans 442
Senecio
  lautus subsp. *dissectifolius* 464
  lautus subsp. *maritimus* 464
  linearifolius 465
  vagus 465
  velleioides 465
Senna
  aciphylla 164
  odorata 164
Seringia arborescens 86
Sloanea australis 82
Smilacaceae 554
Smilax
  australis 555
  glyciphylla 555
Solanaceae 394

Solanum
  aviculare 395
  brownii 396
  celatum 396
  campanulatum 396
  cinereum 396
  prinophyllum 387
  pungetium 397
  stelligerum 398
  vescum 398
Sowerbaea juncea 538
Sparganiaceae 526
Sparganium subglobosum 526
Spartothamnella juncea 411
Sphaerolobium
  minus 203
  vimineum 203
Spinifex sericeus 525
Spiranthes sinensis 589
Sporadanthus gracilis 481
Sprengelia
  incarnata 109
  monticola 109
  sprengelioides 109
Stackhousia
  monogyna 369
  muricata 369
  nuda 369
  spathulata 370
  viminea 370
Stackhousiaceae 369
Stegostyla
  dimorpha 579
  gracilis 579
  testacea 579
Stellaria
  flaccida 67
  pungens 67
Stenocarpus salignus 233
Stephania japonica var. *discolor* 58
Sterculiaceae 82
Sticherus
  flabellatus 24
  lobatus 24
  urceolatus 24
Streblus brunonianus 61
Stylidiaceae 432
Stylidium
  graminifolium 432
  laricifolium 433
  lineare 433
  productum 433

Further reading

Benson, D. and Howell, J. (1995) *Taken for Granted: The bushland of Sydney and its suburbs*, Royal Botanic Gardens, Sydney.
Benson, D. and McDougall, L., 'Ecology of Sydney plant species', Parts 1–10, Cunninghamia, 3(2) 1993–9(1) 2005, National Herbarium of NSW.
Bishop, T. (2000) 2nd edn, *Field Guide to the Orchids of New South Wales and Victoria*, UNSW Press, Sydney.
Costermans, L. (2000) *Native Trees and Shrubs of South-Eastern Australia*, Reed New Holland, Sydney.
Fairley, A. (2004) *Seldom Seen: Rare plants of greater Sydney*, New Holland, Sydney.
Floyd, A.G. (2008), rev. edn, *Rainforest Trees of Mainland South-Eastern Australia*, Terania Rainforest Publishing, Lismore.
Harden, G.J. (1990–1993) *Flora of New South Wales*, vols 1–4, UNSW Press, Sydney.
Howell, J. and Benson, D. (2000) *Sydney's Bushland: More than meets the eye*, Royal Botanic Gardens, Sydney.
Klaphake, V. (2009) *Eucalypts of the Sydney Region*, Klaphake, Sydney.
—— (2006) 4th edn, *Guide to the Grasses of Sydney*, Van Klaphake, Byabarra, NSW.
—— (2004) 4th edn, *Key to the commoner species of sedges and rushes of Sydney and the Blue Mountains*, Van Klaphake, Byabarra, NSW.
Olde, P. and Marriott, N. (1994–1995) *The Grevillea Book*, vols 1–3, Kangaroo Press, Sydney.
Pellow, B., Henwood, M. and Carolin, R. (2009) 5th edn, *Flora of the Sydney Region*, Sydney University Press, Sydney.
Robinson, L. (2003) *Field Guide to the Native Plants of Sydney*, Kangaroo Press, Sydney.
Tame, T. (1992) *Acacias of Southeast Australia*, Kangaroo Press, Sydney.
Wheeler, D. and Jacobs, S. (2002) *Grasses of New South Wales*, University of New England Press.
Williams, J., Harden, G. and McDonald, W. (1984) *Trees and Shrubs in Rainforests of New South Wales and Southern Queensland*, University of New England Press.

SCIENTIFIC NAME INDEX 613

Stypandra glauca 540
Styphelia
  angustifolia 110
  laeta 110
  longifolia 110
  triflora 110
  tubiflora 112
  viridis 112
Suaeda australis 69
Swainsona galegifolia 203
Symphionema
  montanum 233
  paludosum 233
Syncarpia glomulifera 306
Synoum glandulosum 316
Syzygium
  australe 306
  oleosum 307
  paniculatum 307

Tasmannia
  insipida 49
  lanceolata 49
Telopea
  mongaensis 234
  speciosissima 234
Tetrabaculum
  tetragonum 589
Tetragonia tetragonioides 65
Tetratheca
  bauerifolia 345
  decora 345
  ericifolia 345
  glandulosa 346
  juncea 346
  neglecta 346
  rubioides 346
  rupicola 346
  shiressii 349
  thymifolia 349
Teucrium corymbosum 424
Thelionema
  caespitosum 540
  umbellatum 540
Thelychiton speciosus 589
Thelymitra
  carnea 590
  ixioides 590
  media 590
  pauciflora 590
  venosa 591

Thelypteridaceae 32
Themeda
  australis 525
  triandra 525
Thymelaeaceae 124
Thysanotus
  juncifolius 540
  patersonii 542
  tuberosus 542
  virgatus 543
Tmesipteris truncata 16
Todea barbara 20
Toona ciliata 317
Trachymene incisa 357
Trema
  aspera 58
  tomentosa 58
Tremandraceae 345
Tricoryne
  elatior 543
  simplex 543
Triglochin
  microtuberosum 468
  procerum 468
  rheophilum 468
  striatum 468
Tristania neriifolia 307
Tristaniopsis
  collina 308
  laurina 308
Trochocarpa laurina 112
Tropilis aemula 590
Tylophora barbata 384
Typha
  domingengis 526
  orientalis 526
Typhaceae 526
Typhonium
  brownii 471
  eliosurum 471

Ulmaceae 58
Urtica
  incisa 62
  urens 62
Urticaceae 61
Utricularia
  australis 405
  dichotoma 405
  gibba 405
  lateriflora 405
  uliginosa 405
  uniflora 405

Vallisneria gigantea 468
Velleia lyrata 442
Verbenaceae 410
Veronica
  calycina 401
  notabilis 402
  plebeia 402
Villarsia exaltata 386
Viminaria juncea 204
Viola
  banksii 90
  betonicifolia 90
  caleyana 91
  hederacea 91
  sieberiana 91
  silicestris 91
Violaceae 89
Viscaceae 366
Vitaceae 379

Wahlenbergia
  ceracea 427
  communis 426
  gracilis 426
  graniticola 427
  littoricola 427
  luteola 427
  multicaulis 427
  planiflora 427
  stricta 427
Westringia
  eremicola 425
  fruticosa 425
  longifolia 425
Wilkiea huegeliana 51
Wilsonia backhousei 392
Winteraceae 49
Wollemia nobilis 47
Woollsia pungens 113

Xanthorrhoea
  arborea 549
  australis 550
  concava 550
  fulva 550
  latifolia 550
  macronema 552
  media 552
  minor 552
  resinifera 553
Xanthorrhoeaceae 544

Xanthosia
  atkinsoniana 357
  dissecta 358
  pilosa 358
  scopulicola 358
  stellata 358
  tridentata 358
Xerochrysum
  bracteatum 456
  viscosum 456
Xylomelum pyriforme 235
Xyridaceae 474
Xyris
  bracteata 474
  complanata 474
  gracilis 475
  juncea 475
  operculata 477
  ustulata 477

Zamiaceae 42
Zieria
  arborescens 338
  covenyi 338
  cytisoides 339
  fraseri 339
  granulata 339
  laevigata 341
  murphyi 341
  pilosa 341
  smithii 341
Zingiberaceae 527
Zornia dictyocarpa 204